Second Edition

In Situ
Treatment
Technology

Second Edition
In Situ
Treatment
Technology

Evan K. Nyer
Peter L. Palmer
Eric P. Carman
Gary Boettcher
James M. Bedessem
Frank Lenzo
Tom L. Crossman
Gregory J. Rorech
Donald F. Kidd

ARCADIS
GERAGHTY&MILLER

**Environmental Science
and Engineering Series**

LEWIS PUBLISHERS
Boca Raton   London   New York   Washington, D.C.

Library of Congress Cataloging-in-Publication Data

In situ treatment technology / Evan K. Nyer ... [et al.].--2nd ed.
    p. cm.--(Geraghty & Miller environmental science and engineering series)
    Includes bibliographical references and index.
    ISBN 1-56670-528-2 (alk. paper)
    1. In situ remediation. I. Nyer, Evan K. II. Series.

TD192.8 .I5724 2000
628.5—dc21
                                                        00-045045

© 2001 by CRC Press LLC
Lewis Publishers is an imprint of CRC Press LLC

No claim to original U.S. Government works
International Standard Book Number 1-56670-528-2
Library of Congress Card Number 00-045045
Printed in the United States of America 1 2 3 4 5 6 7 8 9 0
Printed on acid-free paper

Preface

Many things have changed, but many things have stayed the same since the first edition of *In Situ Treatment Technology* was published. One thing that has stayed the same is that this is still the most exciting technical area in the remediation field today. Also, many new important technologies have emerged over the past 5 years, and full-scale installations of existing technologies have broadened our knowledge base. The purpose of this book is to provide the reader with a single source that consolidates all of this information on the various *in situ* technologies. The main technology areas of bioremediation (monitored natural attenuation, MNA), vapor extraction, sparging, vacuum enhanced recovery, fracturing, and reactive walls are discussed in individual chapters. New *in situ* technologies like *in situ* reactive zones, and phytoremediation are also discussed in individual chapters. This allows for an in-depth review of the state-of-the-art for each technology including laboratory and pilot plant studies, full-scale design, operation and maintenance, cost analysis, and case histories. We have also added full design sections for the vapor extraction, sparging, and vacuum enhanced recovery chapters. This level of detail will help those new to the field develop the correct design methods for these *in situ* practices.

One chapter has remained for non-*in situ* design considerations. Many of the *in situ* technologies use air movement as part of their applications. The air usually must be collected and brought above ground for treatment. Chapter 6 is devoted to discussing above-ground air treatment.

The book goes beyond discussing individual *in situ* technologies. The authors felt that it was very important for the reader to end up with an understanding of the geologic foundation and limitations of each of the technologies. The first chapter begins by explaining the limitations of pump and treat remediation. Designers have progressed to *in situ* technologies because the pump and treat remediations methods have failed to clean most sites. Chapter 1 provides the technical reasons that the pump and treat systems have had limited success, and how these same reasons may limit the success of *in situ* technologies. The information in Chapter 1 will also provide the reader with a basis to analyze and predict the possible success of any new *in situ* methods that are developed in the future. Chapter 2, Lifecycle Design, shows the importance of the entire life of the design when using an individual technology. Examples of good lifecycle designs are spread throughout the individual chapters. The book is next broken into two sections. Based upon the geological limitations discussed in Chapters 1 and 2, *in situ* technologies are mainly used as either mass removal techniques or to enhance the rate of remediation during the "diffusion limited" portion of the project. The mass removal section includes vapor extraction, sparging, and vacuum enhanced recovery. The diffusion controlled section includes bioremediation, *in situ* reactive zones, and phytoremediation. While none of the technologies are limited to mass removal or enhancement, they tend to have their main uses. The rest of the book covers the remaining technologies and the final chapter tries to prepare the reader for the potential problems we may face when remediating sites.

I have tried to maintain the easy style of writing that my books normally enjoy. However, I felt that it was important to provide the reader the details necessary to be able to implement the *in situ* technologies. This dichotomy is one of the main reasons that I have asked the co-authors to participate in the book. Each of the co-authors work on a daily basis with the technology that he or she wrote about. I reviewed and rewrote each of the chapters, but the co-authors provided the meat. The result is, hopefully, a text that is still easy to read, but provides significant design and operational detail for each technology. The co-authors have their own bylines for the chapters that they wrote so that the reader will know the prime source of the information.

Many people have to be thanked beyond the co-authors. First, ARCADIS Geraghty & Miller has once again provided support and encouragement. There is no way that anyone can write a book today and put food on the table without the support of his or her employer. ARCADIS Geraghty & Miller has allowed me and the co-authors the time required for the book, and provided the staff support from drafting and secretaries. Second, I have to thank the authors from the first edition who decided not to continue with the second edition. Frank J. Johns II, Suthan Suthersan, and Sami Fam were all an important part of the first edition and their efforts continue to be a basis for the quality of the second edition. There are over 200 tables and figures in the book. Brian Herrmann continued his efforts from the first edition to complete the added figures for this edition. Carla Gerstner once again stepped up to furnish the main secretarial support for the book. Without her patience and cool head I am not sure that I would have finished the book, and several of the co-authors would not have remained on speaking terms with me. In the technical area we have to thank Kurt Beil, Steve Brussee, Edmond Buc, Jeff Burdick, Scott Davis, Heidi Dauer, Jennifer Evans, Bill Golla, Mike Hansen, John Horst, Chip Hughes, Dan Jacobs, Gary Keyes, Jack Kratzmeyer, Chris McHale, Jim Morgan, Greg Page, Scott Potter, Eileen Schumacher, Matt Waslewski, and Amy Weinert. There is no way this book would have been finished without their support.

In situ technologies are an important part of being able to clean sites. I hope that the readers will find this book helpful in their applications of these new methodologies.

Evan K. Nyer

The Authors

Evan K. Nyer is Senior Vice President of ARCADIS Geraghty & Miller, Inc. and is responsible for maintaining and expanding the company's technical expertise in geology/hydrogeology, engineering, modeling, risk assessment, and bioremediation. He has extensive experience as a groundwater treatment engineer and has designed and installed more than 400 groundwater treatment systems including biological, *in situ* biological, air stripping, activated carbon, inorganic, advanced oxidation, soil venting, sparging systems, vacuum enhanced remediations, and reactive zones. In addition to being responsible for technical designs and strategies, he has published and presented numerous works on groundwater treatment and other aspects of waste management and remediation.

Mr. Nyer has taught courses on groundwater cleanup and treatment technologies around the world and is the author of four books: *Groundwater and Soil Remediation: Practical Methods and Strategies* (Ann Arbor Press), *Practical Techniques for Groundwater and Soil Remediation* (Lewis Publishers), *Groundwater Treatment Technology* (Van Nostrand Reinhold), and *In Situ Treatment Technology* (Lewis Publishers). He was also a principal author of *Bioremediation* (American Academy of Environmental Engineering) and has written the column "Treatment Technology" for *Groundwater Monitoring and Remediation* since 1987.

Peter L. Palmer, a Senior Vice President in charge of the Remediation Services Business Practice for ARCADIS Geraghty & Miller, has 27 years of experience in providing environmental management services. He has written numerous articles on soil and groundwater remediation strategies. He has extensive experience in performing projects that have encompassed all aspects of hazardous waste management including the evaluation, design, and construction of remedial measures to abate soil and groundwater contamination at RCRA and CERCLA sites. As both a Professional Engineer and a Professional Geologist, he has a unique perspective in developing remedial measures that cost-effectively integrate source controls and plume remediation. He administers ARCADIS Geraghty & Miller's Innovative Technology Development and Training Program to promote the use of creative, cost-effective approaches for solving remedial challenges.

Eric P. Carman, P.G. is a Principal Hydrogeologist and Associate Vice President with ARCADIS Geraghty & Miller. He has more than 15 years of environmental experience and has been a consultant with ARCADIS Geraghty & Miller since 1998. He received his B.S. in Geology from the University of Iowa and M.S. in Hydrogeology from the University of Wisconsin-Milwaukee.

Mr. Carman specializes in implementing and managing innovative cleanup strategies using biotechnologies for industrial and public sector clients. Involved with applications of bioremediation since 1990, he has been working in the field of phytoremediation since 1993. His phytoremediation experience includes projects across the United States, including the first project in Wisconsin to use phytoremediation to address petroleum hydrocarbons. He has published several papers on bioremediation and phytoremediation and has given lectures at many universities.

Mr. Carman is an active member of Society of Military Engineers, American Chemical Society, and Technical Association of the Pulp and Paper Industry.

Gary Boettcher is an Associate Vice President and senior Project Manager for ARCADIS Geraghty & Miller. Mr. Boettcher manages and directs multi-facetted and multi-disciplined projects relating to environmental remediation, property development, and property acquisition. Mr. Boettcher subscribes to the project management concept of *"define, plan, and control"* whereby scope, schedule, budget, and execution progress are clearly and frequently conveyed to his clients such that objectives and expectations are met. Technical project elements have included groundwater investigation, recovery, treatment, and management; soil investigation and treatment; regulatory interface; engineering design, construction, and operations; data review and validation; third-party review, recommendations, and negotiations; and human health and ecological risk management.

Mr. Boettcher has 16 years of environmental experience obtained in the chemical industry, decontamination equipment manufacturing, hazardous waste treatment industry, and environmental consulting. Mr. Boettcher specializes in investigation and remediation of impacted groundwater and soil. He has been a project engineer, scientist, and manager on federal and state Superfund, RCRA, and various industrial projects throughout the United States including Puerto Rico and the Bahamas. He offers his clients broad technical capabilities having served as a consultant, field implementor, and project manager and has consistently provided cost-effective services to clients. Mr. Boettcher's objective is to provide practical solutions to environmental challenges, tailored to client needs and expectations. He has considerable experience in developing, interacting, and negotiating favorable regulatory strategies in California and Texas, designing investigation and remediation strategies to allow property transactions to occur, preparing for potential toxic tort litigation, and implementing projects with the goal of recovering cost from environmental insurance carriers.

Mr. Boettcher has developed added specialization in the area of bioremediation where he has designed, managed, and implemented *in situ* and *ex situ* bioremediation processes to treat hydrocarbons and industrial solvents. These programs were enhanced by his knowledge of chemical properties and their fate in the environment using the literature, experience, and implementation of laboratory treatability studies. Mr. Boettcher has co-authored several papers and textbooks on remediation and contributed to the development of U.S. Environmental Protection Agency (USEPA) guidance documents focusing on use of aerobic biological treatability studies at CERCLA sites.

James M. Bedessem, P.E. has more than 10 years of experience in environmental consulting and currently serves as the manager of engineering for the Tampa and Palm Beach Gardens offices of ARCADIS Geraghty & Miller. During his career, he has served in a wide variety of technical and managerial roles, including technical advisor, project manager, design engineer, construction manager, and operations specialist. He has also evaluated, designed, and implemented numerous treatment technologies for both soil and water, including several innovative technologies. Mr. Bedessem is experienced at leading remedial investigation, design, and implementation efforts at Superfund and RCRA sites.

Donald F. Kidd, P.E. has more than 13 years of experience in the design, installation, and operation of soil and groundwater treatment systems. He is a Professional Civil Engineer in several states and is responsible for technical support and quality assurance of remedial engineering projects conducted within the western region of ARCADIS Geraghty & Miller. His responsibilities include training and professional development for both internal and external staff involved in remediation. Mr. Kidd's technical emphasis is on site data analysis and cleanup strategy development. Since the late 1980s, he has been responsible for planning/coordination of pilot tests, interpretation of results, and design of full-scale systems under a variety of geologic conditions and a wide range of contaminants. Mr. Kidd is a co-author of the first edition of *In Situ Treatment Technology* and is the author of several other published papers covering the practical application of engineering principles within the environmental field.

Frank Lenzo has been involved in the environmental field for more than 18 years, providing expertise in the testing, design, and application of *in situ* and *ex situ* treatment systems for groundwater remediation. He presently serves as a member of ARCADIS' Innovative Strategies Group, providing alternative technical approaches to subsurface remediation problems. He has managed or provided senior support for system designs involving enhanced reductive dechlorination, *in situ* metals precipitation, natural attenuation processes, vacuum enhanced recovery, air sparging, soil vapor extraction, biosparging, and bioventing.

Gregory J. Rorech, P.E. is President of Progressive Engineering & Construction, Inc. He specializes in the evaluation, development, design, and implementation of both innovative and conventional remediation technologies. Mr. Rorech has been utilizing his chemical engineering expertise to assist industrial and municipal clients with environmental and process concerns for more than 15 years. As President of Progressive Engineering & Construction, Inc., Mr. Rorech is responsible for directing the firm's current work at CERCLA, RCRA, hydrocarbon, and consent decree sites throughout the United States. Mr. Rorech's expertise with remedial strategy development, regulatory negotiation, economic analysis, innovative technologies, implementation, and design as well as system operations enables him to develop cost-effective closure strategies for his clients. Remedial technologies recently implemented include monitored natural attenuation, enhanced reductive dechlorination, air sparging, enhanced vacuum extraction, permeable treatment barriers, *in situ* chemical oxidation, ion exchange, reverse osmosis, electrochemical precipitation, phytoremediation, vacuum extraction, and bioremediation. Mr. Rorech has written extensively on groundwater and soil remediation technologies, is a contributing author on six books, and is an instructor for The Princeton Remediation Course.

Tom L. Crossman, Manager Bio/Phytoremediation Services, has a B.S. in Dairy Technology/Microbiology. Mr. Crossman worked in the research and development of fermented food products, focused on accelerated aging and flavor-forming processes in cheeses, yogurts, etc., via enzymes and fermentation technologies. He has applied biotechnology featuring immobilized enzyme and cell reactors resulting in two immobilized enzyme patents via controlled-porosity supports for bioreactor technology. At ARCADIS Geraghty & Miller, Inc., Mr. Crossman's primary focus is on *in situ* intrinsic bioattenuation and intrinsic reductive dechlorination, having evaluated and applied intrinsic remediation in over 60 remedial strategies. His secondary focus is on phytoremediation (use of vegetation) of groundwater for remediation of TCE-contaminated groundwater and weathered hydrocarbons on vadose zone soils. He is a mentor on intrinsic remediation and phytoremediation for the firm. He is the only consultant-member of the ACT 307 Subcommittee in Michigan providing guidelines for bioremediation of chlorinateds and hydrocarbons.

Contents

CHAPTER 1

Limitations of Pump and Treat Remediation Methods

Evan K. Nyer

CONTENTS

During the last 30 years we have been trying to use advective treatment methods to remediate contaminated aquifers and vadose zones. Advective methods rely on a fluid to move through the geology, have the contaminant transfer to the fluid, and then the fluid is brought above-ground for treatment. Water and air have been the two fluids that have been used for remediation. Water is used for the pump and treat remediations, and air is used for several technologies such as vapor extraction and air sparging. Pump and treat methods have been used since the 1970s to try and remediate aquifers. It has now become obvious that pump and treat methods of remediation are not able to reach the required contaminant concentrations in most aquifers. Even with extended operating periods, we will not be able to call the

1-56670-528-2/01/$0.00+$.50

contaminated aquifer 'clean.' Installing wells and forcing water to move past the contaminated portion of an aquifer will not remove the organic and heavy metal compounds at a high enough rate for the project to reach a conclusion in a reasonable amount of time.

Most people who have reached this conclusion have also decided that the next logical step is to treat the contaminants in place or *in situ*. One problem with this thinking is that unless we have a complete understanding of why pump and treat systems cannot clean an aquifer, we will probably have the same problems when we design or evaluate an *in situ* method. A second problem is that many of the *in situ* methods that were first employed are still based on advective methodologies. Instead of water flowing past the contaminated area, the initial *in situ* methods simply used airflowing past the contaminated area. Too often *in situ* methods are solely evaluated upon the rate at which they remove compounds from the aquifer. However, unless the new *in situ* method can overcome the limitations of the advective removal process, there may be limited reason for applying the technology. This means that we must not only analyze the rate at which the *in situ* method can remove contaminants from the aquifer, but also determine whether the new method can reach 'clean' and if so, at what cost.

Therefore, before we can review and discuss the various *in situ* methods, we must have a thorough understanding of the limitations of pump and treat and other advective removal technologies. We have to go beyond the data analysis that shows that the process is limited. We have to understand the macro and micro processes that are occurring in the aquifer and vadose zone. Once the problem has been broken down into its components, we then can analyze an *in situ* method to see if it will overcome all of the individual problems that limit the effectiveness of the original advective method, pump and treat.

This chapter will break down the limitations of pump and treat into its basic components. We will first review water as the carrier of the contaminants and the creation of a contamination zone. Next, we will review how others have described the limitations of pump and treat. Finally, we will break down the limitations of pump and treat into component parts and analyze each one. These components will then become the basis for evaluating advective removal methods in general.

WATER AS A CARRIER

One of the central points that must be made in this chapter is that water is the carrier when we are dealing with the aquifer. We drill a well into an aquifer, collect a water sample, and send it off to a laboratory for analysis to determine if the aquifer is contaminated. The contaminants found in this sample do not directly represent what is in the environment of the aquifer. The concentration of contaminants found in the water sample represents the relationship between the organic compounds located in the aquifer, their adsorption properties in relationship to the aquifer soil, and their solubility in water. If we lose the same amount of acetone and decane to an aquifer, our groundwater sample will show more acetone then decane. The decane will adsorb to the soils in the aquifer and be less soluble in water than the acetone.

The water will carry more acetone than decane to the sampling well. The concentration of acetone will be higher even though the same amounts of acetone and decane are located in the aquifer.

Everything we do in the aquifer relates to water. It is an integral part of the aquifer environment. We have come to rely on the water in the aquifer (groundwater) as our measuring device and as our method of cleaning (pump and treat). Water is also the reason that the contaminants spread from their original release point. The best place to start our understanding of water as the carrier is to review how the contamination plume was originally formed.

THE CONTAMINATION PLUME

Plumes are created when a contaminant comes into contact with the aquifer. Contaminants can be released at the ground surface, into the unsaturated zone or directly into the aquifer. Figure 1 shows a contaminant that was released at the surface. The contaminant travels down through the soil by the force of gravity. The organic compounds will adsorb to the soil as they move through the unsaturated zone. Assuming that all of the organics are not sorbed and that the contaminants do not encounter an impermeable layer, the contaminants will eventually reach the aquifer. We will discuss the unsaturated zone when we review air as a carrier.

As can be seen in Figure 1, the release does not travel a smooth path. There are areas of high flow and many areas that have no flow as the contaminant travels down through the vadose zone. This is due to the fact that very small variations in the geology can cause a major change in the flow direction. Figure 2 shows the results

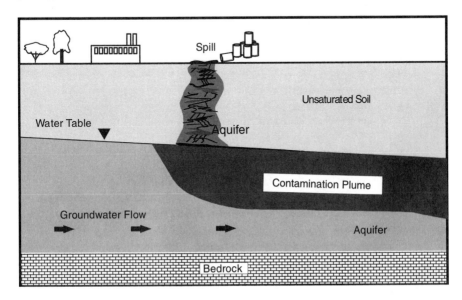

Figure 1   Contamination plume in an aquifer.

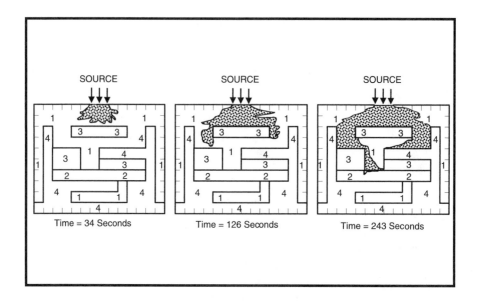

Figure 2   Perchloroethene movement through sand. Feenstra Course Notes 1996, Princeton
Remediation Course, laboratory experiment conducted by Kueper et al. (1989).
These diagrams show the migration of PCE as DNAPL through a sand box having
layer of different K: Layer 1 = 1 x 10^{-1} cm/s, Layer 2 = 2 x 10^{-1} cm/s, Layer 3 = 5
x 10^{-1} cm/s, and Layer 4 = 8 x 10^{-1} cm/s.

of an experiment following the movement of perchloroethene (PCE) through sand
(Kueper, Abbott, and Farquhar, 1989). The three figures show the migration of PCE
through a sandbox having layers of different hydraulic conductivities (K) (see
Darcy's Law below). While all of the sand is the same order of magnitude, the
number in the layers represents the different multipliers for the same orders of
magnitude.

Layer Hydraulic Conductivity
1 = 1 x 10^{-1} cm/s
2 = 2 x 10^{-1} cm/s
3 = 5 x 10^{-1} cm/s
4 = 8 x 10^{-1} cm/s

As can be seen in these figures, a small variation in the makeup of the geology
can act to retard the vertical movement of the PCE. The geologic changes do not
simply slow down the traveling speed of the PCE, they completely change the
direction of movement. This experiment used large blocks of the four different sands.
The same result would occur if the different sand layers were just a few centimeters
thick. In the real world these same, small changes in geology naturally occur all of
the time. The changes are in thin layers, not large blocks. This is why Figure 1
shows the movement of the contaminant going through several directional changes
as it travels down through the vadose zone. The changes shown in Figure 1 are the

result of small changes in the makeup of the geology. Large changes, like sand and clay lenses, will have a more pronounced effect on the movement of the contaminant.

Three things can happen when the contaminants reach the aquifer. Figure 1 shows the contaminant directly entering the aquifer. This will occur if the contaminant is very soluble in water. Table 1 provides the solubilities for 50 organic compounds. These are the organic compounds that are most likely to be found at a contaminated site. This book will provide the other properties of these compounds as that property is discussed in the text. As can be seen in Table 1, acetone is very soluble in water. It is listed as 1×10^6 milligrams per liter (mg/l). This means that the concentration can reach one million parts per million. In other words, acetone has infinite solubility in water. In general, the entire ketone and alcohol families of organic compounds are very soluble in water. If any of them are released to the ground environment, their movement will be represented by Figure 1.

The rest of the organic compounds have a variety of solubilities. The values listed in Table 1 represent the maximum amount of organic that is soluble in water. Do not expect to find these concentrations in a groundwater sample, however. In fact, when the concentration in the groundwater sample reaches 5 to 10 percent of maximum solubility, it is a strong indication that a source of pure compound is in close proximity to the sampling point (EPA, 1992a and Perry, 1984). Several organizations are now promoting that chlorinated hydrocarbon concentrations as low as 1 percent are a strong indication of pure compound. The reader should use the 1 percent level as an indication of pure compound, but not as proof.

The solubility of the compound controls how much or at what rate the organic compound enters the groundwater of the aquifer. Therefore, the solubility also controls the mass rate at which the water can carry the organic away. From Table 1, one liter of water can carry a maximum of 50 mg of hexachloroethane. However, the same liter of water can carry 14,000 mg of 2-hexanone. The actual movement of the organic compound in the groundwater is also affected by the organic compound's affinity for sorption to the soil particles in the aquifer. This is quantified by the retardation factor of the compound, and will be discussed later in this chapter.

Two things can happen when the organic compound has a relatively low solubility in water. If the organic liquid is lighter than water then it will stop its downward movement when it reaches the aquifer. Figure 3 shows the movement of a lighter-than-water organic compound (gasoline in the figure). If the organic liquid is heavier than water, it will continue its downward movement until it has been sorbed by the aquifer soil or it encounters an impermeable layer. In addition, all compounds have some solubility in water, and a small amount of the organic compound will also be left in the aquifer as 'residual saturation' as it moves down through the aquifer. The more soluble the organic compound, the more mass that will be lost to this process. Figure 4 shows the movement of a heavier-than-water organic compound (trichloroethylene). Basically, the organic compounds that are lighter than water will float on the water, and the organic compounds that are heavier than water will sink through the water. As can be seen in Figure 4, the movement of the trichloroethylene (TCE) has the same abrupt changes in direction that the contaminant had traveling through the vadose zone. Even with the geology saturated with water, the movement of a

Table 1   Solubility for Specific Organic Compounds

#	Compound	Solubility[a] (mg/l)	Ref.	#	Compound	Solubility[a] (mg/l)	Ref.
1	Acenaphthene	3.42	2	26	bis(2-Ethylhexyl)phthalate	2.85×10^{-1}	2
2	Acetone	1×10^{6a}	1	27	Heptachlor	1.8×10^{-1}	2
3	Aroclor 1254	1.2×10^{-2}	2	28	Hexachlorobenzene	6×10^{-3}	1a
4	Benzene	1.75×10^{3}	1a	29	Hexachloroethane	5×10^{1}	2
5	Benzo(a)pyrene	1.2×10^{-3}	2	30	2-Hexanone	1.4×10^{4}	2
6	Benzo(g,h,i)perylene	7×10^{-4}	2	31	Isophorone	1.2×10^{4}	2
7	Benzoic Acid	2.7×10^{3}	2	32	Methylene Chloride	2×10^{4}	1
8	Bromodichloromethane	4.4×10^{3}	2	33	Methyl Ethyl Ketone	2.68×10^{5}	1b
9	Bromoform	3.01×10^{3}	1b	34	Methyl Naphthalene	2.54×10^{1}	2a
10	Carbon Tetrachloride	7.57×10^{2}	1a	35	Methyl tert-Butyl Ether	4.8×10^{4}	3
11	Chlorobenzene	4.66×10^{2}	1a	36	Naphthalene	3.2×10^{1}	2
12	Chloroethane	5.74×10^{3}	2	37	Nitrobenzene	1.9×10^{3}	2
13	Chloroform	8.2×10^{3}	1a	38	Pentachlorophenol	1.4×10^{1}	1
14	2-Chlorophenol	2.9×10^{4}	2	39	Phenol	9.3×10^{1}	1a,b
15	p-Dichlorobenzene (1,4)	7.9×10^{1}	2	40	1,1,2,2-Tetrachloroethane	2.9×10^{3}	2
16	1,1-Dichloroethane	5.5×10^{3}	1a	41	Tetrachloroethylene	1.5×10^{2}	1a
17	1,2-Dichloroethane	8.52×10^{3}	1a	42	Tetrahydrofuran	3×10^{-1}	4
18	1,1-Dichloroethylene	2.25×10^{3}	1a	43	Toluene	5.35×10^{2}	1a
19	cis-1,2-Dichloroethylene	3.5×10^{3}	1a	44	1,2,4-Trichlorobenzene	3×10^{1}	2
20	trans-1,2-Dichloroethylene	6.3×10^{3}	1a	45	1,1,1-Trichloroethane	1.5×10^{3}	1a
21	2,4-Dichlorophenoxyacetic Acid	6.2×10^{2}	2	46	1,1,2-Trichloroethane	4.5×10^{3}	1a
22	Dimethyl Phthalate	4.32×10^{3}	2	47	Trichloroethylene	1.1×10^{3}	1a
23	2,6-Dinitrotoluene	1.32×10^{3}	2	48	2,4,6-Trichlorophenol	8×10^{2}	2
24	1,4-Dioxane	4.31×10^{5}	2	49	Vinyl Chloride	2.67×10^{3}	1a
25	Ethylbenzene	1.52×10^{2}	1a	50	o-Xylene	1.75×10^{2}	1c

[a] Solubility of 1,000,000 mg/l assigned because of reported "infinite solubility" in the literature.
1. *Superfund Public Health Evaluation Manual*, Office of Emergency and Remedial Response Office of Solid Waste and Emergency Response, U.S. Environmental Protection Agency (1986).
a. Environmental Criteria and Assessment Office (ECAO), EPA, Health Effects Assessments for Specific Chemicals (1985).
b. Mabey, W.R., Smith, J.H., Rodoll, R.T., Johnson, H.L., Mill, T., Chou, T.W., Gates, J., Patridge, I.W., Jaber, H., and Vanderberg, D., "Aquatic Fate Process Data for Organic Priority Pollutants", EPA Contract Nos. 68-01-3867 and 68-03-2981 by SRI International, for Monitoring and Data Support Division, Office of Water Regulations and Standards, Washington, D.C. (1982).
c. Dawson et al., "Physical/Chemical Properties of Hazardous Waste Constituents," by Southeast Environmental Research Laboratory for USEPA (1980).
2. USEPA, "Basics of Pump-and-Treat Groundwater Remediation Technology" EPA/600/8-901003, Robert S. Kerr Environmental Research Laboratory (March 1990).
3. Manufacturer's data; Texas Petrochemicals Corporation, Gasoline Grade Methyl tert-butyl ether Shipping Specification and Technical Data (1986).
4. *CRC Handbook of Chemistry and Physics*, 71st ed., Boca Raton, FL: CRC Press, Inc., (1990).

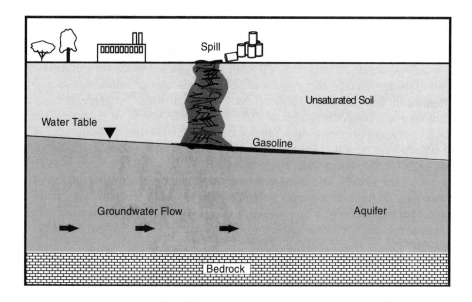

Figure 3   Gasoline spill encountering an aquifer.

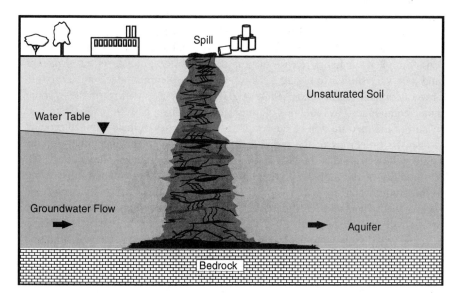

Figure 4   Trichloroethylene spill passing through an aquifer.

nonaqueous phase liquid through the aquifer will change directions when encountering small changes in the geology.

Organic liquids found in the unsaturated zone or the aquifer are referred to as nonaqueous phase liquids or NAPLs. When the NAPL is lighter than water it is referred to as light NAPL or LNAPL. When the NAPL is heavier than water it is referred to as dense NAPL or DNAPL. These are all general terms and can be used for pure compounds or mixtures of organics. If the mixture of organic compounds has a resulting specific gravity greater than water, the entire mixture will sink through the water and can be referred to as a DNAPL. Even if a lighter-than-water organic is part of the mixture, it is the property of the entire mixture that will rule its vertical movement in the aquifer environment. For example, benzene is 5 percent of an organic mixture composed mainly of TCE. The specific gravity of the mixture is 1.25. Even though benzene is lighter than water, it will move with the entire mixture and sink through the water. The same is true for chlorinated compounds. The author is currently working on a project where an LNAPL is composed of 23 percent PCE. The rest of the LNAPL is mineral oils, and the specific gravity of the mixture is less than 1.0. In this case the concentration of the PCE in the groundwater is above 1 percent of the solubility, which would provide strong indication that there is pure compound present. While this is true, the pure compound is in the form of a LNAPL, not a DNAPL.

Table 2 summarizes the specific gravity for 50 organic compounds. In general, petroleum hydrocarbons are lighter than water and chlorinated hydrocarbons are heavier than water. Table 2 does not include many petroleum hydrocarbons because most of the organic compounds that make up the various petroleum products (gasoline, kerosene, fuel oil, etc.) are not hazardous. Table 3 provides the solubility and specific gravity for some of the organic compounds found in petroleum hydrocarbons.

As can be seen in Table 3, a compound like decane is lighter than water, with a specific gravity of 0.73. Decane is also not very soluble in water, with a solubility of 0.009 mg/l. If we had a release of pure decane to the ground, its travel pattern would look very similar to Figure 2. However, all petroleum products are a mixture of several organic compounds. Table 4 shows the concentration of the specific organic compounds, by volume, of three representative gasolines. The fate and transport of the mixture will be the result of the properties of the combined product and also the properties of the individual compounds. The specific gravity of the mixture will determine if the NAPL will float or sink as it encounters the aquifer. The solubility of the individual compounds will control the compound dissolving into the groundwater of the aquifer and forming the plume. For example, gasoline is lighter than water and will float when it reaches the aquifer. The decane in the gasoline will mostly stay with the LNAPL. Benzene, however, is more soluble in water, and will dissolve into the water and be part of the groundwater contamination plume. The same thing would happen with our previous example with benzene being a part of a DNAPL. The rate of release of the benzene to the water of the aquifer would relate to the individual compound and not be controlled by the mixture.

Figures 3 and 4 represent movement of NAPLs that are not soluble in water. These drawings are a simplification of what happens in the real world. As can be seen in Table 1, all organic compounds have some solubility in water. Even hexachlorobenzene is slightly soluble in water. One liter of water will carry a maximum of 6 micrograms of hexachlorobenzene. Therefore, all of the NAPL that is in contact with the groundwater of the aquifer will release organic compounds to the water. Since groundwater is moving, the compounds will be carried away from the original point of contamination. Figures 5 and 6 show the plumes of organics that are

Table 2 Specific Gravity for Specific Organic Compounds

	Compound	Specific Gravity[a]	Ref.		Compound	Specific Gravity[a]	Ref.
1	Acenaphthene	1.069 (95°/95°)	1	26	bis(2-Ethylhexyl)phthalate	.984	1
2	Acetone	.791	1	27	Heptachlor	1.570	5
3	Aroclor 1254	1.5 (25°)	3	28	Hexachlorobenzene	2.044	1
4	Benzene	.87900	1	29	Hexachloroethane	2.090	6
5	Benzo(a)pyrene	1.35 (25°)	4	30	2-Hexanone	.815 (18°/4°)	1
6	Benzo(g,h,i)perylene	NA		31	Isophorone	.921 (25°)	2
7	Benzoic Acid	1.316 (28°/4°)	1	32	Methylene Chloride	1.366	1
8	Bromodichloromethane	2.006 (15°/4°)	1	33	Methyl Ethyl Ketone	.805	1
9	Bromoform	2.903 (15°)	1	34	Methyl Naphthalene	1.025 (14°/4°)	1
10	Carbon Tetrachloride	1.594	1	35	Methyl tert-Butyl Ether	.731	1
11	Chlorobenzene	1.106	1	36	Naphthalene	1.145	1
12	Chloroethane	.903 (10°)	1	37	Nitrobenzene	1.203	1
13	Chloroform	1.49 (20°C liquid)	2	38	Pentachlorophenol	1.978 (22°)	1
14	2-Chlorophenol	1.241 (18.2°/15°)	1	39	Phenol	1.071 (25°/4°)	1
15	p-Dichlorobenzene (1,4)	1.458 (21°)	1	40	1,1,2,2-Tetrachloroethane	1.600	1
16	1,1-Dichloroethane	1.176	1	41	Tetrachloroethylene	1.631 (15°/4°)	1
17	1,2-Dichloroethane	1.253	1	42	Tetrahydrofuran	.888 (21°/4°)	1
18	1,1-Dichloroethylene	1.250 (15°)	1	43	Toluene	.866	1
19	cis-1,2-Dichloroethylene	1.27 (25°C liquid)	2	44	1,2,4-Trichlorobenzene	1.446 (26°)	1
20	trans-1,2-Dichloroethylene	1.27 (25°C liquid)	2	45	1,1,1-Trichloroethane	1.346 (15°/4°)	1
21	2,4-Dichlorophenoxyacetic Acid	1.255	6	46	1,1,2-Trichloroethane	1.441 (25.5°/4°)	1
22	Dimethyl Phthalate	1.189 (25°/25°)	1	47	Trichloroethylene	1.466 (20°/20°)	1
23	2,6-Dinitrotoluene	1.283 (111°)	1	48	2,4,6-Trichlorophenol	1.490 (75°/4°)	1
24	1,4-Dioxane	1.034	1	49	Vinyl Chloride	.908 (25°/25°)	1
25	Ethylbenzene	.867	1	50	o-Xylene	.880	1

[a] Specific gravity of compound at 20°C referred to water at 4°C (20°/4°) unless otherwise specified.
NA = Not Available

1. Dean, J.A. Lange's Handbook of Chemistry, 11th ed. New York: McGraw-Hill Book Co., (1973).
2. Weiss, G. Hazardous Chemicals Data Book, 2nd ed. New York: Noyes Data Corp., (1986).
3. "Draft Toxicological Profile for Selected PCBs" (November 1987). U.S. Public Health Service Agency for Toxic Substances and Disease Registry.
4. "Draft Toxicological Profile for Benzo(a)pyrene" (October 1987). U.S. Public Health Service Agency for Toxic Substances and Disease Registry.
5. Verschueren, K. Handbook of Environmental Data on Organic Chemicals, 2nd ed. New York: Van Nostrand Reinhold Co., (1983).
6. Merck Index, 9th ed. Rahway, NJ: Merck and Co., Inc., (1976).

Table 3   Physical/Chemical Properties of Selected Petroleum Hydrocarbons

Compound	Molecular Weight	Specific Gravity	Solubility mg/l (@°C)	Boiling Point, °C	Vapor Pressure @1 atm and (°C)
Pentane	72.15	0.626	360 (16)	36	430 (20)
Hexane	86.17	0.66	13 (20)	68.7	120 (20)
Decane	142.28	0.73	0.009 (20)	173	2.7 (20)
Cyclopropane	42.08	0.72	37,000	-33	760 (-33)
Cyclopentane	70.14	0.751	<1,000	-	200 (13.8)
Cyclohexane	84.16	0.779	55 (20)	81	77 (20)
Benzene	78.11	0.878	1780 (20)	80.1	76 (20)
Toluene	92.1	0.867	515 (20)	110.8	22 (20)
ortho-Xylene	106.17	0.88	175 (20)	144.4	5 (20)
meta-Xylene	106.17	0.86	175 (20)	139	6 (20)
para-Xylene	106.17	0.86	198 (25)	138.4	6.5 (20)

Note: Compiled from various sources.

Table 4   Some of the Major Constituents of the Gasoline Fraction (b.p. 36-117°C) in Selected Petroleums

Constituent	Volume (%)		
	Conroe, TX	Colinga, CA	Jennings, LA
Alkanes:			
n-Pentane	0.33	0.44	1.12
n-Hexane	6.44	7.75	9.15
n-Heptane	6.90	5.94	8.42
2-Methylpentane	2.89	2.56	3.47
2,3-Dimethylhexane	0.22	1.30	2.39
Cycloalkanes:			
Cyclopentane	0.96	1.76	0.67
Methylcyclopentane	6.51	10.29	5.01
Cyclohexane	10.40	7.63	7.13
Methylcyclohexane	22.00	14.55	18.07
Ethylcyclopentane	2.03	4.38	2.34
Trimethylcyclopentane	3.64	8.12	4.18
Aromatics:			
Benzene	3.27	2.22	3.61
Toluene	16.19	7.94	12.02

Source: Adapted from Perry (1984).

dissolved in the groundwater. These plumes are created from the NAPL's interaction with the moving groundwater.

There are several other ways in which a plume can form. For example, rainwater can move through the vadose zone, dissolve contaminants that are located in that zone, and carry the contaminants to the aquifer. Volatile organic compounds in the

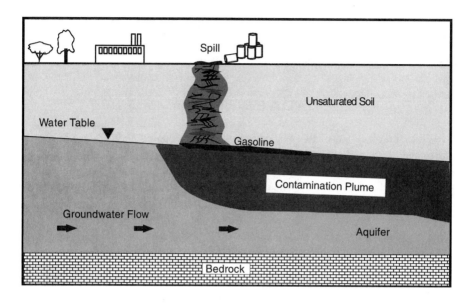

Figure 5   Contamination plume resulting from gasoline spill.

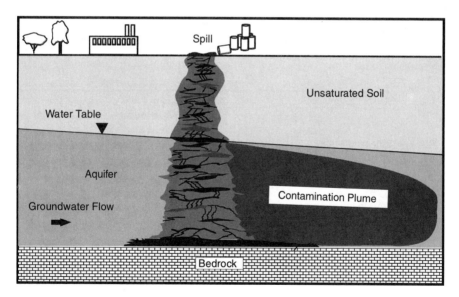

Figure 6   Contamination plume resulting from trichloroethylene spill.

vadose zone can volatilize and move in the gaseous phase (sometimes in a direction opposite to the direction of the groundwater flow). In this phase the organics can come into contact with the aquifer and dissolve into the groundwater. There is no way to list all of the possible methods. The important point is that there is a source of organics, and when this source (all or part of it) comes into contact with the aquifer it will start to dissolve into the groundwater. The groundwater, which is moving, will carry the organics away from the original point of contact. The term 'plume' refers to the dissolved phase of the organics in the aquifer.

PLUME MOVEMENT

While the objective of this book is to provide a detailed analysis of various *in situ* remedial technologies to clean the aquifer, we must first understand how the plume moves in the aquifer. The main processes that affect plume movement are advection, dispersion, retardation, chemical precipitation, and biotransformation. Each compound will be affected differently by these processes.

ADVECTION

Advection is the main process that moves the compounds in the aquifer. Advection is movement by bulk motion, and is quantified by the value of the groundwater velocity. Under most conditions, groundwater is constantly moving, although this movement is usually slow (typically 1–900 feet/year). To determine the flow and direction in an aquifer, basic information is needed. Once we collect or estimate that basic information, the groundwater flow rate may be calculated. The relationship for flow is stated in Darcy's Law

$$V = - \frac{K}{n_e} \frac{\Delta h}{\Delta x}$$

where V = pore water velocity [L^3/T]; K = average hydraulic conductivity, a measure of the ability of the porous media to transmit water [L/T]; n_e = affective porosity for flow; and $\frac{\Delta h}{\Delta x}$ = hydraulic gradient.

To determine the direction and velocity of flow, three or more wells may be drilled into the aquifer and the heads or water levels measured from a datum (typically mean sea level). Groundwater will flow from high head to low head (the negative sign in Darcy's Law keeps the velocity positive as the gradient is always negative). The hydraulic conductivity (K) is a function of the porous medium (aquifer) and the fluid (viscosity and specific weight); finer grained sediments such as silts and clays have relatively low values of K, whereas sand and gravel will have higher values. Other physical factors may affect the hydraulic conductivity including porosity, packing, sorting, solutioning, and fracturing.

This description refers to the perfect aquifer or a very large section of any aquifer. The problem is that water does not move with a uniform velocity in small sections of the aquifer or the microenvironment. In the microenvironment, most aquifers have areas of high water flow, low water flow, and no water flow. All aquifer soils are heterogeneous, some extremely so. The most obvious conditions that create extreme variable velocities are when the aquifer is constructed of material that forms large open areas. The best examples of this are fractured bedrock and karst geology. In both cases the water flows mainly in the large, relatively open channels. Only the compounds that come into contact with one of these flow channels will combine with the water and form the plume. Since the flow channels only cover a small fraction of the cross sectional area of the aquifer, the chances of the contaminant interacting with the water are greatly diminished. This limits the amount of organic that enters the water, but also limits the ability of the water to act as a carrier to clean the aquifer of the contaminant. In the field, the contaminant usually acts as a long-term source in these types of aquifers. The water movement in these aquifers can be quick and the plume length can be substantial.

Another problem created in these situations is in the investigation of the aquifer. We normally drill wells into the aquifer, take soil samples as we drill, and collect samples of the groundwater when the well is finished. In an aquifer in which the main flow is through open areas, if we do not encounter one of those areas, we will not be able to measure the contamination in the aquifer. When we consider the cross sectional area represented by the well in relationship to the overall cross sectional area of the aquifer, we realize the low probability of being able to take a representative sample of the contaminants in these types of aquifers.

The problem with using water as a carrier to clean these aquifers is obvious, and few systems have been designed that use pump and treat as the method to clean these aquifers. Pumping has usually been limited to controlling plume movement. It is beyond the scope of this book to discuss in detail these types of aquifers. Fractured bedrock and karst geology are unique situations. There are many fine books and articles written on the methods used to investigate contamination in these situations. Experts in the field have accepted that certain types of aquifers create flow patterns that cause trouble when we try to measure contamination, predict plume movement, and clean the aquifer. The problem is that we do not recognize these same limitations when the same problems occur on a smaller scale in the aquifer. As stated before, most aquifers have areas in which the water is traveling at a high velocity, areas where water travels at a low velocity, and relatively stagnant areas. When we talk about advection, we must understand on both a macro and micro scale how water travels through the aquifer. Let us review some of the other geology that creates flow patterns in the aquifers.

Many geologic situations can create zones of high and low velocity in an aquifer. The classic descriptions of these two conditions are sand (or gravel) lenses and clay lenses. The soil particles in a sand lens are larger than the surrounding particles in the aquifer. While we are using the term sand lens, we are really describing any aquifer that has an area in which the soil particles are larger than the surrounding particles, and the resultant permeability of the lens is higher than the permeability of the surrounding geology. In the real world, these areas are usually composed of coarse sands and gravel.

A typical sand lens is shown in Figure 7. As can be seen by the flow lines on Figure 7, the water flow in the aquifer will have a preference to move through the area of least resistance. The sand lens will allow most of the water flow in this area

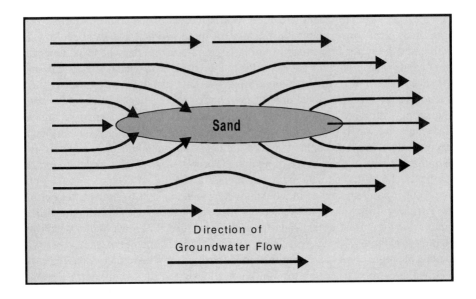

Figure 7   Groundwater flow pattern resulting from a sand lens.

of the aquifer to travel through the sand. While the main path of the water in the aquifer may be the sand, the organic NAPLs may have a different path.

Figure 6 shows the contaminant traveling by gravity down through the aquifer. This flow path is perpendicular to the sand lens. The contaminant will form a plume as soon as it comes into contact with the water of the aquifer. However, while all of the water in the source area will be contaminated, the plume will move out from the source area by the movement of the water. Under the macroenvironment, this movement will be controlled by the sand lens. When the sand lens system is significant in the aquifer, the flow pattern of the plume can be represented by Figure 8. The plume in Figure 8 has 'fingers.' These are areas of high flow that extend the plume faster along those paths.

Figure 8 is two-dimensional. An aquifer is three-dimensional. The same preferential flow paths can occur along the depth of the aquifer. Figure 9 shows the result of sand lenses along the depth of the aquifer. Once again, while all of the water is contaminated in the source area, the plume will mainly move by the movement of the water in the high-flow areas of a sand lens. Figure 9 shows the importance of monitoring well screen placement in determining the contamination in the aquifer.

The opposite of the sand lens is the clay lens. When the soil material in the lens is smaller and the resultant permeability is less than that of the surrounding geology, the water will flow around that area. Figure 10 shows the water flow pattern when the aquifer includes a clay lens. While we refer to these areas as clay lenses, they can be made up of any low-permeability soil material. However, they usually consist of clay or fine silt.

The clay lens blocks water flow through an area of the aquifer. The clay lens is not significant while the plume is spreading. Advection movement of the plume

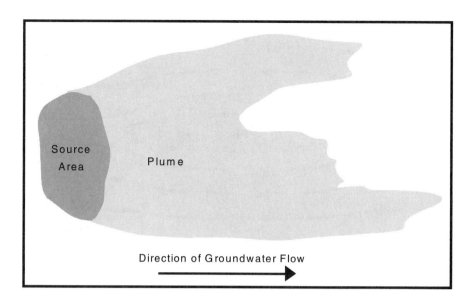

Figure 8   Sand lenses causing "fingers" in the plume, plan view.

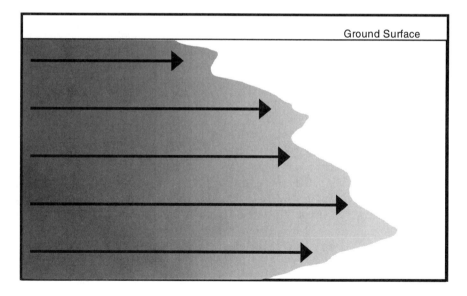

Figure 9   Sand lenses causing variations in flow patterns, section view.

simply moves around that area. However, the clay lens becomes significant if it is contaminated, and we are trying to remove contaminants from that area of the aquifer. Since water has a preference to move around the clay lens, then water will not act as a good carrier to remove the contaminants. The contaminants will remain in the clay lens and act as a source of contamination when we try to remediate the aquifer. We will discuss this further under diffusion later in the chapter.

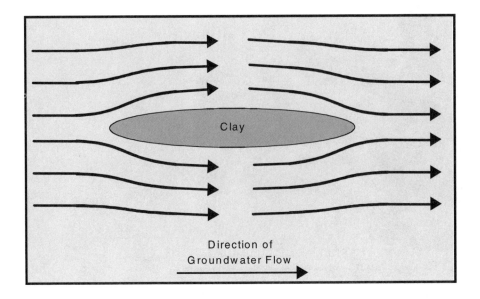

Figure 10 Groundwater flow pattern resulting from a clay lens.

When we discuss plume movement, we usually only consider sand and clay lenses when they significantly affect the flow and pattern of the plume. When the sand lens system is large enough to create the fingers shown in Figure 8, they are included in the description of the aquifer. However, if the sand lenses are relatively small, and other plume movement processes fill in the areas between the lenses, we too often refer to this type of aquifer as homogeneous.

This takes us back to the discussion on Darcy's Law. Darcy's Law is applicable to the perfect aquifer (homogeneous) or to a large enough section of any aquifer. The more heterogeneous factors in the aquifer, the larger the area that is needed to accurately describe the water movement by Darcy's Law and the average bulk hydraulic conductivity. When we have solution channels and/or bedrock fractures, a very large area is required. When we have sand and/or clay lenses, a relatively small area is required, depending upon the size of lenses.

The problem is that no aquifer is perfectly homogeneous. We can always select a small enough section of any aquifer so that the water movement in it cannot be described by Darcy's Law and the average hydraulic conductivity. Figure 11 shows a microenvironment of a homogeneous aquifer. The soil particles in this section are all of similar size. Even with the soil homogeneity, the small section of the aquifer has areas of high water velocity, areas that have low water velocity, and areas that are relatively stagnant. Szecsody described this situation by Figure 12 (Szecsody and Bales, 1989).

This discussion is moot when we are describing plume movement. When we are describing how a plume moves in an aquifer, the micro movement of the water does not matter. The plume moves mainly by advection and Darcy's Law describes that

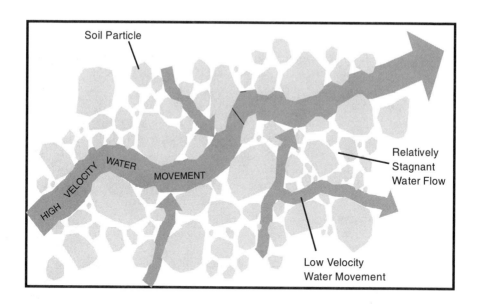

Figure 11   Water flow through soil microcosm.

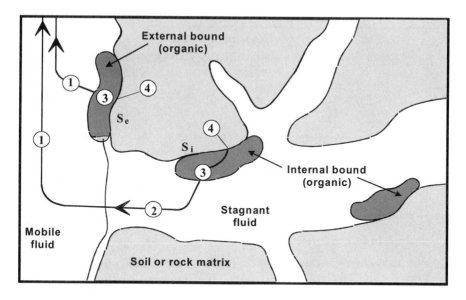

Figure 12   Conceptual model of sorption in porous aggregate.

advection. The microenvironment "fills in" by another method of organic movement, dispersion, which will be discussed in the next section of the chapter. Darcy's Law is only dependent upon selecting a large enough section of the aquifer. The small section that is described in Figure 11 is not significant in that analysis or the advection movement of the plume.

However, when we analyze water as a carrier to move contaminants out of an aquifer, the micro analysis becomes very significant. During pump and treat, we are dependent upon the water coming into contact with the contaminants, picking up the contaminant (solubilizing), and carrying the contaminant to the recovery well. To completely clean an aquifer, the water must get to all of the soil particles. Figure 11 shows that the water will not flow past every particle in the aquifer. The contaminants located in the minor flow and relatively stagnant flow areas will have to move out of those areas by means other than water movement or advection.

The final question becomes how large is the area in the microenvironment that does not have water moving by each soil particle? As part of an EPA study at the Thermo Chem Superfund site in Michigan, a detailed analysis of contaminant movement in the aquifer was performed. Three methods of collecting data were used: (1) standard monitoring wells; (2) monitoring wells set up as a transect; and (3) GeoProbe[7m] wells set up as a transect. Previous studies characterized the site as possessing an upper sand unit underlain by an aquitard. The upper sand unit has a hydraulic conductivity of 1×10^{-3} cm/sec to 1×10^{-4} cm/sec. Figure 13 shows the study area. Previous investigations had constructed 40 monitoring wells over an area of approximately 1000 x 2000 feet. A transect of monitoring wells was set up at the southern end of the site, Figure 13. Figure 14 shows the screening for the 34 transect monitoring wells. The GeoProbe[7m] transect was installed in the same area.

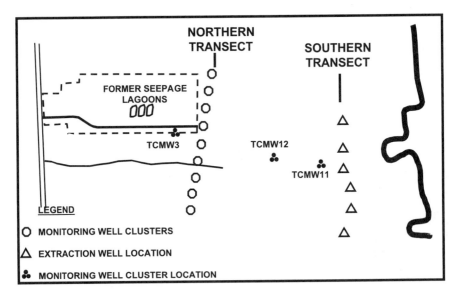

Figure 13   Thermo Chem Superfund site plan.

The first part of the study was to analyze the plume. The monitoring well transect was used to show the contents of the cross section of the aquifer. It was decided that concentrations of the contaminant would not tell the entire story of the plume movement. In order to understand the plume, we have to know the rate of contam-

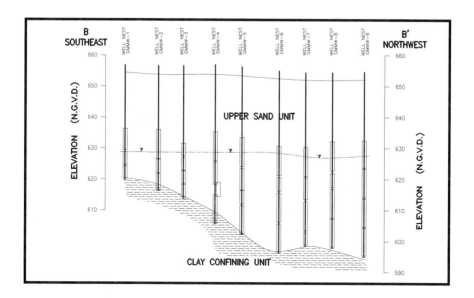

Figure 14 Section view, transect monitoring wells clusters.

inant movement, not just the concentrations. To obtain this data it was necessary to measure the groundwater flow rate and the contaminant concentration for each well. Multiplying the flow rate times the concentration provided the rate of mass movement in that section of the aquifer, the flux. Figures 15 and 16 show the flux rates for TCE and cis-1,2DCE in grams per year (g/yr). The same analysis was performed for all of the contaminants in the aquifer. Looking at Figures 15 and 16 we can see that small areas of the aquifer are responsible for most of the movement of a contaminant mass. We are left with the impression that the contaminants are traveling in a pipeline or a streambed, not a homogeneous sand aquifer. These figures also show that not all of the contaminants are in the same location in the aquifer.

The second part of the study was to compare the three methods of monitoring. An estimate of the flux of the contaminants was calculated based upon the original monitoring wells and the established groundwater flow rate. This flux rate was compared to the flux rate calculated from the groundwater flow and concentration data collected from the two types of transects. Table 5 summarizes the fluxes for four of the chlorinated compounds found in the aquifer: PCE, TCE, cis-DCE, and vinyl chloride (VC). The two types of transect monitoring wells calculated the flux rates at 30 to 1000 times higher than the conventional monitoring wells.

This analysis shows that the placement of the extraction well and its screening interval are critical to optimizing the rate of extraction of the contaminants from the aquifer. It also shows that a single extraction well will not be optimum for all of the contaminants in one area of the aquifer. While we can stop the general movement of the water in the aquifer, and stop the total flux of contaminants, we cannot collect all of the contaminants in the aquifer. This is true even if we used multiple wells in this area of the aquifer. The areas of high flow would empty of their contaminants

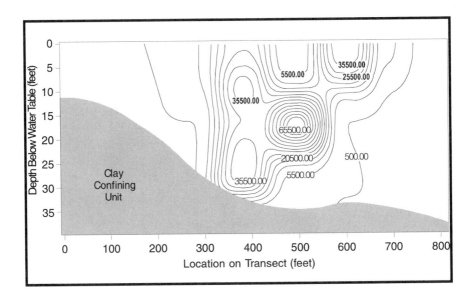

Figure 15   Spatial distribution of TCE flux along northern transect (g/yr), November 1997.

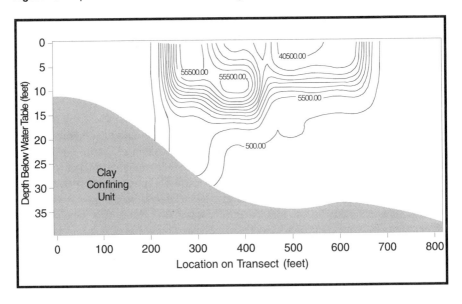

Figure 16   Spatial distribution of cis-1,2DCE flux along northern transect (g/yr), November 1997.

quickly, but the areas of low flow would not be cleaned by advection. The water would mainly travel in the areas of high flow and bypass the areas of low flow.

While advection is the main force in plume movement, is it the main force when we try to remove contaminants from an aquifer? We need to discuss the other types

Table 5   Estimates of Flux Across Transect (kg/yr)

	Permanent Transect	GeoProbe Transect	Conventional Well Array
PCE	55.1	45.9	1.5
TCE	182.5	224.2	8.9
cis-DCE	311.7	918.0	19.0
VC	26.7	53.0	0.05

of movement before we can fully answer these questions. The point to be made is that even though advection is the main process in plume movement, it may not be the controlling process when remediating the plume.

DISPERSION

Dispersion creates two main results in the movement of the plume. Both relate to a spreading of the contaminants. The first result can be shown in Figure 17. Additional spreading beyond the plume movement caused by advection can be the result of dispersion. Keely cautions that permeability differences between strata can be a main cause of plume spreading by dispersion. (Keely, 1989) Figure 18 shows the difference between spreading by advection and spreading by dispersion.

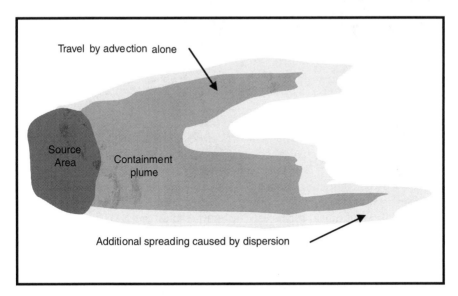

Figure 17   Containment movement by dispersion.

The second movement of the contaminants that can relate to dispersion is filling in the areas between main advection movement. As discussed in the previous section, the main movement of the plume is by advection. The trouble is that there are main flow paths and minor flow paths. The plume would mainly spread along main flow paths and leave the minor flow paths clean. However, the real rate of flow of the water in an aquifer is very slow. Dispersion helps move the contaminants into the areas of relatively slow flow as the plume moves along by advection. Dispersion fills in the areas that do not have direct contact with the main water flow. Even when there are sand or clay lenses, dispersion fills in the area in between the areas of relatively high flow.

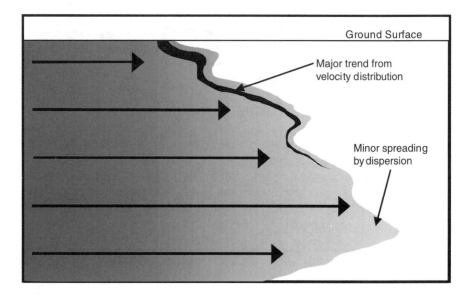

Figure 18   Cross-sectional view of containment spreading. Permeability differences between strata cause comparable differences in advection and, hence, plume spreading.

Dispersion can be divided into two components, (1) mechanical movement, and (2) chemical movement. Mechanical dispersion comes from the difference in velocity from water molecules in contact with soil particles and water molecules moving in between the soil particles. When the water molecule is in contact with the soil particle or when it 'wets' the grain of sand, silt, or clay, its movement is restricted. The molecule in contact will slow its movement relative to the rest of the water molecules. The molecules that move in between the soil particles and have no direct contact with the soil will move at a relatively higher rate. The difference in these velocities creates the mechanical component of dispersion.

The chemical component of dispersion is the result of molecular diffusion. All molecules have independent motion. While the main movement of a contaminant in water will be by the movement of the water carrying the compound, the compound will also have random movement within the water. Diffusion is a minor part of plume spreading. When the water is moving quickly it has a negligible impact. However, when the plume movement is very slow, diffusion can cause significant spreading. What is more important from a remediation point of view is that diffusion can spread the contaminants into areas that have relatively slow flow or are completely stagnant, even when the water is moving fast. This process will place contaminants in areas of the aquifer that have no direct water flow contact. This is not significant when we are watching or trying to predict the plume movement. However, when we reverse the process and use the water to bring back the contaminants, these compounds are out of contact with the flow of water. As the water in the main flow paths gets cleaner, the contaminants that have diffused into the low flow areas reverse course. The compounds naturally diffuse from an area of high concentration (the micro clay lens or other stagnant area) to an area of low concentration (the main water flow path). Thus, these micro areas become a source of low concentration contaminants as we use pump and treat to remediate a site.

RETARDATION

The next major influence on the movement of contaminants in an aquifer is retardation. Retardation is the result of the contaminants being attracted and held to the surface of the aquifer solids. Sorption and ion exchange are the physical and chemical processes that work between the contaminants and the solids. The result is that as the water carries the contaminants through the aquifer, the contaminant comes into contact with the aquifer solids. The solids temporarily hold or retard the contaminant as the water continues to move. Thus the water moves faster than the contaminant. The difference in the rates relates to how "temporary" the aquifer solids hold the contaminant. There are several factors that influence retardation. The properties of the contaminant that affect retardation can be approximated by measurements such as bulk density and partition coefficients. Table 6 provides the octanol-water partition coefficients (K_{ow}) for 50 organic compounds.

Generally, the less soluble and/or more complex the compound, the higher the attraction to the aquifer solids. Carbon tetrachloride has a higher octanol-water coefficient than TCE. It will have a higher attraction to the aquifer solids, and thus move slower than TCE in the aquifer.

The properties of the aquifer solids also have a great effect on the retardation of the contaminants. In general, the higher the organic content which can be measured by total organic content (TOC) of the soil, the more the soil retards the movement of the contaminants. The ion exchange capacity of the aquifer solids can also affect contaminants that exist as ionic species, mainly heavy metals.

The only real way to account for all of these factors in an actual aquifer is by direct measurement. The laboratory can be used by taking soil samples from the site and passing the compounds through a test cell. While these data will be more accurate than the results predicted from partition coefficient and TOC, the best measurement of the effective mobility of each contaminant is made from observing the actual plume composition, and its spreading over time.

Retardation is reported as a ratio of groundwater movement to the movement of the organic compounds, or the velocity of uncontaminated groundwater divided by the velocity of the contaminant. The groundwater is considered the base. If the contaminant travels at one half the speed of the groundwater, then the retardation factor is 2. If the contaminant travels at one fourth the speed of the groundwater, then the retardation factor is 4. Since both the contaminant properties and the aquifer

Table 6 Octanol Water Coefficients (K_{ow}) for Specific Organic Compounds

	Compound	K_{ow}	Ref.		Compound	K_{ow}	Ref.
1	Acenaphthene	1.0×10^4	2	26	bis(2-Ethylhexyl)phthalate	9.5×10^3	2
2	Acetone	6×10^{-1}	1d	27	Heptachlor	2.51×10^4	2
3	Aroclor 1254	1.07×10^6	2	28	Hexachlorobenzene	1.7×10^5	1a
4	Benzene	1.3×10^2	1a	29	Hexachloroethane	3.98×10^4	2
5	Benzo(a)pyrene	1.15×10^6	2	30	2-Hexanone	2.5×10^1	3
6	Benzo(g,h,i)perylene	3.24×10^6	2	31	Isophorone	5.0×10^1	2
7	Benzoic Acid	7.4×10^1	2	32	Methylene Chloride	1.9×10^1	1b
8	Bromodichloromethane	7.6×10^1	2	33	Methyl Ethyl Ketone	1.8	1a
9	Bromoform	2.5×10^2	1b	34	Methyl Naphthalene	1.3×10^4	2
10	Carbon Tetrachloride	4.4×10^2	1a	35	Methyl tert-Butyl Ether	NA	NA
11	Chlorobenzene	6.9×10^2	1a	36	Naphthalene	2.8×10^3	2
12	Chloroethane	3.5×10^1	2	37	Nitrobenzene	7.1×10^1	2
13	Chloroform	9.3×10^1	1a	38	Pentachlorophenol	1.0×10^5	1b
14	2-Chlorophenol	1.5×10^1	2	39	Phenol	2.9×10^1	1a
15	p-Dichlorobenzene (1,4)	3.9×10^3	2	40	1,1,2,2-Tetrachloroethane	2.5×10^2	2
16	1,1-Dichloroethane	6.2×10^1	1a	41	Tetrachloroethylene	3.9×10^2	1a
17	1,2-Dichloroethane	3.0×10^1	1a	42	Tetrahydrofuran	6.6	4
18	1,1-Dichloroethylene	6.9×10^1	1a	43	Toluene	1.3×10^2	1a
19	cis-1,2-Dichloroethylene	5	1a	44	1,2,4-Trichlorobenzene	2.0×10^4	2
20	trans-1,2-Dichloroethylene	3	1a	45	1,1,1-Trichloroethane	3.2×10^2	1b
21	2,4-Dichlorophenoxyacetic Acid	6.5×10^2	2	46	1,1,2-Trichloroethane	2.9×10^2	1a
22	Dimethyl Phthalate	1.3×10^2	2	47	Trichloroethylene	2.4×10^2	1a
23	2,6-Dinitrotoluene	1.0×10^2	2	48	2,4,6-Trichlorophenol	7.4×10^1	2
24	1,4-Dioxane	1.02	2	49	Vinyl Chloride	2.4×10^1	1a
25	Ethylbenzene	1.4×10^3	1a	50	o-Xylene	8.9×10^2	1c

NA = Not Available

1. *Superfund Public Health Evaluation Manual*, Office of Emergency and Remedial Response, U.S. Environmental Protection Agency (1986).
   a. Environmental Criteria and Assessment Office (ECAO), EPA, Health Effects Assessments for Specific Chemicals (1985).
   b. Mabey, W.R., Smith, J.H., Rodoll, R.T., Johnson, H.L., Mill, T., Chou, T.W., Gates, J., Patridge, I.W., Jaber, H., and Vanderberg, D., "Aquatic Fate Process Data for Organic Priority Pollutants", EPA Contract Nos. 68-01-3867 and 68-03-2981 by SRI International, for Monitoring and Data Support Division, Office of Water Regulations and Standards, Washington, D.C. (1982).
   c. Dawson et al., "Physical/Chemical Properties of Hazardous Waste Constituents," by Southeast Environmental Research Laboratory for USEPA (1980).
   d. *Handbook of Environmental Data for Organic Chemicals*, 2nd ed. New York: Van Nostrand Reinhold Co., (1983).
2. USEPA "Basics of Pump-and-Treat Groundwater Remediation Technology" EPA/600/8-901003, Robert S. Kerr Environmental Research Laboratory (March 1990).
3. Lyman, W.J., et al. "Research and Development of Methods for Estimating Physicochemical Properties of Organic Compounds of Environmental Concern" (June 1981).
4. EPA Draft Document "Hazardous Waste Treatment, Storage and Disposal Facilities (TSDF) Air Emissions Model" (April 1989).

solid properties all contribute to the retardation of a specific compound, we cannot provide a table of the retardation factors for each compound. Ranges can be provided for compounds and types of soil, but more precise numbers will have to be generated at each site.

One example of the combined effect of diffusion and retardation is work by Roberts, Goltz, and MacKay (1986). Working on the aquifer at the Canadian Forces Base, Borden, Ontario Canada, Roberts, Goltz, and MacKay studied the movement of organic compounds over a two-year period. They created a pulse of organics in the aquifer and then monitored the concentrations at three locations in the downstream aquifer. The results are shown in Figure 19. As can be seen, the tracer and organic compounds all have a tendency to tail off with time. The higher the retardation factor for the compound, the longer the tail.

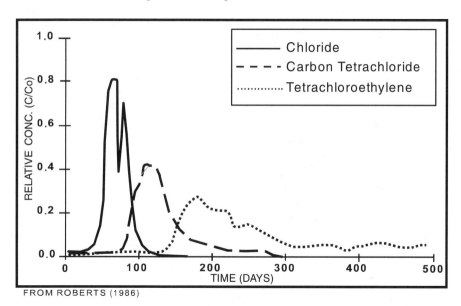

FROM ROBERTS (1986)

Figure 19   Breakthrough responses for chloride, carbon tetrachloride, and tetrachloroethylene at well (X=5.0 M, Y=0.0 M, Z=-3.26 M). (Roberts, Goltz, and MacKay, 1986. With permission.)

The tails in this study were created in a homogeneous, sand aquifer over a 5-meter sampling area. Due to the nature of the aquifer and the short distances, the effect from diffusion will be minor compared to retardation. As the aquifer becomes less homogeneous, the tails of concentration will lengthen.

One final note before we leave retardation. The same properties that can slow the movement of contaminants in an aquifer can also combine to speed the movement of the contaminants. When we discuss aquifer solids, we usually mean relatively large soil particles. However, very small soil particles called colloids also exist in the aquifer. These small particles are not stationary, but can move with the water. The water once again acts as a carrier for these solids. When a contaminant is attached to one of these small particles it travels with the particle. Instead of being retarded, it moves at the rate the groundwater is moving. Sampling methods become important to understand more than how much compound is at a point in the aquifer. How the compound got to that point is also important in understanding what the data really mean.

CHEMICAL PRECIPITATION AND BIOTRANSFORMATION

The last thing that can happen to the contaminants as they move through the aquifer is that they can disappear. The chemicals can change their form, or they can be completely destroyed. These actions, or reactions, can take place chemically or biochemically.

The best known chemical reaction in the aquifer is precipitation. Heavy metals in an ionic or dissolved phase can react with the aquifer soils or anions in the groundwater and precipitate. Once the metal is in a solid phase form, it no longer travels with the groundwater, (except if the precipitate is a colloid that can continue to be carried by the water). If the groundwater does not carry the metal, then it is no longer a part of the plume, and it will not show up at any of the monitoring points.

The solubility of most heavy metals is related to the pH of the water. Figure 20 shows the solubilities for several metals in relationship to the pH of the water. Two points must be made about these curves, (1) they are only relative, and (2) they do not represent exact values. Every water will be slightly different as far as specific solubility and the pH of minimum solubility.

The pH surrounding the immediate area where the dissolved metal comes into initial contact with the aquifer may be low. This could be the interaction of the rest of the chemicals associated with the heavy metal. When industry uses high concentrations of heavy metals, the solutions are usually very acidic. At the low pH the metal would be relatively soluble and move with the groundwater movements. As the groundwater moved from the immediate area, the aquifer solids would interact with the components of the groundwater and tend to neutralize the low pH. As the pH increased, the metal would precipitate as a result of natural water chemistry. The dissolved metal could also directly interact with the aquifer solids and precipitate.

The other major reaction in aquifers is biochemical. Bacteria are ubiquitous. They exist everywhere, including aquifers. The bacteria can also interact with the contaminants and change their form or completely destroy them. Chapter 7 will cover *in situ* biogeochemical reactions in detail. Bacteria can interact with metals, changing their valence state and making them more or less soluble. Iron fouling in wells is an example of bacterial interaction with metals.

A more universal interaction is between bacteria and organic contaminants. The bacteria can use many contaminants as food sources. Bacteria derive energy and building blocks for new bacteria from the carbon molecules in the contaminant. In addition, several of the enzymes that bacteria produce can interact with contaminants that are not considered food. The contaminants can be transformed into new chemicals or can be completely mineralized to CO_2. An example of transformation is that TCE can be transformed into VC in an aquifer. Petroleum hydrocarbons have a tendency for complete mineralization. In both cases, the original compound disap-

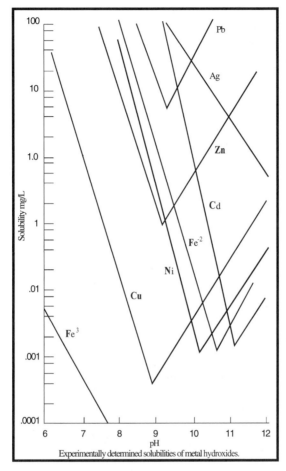

Figure 20   Solubilities of metal hydroxides at various pHs. (Courtesy of Graver Water, Union, NJ.)

pears. In the first case a new compound is created. The new compound is also a contaminant, and its movement in the aquifer will be governed by its chemical properties.

Both chemical and biochemical reactions can change the contaminant. This has to be understood when interpreting the data and plume movement, and is very important to understand when trying to use water as the carrier to remove the contaminant.

NONAQUEOUS PHASE LIQUIDS — NAPLs

The final area we will discuss in this section is the effect of NAPLs on water movement in the aquifer. Most papers and textbooks discuss NAPLs as a source of contamination. We have already discussed this part of NAPLs in a previous section.

It is unusual to consider these pure compounds in a context of water movement. However, as you will see in the following discussion, the NAPLs can also have a significant effect on water as the carrier for our pump and treat remediations.

Most organic contaminants that we encounter in the groundwater exist as liquids when they are in their pure form. Pure TCE, for example, is a free-flowing liquid. If it is lost to the ground, it will flow as a liquid through the ground and aquifer solids. As stated earlier, the liquid will sorb to the soil particles in both the unsaturated zone and the aquifer as it travels down through the ground.

Once the NAPL is in the unsaturated zone or the aquifer it can exist in several physical forms. First, the NAPL can coat the soil particles (Figure 21) or sorb to the soil particles. Second, the NAPL can actually fill all of the pore space between soil particles (Figure 22). Finally, the NAPL can fill fractures or voids in the subsurface materials.

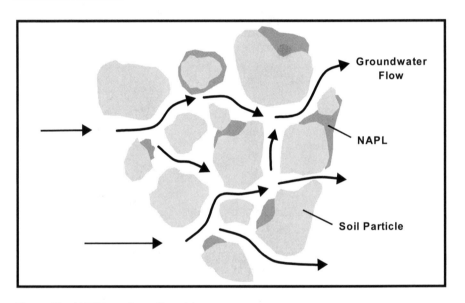

Figure 21   NAPLs coating soil particles.

It is not always easy to find the presence of NAPLs. The easiest method is direct evidence. LNAPLs are normally found by this method. When the well is drilled we find the pure compound floating on top of the water either in measurable amounts or simply as an oil sheen on top of the groundwater. Direct evidence of DNAPLs is much harder to come by. The material sorbed to the soil does not always enter the water and show up in the monitoring well as a separate phase. Also, intercepting where the actual DNAPL traveled down through the ground and trying to find it in core samples is not always reliable. Other methods have been developed to help determine if NAPLs are present in an aquifer. These methods are indirect measurements of the presence of NAPLs. The EPA recommends four indirect methods for determining NAPL presence.

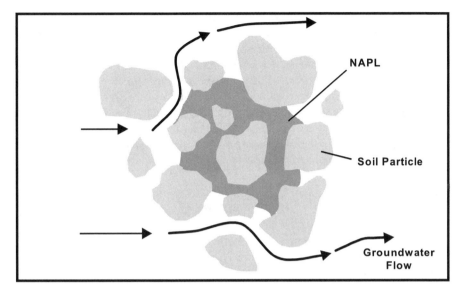

Figure 22   NAPLs completely filling in pores.

1. High concentration of contaminants in the groundwater.
2. Depth of contamination in the aquifer.
3. Persistence of contaminants in a pump and treat system.
4. Contaminant source characteristics (EPA 1992a).

NAPLs can be a significant source of mass of contaminants in an aquifer. Since they are pure compounds, they have a high mass in a relatively small space. What is more significant when we try to remove these compounds is the amount of space that they occupy in the aquifer or vadose zone. When the NAPL coats the soil particle or completely fills the pore space between particles, the movement of other liquids in that same area is severely restricted. Water will not be able to move through an area where the pore spaces are completely filled with NAPL. Water will have severely restricted movement in areas where the NAPL coats the soil particle and reduces the space between the soil particle. All of these restrictions force the water to move around the area where the NAPL is present.

As we discussed before, water must have intimate contact with the contaminants if it is to be used as a carrier to remove the contaminants from underground. The presence of NAPLs in the aquifer causes two problems. First, the water cannot come into direct contact with the NAPL because of movement restrictions. Second, the NAPL represents a huge source of contaminant. Once again, the only method for this contaminant to be removed from the aquifer with water as a carrier is for the contaminant to diffuse over to an area where the water is able to move through the aquifer.

One final property should be discussed before we leave the NAPL discussion. NAPL movement through the ground and water movement through the ground are significantly different. We cannot assume that the main channels that allow water flow will be the same main channels that will help the NAPL flow through the ground. The NAPL will have significantly different chemical and physical properties from water. The interaction with the soil particles will also be very different. In addition, the NAPLs have a tendency to flow vertically in the soil with gravity being the main driving force in the downward movement. Water, in the saturated zones, will mainly flow horizontally. The main flow channels that exist horizontally will not be the same flow channels that exist vertically, although they will intersect. This means that significant sources of contamination can exist in the aquifer in areas that do not receive high flows of water. These NAPLs can then remain stagnant and serve as a source of contamination for years.

PUMP AND TREAT

Now that we have contaminated the aquifer, the objective is to put it back in its original condition or to clean the aquifer. The problem is that our main tool has severe limitations. On a purely logical basis, one would think that in a saturated zone water would be everywhere, and therefore be in contact with all possible points of contamination in the aquifer. This is true. The problem occurs when we try to use the water to carry the contaminants out of the aquifer. All of the restrictions discussed above prevent the total removal of the contaminant. Let us review some real world examples of this problem.

The EPA performed the best known and most complete study of the effectiveness of pump and treat systems. Phase I of the study was completed in 1989 and covered 19 Superfund sites. In 1991, during Phase II of the study, five more sites were included in addition to more data from the original 19 sites. The study is summarized in an EPA report titled, "Evaluation of Ground-Water Extraction Remedies: Phase II," February 1992, Publication 9355.4-05. The conclusions made from that study are as follows:

1. Data collected, both site characterization data prior to system design and subsequent operational data, were not sufficient to fully assess contaminant movement or groundwater system response to extraction.
2. In the majority of cases studies (15 of the 24 sites), the groundwater extraction systems were able to achieve hydraulic containment of the dissolved-phase contaminant plume.
3. Extraction systems were often able to remove a substantial mass of contamination from the aquifer.
4. When extraction systems were started up, contaminant concentrations usually showed a rapid initial decrease, but then tended to level off or decrease at a greatly reduced rate. This may be a result of the type of monitoring data collected as much as a reflection of an actual phenomenon of groundwater extraction systems. For example, it can reflect successful remediation as the contaminated zone shrinks and less-contaminated groundwater is pulled into the extraction system, or poor placement of groundwater monitoring wells.
5. Based on the available information, potential NAPL presence was not addressed during site investigations at 14 of the 24 sites. At five sites they were addressed

because they were encountered unexpectedly during the investigation. As a result, it is difficult to determine NAPL presence conclusively from available site data. Because NAPLs were not addressed in the site investigation, they also were not addressed in the remedial design. Consequently, a groundwater extraction system may be performing as designed (removing dissolved phase contaminants) even though it will not achieve the cleanup goals within the predicted time frame.

6. At 20 of the 24 sites, chemical data collected during remedial operation exhibited trends consistent with the presence of DNAPLs. However, even where substantial soil and water quality data were available, a separate immiscible phase was rarely sampled or observed. This is consistent with DNAPL behavior. In other words, they can move preferentially through discrete pathways that may easily be missed even in thorough sampling schemes. DNAPL was observed at sites where contaminant concentrations in groundwater were less than 15 percent of the respective solubility.

7. The importance of treating groundwater remediation as an iterative process, requiring ongoing evaluation of system design, remediation time frames, and data collection needs, was recognized at all of the sites where remedial action was continuing.

The EPA is careful in the report to state that none of the sites selected had its extraction systems optimized. Even though the contaminant concentrations seem to be stabilized at 17 of the sites, the EPA states, "The apparent stabilization of contaminant concentrations may be due to a number of factors not necessarily related to technical limitations of groundwater extraction. These include non-representative monitoring techniques, other contaminant sources not previously identified, inadequate extraction network design, and/or inefficient operation of the extraction network." In this report, the EPA has not given up on pump and treat as a remediation technology.

The National Research Council reviewed 77 cases for pump and treat and found that it was able to achieve full cleanup at only eight sites. The committee came to the conclusion that pump and treat was ineffective at locations that contained significant amounts of solvents, precipitated metals, contaminants that have diffused into small pore spaces, or those that adhere strongly to soils (National Research Council 1994).

The longest running pump and treat system that ARCADIS Geraghty & Miller has worked on is shown in Figure 23. ARCADIS Geraghty & Miller has been operating a pump and treat system for a client in the northeast since 1978. The system was originally installed to control and remediate a carbon tetrachloride contamination plume. There are several interesting features that can be depicted from Figure 23. First, Well 1 was the downstream well originally constructed to ensure that the carbon tetrachloride plume would not spread further. It has successfully performed this task and the plume has not spread from that well site. In 1982 a second well was installed upstream of Well 1 to help remediate the site faster. Well 2 successfully cut-off all sources of carbon tetrachloride to the area of the aquifer entering Well 1. As can be seen in Figure 23 the concentration in Well 1 seems to go through two periods in which the concentration in the well stabilizes. The first period is prior to the installation of Well 2 when the concentration seemed to stabilize at about 20 micrograms per liter (ʋg/l). The second period is after the source of contamination is prevented from reaching Well 1; the concentration seems to stabilize at about 5 ʋg/l. Even with the original source of carbon tetrachloride being removed from Well 1, the carbon tetrachloride concentration still stabilizes and does not completely clean out of that area of the aquifer. Well 2 also goes through stabilization at approximately 40 ʋg/l. As can be seen, the performance of this pump and treat system coincides with the EPA review of the 24 sites.

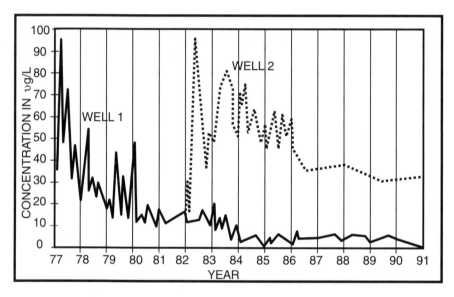

Figure 23   Results from a carbon tetrachloride pump and treat.

Other theoretical and laboratory studies have shown the same problems. Feenstra performed a series of laboratory tests on the influence of DNAPL on the performance of groundwater pump and treat systems (Feenstra 1992). In the laboratory, he had groundwater flow directly through areas contaminated with DNAPL and had water flow parallel to pools of DNAPL. In his studies he found that DNAPLs could continue to act as a source of contamination for over 1,000 pore volumes of water movement. His main two conclusions were that DNAPL zones represent continuing sources and pump and treat are unlikely to accelerate dissolution significantly.

While Feenstra and the EPA have concentrated on DNAPLs as the main source of continuing contamination at sites, the results shown by the ARCADIS Geraghty & Miller pump and treat and others show that geological conditions can also cause these stabilized concentrations in the pump and treat system. Goltz and Roberts conclude, "Relatively small zones of immobile water may result in extensive tailing, if sufficient sorption capacity exists in the zones" (Goltz 1986).

Whatever the reason at a particular site, pump and treat systems are shown to not completely remove contaminants from an aquifer. These results are not surprising if water is viewed as the carrier in pump and treat systems. As discussed in this chapter, the severe, natural limitations of water as a carrier produce the results that we are now finding in full-scale pump and treat systems. No aquifer material is perfectly homogeneous. Therefore, the water cannot come into intimate contact with all portions of the aquifer. While people originally thought and designed for advection controlling the remediation of a plume, advection only controls when the plume is spreading. Because of the nature of the subsurface materials in an aquifer, diffusion controls when we are trying to use water as a carrier to remediate a site. The advection process will allow us to remove large masses of contaminants. The diffusion process will prevent us from cleaning the last residual of the contaminants from the aquifer.

This does not make pump and treat a worthless tool in remediation. Pump and treat is effective, if designed correctly, in stopping the movement of plumes and certain specific functions as in creating a cone of depression to help remove LNAPLs. Remember, the movement of plumes is an advection process and pump and treat is an excellent advection controlling technology. Pump and treat is also successful at removing large masses of contaminants from the aquifer. The failure of pump and treat to get the last residual should not prevent the use of pump and treat in the areas in which it can be successfully applied. But, it must be remembered, that the main limitation of pump and treat is that water is the carrier.

AIR AS THE CARRIER

Some of the most important advances in the remediation field in the last decade have been based upon using air as the carrier for removal of contaminants. Two *in situ* technologies being widely applied in the field today are soil vapor extraction (SVE; also known as vapor extraction systems, VES), and air sparging. Neither of these techniques are, in fact, *in situ* methods. Both of these technologies rely on air movement to remove the contaminants from the ground and aquifer. This does not constitute an "in place" treatment; it is a simple change of carrier. SVE and air sparging use air as the carrier to remove contaminants from the ground and aquifer. Air provides several advantages over water, but still has some of the weaknesses of water.

One of the important advantages in switching from water to air as the carrier is the number of pore volumes that can be processed through the soil or aquifer in a short period of time. Let us compare water and air based upon pore volumes in the same geological setting. The detailed calculations for the following example can be found in Nyer and Schafer's paper.

Figure 24 shows the plume and capture zone boundary for a 10 gpm well located 25 feet from the leading edge of the plume. Analytic element modeling was used to determine the flushing rate by computing the travel time from the up-gradient edge of the plume to the recovery well. Figure 25 shows a streamline approaching the recovery well and includes tick marks at 10-day intervals. The total travel time from the up-gradient edge of the plume to the recovery well is 175 days based on an assumed porosity of 25 percent. Thus, a conservative estimate of pore volume exchange rate is once every 175 days.

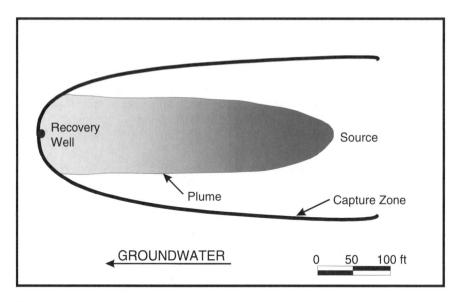

Figure 24   Capture zone for 10 gpm pumping rate.

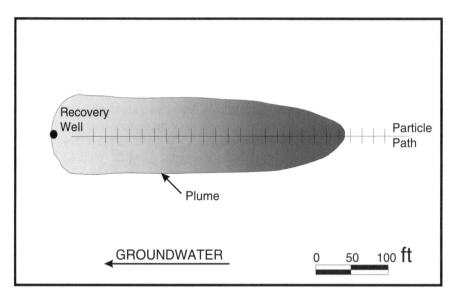

Figure 25   Ten-day time intervals along streamline show 175-day travel time from up-gradient edge of plume to extraction well.

During this time, however, clean water outside the plume, but within the capture zone, converges through the plume to the recovery well. This has the effect of increasing the number of pore volume flushings. To account for this, another way to estimate the number of pore volume exchanges is to simply compare the volume of water in the plume to the extraction rate of 10 gpm. Approximating the plume as an 80 by 300 foot ellipse, the volume of water it contains is expressed as follows

$$V = \pi \cdot \frac{300}{2} \cdot \frac{80}{2} \cdot 40 \cdot 0.25^1$$

$$= 188,495 \text{ ft}^3$$

Dividing by the flow rate of 1,920 cfd (10 gpm) gives an average flushing time of 98 days. We will use this number in comparing pore volume exchanges.

Assume now that we have a similar area of soil contamination above the water table that will be cleaned up using vapor extraction wells. Assume further that the thickness of the vadose zone is 40 feet, the same as the assumed aquifer thickness in the previous example. Figure 26 shows a typical design we might use to vapor extract contaminants using four wells running along the axis of the contaminated area. Often we select flow rates for the vapor extraction wells to produce from one to four pore volume exchanges per day. For this example, we will aim for two pore volume exchanges per day.

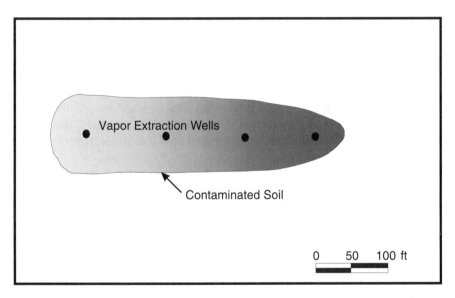

Figure 26   Location of four vapor extraction wells.

Using the same porosity as before, 25 percent, the air volume within the contaminated zone is 188,495 cubic feet. The total airflow rate required for two pore volume exchanges per day is as follows

$$Q = \frac{2 \cdot 188,495}{1,440}$$

$$= 262 \text{ cfm}$$

This requires an average flow rate of 65.5 cfm per well. To see if this is a reasonable expectation, we use the Hantush leaky equation to estimate the drawdown (vacuum) associated with operating vapor extraction wells at this flow rate. Based on this analysis and assuming the vapor extraction well efficiency will range between 20 and 80 percent, the calculated vacuum inside the wells would be expected to range between 15.1 inches of water (80 percent efficiency) and 60.3 inches of water (20 percent efficiency). There would be an additional vacuum component caused by interference by the adjacent wells, but this contribution is minor and may be ignored.

Commercial blowers are readily available to sustain the desired yield at either of the calculated vacuum values, so the projected design is a good one, assuming reasonable well efficiencies are obtained. The Hantush equation can also be used to demonstrate that adequate vacuum is achieved everywhere within the contaminated area.

Thus, the target of two pore volume exchanges per day can be realized under the assumed conditions. Comparing two air exchanges per day with one water exchange per 98 days, it is clear we have a 196-fold increase in the rate of pore volume exchanges with the vapor extraction system.

Many SVE and air sparging systems clean a site in 6 to 24 months, or at least reach equilibrium. We expect pump and treat to take between 5 and 20 years. As can be seen, pore volumes exchange is a significant part of the advantage. Of course, chemical properties can add to the advantage. Sims concluded, "... movement of volatile organic chemicals (VOCs) is generally 10,000 times faster in a gas phase than in a water phase..." (Sims 1990). The 10,000 factor comes from a combination of pore volume exchange rate, solubility versus volatility, and several other factors involved in the dynamic equilibrium of the chemicals in the ground. Hansen summarized the dynamic equilibrium of contaminant partitioning (Hansen, Flavin, and Fam 1994). VOCs partition into four distinct, yet interrelated phases.

1. NAPLs adsorbed to saturated or unsaturated soil particles.
2. Free phase NAPLs.
3. Soluble constituents dissolved in the groundwater.
4. Volatile constituents in the soil pore space of the vadose zone.

All four phases are in equilibrium. Air or water as the carrier must interact with these phases in order to remove the mass from the ground.

The air can come into direct contact with NAPL when it is in the free phase or adsorbed to the soil. The organic compounds can directly volatilize into the air from the NAPL. If the NAPL is located in one of the main flow paths of the air, then the organic will be directly removed from the ground. The rate of removal will be very high since the air will be continuously replaced and the driving gradient of the volatilization will always be at a maximum. Direct volatilization can occur even if the NAPL comes into contact with air that is not in a main flow path. The rate of volatilization will be less as the partial pressure of the organic increases and the driving gradient is reduced. The organic will have to diffuse to an area of high airflow in order to be removed from the ground. Most chlorinated organic compounds and many petroleum-based organic compounds are more volatile than they are soluble. This is a property that will allow the air to carry the organic compounds from the ground faster than water.

The NAPL can also be in direct contact with water when the NAPL is in the free phase or sorbed to the soil. This can be true in the vadose zone as well as the aquifer. The vadose zone is not saturated, but it still contains significant amounts of water in many areas of the country. Of course, desert areas have very little moisture in the vadose zone. The air would be in contact with the water. The organic compounds from the NAPL would dissolve into the water. The compounds would then volatilize from the water to the air when air is the carrier. The rate of transfer to the air would be related to the volatility and the solubility of the compound. The Henry's Law constant would best relate the rate of transfer from the water to the air. The actual removal of the organic compound from the ground would be limited by both the rate of transfer from the NAPL to the water and the rate of transfer from the water to the air. The driving gradient for each transfer would be limited by how far away the organic compound was from a main airflow path. Once again, compounds not in direct contact with a main airflow path would have to diffuse to the airflow path. In the aquifer, the diffusion would occur in the water. In the vadose zone, the diffusion would occur in the air. Compounds that were dissolved in the groundwater, but not from a NAPL source, would behave the same as compounds from a NAPL. The Henry's Law constant would control their transfer to the air, and diffusion would control their removal from non-flow areas.

The other function that air as a carrier can perform is to bring material into the ground. The best example of using air to carry material into the ground is oxygen transfer for biochemical reactions. As we discussed earlier in this chapter, bacteria use many of the organic contaminants as a source of food and energy. The reason that this reaction does not cleanup aquifers by itself is that the rate of reaction is limited. The rate-limiting factor can be oxygen. Moisture content (in the vadose zone), nutrients, temperature, etc. will also limit the rate of biochemical reactions. (Chapter 7 will discuss biochemical reactions and *in situ* biological remediation.) Atmospheric air is 20 percent oxygen. When the air is brought below-ground with an SVE or sparging system, the oxygen can transfer to the water environment of the bacteria and supply the rate-limiting factor to the bacteria. As a comparison, water can only deliver 8-10 mg/l of oxygen. Once again, not only can we provide more pore volumes of air to carry the oxygen into the ground, the air also can carry significantly more oxygen with each pore volume. As we will discuss in Chapter 7, many *in situ* biological designs incorporate air as the method to deliver oxygen to the contaminated zone.

LIMITATIONS

Air is still not the perfect carrier for remediation. There are several factors that limit the use of air. The same geological limitations occur with air as the carrier as occur with water as the carrier. The geology does not change just because the zone is unsaturated. There are still areas of preferred flow and relatively stagnant areas in SVE systems. The compounds still must diffuse over to an area of air movement to be removed from the ground. Diffusion still controls the end of the project.

Air sparging has also been shown to suffer some of the same limitations as pump and treat. Recent tests have proven that the air moves through channels in the aquifer. These channels do not change over the course of the project even if the system is turned off and then on again. The air has preferred paths that it travels. Any contaminants that do not come into direct contact with those channels have to diffuse to a channel in order to be removed from the aquifer by the air carrier. As with water movement, this effect creates the same flattening of the concentration curve after the mass removal phase. Diffusion still controls the process after initial mass removal and helps to determine whether the contaminants reach the cleanup goal within a reasonable time frame. The air is able to remove more mass over a short period of time, but in the end, diffusion still controls the final concentration in the aquifer.

Air also has physical and chemical limitations as a carrier. Not all contaminants can use air as a carrier. If a compound is not volatile, then air will have very limited use. High molecular weight organic compounds and metals will not be able to use air as a carrier. These types of compounds will be left in the ground as the air moves through the contamination zone. Even when compounds are volatile, air may have limitations. If the organic is also very soluble (i.e., ketones and alcohols) then air will not be able to remove the compound from the water. Organic compounds with low Henry's Law constant are not removed from water by air strippers in above-ground treatment systems. Air will also not work below-ground on these compounds when they dissolve in water.

There are also limitations when air is used as a carrier to deliver material to the ground. While oxygen is a limiting factor for biological reactions in the ground, the other factors can also limit reaction rate. Nutrients, nitrogen, and phosphorus cannot be carried by air. There has been some limited work on using ammonia and nitrous oxide in the gas phase to deliver nitrogen through the air, but in general, air is not used for nutrient delivery.

Even when oxygen is the component that needs to be delivered, air has limitations. We have discussed in detail that carriers do not flow past every particle of soil in the ground. The same holds true for using air as a delivery system. Once again, only diffusion is available to transfer the compounds into the low flow and stagnant areas. Oxygen must diffuse through the air or water in order to get to every microenvironment in which bacteria are degrading organic contaminants. Diffusion will limit the rate of oxygen delivery and the subsequent rate of biological degradation.

CONCLUSION

One principle that is consistent throughout all of the *in situ* technologies is the use of carriers for delivery and removal. Mass transfer and diffusion limitations will be a part of every project. The mass transfer portion of the project is usually at the beginning of the remediation. This part of the project can show a significant advantage to switching carriers from water to air. Enhancements to the carrier such as temperature and surfactants will also show significant effect during the mass transfer portion of the project. However, we must consider the entire remediation project when analyzing or designing an *in situ* method of remediation. Diffusion effects must be taken into consideration, and we should address the question of reaching clean by the *in situ* method when we design it. We should not wait 5 to 10 years before we decide if these new *in situ* methods will reach clean.

This book will review each *in situ* technology using the carrier idea to describe how each technique works. This will allow an easier comparison between each technology. The basic comparison for each method will be: What is the carrier? How effective is mass removal with the method? What will limit the method from reaching clean?

We have spent years discovering the limitation of pump and treat. We must understand the replacement technologies better when we install them. We do not want to wait several more years before we realize that there are still significant limitations. This book is orientated toward understanding the limitations now so that we know when and how to install the new *in situ* technologies. We also need to develop a reasonable expectation of what these new technologies can accomplish.

REFERENCES

EPA, Evaluation of Ground-Water Extraction Remedies: Phase II. Volume 1 Summary Report, Publication 9355.4-05. Washington, D.C., EPA, Office of Emergency and Remedial Response, 1992a

Feenstra, S., "Influence of DNAPL on Performance of Groundwater Pump-and-Treat Remedies." Presented at the National Ground Water Association Annual Meeting and Exposition, Las Vegas, Oct. 1992.

Goltz, M. N. and Roberts, P. V., "Interpreting Organic Solute Transport Data from a Field Experiment Using Physical Nonequilibrium Models," Elsevier Science Publishers B.V., 1986.

Hansen, M. A., Flavin, M. D., and Fam, S. A., "Vapor Extraction/Vacuum-Enhanced Groundwater Recovery: A High-Vacuum Approach," Geraghty & Miller, internal paper, 1994.

Keely, J. F., "Performance Evaluations of Pump-and-Treat Remediations," *EPA Ground Water Issue,* Oct. 1989.

Kueper, B. H., Abbott, W. and Farquhar, G., "Experimental Observations of Multiphase Flow in Heterogeneous Porous Media," *Journal of Contaminant Hydrogeology,* 5:83, 1989.

National Research Council, "Alternatives for Ground Water Cleanup," National Academy Press, Washington, D.C., 1994.

Nyer, E. K. and Schafer, D. C., "There are No *In Situ* Methods," *Ground Water Monitoring and Remediation,* Fall 1994.

Perry, J.J., "Microbial Metabolism Of Cyclic Alkanes," *Petroleum Microbiology,* R.M. Atlas, ed. Macmillan Publishing Co., New York, 1984.

Roberts, P. V., Goltz, M. N., and MacKay, D. M., "A Natural Gradient Experiment on Solute Transport in a Sand Aquifer 3. Retardation Estimates and Mass Balances for Organic Solutes," *Water Resources Research,* 2047, 1986.

Sims, R. C., "Soil Remediation Techniques at Uncontrolled Hazardous Waste Sites, A Critical Review," *J. Air & Waste Management Association,* 40, 5, 1990.

Szecsody, J.E. and Bales, R.C., *Journal of Contaminant Hydrogeology,* 4, 181, 1989.

CHAPTER 2

Lifecycle Design

Evan K. Nyer

CONTENTS

The lifecycle concept helps to focus the designer on the main strategies necessary to successfully remediate a site. The concept of lifecycle design and its use on groundwater was first published in 1985 in one of the author's books, *Groundwater Treatment Technology*. The simple basis for the lifecycle concept is that groundwater remediations are unique, and that the requirements for the project will change over the life of the project. One must design for the entire life of the project, not just the conditions found at the beginning. Since 1985, we have continued to use the concept of lifecycle on groundwater treatment designs. However, over the years there have been three major interpretations of the lifecycle. This chapter will review each of the main interpretations of the lifecycle of a groundwater remediation, and within the review, show how design concepts have changed in groundwater remediation over the last two decades.

1-56670-528-2/01/$0.00+$.50

There are two reasons that this book contains an entire chapter on the lifecycle design concept. First, the lifecycle, as originally described in 1985, was an early indicator that we would not be able to reach 'clean' with pump and treat systems. Understanding the lifecycle of a groundwater remediation will help us understand the limitations of pump and treat, and help us understand the possible limitations of any *in situ* treatment method. Second, *in situ* treatment remediations will go through a lifecycle. Once again, the conditions at the beginning of the project will not be the same as the conditions during the middle of the project, and the conditions continue to change as the project progresses to the end. The design of the *in situ* remediation must encompass all of the conditions to be found during the remediation, and not be solely based upon the initial conditions.

LIFECYCLE DESIGN FOR PUMP AND TREAT SYSTEMS

In 1985 the main treatment technology was pump and treat. There was little discussion on remediation methods. The main discussion was on the type of technology used to remove the organics and metals from the water withdrawn as the result of the pump and treat system. The main use of the lifecycle curve was to provide a model that could be used to design the groundwater treatment system. There were three main lessons learned from using the original lifecycle model: concentration changes over time; capital costs were an important consideration because of the limited time each piece of equipment would be used; and operator expenses were a significant part of the treatment costs.

CONCENTRATION CHANGES WITH TIME

There are three patterns that contaminant concentrations follow over the life of the project. These patterns are summarized in Figure 1. First, there is the constant concentration exhibited by a leachate. If we do not remove the source of contamination, then the source will replace the contaminants as fast as they can be removed with the groundwater pumping system. Until the source of contamination is remediated, the concentration will remain the same. We normally think of "mine" leachate or "landfill" leachate, but anytime there is a continuous source of contamination, we are dealing with a leachate. A NAPL or a large clay lens impregnated with dissolved contaminants can also represent a source of contamination. As the water moves around the clay lens and/or the NAPL diffuses into the groundwater, a well downstream of the lens will show a continuous concentration of the contaminant.

Several other examples of sources have developed as we have gained experience with remediation. LNAPLs have been found to be a continual source in two major locations. First, LNAPL can be sorbed in the vadose zone. Rainwater (or other surface water) can cause a vertical migration of the contaminants into the groundwater. Second, the smear zone can act as a source of continual contaminants. The smear zone is created when the change in groundwater levels causes a subsequent change in the level of the LNAPL floating on top of the water. This allows the

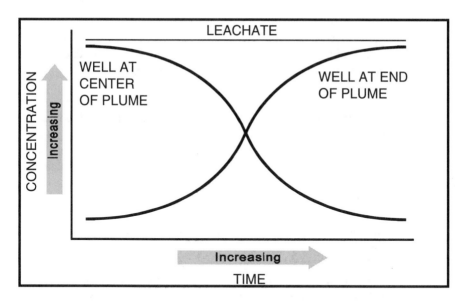

Figure 1   Time effect on concentration.

LNAPL to sorb to the soil over a wide vertical zone. When the water rises again, the sorbed LNAPL does not float out of the soil; it stays sorbed to the soil. This creates an area that has relatively low permeability to both water and air carriers making it difficult to remediate. However, the smear zone is still in contact with both the vadose zone and the aquifer, and the organic compounds making up the LNAPL can diffuse into either area. DNAPLs can cause the same type of effect in the aquifer. As the DNAPL travels down through the aquifer, a portion is sorbed to the soil. While these areas may have a reduced permeability to water movement, water can still move through the affected zone, picking up contaminants. These zones can also act as a diffusion source of organic contaminants.

The second possible pattern arises when the contamination plume is being drawn toward the groundwater removal system. This mainly happens with municipal drinking water wells. In this situation the concentration increases over time. The well is originally clean, but becomes more contaminated as the plume is drawn toward the well. It is important to recognize when this situation will occur. Since the concentration will rise over time, the original treatment system must be overdesigned to allow for increases in concentration. This will allow the treatment system to be designed for the entire life of the project. The curve shown in Figure 1 represents a large plume, or a situation where the source of the contaminant has not been removed. Smaller plumes, which have had their source of contamination controlled, will increase and then decrease. The center of the plume will be drawn toward the low hydraulic head created by the large amount of pumping. Once the center of the plume is pumped, the concentration will start to decrease. This process can take a long period of time, in many cases even decades.

The final pattern is associated with remediation. In this case, the original source of contamination is removed. The pumping system is placed near the center of the plume. This should be the area of highest concentration, and the place where the water will bring the maximum amount of mass of contaminants to the withdrawal point for removal from the aquifer. As the pumping continues, the concentration of the contaminants decreases over time. The rate of decrease is fast at the beginning of the project, slows, and then finally stops decreasing, or reaches an asymptote. The author originally thought that this was the result of just retardation, natural chemical and biochemical reactions, and the dilution of the surrounding groundwater. As discussed in Chapter 1, we now realize that the geology and micro flow patterns play an important role in the lifecycle pattern of remediation. While the beginning part of the lifecycle curve is concerned with the main body of the contamination, further along in the lifecycle minor sources of contaminants control the shape of the curve. When we were designing our first groundwater treatment systems, we were only concerned with the beginning part of the lifecycle curve. In fact, recognizing what occurred during the beginning of the remediation curve was a giant step toward proper design of groundwater treatment systems.

In the early 1980s, the main problem with groundwater treatment designs was that the concentration values used to determine the type of technology and treatment system size were overly conservative. It was common, at the time, to summarize all the concentrations found in the monitoring wells and use the maximum concentration found in the highest concentration well as the initial concentration for the groundwater treatment system. This often led to the incorrect selection of technology. When the pumping wells were installed and the system finally turned on, the actual concentrations found at the influent to the treatment plant were significantly lower than the design concentrations.

Most treatment systems do not get more efficient as the influent concentration decreases. Metal removal and biological treatment systems can have a catastrophic failure if the influent concentration drops below a minimum level. The selection among other technologies can be based on total pounds of contaminant that have to be removed. For example, one of the main costs of carbon adsorption is the replacement of the spent carbon. This is related to the total mass of contaminants that are removed. Carbon adsorption will be skewed as a high cost technology if the wrong concentration is employed in its evaluation. As a result, many treatment systems failed to meet discharge standards, or were not economical on their first day of operation.

Even if the original design did work, this design approach produced treatment systems that were no longer effective after a short period of time. In 1985, the lifecycle concept was introduced so that the designers realized that their treatment system design would have to treat changing concentrations over the life of the equipment and the remedial program.

The concept of the concentration change over the lifecycle of the project was promoted to show that the treatment design would have to be flexible on any groundwater treatment system installation. No matter what the type of contaminant or the geological setting, the lifecycle curve of remediation was consistent. In 1985 we were mainly worried about the beginning portion of the lifecycle curve because we were mainly interested in the design of groundwater treatment systems and the effect of the changing concentration on the actual design. We did not think too much about the later part of the curve. We were not sure if it was a period of slow decrease in concentration and the lifecycle curve would be a straight line if the time was put on a log scale, or if it was a true flattening of the curve and the concentrations had stopped decreasing. While several studies were already available to tell us that the curve was probably flat, they were mainly in the hydrogeological literature. The engineers and hydrogeologists were kept separate at the time, and the design engineers were simply told to design a groundwater treatment system based on the results of the remedial investigation. The first part of the curve was a major advance in the treatment design method; the concentration would decrease as the remediation progressed. The last part of the curve was a simple guess, and we did not realize its importance at the time.

CAPITAL COSTS

Another factor that we faced in the early 1980s was the lack of experience in designing capital equipment for groundwater remediations. Engineers who had designed wastewater treatment systems were the best source of experience at the time. Most of the first designers transferred from the wastewater area. This was similar to many hydrogeologists during the same period who transfered from the oil fields. One problem with wastewater as a background for the groundwater field was the length of time that the project would last. Municipal systems are designed to last up to 50 years. Industrial systems are expected to last at least 20 years. Most equipment used in the field will have a 5 to 20 year life expectancy. Municipal systems switched from steel tanks to concrete tanks in order to extend the life of those unit operations. Pumps and other equipment with moving parts have a lower life expectancy, and tanks and reaction vessels have a longer life expectancy. The cost of equipment in wastewater treatment is figured over the life expectancy of the equipment. However, the cost of equipment on a groundwater cleanup must be based on the time the equipment is used on the project with an upper limitation on the life expectancy of the equipment.

Chapter 1 discussed that the total time for the mass removal portion of a cleanup would probably be much less than the 20 years necessary for an industrial wastewater project. In the previous section of this chapter, we saw that even if the life of the project is 10 years, all of the equipment would probably not be needed for the entire time. As the concentration decreases, some of the equipment would have completed its function. The second part of the lifecycle design switched our thinking from the length of time that the equipment would last to the length of time the project would need the equipment. The difference can be significant.

In 1985, we were mainly interested in equipment associated with groundwater pump and treat systems. The example prepared then was based on a biological treatment system. The lesson still holds true today for the type of equipment that we apply on groundwater remediations. The example below was produced in 1985, but will show the same results as the example that we will provide in a section later in this chapter, Lifecycle Design for *In Situ* Treatment Methods. We have updated the interest rate in the example below, and the daily costs produced will be slightly different from the original 1985 calculations (Nyer and Senz 1995).

Let us assume that the cost of equipment for a submerged, fixed film, biological treatment system is $100,000. If we set the amount of time that we need the equipment and the interest rate that we have to pay for the equipment, then we can calculate the daily cost of the equipment.

One formula for calculating costs would be

$$C = \frac{\text{Cap}}{[1 - (1 + i)^{-n}]/i}$$

where C = cost per time period n; Cap = capital cost ($100,000 in our example); i = the interest rate; and n = the period of time.

We will assume that the interest rate is 9 percent. If the equipment is used for 10 years, the daily cost is $43/day. If the equipment is only needed for 5 years, the daily cost is $70/day. At 2 years, the daily cost is $156/day, and at 1 year, the daily cost is $299/day. All of these figures assume that we have no use for the equipment after its usefulness is finished on this project. Figure 2 summarizes the daily cost of equipment when used for various periods of time.

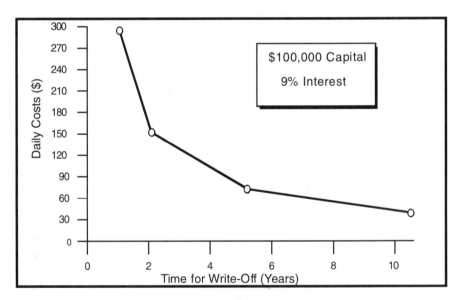

Figure 2   Capital cost as a function of time.

As can be seen, the cost of equipment gets significantly higher as the time of use decreases. The normal method of comparing the cost of treatment by different technologies is to base the comparison on cost of treatment per 1,000 gallons of water treated. At a flow of 25,000 gpd, the cost of treatment goes from $1.72/1000 gallons at 10 years to $11.96/1000 gallons at 1 year. Using the treatment equipment for 1 year will cost six times as much per gallon treated as using the same equipment for 10 years.

A great many groundwater cleanups will be completed in under 10 years, and many more will not use all of the equipment for the entire life of the project. This makes the cost of equipment over time another part of the lifecycle design. The design engineer will have a problem on the shorter projects and on the longer projects in which a particular piece of equipment is only needed for a short period of time. An obvious solution to short-term use is to rent the equipment, or to use it over several different projects. This would allow the equipment to be capitalized over 10 years even though it was only required for 1 year on a particular project.

Of course, any equipment that is to be used for more than one project will have to be transported from one site to the next. The equipment will have to be portable. For example, the design engineer needs a 15,000 gallon storage tank. They have a choice of one tank 17 feet in diameter and 10 feet in height, or two tanks 12 feet in diameter and 10 feet in height. If the equipment is to be used only a short period of time, the proper choice is the two 12 foot diameter tanks. The legal limit for a wide load on a truck is 12 feet. In general, to be transported by truck, the treatment equipment should also be less than 10 feet in height and 60 feet in length. Rail transport can take slightly wider, higher, and longer units, but to be able to reach most destinations in the United States, shipment by truck should be assumed in the design.

Most of the equipment used today on groundwater pump and treat systems, and in fact, on all remediation systems, are portable. Most of the pump and treat systems are for very small flows. A 100 gpm unit is considered a medium to high flow system. Even for larger flow systems it is not hard to make an air stripper portable. A 750 gpm packed tower air stripper would be significantly less than 12 feet in diameter. Biological units have been designed in rectangular tanks to be able to fit on trucks. All carbon adsorption units are portable. Other equipment has also taken on the shapes and limitations necessary to make them portable. In 1985 portability was introduced as part of the lifecycle design requirements for groundwater treatment systems. Today we accept portability as part of the unique requirements on most groundwater remediation systems.

In the mid-1990s, a new practice started to become acceptable. Many small remediation projects, such as gasoline stations, are starting to use equipment that no longer has a long life expectancy. Small systems can cost more to move than they are worth. Plastics and other less expensive materials are being used for construction. The life expectancy of the equipment in these systems is on the order of 5 years. The equipment is thrown away after it is used at the site.

As will be discussed later in the chapter, one of the main problems with using air as the carrier is that the lifecycle occurs over a short period of time. This can create situations that have a large variation of organic concentration in the air stream over a 6 month to 2 year period. Chapter 6 provides a case history for the lifecycle design of an air treatment system. In this case the site was divided into three sections. As the VES in each section was brought on line, its air stream was sent to a high concentration treatment system. When the concentration reached lower levels, the air stream was switched to a low concentration treatment system, and the next section was brought on line and sent to the high concentration treatment system. This lifecycle design allowed the high concentration system to be used over a longer period of time on the project and to be designed at a lower flow rate.

OPERATOR EXPENSES

One final area that has to be discussed under lifecycle design is operator expenses. Any system that requires operator attention will cost more to operate than a system that does not require operators. All wastewater treatment systems should have operator expenses factored into the design. With groundwater treatment systems, this factor takes on added importance. The main reasons for this importance are the relative size of a groundwater treatment system, the remote locations of many sites, and the many remediations that occur at properties no longer active or sold to new owners. Once again, the engineer cannot take a design developed for wastewater treatment systems and reduce its size for groundwater treatment. Most groundwater treatment systems will be very small in comparison to wastewater treatment systems, and most wastewater systems are associated with an active industrial plant. The operator costs, therefore, become more significant when dealing with groundwater treatment system designs.

In the 1980s we thought that most groundwater treatment systems would require regular operator attention. Current systems are designed so that they work without regular operators. These new system designs use various automatic analyzers and telemetry systems in order to inform a central office if the system needs attention. Remote locations and inactive sites have forced this change in approach for many remediations.

The significance of the cost of operator attention can still be shown by analyzing the relative cost of regular operator attention. Let us look at the biological treatment system example once again. Assume that a 15 hp blower is required for the system at $0.06/kwhr. In addition, chemicals and miscellaneous costs are $3.00/day. At a 10 year life expectancy for the equipment, the daily costs would be:

Equipment	$43.00
Power	$29.00
Chemicals	$ 3.00
Total	$75.00

Figure 3 summarizes the relative costs for each category. Without any operator attention, the equipment represents about 60 percent of the daily cost of operation. The power is about 35 percent and the chemicals are about 5 percent of the daily costs. Figure 4 shows what happens to this relationship if one operator is required for one 8 hour shift per day and is paid, including benefits, $10.00/hr. Now 50 percent of the daily cost is represented by operator costs. Equipment drops down to 30 percent, power to 18 percent and chemicals to 2 percent. At just one shift per day, the operator is now the main expense of the treatment system.

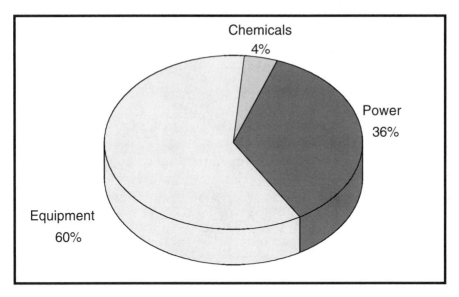

Figure 3   Ratio of daily cost with no operator.

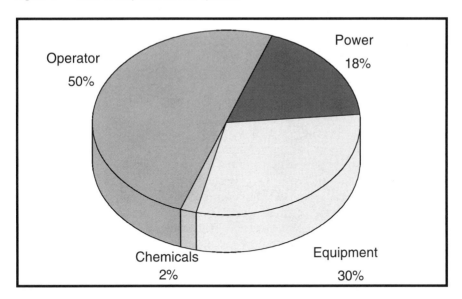

Figure 4   Ratio of daily cost with 8 hr/day operator attention.

If the treatment system requires full time observation, the operator costs become even more important. Figure 5 shows the relative costs when an operator is required 24 hours per day and paid $10.00/hr. Now, the operator represents 75 percent of the cost of operation. Three out of every four dollars spent on the project would go to personnel.

Daily costs for the project double if an operator is required for an 8 hour day when compared to operating with no personnel. The costs triple at two shifts per

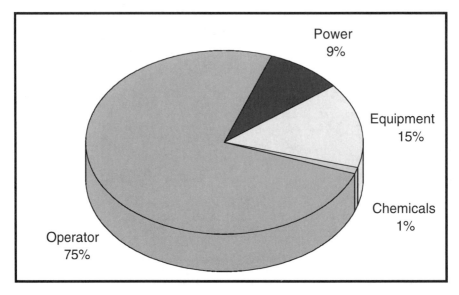

Figure 5   Ratio of daily cost with 24-hr/day operator attention.

day, and costs quadruple when around-the-clock attention is required. These costs are summarized in Figure 6. As can be seen from these data, the design engineer cannot ignore the effect of the operator on treatment system costs. Even when we extend the curve below the 8 hour day operator attention data point, we see that small amounts of operator attention can still add significant costs to the remediation. The designer should spend a significant amount of effort on minimizing the operator time required for a particular design.

The effect of the operator does not decrease even as the size of the equipment increases significantly. Figure 7 represents the relative costs from a treatment system five times the size of the present example and requiring 24 hours per day of operator attention. The operator still represents over one-third the cost of treatment. Even as the total cost of the treatment system approaches $500,000, the design engineer must take special precaution to keep the required operator attention to a minimum.

In summary, there are three main factors that must be considered when performing a lifecycle design for a groundwater treatment system. First, the concentration may change over time. The treatment design must meet the requirements at the beginning of the project, at the middle of the project, and at the end of the project. Second, because of the relatively short time that equipment is needed on groundwater projects, portable or inexpensive equipment should be considered. Finally, due to the relatively small size of groundwater equipment and the strong possibility of operation at an inactive site, manpower costs from operators become a significant, if not controlling, factor in equipment design (Nyer 1989).

As you can see, all of the above discussions relate to groundwater treatment systems. It was a given that the groundwater was going to be pumped above-ground and treated. It was assumed that this would clean the aquifer. The only question was

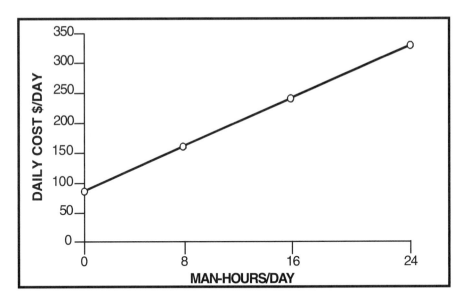

Figure 6 Daily cost of treatment with variable operator attention.

the most cost effective method of removing the contaminants once they were above-ground. In the late 1980s we realized that the pump and treat systems were not going to be able to complete the job. The next use of the lifecycle curve was to try to understand the end of the project, and define it in a way that would allow the pumping system to be turned off.

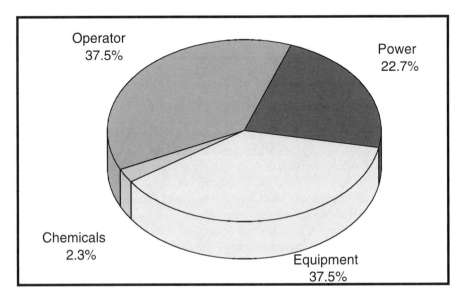

Figure 7 Ratio of daily cost for a $500,000 treatment system with 24-hr/day operator attention.

USING LIFECYCLE DESIGN TO DESCRIBE
THE END OF THE PROJECT

In the late 1980s many pumping systems had been running for several years. A few things started to become obvious. First, we were right about the concentration decreasing as the project progressed. The concentration in the pumping wells and in the aquifer itself decreased as the pump and treat system operated. Second, the concentration curve flattened, and the concentration stopped decreasing after the systems had been running for extended periods of time. This caused two problems, (1) the pumping systems were no longer removing significant amounts of contaminants; however, operational costs generally remained the same; and (2) the concentrations were not low enough to declare the aquifer clean and shut off the treatment system.

The lifecycle curve was once again used to describe the change, or lack of change, in the concentration. In addition, the lifecycle was used as a basis to show a method to turn off the treatment equipment, but still continue to make progress in the remediation of the aquifer. While the lifecycle curve was still mainly concerned with above-ground pump and treat systems, our better understanding of *in situ* processes started to affect our interpretation of the curve.

Figure 8 shows the normal lifecycle concentration of a remediation project. This is the same as the *well at center of the plume* curve found in Figure 1. We have already discussed the beginning portion of the curve. We now will consider the bottom portion of the curve. As the project progresses, the rate of removal decreases. As you can see in Figure 8, the curve tends to become parallel to the horizontal axis. There are several factors that contribute to the flattening of this curve, and they were discussed in Chapter 1. In this chapter, let us build a framework to better understand how the lifecycle curve can help us reach the end of the project.

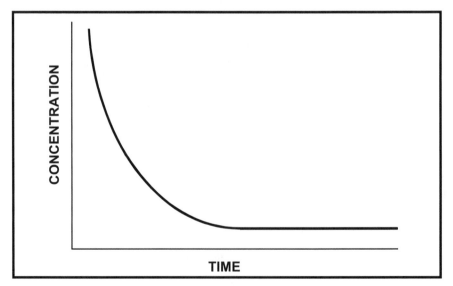

Figure 8   Lifecycle concentration during remediation.

WHAT IS CLEAN?

The objective of most remediations is to clean the site. This raises two questions: what is clean, and how do we define clean? The three main, conservative methods used to establish clean are, (1) risk assessment; (2) government regulations (federal, state, and/or local); and (3) analytical detection limits. All of these methods have advantages and disadvantages.

1. Risk assessment can be uniquely designed for the specific site. Any unusual paths for human contact, and any design method used to prevent human contact, can be incorporated into the risk assessment. However, there are no official standard methods to develop the risk formulations. Different basic assumptions will result in different specific numbers calculated for clean. Some states will not allow risk assessment to be used because of the possible variable results from the same data. While risk assessment may be the only method that can consider local anomalies, the variable output may cause long discussions on the reality of the numbers.
2. Federal, state, and/or local regulations are another source of numbers that can be used to establish what is clean. The Clean Water Act has established maximum concentration levels (MCLs) for many compounds. Table 1 provides a list of current federal MCLs. These numbers are set and specific. Their bases were published and discussed before the final figures were made official. Over time, more compounds will be included on the list. The only problem for remediation sites is to establish the relationship between the soil and the groundwater. For example, all of the drinking water standards are based on concentrations in water. The organics sorbed to the soil particles in the vadose zone and the aquifer will slowly be released into the groundwater. While drinking water standards have a strong technical basis, this method does not directly address the sorbed material.
3. Detection limits are the third method. For compounds that are highly toxic, or in situations in which the numbers developed during a risk assessment are less than the detection limit of the compound, the analytical detection limit can be used to establish what is clean. The main problem with this method is that the ability to detect a compound continuously improves. No one should accept detection limits as a definition of clean. Instead, the detection limits should be used as a basis for a specific concentration. Without a specific number, what is clean will change over the life of the project.

Any of these methods can be used to determine what is clean. However, if we look at the basis for each of these methods, we are basing these concentrations on contact with human beings. In other words, clean is when the groundwater or soil is safe for human consumption. The end of the remediation occurs when the site is safe for people.

A fourth method of determining clean started to gain ground in the 1990s. Ecological risk assessments have been performed to determine the risk of harming

Table 1 Federal Primary Drinking Water Standards

Organic Primary Standards		Volatile Organic Primary Standards		Microbiological Primary Standards		Physical Primary Standards		Radionuclides Primary Standards		Inorganic Primary Standards	
Contaminant	MCL (ppb)	Contaminant	MCL (ppb)	Contaminant	MCL	Contaminant	MCL	Contaminant	MCL	Contaminant	MCL (ppb)
Alachor	2	Benzene	5	Total Coliform Bacteria	(systems that collect 40 or more samples per month) 5% of monthly samples are positive; (systems that collect fewer than 40 samples per month) 1 positive monthly sample.	Turbidity	1TU monthly ave.	Gross alpha	15 pCi/L	Antimony	6
Atrazine	3	Carbon tetrachloride	5									
Benzo(a)pyrene	0.2	1,2-Dichloroethane	5	Fecol Coliform & E. coli	A routine sample and a repeat sample are total coliform positive, and one is also fecal coliform or E. coli positive.			Manmade Beta	4 milli-rem/yr	Arsenic	50
								Radium 226 and 228	5 pCi/L	Asbestos	7.0 MFL
Carbofuran	40	1,1-Dichloroethylene	7							Barium	2000
Chlordane	2	cis-1,2 Dichloroethylene	70							Beryllium	4
2,4-D	70	trans-1,2- Dichloroethylene	100							Cadmium	5
Dalapon	200	Dichloromethane	5							Chromium	100
										Copper[a]	1300
o-Dichlorobenzene	600	1,2-Dichloropropane	5							Cyanide	200
p-Dichlorobenzene	75	Ethylbenzene	700							Fluoride	4000
										Lead[a]	15
Dibromochloropropane	0.2	Monochlorobenzene	100							Mercury	2
Di(2-ethylhexyl)adipate	400	Styrene	100							Nickel[b]	100
Di(2-ethylhexyl)phthalate	6	Tetrachloroethylene	5							Nitrate (as N)	10,000
Dinoseb	7	Toluene	1000							Nitrite (as N)	1000
Diquat	20	1,1,1-Trichloroethane	200							Selenium	50

Table 1 Federal Primary Drinking Water Standards (continued)

Organic Primary Standards		Volatile Organic Primary Standards		Microbiological Primary Standards		Physical Primary Standards		Radionuclides Primary Standards		Inorganic Primary Standards	
Contaminant	MCL (ppb)	Contaminant	MCL (ppb)	Contaminant	MCL	Contaminant	MCL	Contaminant	MCL	Contaminant	MCL (ppb)
Endothall	100	1,1,2-Trichloroethane	5							Thallium	2
Endrin	2	Trichloroethylene	5								
Ethylenedibromide	0.05	Trihalomethanes (total)	100								
Glyphosate	700	Vinyl chloride	2								
Heptachlor	0.4	Xylenes (total)	10,000								
Heptachlorepoxide	0.2										
Hexachloro-Benzene	1										
Hexachlorocyclopen-tadiene	50										
Lindane	0.2										
Methoxychlor	40										
Oxamyl (Vydate)	200										
Pentachlorophenol	1										
Picloram	500										
Polychlorinatedbiphenyl	0.5										
Simazine	4										
2,3,7,8-TCDD (Dioxin)	0.00003										
Toxaphene	3										
2,4,5-TP (Silvex)	50										
1,2,4-Trichlorobenzene	70										

[a] Action level
[b] The status of the nickel primary MCL is discussed in USEPA, 1998. Announcement of the Drinking Water Contaminant Candidate List. Federal Register: 63 FR No. 40, 10273-10287, March 2.

some portion of the ecology when its environment is a receptor of the plume movement. For example, if the groundwater discharges into a surface stream, then the ecological risk assessment would determine if any of the fish or other fauna would be harmed by the contaminants. On a recent project in Michigan the responding parties were forced to run fish toxicity tests directly on the groundwater sampled from wells along the river. There is much controversy over whether this direct test is appropriate, but as of the year 2000, the testing requirement was still part of Michigan's regulations.

The problem with any of these definitions of clean is shown in Figure 9. As the site gets closer to clean, the contaminant concentration reaches its asymptote. Figure 9 represents a worst case scenario in which the site never reaches clean. Even in cases where the site does reach clean, it can take many years.

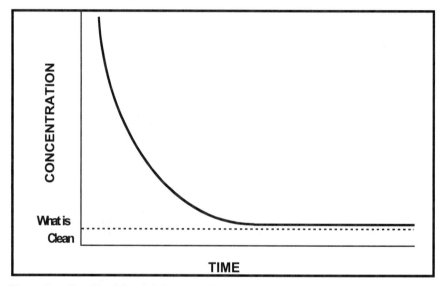

Figure 9   Reaching "clean" during remediation.

During the last years of the project, the treatment system suffers from diminishing returns. The treatment system continues to run, but the amount of material removed is minimal. The money is still being spent on the site, but the benefits are minimal for this financial outlay.

RETARDATION vs. BIOCHEMICAL ACTIVITY

A second factor comes into consideration when we approach the asymptote of the lifecycle concentration. Natural biochemical reactions may be occurring at the same rate as the natural release of the contaminants due to sorption and geological factors. This will be the diffusion controlled portion of the project as discussed in Chapter 1.

As we will discuss in Chapter 7, natural bacteria exist throughout the soil, vadose zone, and aquifer. If no toxic conditions exist, then the natural bacteria are already degrading the contaminants. Their rate of degradation is limited by the presence of the proper bacteria, a final electron acceptor, nutrients, and the degradability of the organic compound. Degradation in the vadose zone is usually by aerobic bacteria, and therefore, the presence of oxygen is the limiting factor in the rate of biodegradation. In groundwater, both aerobic and anaerobic degradation will occur. Some compounds will required cometabolites or specific environmental conditions in order for the bacteria to be able to degrade them.

The natural rate of degradation is slow compared to an above-ground reactor or to enhanced *in situ* biological reactions. The vadose zone or aquifer have a limited capability to replenish oxygen or other final electron acceptors used by the bacteria. But when the contaminant concentration reaches very low levels, the vadose zone and aquifer can naturally supply what is needed. At the same time, all of the contaminants do not release to the water at the same rate. The contaminants are sorbed to the soil. Depending on the organic content of the soil and the chemical properties of the contaminant, the individual compounds will release to the water at a different rate. In addition, some of the contaminants will have to diffuse from their location to a place where they will be part of the main flow of the aquifer.

Other references (Rafai et al. 1988, Nyer, Mayfield, and Hughes 1998, Rice et al. 1995, Wiedemeier et al. 1995, and Wiedemeier et al. 1996) showed that pumping would not improve the remediation of petroleum hydrocarbons. In fact, more compounds would be exposed to humans from a pumping system than by allowing the natural bacteria to degrade the compounds *in situ*. We can project a complete natural elimination of the plume based on the same data.

Recent data (Nyer 1996 and Wiedemeier et al. 1996) have shown that chlorinated hydrocarbons also may be eliminated by natural attenuation. The main difference between natural attenuation for petroleum hydrocarbons and chlorinated hydrocarbons is that chlorinated hydrocarbons require anaerobic conditions and an electron donor. (See Chapter 7 for a full discussion.) When these conditions are met, the biochemical activity at these sites will remove chlorinated compounds and these reactions must be incorporated into the lifecycle design.

At some point, the rate at which the compounds can be removed from the aquifer by pumping will be equal to the natural biological degradation rate of the aquifer. At this point, we will not be able to speed the cleanup no matter how fast we pump the well. In fact, for petroleum hydrocarbons, if the aquifer can naturally replenish the final electron acceptor demand from the contaminants, then pumping will not increase the rate that we reach clean. Once we reach this point, the money spent on pumping and treating this water is wasted. The pumps could be turned off and all of the equipment removed, and we would still reach clean at the time that we would have by leaving the system running.

ACTIVE MANAGEMENT

In the late 1980s we tried to develop another point in the lifecycle of the project that determines when we can turn off the treatment system. This is the point at which active management ceases, and we can stop spending most of the money. Clean occurs when the contaminant concentrations have reached the point where they satisfy the regulator's cleanup criteria. The active management line in Figure 10 represents when we can turn off the remediation equipment.

Figure 10 also shows the lifecycle curve with the clean line and the active management line. The active management line represents the point in the lifecycle where pumping will no longer speed the cleanup of the site. We should stop spending money at this point. The clean line represents when we can use the water and the site for human consumption.

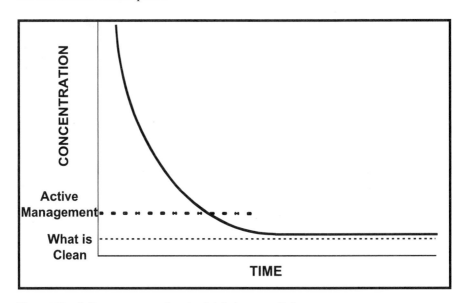

Figure 10   Active management end point during remediation.

This is similar to the situations that occurred in the 1970s when we cleaned up rivers and lakes. We installed wastewater treatment systems on municipal and industrial wastewater projects. This reduced the levels of contaminants entering the water body (lake or river), and allowed the river or lake to clean itself. This took time, and we did not use the water body the day after we installed the treatment system. We waited until the water was clean, or safe to use.

Rivers and lakes can remediate themselves faster than can groundwater. One reason for this is that oxygen transfers into these water bodies faster than it can into groundwater. However, groundwater can remediate itself. The problem is that the rate is so slow that we normally find the time frame unacceptable. If we remove most of the contaminants, then the aquifer can finish the job on its own. If this rate is the same as the rate from pump and treat, then there is no reason to continue the pump and treat. We are at the point in the lifecycle of the project when we must wait before we can use the water. Active management of the site should stop, and we should only monitor the site while we wait for clean.

The only problem left is to determine when we reach the active management point of the lifecycle of the project. For degradable organics this point is basically a comparison of the normal fate and transport of chemicals to the biochemical reaction rate of those same chemicals. Some type of biomodeling will be required in conjunction with the solute transport materials. Chlorinated hydrocarbons and metals will also undergo natural reactions that will contribute to their disappearance from the water. Modeling methods must include these rates of disappearance in order for the design to be able to determine when active management should end. One simple model that can be used to assist in this evaluation is EPA's Bioscreen model. This model incorporates the degradation of a compound into the predicted fate and transport of that compound. The model also shows the fate and transport of the compound if it did not degrade. The model is available for download at www.epa.gov.

Lifecycle design was still a tool of the pump and treat system in the late 1980s. As we started to recognize the limitations of water as a carrier, the importance of natural reactions occurring below-ground, and the benefits of enhancing natural reactions, we started to incorporate *in situ* methods into our planning for the lifecycle of the project.

LIFECYCLE DESIGN FOR *IN SITU* TREATMENT METHODS

The current interpretation of the lifecycle curve completes the incorporation of the *in situ* reactions into the planning of a remediation. We now separate the *in situ* technologies that are based on the use of water or air as the carrier to remove mass from the *in situ* technologies that are directed toward the bottom of the curve. The first part of the curve represents the mass removal technologies. Figure 11 represents the change in concentrations over the life of a vapor extraction system, sparging, or vacuum-enhanced recovery. The first section of this book is dedicated to the mass removal technologies. Each technology has its own chapter. For this discussion, we just have to understand that all three technologies use air as the carrier.

As discussed in Chapter 1, air has many advantages over water as the carrier in a remediation. Air will be able to remove the organic compounds from the ground at a much faster rate than techniques that use water as the carrier. The simple advantage of air being able to move more pore volumes in the same amount of time would have the effect of shortening the amount of time to remove an organic compound. Assuming that the comparison is based on two projects that had the same amount of volatile organic originally in the ground, the concentration would decrease much faster with air as the carrier compared to water. The left side of the curve will have a steeper slope. (Of course, if we used pore volumes for the x axis then the curves would be the same for water or air.) If the organic compound is volatile, then the curve would be even steeper with air as the carrier.

However, the right side of the curve would not automatically be affected with a switch from water to air as the carrier. Diffusion and the geology of the site control the right side of the curve. Air has many of the same limitations as water. The material that is not in direct contact with the main flow path of the carrier must diffuse over to the main path in order to be removed from the ground. The right side of the curve would look the same for water or air. So while the slopes would change, the overall curve would still remain an accurate description of the change in concentration over the life of the project.

The lifecycle curve has always been an important tool in the design of a remediation system. This is still true for *in situ* technologies that rely on one of the carriers. When we analyze the change in the shape of the lifecycle curve between the two carriers and the various enhancements to the carriers, we realize that the left and right sides of the curve must be treated separately.

Figure 11 is the best way to think of the lifecycle curve when designing a remediation. While the curve is the same, we have completely separated the left half and the right half of the lifecycle line. The left half represents the mass removal portion of the remediation. The right half represents the diffusion limited portion of the remediation.

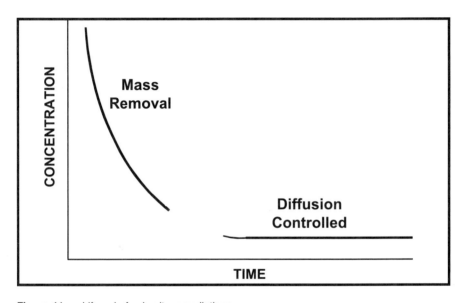

Figure 11   Lifecycle for *in situ* remediations.

There are really two separate projects in any remediation. The first project is the removal of the maximum amount of mass of contaminants. The *in situ* technologies that rely on air as the carrier have been shown to have a significant advantage in this type of project. (Sims reported a factor of 10,000.) Most of the published information that we are currently receiving on *in situ* processes are really an analysis of the mass removal capability of these techniques. The second project is reaching the mandated concentration required to declare the site clean.

Pump and treat never failed as a technology for the first type of project. Water was capable of removing a significant portion of the mass of contaminants from the aquifer. Air may cost less and remove the compounds faster because of its advantages, but that does not mean that water failed. The EPA and the NRC reviews were based on the diffusion limited part of the remediation. When they stated that pump and treat was not capable of remediating an aquifer, they really meant that pump and treat failed to reach the concentration that would allow the aquifer to be designated as clean. (Both reports made it very clear that they believed pump and treat had been misapplied and, that it is a successful technique for controlling plume movement.)

The real problem with pump and treat systems is that the designers expected that the systems would be able to clean the aquifer. They did not include the geology and diffusion limitations of using a carrier when they designed their remediation. Because of the slow pace of pore volume exchange with water as the carrier, the results did not show up for many years. This delay allowed thousands of remediation systems to be installed based on the assumption that pump and treat could reach clean.

Air as a carrier has the same geology and diffusion limitations. However, we are now installing the *in situ* technologies based on air and expecting them to reach clean. We are taking the results from the first part of the remediation and declaring that since the compounds are being removed much faster, then obviously the site will reach clean.

This is why it is very important to separate the two projects. Mass removal is completely separate from diffusion controlled. A remediation design should be analyzed as two separate projects. What will the technique do for mass removal? What will the technique do for diffusion controlled? Separating the remediation into two projects does two major things for the designer. First, we are able to predict at the beginning of the project what will happen over the entire life of the remediation. Second, we are able to realize that different technologies may have to be applied to the site in order to complete the remediation. One technology may have to be used for the mass removal, and a different technology used for diffusion controlled.

Figure 11 is the representation of the lifecycle of the remediation that we should be using. It is still important to use the lifecycle curve when developing a strategy for a remediation. Figure 11 incorporates our latest understanding of the different factors that affect the entire remediation project.

DETERMINING THE TIME REQUIRED TO COMPLETE A LIFECYCLE IN GROUNDWATER REMEDIATION

The last two questions that must be answered to complete the lifecycle design are, (1) when do we switch from mass removal to diffusion controlled; and (2) when is the project finally over? Both are very difficult questions. Too many people want to answer, "you'll know it when you see it" and just start the project. Good design requires that the designer have some idea about when these two events will happen, and at the least, a method of measuring when the events do occur.

The first question is easy to define on paper, but difficult to define in terms of a remediation project. The main requirement from the mass removal portion of the remediation is to remove enough mass from the ground so that the method applied during the reaching clean portion of the remediation can function. For example, if we were using natural attenuation for the diffusion controlled portion of the project, then natural rate of biodegradation becomes the key to the first portion of the project.

We must remove enough organics during the mass removal so that the natural bacteria can completely consume the remaining organics left in the aquifer. The rate of degradation must not only prevent further spread of the organics, but actually consume the organics fast enough to shrink the plume. The level of reduction can be defined at the point at which the diffusion controlled portion of the project can function without mechanical assistance. The natural bacteria will degrade the organics remaining in the ground.

While this example seems very simple, the concept is still relatively new. One of the authors was working on a project that tried to use this concept as the method to design and control the remediation at a site. While the overall lifecycle concept of two separate projects was accepted, the problem occurred when we tried to define how to switch from the mass removal to the diffusion controlled portions of the remediation. The technical support for the regulators was of the opinion that the mass removal portion of the project should be run until it no longer was removing a significant amount of material. The author tried to use the above description as a method from switching from mass removal to reaching clean.

The regulators suggested two ways to define the point at which the mass removal was no longer removing a significant amount of material. The first was a simple pounds per day or concentration per well that was being removed with the VES system. The second method was to do a statistical analysis of the concentration coming out of the VES well and run the system until an asymptote was reached. The first method can suffer from background organic concentrations that show up in the VES system. These background concentrations, especially near gasoline stations, can make set concentrations or mass removal per day values very difficult to delineate. If the concentration is set too low, this can add years to the operation of the mass removal portion of the project without any significant decrease in mass in the ground.

The second method suffers from analytical costs. Developing enough data over a large site with multiple wells is always difficult. The number of data points and sampling events that would be required to statistically show that the data had reached an asymptote over a period of time can be expensive. On this one project, the analytical costs of determining the asymptote were more expensive than the actual operations of the mass removal project.

This project pointed out that the first step in determining the lifecycle design of the mass removal portion of the project is to determine the objectives of that portion of the project. If the stated objective is to remove as much as possible then the two methods suggested by the regulatory agency, minimum mass per day or asymptote, are two methods that can be used to determine the end of the project. However, if the objective of the mass removal portion of the project is to establish a favorable subsurface environment so that the reaching clean portion of the project can be achieved naturally, then these two methods may extend the operation of the mass removal project unnecessarily.

Simple monitoring of the system may show that the natural attenuation of even non-degradable compounds is sufficient after most of the mass has been removed. One operation method that may satisfy the regulators is to not remove the mass removal equipment once the end of that project is thought to have been reached.

The equipment can simply be turned off and the vadose zone and the groundwater monitored for the next several quarters. If the monitoring data shows that enough mass had not yet been removed, then the mass removal equipment can simply be turned on. This method allows flexibility with the design. Because of all the variables associated with the organics and the geology, it is difficult to make exact predictions of when the project should be changed from one portion to the next portion of the remediation. The simple testing of the actual results while the equipment is still in place is probably the best method at this time of determining when to switch from one project to the other. Hopefully, with time and experience we will become much better at knowing when to switch from one project to the next. Of course, all parties will first have to agree on the original objective of each portion of the project.

The second question is much more difficult to answer. When will the project finally be over and the site declared clean? There are two problems with declaring a site clean. The first is point of compliance, and the second is understanding and modeling the natural reactions of the compounds in the vadose zone and aquifer. Point of compliance is extremely important when trying to determine if the site has reached clean. Monitoring wells represent the concentration of the aquifer or vadose zone in the immediate area of that monitoring well. As we will discuss in Chapter 7, there are several biochemical reactions that occur underground. If these reactions are fast enough, or at least as fast as the diffusion rate of the compounds, then material located in one portion of the aquifer may not reach the monitoring wells. Is this site still contaminated? Other work has fully explored this question (Nyer and Senz 1995). The reader should look to these other publications for full discussion on this subject. For this chapter, we will say that the point of compliance is extremely important as far as determining whether the site is determined to be clean.

The second portion of the question becomes when will a non-degradable compound reach clean? From the work that the authors and others have done, especially if NAPLs are present, many sites may never reach clean. This is true for the center of the site where the contamination originally occurred. The term "never" refers to any site that takes more than 50 to 100 years to remediate. Under these conditions the project really turns into a removal of mass and control of the site. The answer to the second question then is relatively black and white. Does the site have a chance to reach clean, or does the nature of the contaminants and the aquifer force us to believe that the compounds will be slowly released to the aquifer over an extended time?

One of the exciting areas that will be discussed in this book are the new *in situ* technologies that have been developed for the diffusion controlled portion of the project. Chapter 8, Reactive Zone Remediation, and Chapter 9, Phytoremediation, discuss two important technologies that are applied to the diffusion controlled portion of the project. Both technologies enhance natural reactions that occur in the ground. These techniques will allow us to remediated sites in a reasonable amount of time that previously we thought would take more than 50 years.

If the compounds are of a nature that the aquifer can reach clean, then the second part of the project is usually set up for natural attenuation and monitoring. Bio-modeling may be able to show, or at least give a range of the time it will take to finally reach the required concentration to declare the site clean. Monitoring is the only way to confirm that the actual conditions have been reached. When it is decided that the compounds cannot be removed successfully by natural or enhanced means, then the project switches over to a control project. The design objective switches from one of complete removal to one of complete control. One must design the system so that no compounds leave the site. Pump and treat is a very good technology for this control in an aquifer. New technologies such as treatment walls may be more cost effective for this control. While the vadose zone and aquifer are both contaminated, the decision for removal or control will have to be separate for each zone.

Even the *in situ* technologies cannot escape the lifecycle of a remediation. The lifecycle concepts have taught us over the years how to apply technologies and design remediation. It is important to use Figure 11 when trying to design an *in situ* project. Separate the project into two major portions, the mass removal and the diffusion controlled. Using this technique, the designer can incorporate the entire project into the original concept. The designer must decide in the beginning of the project what the objectives are and what can be accomplished at this particular site. The lifecycle concept helps to focus the designer on the main strategies necessary to successfully remediate a site.

REFERENCES

Nyer, E. K., "The Effect of Time on Treatment Economics," *Groundwater Monitoring Review,* Spring 1989.

Nyer, E. K., "Biochemical Effects on Contaminant Fate and Transport," *Groundwater Monitoring Review,* Spring 1991.

Nyer, E. K., Senz, C. D., "Is This Site Contaminated?," *Groundwater Monitoring and Remediation*, Winter 1995.

Nyer, E. K., Mayfield, P., and Hughes, J., "Beyond the AFCEE Protocol for Natural Attenuation," *Groundwater Monitoring and Remediation,* Ground Water Publishing Company, Summer 1998.

Piwoni, M.D. and Bannerjee, P., "Sorption of Volatile Organic Solvents from Aqueous Solution onto Subsurface Solids," *J. Contam. Hydrogeology*, 4, 163-179, 1989.

Rafai, H. et al., "Biodegradation Modeling at Aviation Fuel Spill Site," *Journal of Environmental Engineering,* 114(5), October 1988.

Rice, D.W., Dooher, B.P., Cullen, S.J., Everett, L.G., Kastenberg, W.E., Grose, R.D., and Marino, M.A., "Recommendations To Improve the Cleanup Process for California's Leading Underground Fuel Tanks," Lawrence Livermore National Laboratory, University of California, October 16, 1995.

Wiedemeier, T.H., Wilson, J.T., Kampbell, D.H., Miller, R.N., and Hansen, J.E., "Technical Protocol for Implementing Intrinsic Remediation with Long-term Monitoring for Natural Attenuation of Fuel Contamination Dissolved in Groundwater," U.S. Air Force Center for Environmental Excellence, San Antonio, 1995.

Wiedemeier, T.H., Swanson, M.A., Moutoux, D.E., Wilson, J.T., Kampbell, D.H., Hansen, J.E., and Hass, P. "Overview of the Technical Protocol for Natural Attenuation of Chlorinated Aliphatic Hydrocarbons in Ground Water Under Development for the Air Force Center for Environmental Excellence," In: Symposium on Natural Attenuation of Chlorinated Organics in Ground Water, USEPA 540/R-96/509, 35-59, 1996.

CHAPTER 3

Vapor Extraction and Bioventing

Gregory J. Rorech

CONTENTS

1-56670-528-2/01/$0.00+$.50
© 2001 by CRC Press LLC

INTRODUCTION

The vapor extraction and bioventing technologies induce airflow in the subsurface using an above-ground vacuum blower/pump system. Adequate air movement within the contaminated zones is of primary importance to the success of the VES. The induced airflow brings clean air in contact with the contaminated soil, NAPL, and soil moisture. The contaminated soil gas is drawn off by the VES and the air in the soil matrix becomes recharged with new vapor phase contamination as the soil/pore water/soil gas/NAPL partitioning is re-established.

Bioventing, or bioenhanced vapor extraction, is a remedial method similar to vapor extraction in that it relies upon an increase in the flow of air through the vadose zone. Vapor extraction is performed to volatilize the volatile organic constituents *in situ*. In bioventing, the increase in the flow of air is to provide oxygen in the subsurface to optimize natural aerobic biodegradation, and this becomes the dominant remedial process. While the design criteria for vapor extraction and bioventing are different, once the physical system is in operation both processes occur. Compounds that are volatile move with the air, and compounds that are degradable

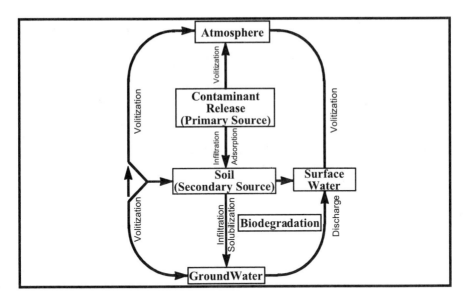

Figure 2   Environmental compartment model for VES.

When a volatile NAPL is present on the soil, the bulk of the mass removal by VES will come from direct volatilization of the NAPL. This would be similar to a fan blowing past a pool of gasoline. Research workers (Hoag et al. 1989) have shown that the bulk (more than 95 percent) of the NAPL can be removed within passage of several hundred pore volumes of air through the experimental soil columns (NAPL within the dry soil void space). In field applications, where airflow is usually over the NAPL layer rather than through it, VES still often recovers the bulk of the NAPL with passage of several hundred pore volumes of air. Under conditions of NAPL presence, mass removal rates are often linearly correlated with airflow rates. When NAPL is not present in the subsurface, airflow requirements become very different, and are often governed by nonequilibrium rate limiting conditions.

The following section describes the partitioning of contaminants in the subsurface. These equations are the basis of numerical simulation models that attempt to predict the remediation process of vapor extraction. Models achieve this prediction by repeatedly evaluating the new partitioning relationships after passage of clean air past the contaminated soil. Modeling is discussed later in this chapter.

Under moist soil conditions, contaminant partitioning in the vadose zone (no residual NAPL) can be described by the following equation (1). Figure 3 is an illustration of the equation

$$C_T = p_b C_A + \theta_L C_L + \theta_G C_G \tag{1}$$

where C_T = Total quantity of chemical per unit soil volume; C_A = Adsorbed chemical concentration; C_L = Dissolved chemical concentration; C_G = Vapor concentration; p_b = Soil bulk density; θ_L = Volumetric water content; and θ_G = Volumetric air content.

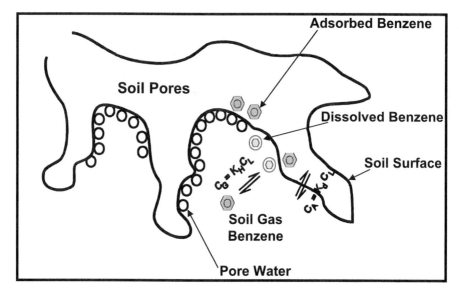

Figure 3   Equation schematic.

The equilibrium relationship between vapor concentration (C_G) and the associated pore water concentration (C_L) is given by Henry's Law

$$C_G = K_H C_L \qquad\qquad (2)$$

where K_H = Henry's Law constant.

Henry's Law is often thought of in relation to air stripping. In air stripping, air removes dissolved VOCs from the water stream. The efficiency of this removal process is related to the compound's Henry's constant. Under moist soil conditions, the extracted vapors during vapor extraction similarly remove VOCs from water. The success of this removal process is similarly related to the compound's Henry's constant.

Likewise, the relationship between equilibrium solution concentration and adsorbed concentration is given by

$$C_A = K_d C_L \qquad\qquad (3)$$

where K_d (L^3/M) = the distribution coefficient expressed as $K_d = f_{oc} K_{oc}$, and where f_{oc} = the mass fraction of organic carbon, and K_{oc} = the organic carbon partitioning coefficient.

Equation (3) is generally considered to be valid for soils with high organic content ($f_{oc} > 0.1$ percent solids). For soils with lower organic carbon content ($f_{oc} < 0.1$ percent solids), sorption to mineral grains may be dominant (Piwoni and Bannerjee 1989 and Brusseau, Jessup, and Rao 1991). Adsorption is further discussed later in the chapter. There are several good sources for organic carbon partitioning coefficient

(K_{oc}) values (U.S. EPA 1996 and Fetter 1994). The EPA document checked various literature sources and then calculated the geometric mean of what they considered the most reliable sources. This document also provided some relationships for various types of compounds to estimate K_{oc}, if K_{ow} is known, as presented below in Equations (4) and (5). Other relationships for these and other types of compounds are also available in the literature with some slight variation in each source.

For polyaromatic hydrocarbons, polychlorinated biphenols and phthalates

$$\log K_{oc} = \log K_{ow} - 0.094 \tag{4}$$

For volatile organic compounds and chlorinated pesticides

$$\log K_{oc} = 0.78 \log K_{ow} + 0.151 \tag{5}$$

Mathematical relationships can also be developed to quantify biological and/or other reaction transformations of the chemical contaminants. These relationships can subsequently be used to develop mathematical models of the subsurface conditions under advective air movement conditions. Mathematical models can be used to simulate subsurface changes caused by the VES (perturbation) and allow the user to select the most efficient perturbation that leads to the most contaminant mass removal.

The above equations have defined partitioning of contaminants in the subsurface without air movement. Partitioning without advective air movement occurs via diffusion. The VES induces airflow (advective) past the contaminated zone; therefore, under VES operating conditions, both diffusive and advective transport are occurring.

Under the assumptions of uniform moisture distribution across the soil and incompressible air phase, the advective-dispersive transport equation in cartesian coordinates can be written as (Armstrong, Frind, and McClennan 1994, and Gierke, Hutzler, and McKenzie 1989)

$$L(C_G) = \theta_G D_{ij} \frac{\partial}{\partial x_i} \frac{(\partial C_G)}{(\partial x_j)} - \theta_G v_i \frac{\partial C_G}{\partial x_i} = \theta_G \frac{\partial C_G}{\partial t} + \theta_L \frac{\partial C_L}{\partial t} + P_b \frac{\partial C_A}{\partial t} \tag{6}$$

$$i, j = x, z$$

where subscripts G, L, and A designate the gaseous, dissolved, and sorbed phases of the contaminant. C_A, C_L, C_G, P_b, θ_G, and θ_L are as defined above. The air velocity component is derived from the air continuity equation and the subsequent application of the Darcy equation to pressure. The continuity equation states that the same mass of material entering a unit volume of space must also exit that volume space without biodegradability. The Darcy equation relates groundwater velocity to hydraulic gradient. In this application, the Darcy equation ($V = k/\phi \cdot dh/dl$) is applied for air movement rather than groundwater. The term D_{ij} is the dispersion tensor, defined in terms of longitudinal and transverse dispersivity and the diffusion coefficient. Equations (1), (2), and (3) can be substituted into Equation (6) in order to represent the transport equation in terms of the gaseous phase only to yield

$$K_d = f_{oc} K_{oc} \qquad (7)$$

$$\theta_a D_{ij} \frac{\partial}{\partial x_i} \frac{(\partial C_G)}{(\partial x_j)} - \theta_G v_i \frac{\partial C_G}{\partial x_i} = R\theta_G \frac{\partial C_G}{\partial t} \qquad (8)$$

where R is the retardation factor

$$R = 1 + \frac{\theta_L}{\theta_G H} + \frac{P_b K_d}{\theta_G H}$$

Equations (7) and (8) therefore, are a mathematical representation of partitioning under diffusive and advective conditions. Actual field observations, however, indicate that the equations are only valid for diffusion dominated or weakly advective conditions.

Under strongly advective conditions that can be found while operating a VES, the above equations do not account for the asymptote that is observed in field applications. Tailing is the phenomenon that is often observed in VES applications whereby the contaminant mass removal rates are slower and the residual mass of contaminant adsorbed to the soil after vapor extraction is greater than what would be predicted by the equilibrium equation, Equation (8). Equilibrium predicted nontailing and nonequilibrium type tailing effects are shown schematically in Figure 4.

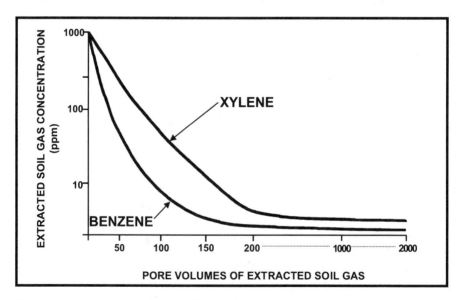

Figure 4   Equilibrium asymptotic tailing effect.

As discussed in Chapter 2, the performance of a system in removing contaminants changes with time. In most VES applications, the initial stages of the project yield the highest mass removal. The tailing effect implies that efficiency is decreasing with system lifetime and that closure goals or system operational modifications should account for this temporal change.

Several authors (Armstrong, Frind, and McClennan 1994; Gierke, Hutzler, and McKenzie 1989; Sleep and Syikey 1989; and Brusseau, Jessup, and Rao 1991) have shown that the tailing effects can be represented by first order mass transfer, nonequilibrium, physical, and/or chemical processes. These nonequilibrium modifications to the equilibrium relationships (Equations (1)-(3), (6)-(8)) will not be presented in this chapter; however, the reader is referenced to the appropriate research publications for details. In brief, the nonequilibrium notions relate to the existence of a rate limiting criteria governing the mass transfer process for VES. Under fully wetted soil conditions without NAPL, the rate limiting step may be transfer across the air/water interphase, the soil/water interphase, dead end micropore effects, and/or a combination of all effects. Simply stated, these nonequilibrium effects slow down the vapor extraction process once the bulk of the contamination has been removed. (Armstrong, Frind, and McClennan 1994) have conducted a sensitivity analysis on several of these nonequilibrium rate limiting conditions.

A sensitivity analysis or demonstration of the physical limitation of the VES ability is a powerful tool in ascertaining the system's capability and achievable closure goals. This can be utilized to negotiate reasonable closure criteria with regulatory agencies and/or modify system operation during the project's lifetime to minimize expenses while maximizing mass removal. VES is a very powerful mass removal technique. However, as the historic research and long-term operations have shown, VES can not overcome the natural limitations of the geology. Once most of the mass has been removed by the VES, an alternative treatment technique may have to finish the removal.

AIRFLOW REQUIREMENTS AND CAPABILITIES

The need to understand and predict the subsurface mass transfer relationships relates to the practical need to deliver the required airflow to achieve the remedial goals. Often the designer will only want the minimum subsurface air movement to achieve the remedial goals, since excessive airflow results in larger, expensive off gas treatment equipment as well as higher operating costs.

In instances where NAPL is present in pockets, pools, or as a layer atop the groundwater, mass removal will often be linearly or semi-linearly rated to the airflow. This does not imply that if NAPL is present, high airflow is required, since often the NAPL is removed rapidly, leading to site conditions that may not require further high airflow. Airflow generation capability (airflow that can be generated based upon subsurface soil conditions) and airflow requirement (to achieve remediation) needs must be met in order to appropriately install a VES.

Airflow Capability

Figure 5 presents predicted airflow rates per unit well screen (from an extraction well) depth for a 4-inch diameter extraction well and a wide range of soil permeabilities and applied vacuums (Johnson et al. 1990, and Johnson et al. 1991). The graph was generated by solving the logarithmic Jacob's equation and assuming a 40 foot radius of air collection (i.e., zone of influence). The Jacob's equation is the same Jacob's equation used in groundwater applications to determine zone of influence of a pumping test. In this instance, however, it is utilized for subsurface airflow rather than groundwater flow. This figure provides an excellent screening tool to determine the necessary vacuum equipment to generate the required airflow. This will be further discussed later in the chapter.

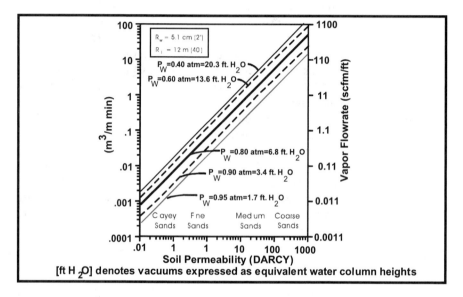

Figure 5   Airflow generation plot.

The measurement of vacuum at locations away from the vapor extraction well implies that subsurface airflow is present at that location. Subsurface vacuum is easy to measure in the field, therefore, it is often measured in place of subsurface airflow during VES pilot testing and application.

Figure 6 illustrates how quickly vacuum measurement profiles die off away from the extraction well point. This quick decay of induced vacuum readings in turn implies that subsurface airflow quickly dies off at increasing distances from the extraction well. If the airflow was perfectly radial from the well, then the airflow velocity would decrease at a rate proportional to the square root of the distance from the well. Several design techniques will be discussed later in the chapter that focus on maintaining the airflow across the contaminated area of the vadose zone and not simply in a radial flow from the extraction well. This will help produce the required airflow with a minimum of equipment.

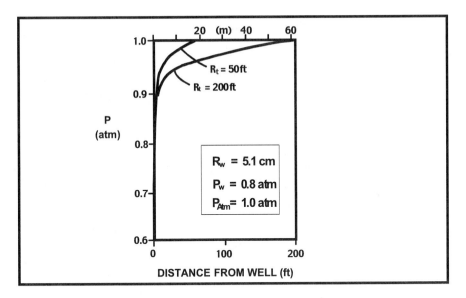

Figure 6   Vacuum tailing effect.

Airflow Requirements

Delivering the required airflow requirements to achieve the cleanup criteria is the basic design goal for VES installation. This basic design goal, however, remains the most difficult to predict due to our limited understanding of subsurface conditions. This understanding is required to quantify the mathematical relationships used in formulating the simulation models. The most distant location from the extraction well should receive sufficient airflow to achieve remediation. The most distant location from the extraction well is termed the zone of influence. Sufficient wells are spaced in the contaminated area to deliver the minimally acceptable airflows across the entire site.

The required airflow at the most distant location can be determined by (1) comparison of published literature values for similar soil types and contaminants; (2) by conducting bench scale column tests to determine this required airflow for contaminants and soil conditions (discussed later in the chapter); or (3) by conducting computer simulation modeling (discussed later in the chapter).

EVALUATION OF CONDITIONS WHERE VES IS APPLICABLE

Vapor extraction system efficiency is affected by parameters relating to the contaminants to be removed and by variables relating to the site to be remediated. Contaminant properties that affect VES are vapor pressure, solubility, Henry's Law constant, biodegradability, and other molecular structure properties (Jury et al. 1990). Vapor pressure and solubility tend to be the most important chemical parameters affecting VES performance. Soil properties that affect VES performance include soil porosity, soil adsorption, soil moisture, site topography, depth to water, and site homogeneity. Of all the soil properties, permeability tends to be the most critical parameter relating to system success.

CONTAMINANT PROPERTIES

Vapor Pressure

Vapor pressure is the parameter that can be used to estimate a compound's tendency to volatilize and partition into the gaseous state. The vapor pressure of a compound is defined as the pressure exerted by the vapor at equilibrium with the liquid phase of the compound in the system at a given temperature. Vapor pressure is often expressed in millimeters of mercury. Table 1 provides a listing of vapor pressures of some common environmental contaminants. When chemicals exist in pure form, the vapor pressure of the contaminant is very important to its removal efficiency by a VES. The higher the vapor pressure, the more it is amenable to vapor extraction. In turn, the lower the contaminant's vapor pressure, the less likely it will be volatilized and the greater reliance on bioventing (if the contaminant is biodegradable) for successful remediation. For mixtures of compounds (i.e., gasoline), the composition of the mixture also has a bearing on the vapor pressure according to the following relationship

$$P_i^* = X_i A_i P_i^o \tag{9}$$

where P_i^* = Equilibrium partial pressure of component i in the organic mixture; X_i = Mole fraction of component i in the organic compound mixture; A_i = Activity coefficient of component i in the organic compound mixture; and P_i^o = Vapor pressure of component i as a pure compound.

Table 1   Contaminant Properties—Vapor
            Pressure Parameters

Compound	Vapor Pressure (mmHg)
Acetone	89 at 5°C
Benzene	76 at 20°C
Toluene	10 at 6.4°C
Vinyl chloride	240 at -40°C
o-xylene	5 at 20°C
Ethylbenzene	7 at 20°C
Methylene chloride	349 at 20°C
Methyl ethyl ketone	77.5 at 20°C
Trichloroethylene	20 at 0°C
Tetrachloroethylene	14 at 20°C

For example, the above relationship states that the vapor pressure of benzene is related to the percentage (mole fraction) of component benzene in gasoline.

If NAPL is not present in the soil, vapor pressure becomes a less accurate predictor of VES efficiency since other relationships (adsorption to soil, moisture content, etc.) become more important in governing system success. It should be noted, however, that even under conditions of no NAPL presence, a compound must be volatile enough to be removed by VES. Sufficiently high vapor pressure can therefore be viewed as a prerequisite for successful vapor extraction. Although the definition of sufficiently high vapor pressure is rather subjective, one to two millimeters of mercury (mm Hg) should be used as a guideline. Compounds with lower vapor pressures will likely be more slowly removed and/or a greater reliance will be required on *in situ* biological breakdown of the compounds.

Solubility

Aqueous solubility is one of the most important parameters governing the partitioning, transport, fate, and therefore, the ultimate remediation of site contaminants. Solubility can be defined as the maximum amount of a constituent that will dissolve in pure water at a specified temperature. For organic mixtures such as gasoline, solubility is additionally a function of the mole fraction of each individual constituent in the mixture according to Equation (10). Chapter 1 contains a table listing pure water solubilities for some common environmental contaminants.

$$C_i^* = X_i A_i C_i^o \tag{10}$$

where C_i^* = Equilibrium concentration of component i in the organic mixture; X_i = Mole fraction of component i in the organic compound mixture; A_i = Activity coefficient of component i in the organic compound mixture; and C_i^o = Equilibrium solute concentration of component i as a pure compound.

Under most vapor extraction scenarios, the vadose soil is relatively moist (10 to 14 percent by weight) and contaminants are generally dissolved in the soil pore water. Solubility is also a critical factor for the bioventing of contaminants, since biological degradation is enhanced and simplified if the contaminants are more available for microbial uptake by being dissolved in the pore water. A soil moisture of 12 percent by weight is generally required for adequate bioventing.

Henry's Law

The interaction of solubility and vapor pressure produces a behavioral modification that renders the additive effects of solubility and vapor pressure nonlinear. This interaction has particular impact on volatilization of organics from water. The reader is referenced to basic chemistry textbooks for the derivation of Henry's Law (Mahan 1966). Henry's constant is functionally defined as the ratio of saturated vapor density to chemical solubility for a given compound

$$H = C_{sg}/C_{si} \tag{11}$$

where H = Henry's Law constant; C_{sg}= Compound concentration in the vapor phase at the water/vapor interphase; and C_{si} = Compound concentration in the water phase at the water/vapor interphase.

The Henry's Law relationship is depicted graphically in Figure 7. Table 2 presents a list of various compounds Henry's Law constants. Under moist soil conditions, therefore, VES efficiency is Henry's Law dependent much like VOC removal by air stripping. For example, although acetone is very volatile, it is not well remediated by a VES due to its high water solubility. Acetone tends to biodegrade readily, however, and is therefore amenable to bioventing.

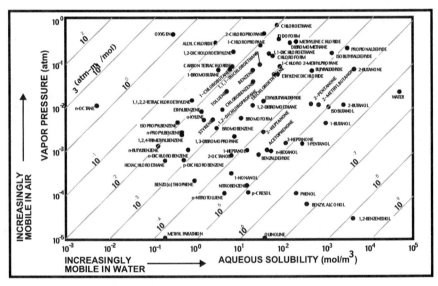

Figure 7   Henry's Law.

Much like air stripping efficiency is temperature dependent, VES efficiency has a temperature dependence. The temperature relationship is more complex for VES due to the existence of multiple system variables such as biodegradation, adsorption, etc. In general, higher temperatures in the vadose zone enhance volatilization which improves operation of a VES. Increased temperature also enhances biodegradation, increases the rate of desorption, and weakens the adsorption binding.

Other Molecular Properties

There are several other molecular properties of the contaminant that influence the success of vapor extraction or bioventing. Although these properties are not as significant as vapor pressure, solubility, and Henry's Law, and hence may be considered secondary, these properties often may be the rate-limiting criteria to site remediation. Compound size, molecular weight, electronegativity, and polarity affect

Table 2   Henry's Law Constants

Compound	Henry's Law Constant[a] atm	Reference
1 Acetone	0	1
2 Benzene	230	1
3 Bromodichloromethane	127	1
4 Bromoform	35	3
5 Carbon tetrachloride	1282	1
6 Chlorobenzene	145	2
7 Chloroform	171	1
8 2-Chlorophenol	0.93	2
9 p-Dichlorobenzene (1,4)	104	4
10 1,1-Dichloroethane	240	1
11 1,2-Dichloroethane	51	1
12 1,1-Dichloroethylene	1841	1
13 Cis-1,2-Dichloroethylene	160	1
14 Trans-1,2-Dichloroethylene	429	1
15 Ethylbenzene	359	1
16 Hexachlorobenzene	37.8	2
17 Methylene chloride	89	1
18 Methylethylketone	1.16	2
19 Methyl naphthalene	3.2	2
20 Methyl tert-butyl-ether	32.6	1
21 Naphthalene	20	4
22 Pentachlorophenol	0.15	2
23 Phenol	0.017	2
24 Tetrachloroethylene	1035	1
25 Toluene	217	1
26 1,1,1-Trichloroethane	390	1
27 1,1,2-Trichloroethane	41	2
28 Trichloroethylene	544	1
29 Vinyl chloride	355000	3
30 o-Xylene	266	1

[a]At water temperature 68°F

1. Per Hydro Group, Inc., 1990.
2. Solubility and vapor phase pressure data from *Handbook of Environmental Data on Organic Chemicals*, 2nd edition, by Karel Verschueren, Van Nostrand Reinhold, New York 1983.
3. Michael C. Kavanaugh and R. Rhodes Trussel, "Design of aeration towers to strip volatile contaminants from drinking water," *Journal AWWA*, p. 685, December 1980.
4. Coskun Yurteri, David F. Ryan, John J. Callow, Mirat D. Gurol, "The effect of chemical composition of water on Henry's law constant," *Journal WPCE*, Volume 59, Number 11, p. 954, November 1987.

its adsorption to soil particles and its travel through soil micropores. Larger, bulkier (branched chains) molecules travel more slowly within soil micropores and tend to adsorb more strongly to soil surfaces. Once the bulk of the more accessible (large pores, not directly adsorbed to soil) contaminant is removed, the final molecules removal tends to be rate limiting. Polarity and electronegativity relate to a compound's effective charge, and its interaction with the soil's surface charge. These topics are further discussed below.

Summary

Vapor pressure and solubility are the most dominant contaminant properties affecting the success of a VES. Secondary contaminant properties include molecular weight, compound size and structure, and surface charge. The properties of the compounds are only part of the controlling factors for VES. The compounds interact with the soil. The stronger the interaction, the less effective the VES. The degree of interaction is dependent upon the properties of the compounds discussed above and the properties of the soil.

PROPERTIES OF THE SOIL

The typical volumetric composition of soil is depicted in Figure 8.

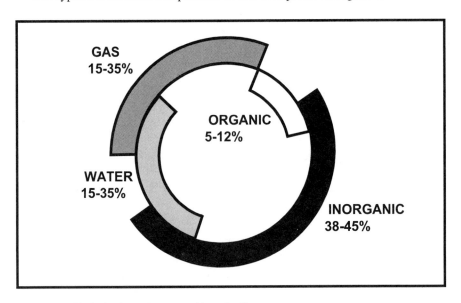

Figure 8   Typical volumetric composition of soil.

Bulk Density/Soil Porosity

Decreasing soil permeability will generally reduce the efficiency of a VES because the diffusive transport from the soil matrix to the soil surface and in turn to the soil gas. Also, the increased path length (path to convective airflow) and decrease in cross sectional area for airflow will reduce transport. A secondary influence of decreasing soil porosity is the increase in soil surface areas available for contaminant binding.

The permeability will also have an economic affect on the VES. The less permeable the soil, the higher the vacuum that is required to maintain the same airflow rate. The zone of influence will also be affected, requiring an increasing number of wells to cover a volume of soil.

Soil Adsorption

There are two methods by which the soil can adsorb the organic contaminant. The soil organic content or its mineral adsorptive surface sites are both capable of contaminant adsorption. The adsorption of contaminants to the soil organic or mineral clay surfaces tends to increase the immobile fraction of the contaminant (adsorbed to the soil) and to decrease the vacuum extraction system efficiency. Soil adsorptive interactions become particularly important (rate limiting) under drier soil moisture conditions.

The clay soil composition is strongly correlated to many water transport and retention properties which influence the volatilization process. Clay holds onto water tightly and is poorly transmissive. This presence of water in the soil pores reduces the available space for vapor transport. The effects of soil moisture or pore water on VES are further described below. Clay soils also tend to have small micropores making transport path lengths longer. The longer transport path lengths reduce the efficiency of vapor extraction.

Clay mineral surfaces tend to have a net negative charge. They will influence the adsorption reaction of most compounds in some fashion. Clays are an excellent adsorbent of positively charged molecules (such as metals) or very polar organic compounds.

Soil total organic carbon (TOC) matter content is strongly correlated with the binding capacity for organic chemicals. This general correlation is valid as a general guidance despite the wide variation in the makeup of soil organic content, its state of decomposition, and consequently its binding capacity. The soil adsorption correlation coefficient is likely to be highest when the soil organic content is high, but observations of strong binding have been documented for soil TOC as low as 0.1 percent. A relationship describing the soil's binding capacity to organics was defined in Equation (3). The parameter K_{oc} in Equation (3) is defined as

$$K_{oc} = \text{ug of contaminant adsorbed/gram of soil TOC/ug/ml solution} \quad (12)$$

Due to the preponderance of other soil organic matter surfaces and the nonpolar nature of most organic contaminants, there is usually little correlation between clay content and VOC adsorption (Jury et al. 1990).

Most organic contaminants are more easily adsorbed to the soil than they are desorbed. It therefore takes much longer and requires more energy to remove the contaminants from the subsurface than it does to distribute them. This phenomenon, known as hysterisis, tends to slow down the vapor extraction process more than would be predicted by simple adsorption isotherm data.

Soil Moisture

Soil moisture is a very important parameter for VES success. High soil moisture content limits air advection travel pathways by occupying void space. Since movement of VOCs is much faster in the gas phase than in the liquid phase, it would be expected that VOC removal by vacuum extraction would be enhanced by decreasing soil moisture. This trend is not always observed. The lack of soil moisture allows for contaminant adsorption to soil surfaces to play a more prominent role in mass transfer as the water particles are removed from the surface. If the soil adsorptive capacities are strong, the benefits of soil dewatering (increased air travel pathways) may be partially offset by this increased soil binding capacity (Sims 1990 and Thibaud, Erkey, and Akgerman 1993). The moisture content at which a decrease in vapor concentration during VES operation is observed is often termed the critical moisture content and is often empirically defined as one monolayer of water molecules coating the soil surface. Recent research observations hypothesize that water particles may act to kick out adsorbed organics, thereby enhancing VES operation under certain conditions. Figure 9 provides a schematic of this concept. If soil adsorptive capacities are very weak (low TOC), it may be advantageous to conduct vapor extraction under drier conditions.

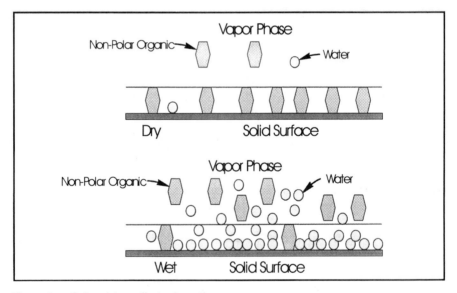

Figure 9   Soil moisture effect schematic.

The notion that an optimal moisture content exists for a given contaminant, based upon its Henry's constant and the soil binding capacity, should allow for some process control and optimization of the VES performance. Although theoretically possible, this notion is rarely applied in the industry due to limited modeling resources and incomplete understanding of the site's partitioning coefficients. Also, it is not practical to regulate soil moisture. It would be expensive to try to dry the air that is entering the vadose zone for the VES. Even with injection wells it may not be possible to completely eliminate the natural moisture of that climate. Moisture can be added in desert environments. This is usually only considered for bioventing projects.

Site Surface Topography

Site surface topography can greatly influence the success of a VES. Ideally the site should be covered by an impermeable surface such as a pavement or concrete. The covered surface serves two functions. First, it minimizes the infiltration of rainwater to the vadose soils and consequently allows some control over soil moisture. Second, the covered surface eliminates the possibility of extraction well short circuiting (Figure 10), where the majority of the extracted volume of air is coming from near the ground surface, thereby more distant locations from the extraction wells receive minimal airflow. The basic design premise of any VES is to create an airflow pattern across the contaminated zone, and short circuiting to the surface may prevent that pattern.

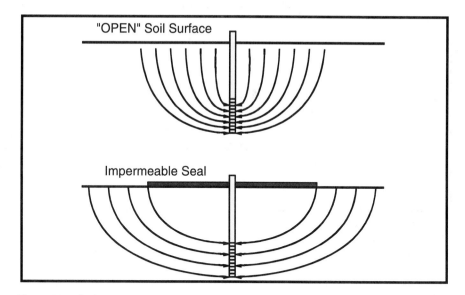

Figure 10   Surface seal effects schematic.

Short circuiting may also be due to the presence of higher permeability zones such as utility trenches. If operation of a VES is conducted in a zone that is prone to short circuiting, a higher number of VES wells will generally be required. This will, in turn, lead to higher airflows and higher capital costs for air treatment devices.

In order to minimize the effects of surface short circuiting, wells should not be screened within 5 feet of the surface. In instances where a surface seal is not available, plastic sheeting can be applied, preferably buried under one foot of cover to enhance system performance.

Depth to Water Table

If the VES well penetrates the water table (use of a converted monitoring well for vapor extraction) and vacuum is applied to the well, the water table within the well will rise by an amount equal to the level of applied vacuum. A 60 inch water column vacuum at the well head will therefore result in a water table rise of 5 feet (Figure 11). If there are only 5 feet of well screen above the water table, the water table rise may clog the available well screen. This scenario may be encountered in a situation where the water table is shallow or when the well is inappropriately designed. Horizontal wells can be utilized in shallow water table situations thereby enlarging the available screen length and reducing well head vacuum, thus minimizing water uplift.

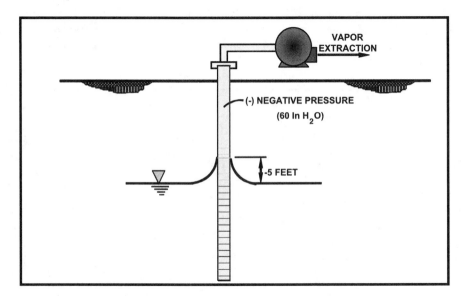

Figure 11   Water uplift effects schematic.

Installing the VES well with well screen above the water table will minimize water table uplift as the vacuum at the water surface will be less than at the wellhead. As Figure 6 points out, wellhead vacuum dissipates quickly away from the extraction well. As a general rule, the bottom of the VES well should be a minimum of 2 to 3 feet above the water table, if possible, to prevent the effect.

Site Homogeneity

Site homogeneity is important to ensure that airflow reaches all areas requiring remediation. The air carrier must flow past the contaminants if they are to be removed. Transport of the contaminants by the carrier airflow minimizes diffusion requirements for mass removal. This reduces the travel path length to remediation and expedites the cleanup. As an example, NAPL floating on top of the water table takes longer to be volatilized because airflows over it rather than through it. Lab experiments where air is drawn through NAPL saturated soils usually result in NAPL volatilization that is much faster than volatilization of NAPL floating on the water table.

Site nonhomogeneity can be partially alleviated by varying well screen designs to maximize air movement in contaminated zones, fully opening some extraction wells in low permeability zones and closing others in high permeability zones, and by possibly breaking the area into more than one zone based upon permeability. Wells in high permeability zones can be connected to a moderate vacuum blower and wells screened in low permeability zones can be connected to high vacuum liquid ring type pumps (Figure 12).

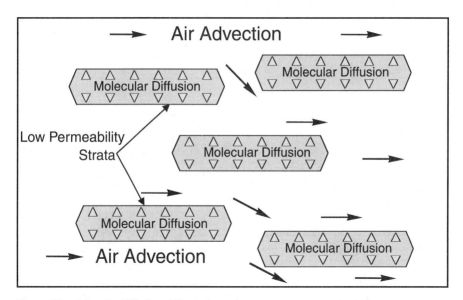

Figure 12   Advective/diffusive airflow schematic.

The presence of micro lenses of highly adsorptive, low permeability soils will often be rate limiting during VES operation (Figure 13). These highly adsorptive lenses are not accessed by the carrier air and rely on concentration gradient driven diffusion.

At a VES site the presence of utility trenches (generally constructed of high permeability fill material) or other high permeability airflow paths may also provide short-circuit pathways. In these instances, well screening and/or placement may require adjustment to accommodate the contaminant distribution. This adjustment usually requires deeper well screens and a higher density of extraction wells.

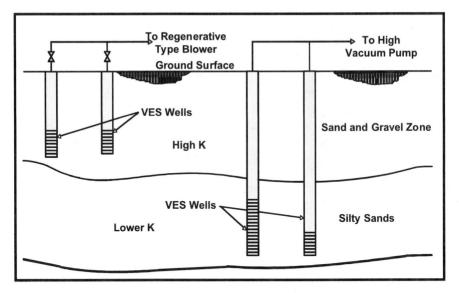

Figure 13   Two-zone venting system.

MODELING TOOLS FOR VAPOR EXTRACTION SYSTEM DESIGN

Computer simulation modeling is utilized to better design vapor extraction systems. This enhancement is due to the simulation of complex subsurface processes. This allows the model user to vary operating parameters and observe a simulation of the result. For example, the model allows for comparing subsurface flow regimes with 10 or 20 extraction wells. The modeler can therefore evaluate the benefits to be gained by installing the additional 10 wells and decide whether the added benefit is worth the additional installation and operational costs.

For the purposes of this discussion, three classes of modeling tools for use in VES design are described. This is a broad, crude model classification, but is useful in presenting the basic requirements for VES design. The following three different types of models will be discussed.

1. Engineering design model
2. Airflow models
3. Multiphase transport models

Engineering Design Model

Models are available that simulate vacuum losses in piping networks, vacuum equipment, well design, valving networks, and above-ground air treatment equipment. These models are particularly useful for the design of larger systems (greater than 10 extraction points) where these calculations may become cumbersome. These models are often based on solution of Hardy Cross type of pneumatic flow equations.

Flow Models

Flow type models Airflow SVE (Waterloo Hydrogeologic Software, Waterloo, Ontario Canada), HYPER VENTILATE (U.S. Environmental Protection Agency), and AIR 3D (ARCADIS Geraghty & Miller) simulate airflow. Airflow SVE is also a transport model while the others do not simulate mass transfer among the various compartments. These models are simple to operate and allow the user to locate wells in order to achieve a predetermined vacuum or conversely (Airflow and AIR 3D) allow the user to translate vacuum influence readings into cross sectional airflow at a given remote location. In either case the models allow for vacuum and airflow profile distributions across the generating two or three (AIR 3D) dimensional profiles.

Flow models do not account for volatilization from pore water, soil surfaces, or any equilibrium partitioning. If flow models are utilized, the user must select the minimally acceptable airflow across the site based on published literature values for the contaminant of concern within similar soils. For example, if TCE is to be extracted from sandy soils, a minimally acceptable airflow across the most distant location from the extraction well must be selected. A limited but useful database (minimally acceptable airflows) exists in the published literature for this purpose. Airflow models therefore allow for selection and optimization of well placement and well screening within the contaminated zone. This selection is based on a predetermined minimal acceptable airflow.

Multiphase Transport Models

Multiphase transport modeling is usually not required for the majority of simple VES applications. Vapor extraction at a typical quarter-acre sandy soil, shallow water table, service station does not require multiphase transport modeling. The installed systems are usually small, overdesigned, and to a large extent rely on bioventing to supplement any shortcomings of VES design, since the majority of contaminants are biodegradable.

At more complex sites, however, where any or all of the following is desired, multiphase transport modeling can be very useful.

1. System layout and sizing optimization is desired due to the site's large size (too expensive to overdesign)
2. System operating parameter optimization is desired
3. Vapor extraction feasibility requires demonstration
4. Contaminants may not be biodegradable
5. Impacts to groundwater may be significant
6. Closure time-frame prediction is required

Solution of Equation (8) with the requisite modifications to account for nonequilibrium, nonlinear behavior is the basis for the multiphase transport models. Several models have been developed by researchers (Armstrong, Frind, and McClennan 1994), but few models that can adequately describe multiphase transport are commercially available.

Multiphase transport modeling allows for simulating the remediation process over time based on the selected well layout. The multiphase transport models are therefore often preceded by simple airflow models that locate extraction wells. The multiphase transport models simulate the VES perturbation (inducement of airflow) to the contaminant partitioning and predict contaminant concentrations within the various media at future points in time. The modeler can vary the site's soil moisture, airflow, or any other parameter and observe the predicted effects.

A numerical model is therefore the best means of understanding the mass transfer between the liquid (soil pore water) and gas phase, and the degradation of constituents into different species. MOTRANS, developed for EPA by Parker and Kaluarachchi at The Virginia Polytechnic Institute and State University in Blacksburg, Virginia is commercially available. MOTRANS can be used to simulate either two-phase flow of water and NAPL in a system with gas present at a constant pressure, or three-phase flow of water, NAPL, and gas at variable pressure. Systems with no NAPL present or with immobile NAPL at a residual saturation may also be modeled by an option that enables elimination of the NAPL flow equation. The transport module can handle up to five components that partition among water, NAPL, gas, or solid phases assuming either a local equilibrium interphase mass transfer or first-order kinetically controlled mass transfer.

The flow of water and vapor and the transport of constituents in the vadose zone are highly complex processes. The equations governing these processes are strongly nonlinear, difficult to solve, and require extensive data input to characterize the physical properties of both the media and the fluids. In general, the principal limitation in applying modeling codes is characterization of the problem. Migrations of constituents in the vadose zone are controlled by local heterogeneities, which may be difficult to define.

In addition, the physical properties characterizing the relative permeabilities and fluid retention characteristics are rarely collected. Multiphase flow and multicomponent transport require specification of permeability/saturation/capillary pressure relationships, air-water capillary retention function parameters, NAPL surface tension and interfacial tension with water, NAPL viscosity, NAPL density, maximum residual NAPL saturation, soil permeabilities and dispersivities, initial phase concentrations, equilibrium partition coefficients, component densities, diffusion coefficients, decay coefficients, mass transfer coefficients, and boundary conditions. These relationships and parameters can be determined from direct measurements in laboratory treatability tests that can accompany modeling efforts or found through a literature search. Multi-phase models can therefore potentially account for:

1. Advection
2. Dispersion
3. Sink/source mixing
4. Chemical and equilibrium partitioning

The models can potentially simulate the removal of the contaminants from the subsurface under a variety of conditions (different flow velocity, different well screen positions, different moisture levels, different extraction and passive well locations, different concentration profiles, etc.). The models can allow the user to optimally choose:

1. Well screening positions
2. Well locations
3. Well positioning
4. Vapor flow rate
5. Applied vacuum
6. Soil moisture content

PILOT STUDIES

Pilot studies are generally conducted in order to gather relevant information to design a full-scale VES. Field pilot studies gather information regarding the pneumatic flow characteristics of the vadose zones and the extracted air quality. Laboratory soil column and soil cube tests allow for simulation of the vapor extraction remediation process by passing air through a small amount of soil. This also allows for selection of the minimally acceptable airflow velocity for removal of the contaminants.

LABORATORY STUDIES

Laboratory studies are often the best method to optimize the required airflow and moisture content, as well as other control parameters for a VES. Laboratory studies allow for manipulation of these parameters under controlled conditions and can be done in conjunction with modeling and field pilot studies to optimize system performance. Due to the expense associated with these activities, their implementation is rare except at large sites contaminated with nonbiodegradable VOCs where volatilization must be optimized.

Lab studies can be conducted using columns or soil cubes. Figure 14 presents a schematic of the laboratory setup. Soil cubes offer the benefit of providing better simulation of actual airflow profiles in the subsurface. However, soil cubes will not be able to simulate the macrogeological conditions of a sand lens, clay lens, or any other major change in the geology.

Researchers have increasingly concentrated on evaluating the minimally acceptable airflow velocity for successful vapor extraction (Armstrong, Frind, and McClennan 1994, Gierke, Hutzler, and McKenzie 1989, Sleep and Syikey 1989 and Brusseau, Jessup, and Rao 1991). It is anticipated that upon completion of sufficient lab studies, a matrix of contaminant type, soil type, moisture content, and minimal airflow velocity can be compiled. This matrix would greatly improve the design of VES by practitioners.

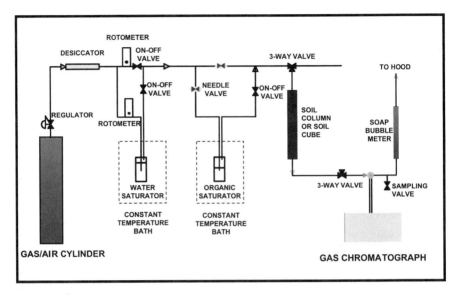

Figure 14   Lab columns/cubes.

FIELD PILOT STUDIES

This section will describe the test wells, procedures, and equipment required to conduct a pilot or feasibility test. A pilot test will allow determination of the subsurface air permeability, the expected mass removal rates of contaminants at system start-up, and allow for the selection of appropriate vapor treatment equipment. The pilot test should also determine the required vacuum and number of extraction points to achieve the desired airflows during VES application. A pilot test should be conducted prior to installation of most VES applications. Only systems whose size does not greatly exceed (two to three times) the size of the pilot system should be installed without pilot testing. For small systems, the cost of the pilot test can sometimes not be justified.

A typical pilot test set up is shown in Figure 15. The pilot test will include installation of a minimum of one extraction well, several observation wells (or points), and hook up of the extraction well to the vacuum equipment. Upon start up of the vacuum pump, several field measurements from the wells and the extraction vacuum pump are taken prior to the conclusion of the test.

Vapor Extraction Testing Well

The extraction well should be located near the center of the contaminated zone in order to ensure gathering of data that may be representative of start up conditions. The vapor extraction well should be 2 to 4 inches in diameter. At most locations, 2 inch diameter construction is adequate; 4 inch diameter construction is advisable in

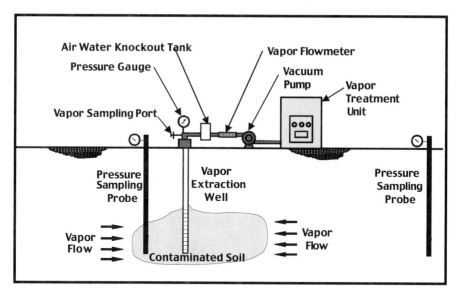

Figure 15   Pilot test set up.

instances where groundwater pumping is considered from the vapor extraction well (vacuum-enhanced recovery or dual extraction) or for deeper contaminated zones where significant airflow may be generated.

In some instances, existing monitor wells can also be used for vapor extraction; however, water uplifting should be expected to reduce the available screen length. Should uplifting significantly reduce the available screen length, installation of a proper vadose zone extraction well should be undertaken.

The screened interval of the extraction well should be within the contaminated zone. Since air circulation below the screened interval is generally not significant, the screen should be placed accordingly to attain the desired air distribution in the contaminated zone. Where contamination is deep and permeability is high throughout the soil column, the screened section should be extended to the maximum depth possible and slotted only at the bottom to maximize the area of treatment (rather than slotted fully vertically) (Sims 1990). At best, the slots outside of the contaminated area will decrease the efficiency of the VES system. At worst, the slots will connect into a highly permeable layer of the vadose zone, and none of the airflow will be through the area of contamination.

The design of a typical VES vertical well is shown in Figure 16. Horizontal wells may also be considered in shallow water table situations. Horizontal wells allow for vacuum distribution over a larger screened area to minimize water uplift. The length of horizontal wells should not be more than 10 feet to minimize preferential flow distribution. Short circuiting concerns are generally amplified with shallow horizontal wells and therefore extraction well vaults are generally not installed at the well. Piping is generally run back to a point further away from the extraction lateral where the vault can be safely installed.

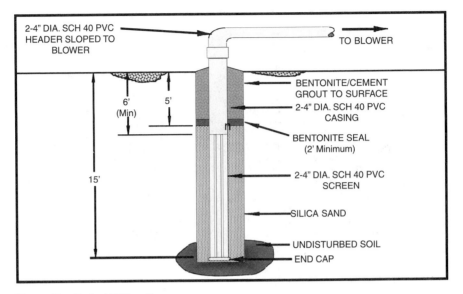

Figure 16 Typical VES well.

Vapor Extraction Monitoring Wells

A pilot test will generally require three to four monitoring points in a homogeneous setting to assess the zone of airflow influence. The number of required monitoring points increases if the site is nonhomogeneous and nonisotropic. The increased number of points depends on the complexity of the site. It is best to locate at least one point close to the VES well in order to ensure one positive result point. The monitoring points should be radially distributed at various distances from the extraction well. Other points are located at increasing horizontal (and vertical if required) distances. Often one tries to locate the test in an area where existing monitoring wells can be utilized for observation.

If a model is going to be utilized to design the VES, the use of monitoring points at various depths should also be considered. This data will allow determination of a vertical air permeability in addition to the usually obtained horizontal air permeability. Data collected from vacuum measuring points at various elevations will also enable the designer to identify microlenses of less permeable material that will control the tailing effect, as noted in Figure 4. This is a reason for also collecting data from various elevations even if a model is not going to be utilized. This information will help the designer to strategically place the extraction wells.

Monitoring points can often be existing groundwater monitoring wells if they are adequately located and screened. Alternatively, monitoring wells of similar construction to the extraction well can be installed. If the observation wells will only be used for vacuum measurement, they may be of small diameter (5/8 inch to 1 inch); however, if one desires to gather *in situ* air velocity profiles with a down-well anemometer, appropriately sized wells are required. In shallow conditions, and instances where the soils are sandy and not silty, the observation points may be driven (5/8 inch soil vapor probes can be used). Driving points in silty conditions may lead to clogging of the drive point screen.

The pilot test can be subdivided into three stages: pilot test planning, conducting the pilot test, and data evaluation. These stages are described below.

Stage 1: Pilot Test Planning

In addition to considering the contaminant and site characteristics prior to conducting the VES pilot test, there are several issues which need to be addressed.

1. *Well location and construction*: The extraction well must be adequately screened and located to achieve the desired results. The screen should be in the contaminated zone and groundwater uplift must be given consideration. It is generally advisable to locate the well in the middle of the contaminated zone. Wells located outside of the contaminated area can save money by not requiring air treatment during the pilot test. However, experience has shown that the money for the air treatment is worth spending to prevent the possible problems from a change in the geology from the test location to the actual contaminated site, even if the distance is relatively small.

2. *Short circuit pathways*: Short circuit pathways should be minimized. Wells should be located and screened to minimize the short circuiting effects of high permeability utility trenches, open surfaces, or other low resistance flow pathways. For example, a well screen which spans two geologic sections will collect most of the airflow from the more permeable section. If the airflow is desired in the low permeability formation, the high permeability zone will act as a short circuit pathway. Once again, shallow VES have to pay particular attention to short circuiting to the surface. Pilot studies may require some type of surface coating to prevent this type of short circuiting.

3. *Required test duration*: The pilot test must evacuate a minimum of 1.5 to 2.0 pore volumes of contaminated vapors in order to gather vapor quality that would be representative of VES startup conditions. A pore volume can be roughly calculated by assuming a radius of capture (40 to 50 feet typical), extraction rate, assuming a soil porosity, and evaluating the time requirement to evacuate that volume of air, as presented below

$$t_p = n\Pi R_i^2 h / Q \tag{13}$$

where t_p = Time required to evacuate one pore volume, min; n = Porosity of the vadose zone; R_i = Radius of influence (estimated), feet; h = Thickness of impacted vadose zone, feet; and Q = Volumetric flow rate, cubic feet per minute.

Upon removing the first pore volume, the VOC concentrations will typically drop and will be more indicative of VOC levels to be observed upon system startup. Soil vapor concentration levels of the first pore volume are indicative of equilibrium conditions and tend to be higher than observed during VES operation. This is true even when NAPLs are present.

4. *Vacuum equipment needs*: You should check Figure 5 for expected vacuum needs based upon the soil conditions of the site and select a blower/vacuum pump that will yield the required airflow at the expected subsurface resistance. The investi-

gation data is utilized to estimate a soil permeability. The soil permeability estimation is then utilized with an assumed required vacuum to induce airflow, which will yield an estimated extraction rate per foot of vadose zone. Most wells are not 100 percent efficient and higher vacuum requirements should therefore be anticipated. A 25 percent decrease in well efficiency can often be expected, resulting in higher vacuum requirements. Inducing airflow at the pilot test is a requirement; therefore, one must utilize a blower/vacuum pump that can induce airflow in the subsurface at the expected resistance. In the case of uncertainty regarding the vacuum requirements, it is best to be conservative and utilize a high vacuum pump. If the high vacuum is not required, the well head vacuum during the pilot test will be low, and full-scale system design can include a low vacuum blower. Another method of evaluating various vacuums and flow rates is to utilize two blower/vacuum pumps that can be operated either in series or parallel. This type of arrangement maximizes flexibility during the test and enables one to collect various operational data that can be evaluated to optimize VES system design.

5. *Type of monitoring points*: Existing groundwater monitoring wells can be utilized, if they are screened in vadose zone. Alternatively, new monitoring points can be installed. Soil vapor probes (5/8 inch) can be driven at various depths (15 feet practical limit), but risk screen clogging in silty conditions. Checking the vacuum rebound on a probe is a good test to see if it is clogged. The vacuum rebound test is done by applying a vacuum to the probe with a small vacuum pump and seeing how quickly the gauge rebounds after the vacuum is turned off. A clogged screen rebounds slowly whereas a unclogged screen bounces back instantly. Probes offer the flexibility of varying locations and depths during the test. The use of probes precludes the use of a down-well anemometer to record *in situ* airflow velocity.

6. *Off gas treatment requirements*: Since the pilot test is of short duration, the choice is generally between granular activated carbon (GAC) or rental of other equipment (catalytic/thermal oxidizers). GAC consumption and cost needs to be evaluated versus other equipment rental costs to make the selection. Many times the regulatory requirements for air discharges are based upon a total monthly discharge mass. If the state regulatory agency allows special consideration for pilot studies and the total mass is below the regulations, then no off gas treatment may be necessary for the pilot test.

Stage 2: Conducting the Pilot Test

The VES pilot test will include the following measurements/activities.

1. Measure the flow rate of extracted air and well-head vacuum. Vacuum and flow rates can be adjusted by opening and closing a dilution valve on the influent side of the blower/vacuum pump. Opening the valve reduces the subsurface airflow/vacuum application. This variation and the monitoring of the subsurface response is useful for computer model verification/calibration as well as providing multiple data points for field parameter (permeability) evaluation. Make sure the flow meter is measuring the extraction rate from the vadose zone and not the pumped rate, if a dilution valve is utilized.

2. Measure vacuum influence in the soil probes or monitoring wells that is induced by the extraction well using magnehelic gauges. If the extraction flow rates are

high enough, down-well anemometers may be able to monitor well airflow velocities. Anemometers have a limited flow velocity range (should be used closer to
the extraction wells) and are sensitive to moisture build up. Readings should be
taken as quickly as possible to minimize the impact of moisture build up. The
measurement of *in situ* air velocities is not widely practiced and limited information
is available on the best methods for measurement. Figure 17 provides a suggested
test measurement set up.

3. Measure the vapor emissions before and after on-site treatment using a field
   instrument and collect lab samples for verification. This data should be collected
   at the beginning, middle, and end of the test. This data will be essential for
   evaluating vapor treatment options. It is important that a minimum of one pore
   volume be extracted and preferably more.
4. Measure the oxygen and carbon dioxide content of the withdrawn air. As will be
   discussed in the bioventing section, this data is used to assess subsurface biodegradation rates.

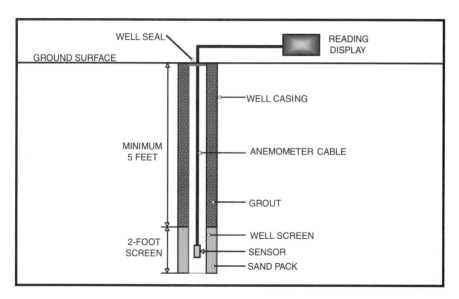

Figure 17   Anemometer set up.

Stage 3: Evaluating the Data

The field data is evaluated to determine or select the following.

1. The pneumatic permeability of the site.
2. The location of extraction points in the full-scale remedial design to achieve the
   required airflow distribution in the subsurface.
3. The appropriate extraction blower/vacuum pump to provide the required vacuum
   and airflow rates.
4. The appropriate air treatment technology.

The first two analyses are discussed below. Chapter 6 of this book discusses vapor treatment, and vacuum equipment selection is discussed later in this chapter.

Evaluation of Pneumatic Permeability

A determination of the soil's permeability is an important element for full-scale system design, evaluation of operating results, and/or modification to an existing system. There are several methods of determining a soil's permeability. One of more common methods, first presented by Johnson, Kemblowski, and Colthart (1990) utilized the pneumatic form of the Jacob equation as presented below

$$Q/H = (k/u)P_w II\{1 - (P_{atm}/P_w)^2\}/1n(R_w/R_i) \qquad (14)$$

where k = soil permeability to airflow (cm²) or Darcy; u = viscosity of air = 1.8 x 10^{-4} g/cm-sec or 0.018 cp; P_w = absolute pressure at extraction well (g/cm sec²) or (atm); P_{atm} = absolute ambient pressure = 1.01 x 10^6 g/cm-sec²or 1 atm; R_w = radius of vapor extraction well (cm); and R_i = radius of vacuum influence of the vapor extraction well (cm).

Rearranged to determine, k

$$k = (Q/H)(u/P_w II)1n(R_w/R_i)/\{1 - (P_{atm}/P_w)^2\} \qquad (15)$$

It should be noted that this evaluation for k is only valid if the vacuum distribution in the vadose zone develops essentially instantaneously. In other words, there is not a vacuum distribution with time as the pilot study continues. This is usually the case for vadose zones 10 feet thick or less.

For vadose zones thicker than 10 feet, where the vacuum distribution varies as the pilot test progresses, Johnson, Kemblowski, and Colthart (1990) presented the transient soil permeability calculation as noted below

$$P' = Q/4IIm(k/u)\{-0.5772 - 1n(r^2nu/4kP_{atm}) + 1n(t)\} \qquad (16)$$

where P' = measured vacuum at a distance r and time t, g/cm sec²; m = stratum thickness, cm; r = radial distance from vapor extraction well, cm; Q = volumetric vapor flow rate, cm³/sec; and P_{atm} = ambient atmospheric pressure = 1.01 x 10^6 g/cm sec².

The above equation predicts a plot of P' vs $1n(t)$ should be a straight line with slope A and y-intercept B where

$$A = Q/4IIm(k/u)$$

$$B = Q/4IIm(k/u)\{-0.5772 - 1n(r^2nu/4kP_{atm})\}$$

When Q and m are known

$$k = Qu/4AIIm \qquad (17)$$

When Q and m are not known

$$k = (r^2 nu/4P_{atm})e^{[B/A + 0.5772]} \qquad (18)$$

Equation (15) is not sensitive to values of R_i; therefore, the subjective nature of the value does not alter the evaluation significantly. Evaluation of the pneumatic permeability using Equation (15) allows for calculation of airflow velocities across the site as discussed in the next section.

Extraction Well Placement

Using the pneumatic permeability (k) value, vacuum profiles can be converted into velocity profiles across the site by substitution of (k) into the Darcy and continuity equations for airflow. The Darcy equation is the same as the equation for groundwater flow except that vacuum pressure is the gradient rather than water elevation. The continuity equation simply states that mass entering a unit volume must also exit the unit volume (if biodegradation does not occur in the unit volume). This task is best accomplished using numerical modeling to provide velocity profiles across the entire site (using airflow models).

The velocity profile is used to locate wells across the site to achieve the minimally acceptable velocity in the contaminated zone. Numerical modeling is often beneficial to simulate the effects of multiple extraction points. Minimally acceptable airflow velocities can be determined based on literature reviews for the particular soil type and contaminant, or can be based on laboratory studies that simulate the subsurface flow/contaminant conditions. Laboratory studies are typically not conducted for small applications and or applications where the contaminants are biodegradable (petroleum contamination), since it is the general belief that the inducement of airflow will eventually lead to biological breakdown of the contaminants.

Reliance on minimally acceptable vacuum readings (0.5 inches to one inch of water) or minimally acceptable percentage of applied well head vacuum (1 percent of applied vacuum) to determine the VES radius of influence is generally not the most desirable method of data analysis for large sites or sites contaminated with non-biodegradable compounds. These methods fail to incorporate the interaction between multiple wells. This practice, however, is often conducted at small sites such as service stations where well spacing is generally conservative (typically 20 to 40 feet) and is often limited by physical barriers (tank, piping, and building location) and therefore cannot be reasonably optimized. Most of these VES installations have a significant reliance on bioventing, where volatilization velocities are not optimized.

SYSTEM DESIGN

Vapor extraction system design can be categorized as being performed with the aid of modeling for larger more complicated projects, or more empirical design without modeling for smaller type projects. The design considerations for both approaches are similar and are enumerated below. The use of airflow and/or multiphase transport models allows for significantly improved abilities to locate extraction wells and select the most appropriate subsurface airflow velocities. Models allow the designer to design a system with greater confidence that it will achieve its remedial goals. Empirical design methods rely on past experience or on achieving airflows based on predetermined vacuum levels in the subsurface. This approach generally yields lower confidence levels that remedial goals will be achieved. The designer must decide whether the higher confidence levels warrant the added modeling cost.

The following design considerations are enumerated as a practical checklist guide for the designer:

1. *Airflow rates*: Airflow rates to achieve remedial goals within the required time frame: Airflow requirements will dictate vacuum equipment selection and air emission control equipment sizing. Airflow requirements are obtained by modeling, pilot testing evaluations, empirical knowledge, and/or literature review to provide adequate airflow across the entire site. Higher airflow rates are not always better since it will increase the blower/vacuum size and associated increased power requirements, and more importantly air emission control costs. Not only will the capital costs increase but also the operation and maintenance costs. Also, the higher airflow rates may not necessarily cleanup the site faster, particularly in a non-homogeneous vadose zone, where diffusion will control cleanup time.

2. *Vacuum equipment selection*: The required vacuum to induce the desired airflow rates dictates the type and size of blower/vacuum pumps that are required. This information is generally based on a scaling up of the field pilot test results, modeling analysis or empirical design methods. Regenerative type blowers are typically used in low vacuum (to 8 inches Hg) and high flow applications; positive displacement blowers are used in medium vacuum applications (to 12 inches Hg), and liquid ring pumps are used to induce flow at high resistance applications (to 25 inches Hg). Other types of blower/vacuum pumps such as rotary vane and gear pumps have been successfully used for VES but tend to be less common. Vacuum equipment systems should always be designed with dilution valves in order to adjust the applied vacuum, and/or dilute the recovered vapors for vapor treatment if required.

3. *Knock out tanks/filters*: Blower/vacuum pumps should be protected from suction of water and/or particulates by using in-line filters and moisture knock outs. In instances where significant moisture is accumulated (high vacuum, high airflow rates), the accumulated water in the knock out tank is often transferred to the groundwater treatment system using a transfer pump. If a groundwater treatment system is not present at the site, other treatment/disposal methods need to be considered. This is another reason why higher airflow rates are not necessarily better since they may lead to excessive moisture that needs to be handled.

4. *Measurement devices*: VES should be equipped with both flow measurement and vacuum measurement devices. Vacuum gauges should be placed on the suction

side of the blower and at the well heads. Since the well head vaults can often be moist and hot and not a particularly conducive environment for instrumentation, use of a portable vacuum gauge should be considered, particularly if many wells are part of the system. Piping arrangements would have to made so that the vacuum gauge could easily be attached, a reading obtained, and the gauge then utilized on the next well. Airflow measurements are usually obtained with either a Pitot tube, which measures airflow velocity from pressure differentials, or a large rotometer.

5. *VES well design*: VES wells should be vertically installed as shown in Figure 18. Lateral well installation can be considered for high water table applications, or when surface structures do not allow the installation of multiple vertical wells. Wells should always be equipped with flow adjustment valves and vacuum gauges or the ability to obtain readings. Valves allow for process control by fully or partially utilizing the extraction point during the remedial process. The use of trenches for vapor extraction is generally a poor choice due to the limited ability to control flow pathways and the high risk of short circuit pathways within trenches. Unlike wells, trenches do not allow for valving down/closing certain areas of the site during the cleanup process.

6. *Number and location of VES wells*: The number of VES wells will be dictated by the need to maintain adequate airflow in the subsurface. Pilot test data, empirical knowledge, and computer simulations will determine the number of required wells and their ideal locations.

7. *Passive well placement*: Passive wells provide influent air and should be located to minimize dead (no flow) zones that may develop in multiple extraction point configurations. Passive wells can also be utilized to provide engineered short circuit pathways so as to eliminate migration of contaminants from certain zones as shown in Figure 18. Passive wells are also a good design method for deep VES systems. Passive wells can help ensure that the airflow pattern is optimized to flow through the contaminated area of the vadose zone.

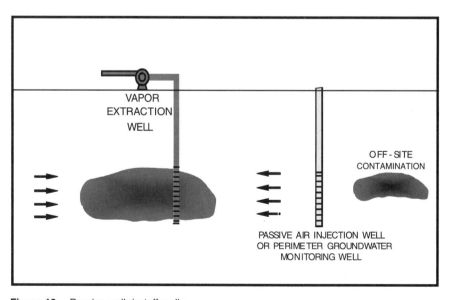

Figure 18   Passive wells/cutoff wells.

8. *Well screen positioning*: Screened intervals are particularly important in ensuring that advective flow is induced in the most contaminated zones. During vapor extraction from thick vadose zones, the screen should placed near the bottom of the vadose zone rather than through the entire zone. One way to ensure that the screens are in the proper location is to monitor the well installation with field testing equipment. This will help keep the screens in the contaminated portion of the vadose zone.

9. *Soil moisture content*: Soil moisture controls air permeability and may control contaminant partitioning and VES rate limitation. Its proper manipulation may enhance system performance. If one cannot induce sufficient airflow for volatilization, one may want to consider keeping the moisture content high and enhancing bioventing for biodegradable compounds. While it is difficult to change the natural soil moisture, steps can be taken to adjust the moisture content in some instances. Surface covers can be used to prevent rain moisture from entering the vadose zone. The airflow will then be able to slightly dry the formation, clear the water out of the pores, and allow greater airflow through the system. Moisture can also be added to dry areas by injecting humidified air.

10. *Off gas treatment technology selection*: Pilot test data, cost analysis, and operation and maintenance logistics will enable the appropriate selection of vapor treatment technologies (see Chapter 6). It is important to remember that the concentration in the off gas will change rapidly for a VES system. The lifecycle concentration will have a large swing in a relatively short period of time. More than one type of treatment technology may have to be used to be cost effective over the life of the project.

11. *Piping*: All underground piping should be vacuum or pressure tested prior to burial. Schedule 40 PVC is the most common and is generally an acceptable material of construction for most underground piping systems. If greater strength is required then the use of Schedule 80 PVC or steel piping should be considered. The piping between the blower and the treatment system will be pressure not vacuum piping. Local regulations may require steel piping for any pressure service. Also, the air will heat up as it passes through the blower and could melt standard PVC, so either steel piping or CPVC should be utilized.

In many respects the design of a VES is quite complex. Figure 19 shows a schematic of a typical VES design. The designer must achieve the optimal airflow velocities in a complex subsurface that is likely to be nonhomogeneous. In many other respects, the above-ground components of a VES are so simple that the designer can forget about the many subsurface uncertainties. The above listed design considerations simply provide the issues to be considered but, unfortunately, cannot shed light into the nonhomogeneous subsurface.

BIOVENTING

Introduction

Bioventing, or bioenhanced soil venting, is a remedial method similar to soil venting in that it relies on an increase in the flow of air through the vadose zone

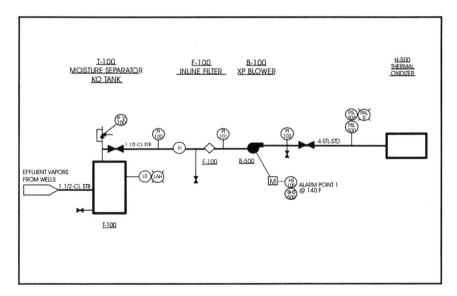

Figure 19   Typical VES system.

by pumping soil air from a well completed in the vadose and capillary fringe zone. Soil venting is performed to volatilize the constituents *in situ*. In bioventing, the increase in the flow of air is to provide oxygen in the subsurface to optimize natural aerobic biodegradation, and this becomes the dominant remedial process. Bioventing has the advantages of being one of the few *in situ* technologies for remediating both volatile and nonvolatile fractions as long as the compounds are biodegradable. In addition, bioventing can minimize air treatment prior to reinjection, if used and has been shown to reduce the contaminant concentrations more rapidly than other methods.

The bioventing technology optimizes the airflow rate to minimize volatilization while providing sufficient airflow to enhance biodegradation. This will often eliminate or minimize air treatment requirements. During implementation of bioventing, soil gases (O_2, CO_2, CH_4, etc.; not just system vent gas composition) are generally monitored to ensure the presence of aerobic conditions. Moisture and/or nutrient addition can be considered to enhance system performance. Both moisture and nutrients must be delivered in the aqueous state. Providing coverage of the vadose zone with solutions is difficult and requires a sophisticated subsurface distribution system. Ammonia delivery can theoretically be delivered in the gaseous state, although this has not been commonly practiced. Some recent work has also looked at providing phosphorous in a gaseous form. Both the ammonia and the phosphorous compounds are soluble and do not travel beyond the soil moisture in the vicinity of the injection well. This creates high concentrations in local areas (possibly toxic levels) and small amounts of nutrients in most of the contaminated area.

Advantages of Vapor Phase Biotreatment

The bioventing system has significant advantages in comparison with ground-water based aerobic biodegradation systems. By using air as the oxygen source, the minimum ratio of air pumped to the contaminants is approximately 13 pounds of air per pound of contaminant for typical petroleum contamination. This compares to a requirement of delivering over 1,000 gallons of groundwater to deliver the same amount of oxygen to the contamination. This is because oxygen saturation in water is roughly 8 mg/l, whereas air is 20 percent oxygen. Another major advantage is that gases have much greater diffusivities than liquids.

Geological heterogeneities present a particular problem for water-borne oxygen flow, since the groundwater will be channeled to the more permeable pathways/channels. As a result, oxygen delivery to the less permeable zone must occur by diffusion. If air is the oxygen carrier, the diffusion can take place several orders of magnitude faster than in the liquid phase. By dewatering and exposing additional vadose zone, the bioventing process can be conducted in the once saturated zone, significantly expediating the remedial process. Chapter 4 shall discuss the benefits of transforming the contaminant carrier from water to vapor in greater detail.

PERFORMANCE CRITERIA AND BIOVENTING PLAN PROTOCOLS

Bioremediation projects have often been evaluated in the laboratory by determining whether the contaminant levels are decreasing and whether the microorganisms from the site have the capability to metabolize the contaminant. This simplistic evaluation is inadequate for the following reasons:

- Field microbial activity may not behave in the same manner as the laboratory cultures. In the laboratory the field microorganisms are put into contact with the contaminant of interest under laboratory conditions which may not simulate natural conditions
- Biotransformation and/or other abiotic processes (such as volatilization) may cause reduced contaminant concentrations without actually resulting in biological breakdown of the contaminant mass

In order to demonstrate that bioventing or biological activity is occurring at a site, sound scientific logic must be demonstrated. This is particularly true due to the difficult credibility route *in situ* bioremediation has suffered in the industry. According to some researchers the following three performance objectives must be met (Rittman and McDonald 1993).

1. Documented loss of contaminants from the site by sampling and chemical analysis.
2. Testing that shows microorganisms in the laboratory assays have the potential to transform the contaminants under the expected site conditions.
3. One or more pieces of evidence showing that the biodegradation potential is realized in the field. The simplest test to conduct in the field is to measure the

electron acceptor uptake rate (oxygen, under aerobic conditions), and to measure the subsurface inorganic carbon production rate (respirometry test).

The laboratory and field biotreatability testing are integral components to demonstrating that bioventing is a viable technology at a given site. The following sections outline the procedures to be followed for conducting the various demonstration tests. The field respirometry testing satisfies the third performance objective. Laboratory treatability testing for the contaminants of interest satisfies the first objective.

LABORATORY TESTING

A series of analyses are generally performed on site soil samples to evaluate the potential for biological degradation of the contaminants. The evaluations are conducted to determine if the respective soil samples harbor microbial populations capable of using the components of contaminants as carbon sources with the possibility of enhancing the populations to remediate the source material. Both impacted and nonimpacted samples from the site are often used in the evaluation.

The objectives of the lab studies are, (1) determine if aerobic microbes are present in the samples; (2) determine if the microbes have adapted to degrade the selected organic compounds; (3) determine if the environmental conditions (pH and moisture content) are conducive to support microbes; and (4) determine if soluble inorganic nutrients such as ammonia and phosphate are present in sufficient quantities for bioremediation of the contaminants.

Total heterotrophic aerobic microbes count will generally be performed using spread plate procedure. Heterotrophic aerobic microbes capable of degrading specific contaminants can be determined using modifications/variations of the above procedure. Soil respiration testing can subsequently be determined using a respirometer. The concentration of oxygen and carbon dioxide in the soil chamber headspace are generally measured periodically during a one day period. The difference in oxygen and carbon dioxide concentrations is subsequently graphed and may be correlated with the rate of respiration.

The soil pH, soil moisture, soluble ammonia, and ortho-phosphate concentrations are also generally determined according to accepted standard methods. The contaminant concentrations in the soils are also analyzed according to the relevant and accepted analytical methods in order to quantify the observed degradation in terms of contaminant mass.

If the contaminants in the soil are not known to be biodegradable and/or system parameter (soil moisture/nutrients, etc.) manipulation is considered, a more sophisticated treatability test can be conducted using soil columns/microcosms. Lab studies involve utilizing multiple microcosms of soil samples in order to vary the parameters of interest (moisture, airflow, nutrients, etc.). The soil within the microcosms is analyzed at various points in time to assess degradation of the contaminants of interest. Parameters such as moisture, pH, and nutrient levels are also generally monitored during the test period. Analytical costs for these tests can be in the range of $10,000 to $50,000. These treatability tests are therefore not generally conducted for compounds known to be biodegradable, or at sites with reasonably good nutrient/moisture conditions.

FIELD RESPIROMETRY TESTING

The respirometry test consists of ventilation (introduction of oxygen) of the contaminated area and periodically monitoring the depletion of oxygen and the production of carbon dioxide for a period of time (three to five days) after the air source is turned off. Based on the results of the respirometry test, oxygen uptake rates and biodegradation rates can be approximated. The oxygen uptake rate can subsequently be utilized to optimize airflows in the subsurface.

The typical test setup is shown in Figure 20. The respirometry test procedures have been best documented by Hinchee and Ong for several demonstrations at United States Air Force bases (Hinchee and Ong 1992a; Hinchee and Ong 1992b). The monitoring points are typically narrowly screened in the zone of interest (contaminated area). This is because oxygen concentrations in the typical monitoring well may not be representative of local conditions. The narrow screening ensures measurement at precise locations. Air is typically injected in one to five points at flow rates in the range of 1 cfm for a period of 24 hours for a typical 10 to 20 foot vadose zone. Subsequently, monitoring point gas composition is analyzed for two to four days. The monitoring points can be as simple as soil gas monitoring probes (5/8 inch diameter steel with a screened 6 inch section) or slightly larger diameter (2 inch) monitoring wells. It is generally advisable to mix 1 to 2 percent helium with the injected air as a tracer gas. Helium can be easily monitored and detected with

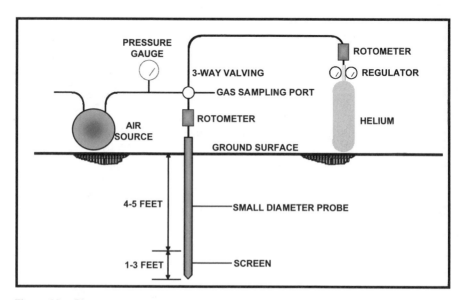

Figure 20   Biovent test set up.

field instruments (with accuracy to 0.01 percent). Detection of helium implies that the gas sampled is the same gas that was injected and that changes in its make-up (O_2/CO_2/contaminant distribution) are attributable to bioactivity. A relatively constant detection of helium concentrations over the monitoring period implies that the gas injected is the gas sampled.

After injection of the air/helium mixture, the soil gas is periodically monitored at a frequency consistent with the oxygen uptake rate (every 2 to 12 hours). The test is normally terminated when the oxygen concentration drops to 5 percent and/or after three to five days. Oxygen utilization rates can be determined from the slope of the O_2 percent vs. time graph, if a zero order respiration rate is assumed. Zero order rates have generally been observed for jet fuel degradation (Hinchee and Ong 1992a; and Hinchee and Ong 1992b). As a first approximation to estimate the biodegradation rate, the stoichiometric relationship for contaminant consumption can be formulated. The following example illustrates the calculations for consumption of benzene:

$$C_6H_6 + 7.5 \ O_2 \rightarrow 6 \ CO_2 + 3 \ H_2O \quad\quad\quad (19)$$

Using the oxygen consumption rate calculated from the oxygen concentration/time graph, the biodegradation rate in terms of mg of benzene-equivalent per Kg of soil/day can be estimated using the following relationship:

$$K_D = -K_R V D_o C / 100 \quad\quad\quad (20)$$

where K_D = biodegradation rate (mg/Kg/day); K_R = respiration/oxygen utilization rate; V = volume of air per Kg of soil (L/Kg); D_o = density of oxygen (1,330 mg/l; temperature, pressure specific); and C = mass ratio of benzene to oxygen required for degradation (78/240=0.32).

Based on a porosity of 0.3, a soil density of 1,440 Kg/m^3, K_D can be estimated to be roughly 0.89 times the measured respiration rate (K_R) for benzene. Alternate methods that evaluate biodegradation rates based upon CO_2 generation may be less accurate than oxygen consumption based calculations, if carbonate precipitates may be formed from the gaseous CO_2 production. Formation of these precipitates is dependent on subsurface pH and alkalinity.

SOIL GAS PERMEABILITY TESTING

The pneumatic permeability testing procedure for bioventing is similar to the pilot testing procedures for VES design. The analysis for zone of influence is based on providing sufficient oxygen delivery to the most distant location from the extraction/injection point rather than providing adequate airflow for volatilization. Airflow requirements are generally much lower for bioventing than for vapor extraction, and therefore the notion of using vacuum/pressure zone of influence for well spacing becomes more acceptable and less prone to error than for VES design. If a vacuum is observed, then some airflow is being delivered to that location. Despite this simplistic assessment, a calculation to evaluate whether this vacuum/pressure delivers the required oxygen should be conducted to ensure optimal system performance. This will require translation of the vacuum profiles across the site into airflow velocities. The airflow velocity can in turn lead to an oxygen delivery calculation based on 20 percent oxygen content in air. This delivery rate can be compared to the respiration rate. Since many areas in the site will receive oxygen depleted air, the oxygen delivery calculation yields best case oxygen delivery information. A pore volume calculation for the zone of influence (based on vacuum influence), and the subsequent translation of the oxygen delivery through the zone of influence, provides an alternate method of calculating whether sufficient oxygen is being provided. For example, passage of one pore volume per week can lead to evaluation of the oxygen content of this pore volume. Subsequently, one can estimate the required number of pore volumes of air to deliver sufficient oxygen to degrade the adsorbed contamination. Since only some of the delivered oxygen is consumed, more than the stoichiometric oxygen needs will require delivery.

BIOVENTING SYSTEM CONFIGURATIONS

Bioventing systems can be operated to extract, inject, or extract and inject air. Some researchers have noted that high vacuums (extraction mode) affect microbial growth and therefore some practitioners prefer the air injection mode of operation. Since high vacuums (5 inches water) are generally only observed near the extraction well, this is not likely to be a significant issue. Alternating system well operation from injection/extraction mode is also often used to overcome this concern. Injection systems have the advantage of delivering the full 20 percent oxygen to the highest contaminated zone. If oxygen is limiting, this delivery system will reduce the time required for biodegradation. Injection systems can also be designed with no point discharges for air. This can save on above-ground treatment equipment, but the designer must make sure that the volatile gases are not being spread by the operation. The injection rate has to create an airflow rate that is less than the rate of biodestruction. Figure 21 illustrates several system configurations.

CLEANUP GOALS/COSTS

Although vapor extraction allows for significant mass removal of VOCs from the subsurface and can be much more efficient than pump and treat remediation, due to the higher transport abilities of the air carrier, reasonable cleanup goals still must be established. Vapor extraction systems reach asymptotic cleanup levels due to nonequilibrium partitioning such as desorption, pore diffusion, or other rate limiting transport steps that eventually render the system diffusion process limited (Figure 4). In order to reach the cleanup criteria, the end points must be realistically determined. Due to its ability to rapidly remove large amounts of VOCs, vapor extraction is an excellent source control remedial strategy. As has been previously

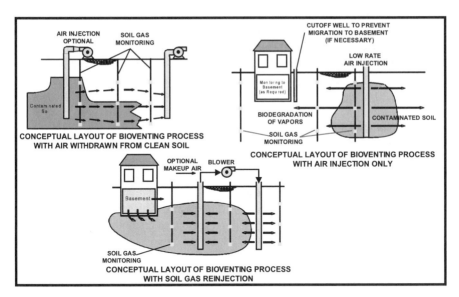

Figure 21   Bioventing system configurations.

discussed, VES efficiency declines as the remaining contamination mass declines. This tailing effect implies that alternate strategies should be considered during the end of the project's lifecycle. The alternate strategies may include a reduction in the extracted air volume and a greater reliance of passive venting or bioventing. Alternately, upon completion of the achievable source control (asymptotic levels), natural attenuation of the receptor groundwater may become an acceptable remedy.

Realistic expectations of system achievements are essential in order to minimize operation and maintenance costs. Table 3 provides typical costs to be expected with installation of a VES system. As can be seen, minimizing O&M lifetimes leads to significant reduction in project costs. Systems should be operated to the point where the total mass minimizes health and environmental impacts.

CASE STUDY

A one and a half acre Superfund site in the midwest was contaminated by past operations at the site which included storage and repackaging of solvents. The site also contained two below-ground and two above-ground storage tanks. The site was in operation from the early 1960s to mid 1980s. Soil contamination was identified throughout most of the site, consisting of a variety of chlorinated and nonchlorinated compounds. Triachloroethene was the most prevalent compound detected in the soils at the site, which also impacted groundwater. Based on the remedial investigation soil sample results, the estimated volume and contaminant mass of impacted soils was 26,000 cubic yards and approximately 10,000 pounds, respectively.

The average depth to groundwater was approximately 12 feet below land surface (ft bls). The geology of the vadose zone consists of relatively permeable Pleistocene

Table 3   Typical System Component Costs for Vapor Extraction

Item	Cost ($)
1. Blowers/vacuum pumps skid mounted system (100 cfm)	
a. Regenerative type	5,000
b. Positive displacement type	8,000
c. Liquid ring type	20,000
2. Extraction wells	
a. 20 foot vertical 2 inch PVC well (does not include drill cuttings disposal)	1,200
b. Shallow (6 foot) horizontal extraction well (4 inch PVC)	1,000
3. Trenching and underground work	
a. Costs per linear foot (4 inch PVC header), oversight costs included, soil disposal not	20
b. Protective wellhead, beneath ground (2 feet x 2 feet)	400
4. Other internal above-ground components (100 cfm)	
a. Piping/valve (PVC)	2,000
b. Gauges and flow meters	1,000
c. 10 foot x12 foot heated and insulated shed	4,000
d. Controls and electrical	5,000

and recent glacial and alluvial sand. A particle size distribution as determined by the Unified Soil Classification System for a soil boring at the site 10 ft bls is shown in Table 4.

Table 4   Particle Size Distribution

Soil Type	%
Gravel	5.70%
Coarse Sand	4.50%
Medium Sand	21.20%
Fine Sand	64.00%
Silt and Clay	4.60%

A pilot study was performed with four wells at individual extraction rates ranging from 60 to 165 standard cubic feet per minute (scfm) and wellhead vacuums of 40 to 55 inches of water. The extraction wells were operated independently to determine their radius of influence and vapor flow rate/vacuum pressure relationship. In addition VOC loading rates from the individual extraction wells at the various operating conditions were also evaluated. The estimated radius of influence determined from vacuum measurements at various depths and distances from the extraction wells was 50 feet. This information along with design/operating criteria presented below were utilized to design the full-scale system.

Design/Operating Criteria

Based upon the pilot study data and vapor extraction modeling, the following design criteria were used for the full-scale system.

- 120 scfm per VES well
- 40 inch water vacuum at each VES well
- 1 inch water vacuum throughout impacted area
- 50 percent overlap of VES well zones of influence

The full-scale system consisted of eighteen VES wells and seven piezometer clusters. The piezometers were installed in clusters to demonstrate that the 1 inch water vacuum was achieved not only at the surface but with depth. The VES wells were constructed of 4 inch Schedule 40 PVC well casing extending 20 ft bls with the bottom 15 feet consisting of 0.010 slot well screen. Each VES well was outfitted with a flow control valve, vacuum gauge, thermometer port, flow meter port and sample valve. The piping system was mounted above grade and consisted of 4, 6, 8, and 10 inch Schedule 40 PVC pipe mounted on unistruts. Each piezometer cluster was constructed of 1 inch Schedule 40 PVC with 3 foot 0.01 slot screen sections. The top portion of the piezometer clusters were placed with the screen section starting at 3.5 to 5 feet bls and bottom screen section ending at the water table. Figure 22 presents a layout of the VES system.

The equipment consisted of a 2,000 gallon knock-out tank (air/water separator), particulate filter, blower, and carbon adsorption system. The 2,000 gallon knock-out tank was constructed of steel and outfitted with a sight gauge and high level switch to shut the system down at high water level. The blower was a 2,250 scfm multi-stage direct drive centrifugal blower powered by 75 hp explosion-proof motor. The carbon adsorption system consisted of two banks of four vessels, each vessel capable

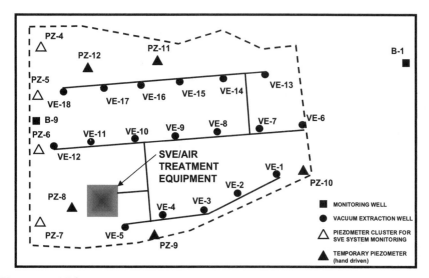

Figure 22   VES system layout.

of handling 600 scfm (or 2,400 scfm per bank). The blower and controls were housed in a temporary wooden shed. Figure 23 presents a process flow diagram for the system.

The system operated for less than one year since the VOC concentrations in the VES wells were not measurable after approximately 150 days of operation. Figure 24 presents the total VOCs removed with the VES system based on air sampling

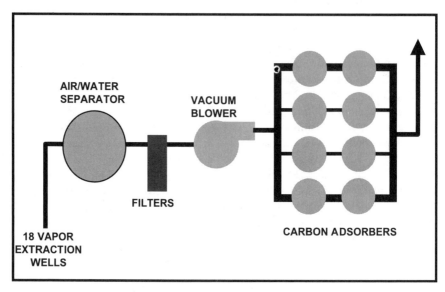

Figure 23   Process flow diagram for case history.

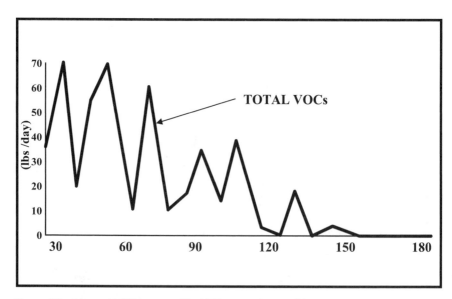

Figure 24   Mass of VOCs removed by VES system in case history.

results from each VES well. As can be seen, there was a rapid decrease in concentration. Based on these results and the costly operation of the VES system, it was shut down and soil samples were taken to demonstrate cleanup. Table 5 presents the sample results from the most highly impacted area of the site. It is interesting to note that even though the VES system extracted air concentrations were below detection levels some contamination still remained. The VES system removed a significant amount of mass but small isolated areas may still remain that are above soil cleanup criteria.

Table 5   Soil Sample Results

Parameter	Cleanup Criteria (mg/kg)	8-10 ft bls (mg/kg) Pre VES	Post VES
Ethylbenzene	1,400	35,000	BDL
Toluene	16,000	140,000	BDL
Xylenes	6,000	11,000	BDL
1,1,1-Trichloroethane	4,000	200,000	BDL
Tetrachloroethene	10	33,000	240
Trichloroethene	60	230,000	4

In summary, the VES system operated for approximately 180 days, removing 4,650 pounds of VOCs. The carbon adsorption air treatment system required two change-outs, utilizing 28,000 pounds of carbon. The total project cost, not including soil cleanup verification sampling, was $375,000.

This case history demonstrates the rapid success of vapor extraction and the diminishing mass removal effects of continued operation. The slower pace of groundwater concentration decline demonstrates that groundwater flushing is slower in reaching this asymptotic level as well, being less efficient in total mass removal. Chapter 4 provides greater detail on comparison of mass removal by airflow versus groundwater carriers.

DESIGN EXAMPLE

Problem

A former manufacturing facility has been impacted from past material handling and storage practices. Site investigation results indicate that both soil and groundwater have been impacted by hydrocarbon compounds and to a lesser extent chlorinated VOCs. The site is approximately 1.5 acres in size and about a third of the site soils have been impacted. There is no surface seal; the site is overgrown with weeds. Soil boring data indicate that the vadose zone soils are impacted primarily by benzene, toluene, ethyl-benzene, xylenes, and to a lesser extent tetrachloroethene and trichloroethene. The vadose zone is approximately 10 feet thick and consists of fine to medium grain sands.

Solution

First, we need to determine if the contaminants are amenable to VES. Using Table 2, the contaminants are either volatile and/or biodegradable, and VES should be considered. Secondly, we need to determine if the geologic and site surface features are acceptable for VES. Classification of the soils (fine to medium grained sand) indicates that a standard VES should suffice. Even though the site does not have a surface seal, we can still proceed and evaluate whether any modifications are warranted as we proceed with the design process.

Pilot Test Planning

A review of the well construction details indicates that only one of the existing monitor wells can be utilized for the pilot test. In order to have a good distribution of data collection points more wells will be required. Therefore, it was decided to install a VES test well and five monitoring points. The VES test will also be performed in the impacted area to collect as much representative data as possible, including air emissions data. Figure 25 is a site plan depicting the monitoring well locations and approximate extent of impacted soil, based on soil boring testing results.

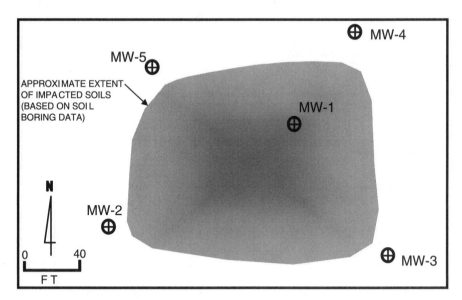

Figure 25   Site plan for design example.

Short circuiting of the vapors is a concern and plastic sheeting (20 mil visqueen) will be available in the event that unacceptable results are obtained. However, it is difficult to get good surface seal with plastic sheeting and this will only be tested as a last resort.

Utilizing Equation (13) we will estimate the required test duration

$$t_p = n \Pi R_i^2 h / Q \qquad (13)$$

assume a porosity, $n = 0.3$, and estimate the radius of influence, $R_i = 20$ feet.

To estimate the volumetric flow rate, Figure 5 will be utilized, as shown in Figure 26. The middle and lower well VES well vacuums were selected since these vacuums can be accomplished with a regenerative type of blower, not a liquid ring as would be required for the higher vacuums. Also, based on the geologic conditions of the site (fine to medium grain sands) vacuums in 20 to 80 inches of water range are all that is likely to be required to induce airflow. Therefore, the volumetric flow rates for the pilot test should range from 8 to 32 scfm.

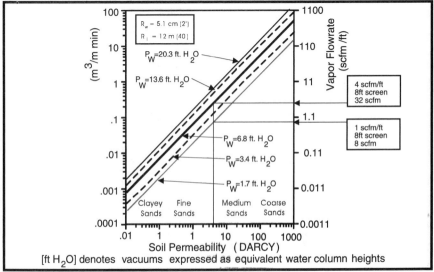

Source: Johnson et. al., 1990(modified)

Figure 26   Predicted steady-state flow rates (per unit well screen depth) for a range of soil permeabilities and applied vacuums (Pw).

Utilizing an estimated volumetric flow rate of $Q = 20$ scfm, yields

$$t_p = (0.3)(3.14)(20 \text{ ft})^2(10 \text{ ft})/(20 \text{ ft}^3/\text{min})$$
$$t_p = 188 \text{ minutes} = 3.14 \text{ hours}$$

Therefore, the pilot test should run for a minimum of 4 to 5 hours to withdraw a minimum of one pore volume and account for any surface leakage.

As shown in Figure 26, the well vacuums estimated to induce airflow and utilized to estimate duration of pilot test ranged from 20 to 80 inches of water. Since a range of vacuums and flow rates should be tested, the equipment brought to the field should be able to accommodate the possible variables. In order to maximize flexibility a good VES pilot test set up could consist of two blowers that could either be operated in series or parallel.

Since the pilot test is to be conducted in an impacted area, a temporary carbon canister will be utilized to treat the off-gas.

Conducting the Pilot Test

Figure 27 is a depiction of the pilot study layout. As you will notice, the monitoring points are radially distributed at various distances from the VES well. Radial distribution should minimize any potential interference from site heterogeneities and obtain the data necessary for full-scale design. The pilot test was conducted at various vacuums and corresponding flow rates, as presented in Table 6. Various ways of determining the radius of influence from a pilot test have been presented in the literature. The majority of methods base radius of influence as a function of vacuum in monitoring points with distance from the VES well on a linear, semi-log, or log-log plot. The best fit line of the plotted data is then utilized to choose a radius of influence based on either a specified vacuum or 1 percent of the applied vacuum. Figures 28 and 29 present a couple of ways of estimating the radius of influence from a VES well. Figure 29 (based upon Table 7) normalizes the

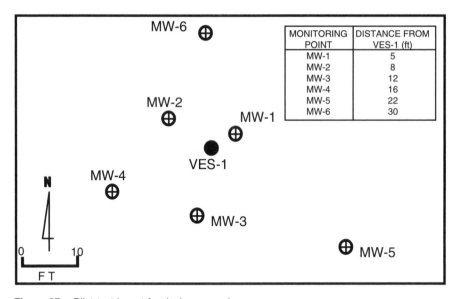

MONITORING POINT	DISTANCE FROM VES-1 (ft)
MW-1	5
MW-2	8
MW-3	12
MW-4	16
MW-5	22
MW-6	30

Figure 27   Pilot test layout for design example.

Table 6   Pilot Test Results

Test	VES-1 Flow Rate (scfm)	Vacuum (inches water)						
		VES-1	MW-1	MW-2	MW-3	MW-4	MW-5	MW-6
A	40	104	6.2	4.4	3.8	1.9	1.1	0.2
B	30	91	3.9	3.0	2.4	2.0	1.2	0
C	20	78	3.6	2.9	2.1	1.6	1.0	0.1
D	10	54	1.8	1.5	1.1	0.5	0.2	0

data to enable various test results to be evaluated simultaneously. This method is also useful since the full-scale system will rarely operate at the exact flow rate and vacuum of the pilot test. Other methods utilize an estimation of mass to be removed based on time desired for cleanup.

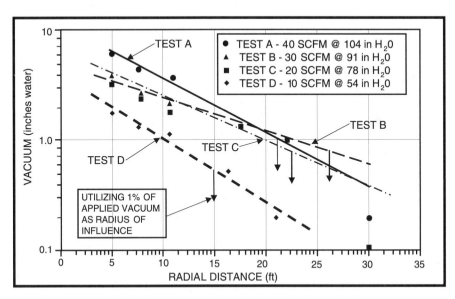

Figure 28   Pilot Test results from design example.

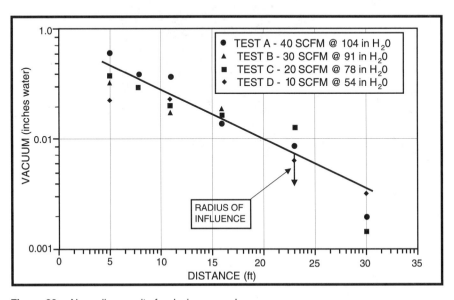

Figure 29   Normalize results for design example.

Table 7   Pilot Test Results, Normalized

Test	VES-1 Flow Rate (scfm)	Vacuum at Monitoring Well/Vacuum at VES-1						
		VES-1	MW-1	MW-2	MW-3	MW-4	MW-5	MW-6
A	40	1	0.060	0.042	0.037	0.018	0.011	0.002
B	30	1	0.043	0.033	0.026	0.022	0.013	0
C	20	1	0.046	0.037	0.027	0.021	0.013	0.0013
D	10	1	0.033	0.028	0.028	0.009	0.004	0

Evaluating the Data

Utilizing an approximate radius of influence of 24 feet, a layout of ten soil vapor extraction wells were distributed within the impacted soil zone, as shown on Figure 30. Some overlapping of the well's influence is provided to try to ensure that some air movement is occurring throughout the impacted area. It must be remembered that the purpose of an VES system is to induce airflow and that the use of vacuum data from the pilot test is just an indication that air will flow from outside the zone where an air vacuum exists. Some researchers have reported that a minimum airflow velocity of 0.01 feet/minute should be maintained to produce good removal results.

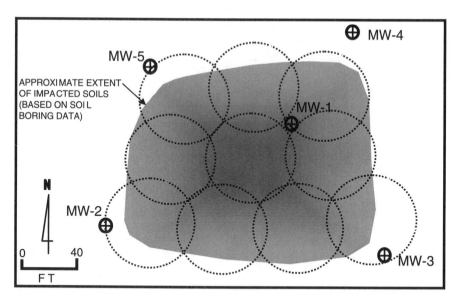

Figure 30   Site plan based on pilot data on design example.

As evident from Figure 28 the use of airflow rates from 10 to 30 cfm per well point should produce the desired radius of influence. Before we look at blower curves to select a blower let's check to see if the 10 to 30 cfm flow range with a radius of influence of 24 feet will yield an acceptable air velocity. Assuming a cylinder and no surface leakage, the velocity can be evaluated as presented below.

V = $Q/(2)(\Pi)(R_i)(h)(n)$ = 10 ft³/min / (2)(3.14)(24 ft)(10 ft)(0.3)
= 0.02 ft/min which is greater than 0.01 ft/min, therefore this extraction
rate should produce an acceptable airflow.

At this point the manufacturer's blower curves need to be evaluated to see the range of possibilities available. It is not always the case that more airflow is better, since more airflow will require additional power and increased air phase treatment costs. In addition, since the vapor extraction system will cleanup the more permeable zones first and rely upon diffusion for the remaining impacts there may not be any benefit to an increased airflow. Therefore, each site has to be independently evaluated as to the objectives, geologic conditions, types of blowers, and air phase treatment. Once the blowers have been selected, the designer can furnish the design by selecting pipe sizes and layouts, valves, and equipment locations and housing. While this chapter provides good details and examples, the author highly suggests that one other component be included: experience. Get experienced help on your first real installation.

REFERENCES

AIR 3D, Geraghty & Miller, Inc., 1070 Parkridge Blvd., Reston, VA.

Armstrong, J.E., Frind, E.O., McClennan, R.D., "Non-equilibrium Mass Transfer Between the Vapor, Aqueous, and Solid Phases in Unsaturated Soils During Vapor Extraction," *Water Resources Research*, 30, 355, 1994.

Brusseau, M.L., Jessup, R.I., Rao, P.S.C., "Transport of Organic Chemicals by Gas Advection in Structured or Heterogeneous Porous Media. Development of a Model and Application to Column Experiments," *Water Resources Research,* 27, 3189, 1991.

Buscheck, T.E., Peargin, R.G., "Summary of A Nation-Wide Vapor Extraction System Performance Study," Chevron Research and Technology Company, Richmond, CA, 1992.

Farrell, J., Reinhard, M., "Desorption of Halogenated Organics From Model Solids, Sediments, and Soil Under Unsaturated Conditions, 1. Isotherms," *ES&T,* 28, 53, 1994.

Farrell, J., Reinhard, M., "Desorption of Halogenated Organics From Model Solids, Sediments, and Soil Under Unsaturated Conditions, 2. Kinetics," *ES&T,* 28, 63, 1994.

Fetter, C.W. Jr., *Applied Hydrogeology, 3rd Edition,* Macmillian Publishing Co., New York, N.Y., 1994.

Gierke, J.S., Hutzler, N.J., McKenzie, D.B., "Vapor Transport in Columns of Unsaturated Soil and Implications For Vapor Extraction," *Water Resources Research*, 25(1), 81, 1989.

Grathwohl, P., Reinhard, M., "Desorption of Trichloroethylene in Aquifer Material: Rate Limitation at the Grain Scale," *ES&T,* 27, 2360, 1993.

Hansen, M.A., Flavin, M.D., Fam, S.A., "Vapor Extraction/Vacuum-Enhanced Groundwater Recovery: A High Vacuum Approach," Proceedings of Purdue Industrial Waste Conference, Lafayette, Inc., 1994.

Hinchee, R.E., Ong, S.K., "Test Plan and Technical Protocol for A Field Treatability Test for Bioventing," Document prepared for US Air Force Center for Environmental Excellence, Brooks Air Force Base, TX, 1992.

Hinchee, R.E., Ong, S.K., "A Rapid *In Situ* Respiration Test for Measuring Aerobic Biodegradation Rates of Hydrocarbons in Soil," *J. Air Waste Management Association,* 42, 1305, 1992.

Hoag, G.E., Bruell, C.J., Marley, M.C., "Induced Soil Venting For Recovery/Restoration of Gasoline Hydrocarbons in the Vadose Zone," *Oil in Freshwater,* 176, 1989.

Johnson, P.C., Kemblowski, M.W., Colthart, J.D., "Quantitative Analysis for the Cleanup of Hydrocarbon-Contaminated Soils by *In Situ* Soil Venting," *Groundwater Monitoring Review,* 413, 1990.

Johnson, P.C., Kemblowski, M.W., Colthart, J.D., Byers, D.L., "A Practical Approach to the Design Operation and Monitoring of *In Situ* Soil Venting Systems," *Groundwater Monitoring Review,* 1991.

Jury, W.A., Russo, D., Streile, G., Abd, H.E., "Evaluation of Volatilization of Organic Chemicals Residing Below the Surface," *Water Resources Research,* 26, 13, 1990.

Rittman, B. McDonald, J.A., "Performance Standards for Bioremediation," *ES&T,* No.10, 27, 1993.

Sims, R.C., "Soil Remediation Techniques at Uncontrolled Hazardous Waste Sites, A Critical Review," *J. Air Waste Management Association,* 40, 704, 1990.

Sleep, B.E., Syikey, J.F., "Modeling of Transport of Volatile Organics in Variable Saturated Media," *Water Resources Research,* 25 (1), 81, 1989.

Thibaud, C., Erkey, C., Akgerman, A., "Investigation of the Effect of Moisture on the Sorption and Desorption of Chlorobenzene and Toluene From Soil," *Environmental Science Technology,* 27, 2373, 1993.

U.S.EPA, Superfund Chemical Data Matrix Office of Emergency and Remedial Response, Washington, D.C., EPA/540/R-96/028, 1996.

CHAPTER 4

Vacuum-Enhanced Recovery

Peter L. Palmer and Evan K. Nyer

CONTENTS

1-56670-528-2/01/$0.00+$.50
© 2001 by CRC Press LLC

INTRODUCTION

Although vacuum-enhanced recovery has been used for decades as a standard approach for dewatering and stabilization of low permeability sediments or to speed dewatering of more permeable sediments, it hasn't been until recently that it has been incorporated into groundwater remediation applications. The use of vacuum-enhanced recovery systems in environmental remediation is unique because whereas most remediation methods rely on either water or air as the carrier, vacuum-enhanced recovery relies on a combination of both as carriers. Because of this characteristic to remove both liquids and vapors, vacuum-enhanced recovery is also referred to as dual-phase extraction. Vacuum-enhanced recovery uses a combination of two forces, gravity and pressure differential, to move the water. This can be beneficial in enhancing cleanups when used in the proper hydrogeologic setting.

In Chapter 1, we discussed how gravity is used as a main force to move water in all methods that use water as a carrier. The main method to control the direction of water movement is to use a well to remove water from the aquifer. This creates a drawdown of the water in the aquifer and water travels from a place of high head (high water level) to a place of low head (low water level at the pumping well). The liquid state of water allows this vertical force to create a horizontal movement of the water. This principal has been successfully practiced as a remedial technique in the medium and high permeability geologic formations. The success of this technique (using gravity alone as the main force), however, can be severely restricted in lower permeability formations due to the diminished groundwater flows that can be achieved by standard remedial recovery equipment. Vacuum-enhanced recovery systems overcome this limitation by using a second force, pressure differential, to help the movement of the water when gravity movement is limited by the geology.

Vacuum is applied, in addition to pumping, to move the water. This combination allows us to move air and water in geologic formations that were inaccessible before.

Within this chapter we will demonstrate the value of applying a vacuum to improve the performance of a well in moving water and/or NAPL in lower permeability geologic settings. However, vacuum-enhanced recovery systems are also advantageous in lower permeability formations in using air as a carrier for many organic contaminants. Lower permeability formations generally have silts and clays incorporated into the matrix. These are generally very adsorptive of organic contaminants, and use of water alone as the carrier makes it difficult to achieve removal of these contaminants. However, air is an effective carrier of many of these organic constituents, and vapor enhanced recovery provides a mechanism to incorporate air as a carrier in these lower permeability formations.

To overcome the air and groundwater flow restrictions of low permeability formations, high vacuums are created at a well by liquid-ring pumps or other specialty pumps and these, when coupled with recovery wells, are collectively referred to as vacuum-enhanced recovery systems. These specialty pumps are used to create high vacuums (up to 24 inches of mercury as opposed to 3 to 6 inches of mercury for conventional vacuum blowers) which results in a much greater driving force (pressure differential) for airflow in the unsaturated zone. When combined with gravity, this increases the rate of groundwater and/or NAPL recovery and the size of the groundwater and/or NAPL recovery capture zone. During this chapter we will discuss the geologic setting and constituent types that are most suitable for vacuum-enhanced systems, and we will take you through the steps needed from evaluation of system applicability to system design.

MASS BALANCE APPROACH TO SITE REMEDIATION

Since vacuum-enhanced recovery uses both water and air as a carrier, it is particularly important to understand which phase the constituents are in so that vacuum-enhanced recovery can be used to its fullest benefits. In a general sense, the transfer of the constituents between phases (vapor, dissolved, free phase, and adsorbed) is affected by the relative affinity of the constituents to each phase which can be evaluated using the constituent partitioning coefficients. The most effective remediation will create subsurface conditions that will drive the interphase transfer towards the phase(s) that allow for the most efficient mass removal. Site remediation can therefore be viewed as implementing perturbations to the subsurface that will drive physical, chemical, and biological processes towards the site remediation goals. For instance, at sites contaminated with petroleum-derived volatile organic compounds (VOCs), analysis of the equilibrium relationships is greatly simplified. Gasoline derived compounds are generally biodegradable, volatile, and relatively water insoluble. Mass transfer and removal of gasoline constituents can be 10,000 times more efficient in the vapor phase than in the soluble water phase. Perturbations that alter the equilibrium in the subsurface in order to drive interphase mass transfer from the free, dissolved, and adsorbed phases to the vapor phase would therefore be viewed as beneficial to site remediation. As always, each site is different and must be looked at on a site-specific basis. Although it may be desirable to lower groundwater levels in order to remove LNAPLs using air as a carrier, there are situations where the most effective remedy is to use a vacuum system to enhance removal of LNAPL with minimal drawdowns in groundwater levels. This minimizes the smear zone and uses the vacuum system to remove residual mass from the vadose zone.

One way to shift from dissolved phase remediation (pump and treat) to the more efficient vapor phase remediation is to dewater the contaminated sediments and remove easily accessible LNAPLs, and then remove the VOCs by vapor extraction. In high permeability formations, site dewatering can create extremely large volumes of water. Therefore, this approach is generally not practical due to cost considerations and you would likely rely on water alone as your carrier. However, lower permeability formations (hydraulic conductivity ranges from 1×10^{-3} to 1×10^{-6} cm/sec) do not deal with large volumes of water. These can be dewatered to expose formerly saturated soils to vapor extraction. Fine-grained sediments associated with lower permeability sediments also possess greater specific retention due to increased capillary pressure, which can trap water in the soil pore space and impede the needed airflow. Application of high vacuum via a vacuum-enhanced recovery system can generally overcome most soil capillary pressures, thereby allowing for the removal of the water, which results in increased airflow and increased constituent removal.

In summary, vacuum-enhanced recovery can be beneficial in aquifer remediation when properly applied. In the water carrier phase, it can be used to enhance the physical removal of NAPLs and/or dissolved constituents. It can also be used to dewater sediments and allow the air carrier phase to, (1) remove constituents absorbed onto fine grained sediments, and/or (2) provide oxygen to enhance *in situ* degradation of biodegradable compounds.

GROUNDWATER RECOVERY ENHANCEMENT

Let us put aside our focus on the advantages of air being a carrier and first focus on enhancing the water carrier aspect. The basic premise behind extraction wells is that the withdrawal of groundwater from a central point causes a decline in groundwater levels (drawdown) in the vicinity of the well which is referred to as the cone of depression; the drawdown induces groundwater flow towards the well. The area within which all of the water flows to the extraction point or well is referred to as the capture zone (see Figure 1). The point on the down-gradient side of the well where water is no longer captured is referred to as the stagnation point. The capture zone is elliptical in shape as a result of the effect of the existing groundwater gradient. It is important to note that the capture zone and cone of depression are not the same due to the existing gradient. The cone of depression is circular and will continue to expand as pumping time increases until equilibrium conditions are reached. The capture zone is determined by superimposing the cone of depression on the water table (groundwater levels) and as shown in Figure 2. It is located in the immediate vicinity of the recovery well and up-gradient from it. The capture zone determines the area that will be contained and/or remediated by a given extraction system.

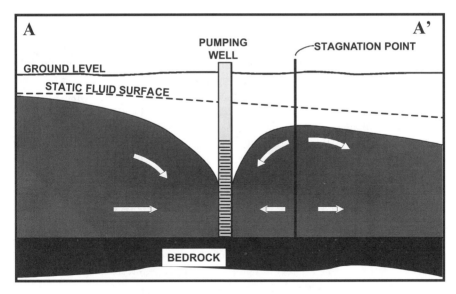

Figure 1   Cross section showing capture zone.

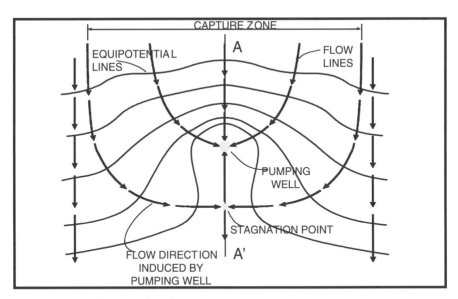

Figure 2   Plan view showing capture zone.

Low aquifer transmissivities limit the capture zone. Low transmissivities result from either a thin saturated thickness or low permeability deposits. In low permeability environments, there may be available drawdown; however, the low permeability prevents the formation of a significant capture zone under the influence of gravity alone. In an aquifer with a thin saturated thickness, there is little drawdown available to create a significant capture zone. In these situations, the capture zones are extremely small and the number of wells required using conventional approaches to contain or remediate an area is large.

The capture zone in an aquifer is proportional to the discharge that can be obtained which in turn is proportional to the aquifer transmissivity. Discharge for a given transmissivity is limited by the drawdown that can be produced at the well. Under normal water table conditions, the maximum drawdown is the available saturated thickness and it corresponds to the maximum achievable discharge. If the gradient (drawdown) can be increased beyond the saturated thickness, then the discharge and the capture zone could be increased. This is the basic idea behind vacuum-enhanced recovery—increase the gradient beyond that which can be achieved by pumping alone.

To get this point across, let us first observe the effects of a vacuum on static well conditions. When a vacuum (negative pressure) is applied to the well, the fluid level in the well rises and a cone of impression is formed around the well as a result of the negative pressure. This is shown in Figure 3 which shows the water level in the well rising 2 feet above the static water table in response to a vacuum applied at the well. This is a key concept since the negative pressure (represented by a rise in water levels) increases the available drawdown.

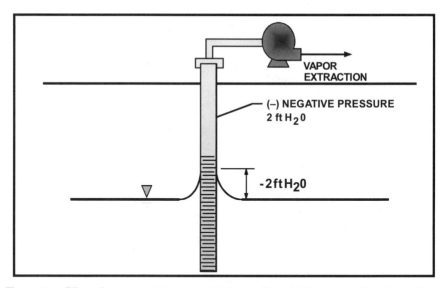

Figure 3   Effect of vapor extraction on water levels. Sawyer, S.W. 1984, "State Water Conservation Strengths and Strategies," Water Resources Bulletin, 20(5): 683. With permission.

Now let us look at its effect under pumping conditions. To do this let us first look at the effect on water levels from pumping alone. This is shown in Figure 4, which is a cone of depression formed by an extraction well. Note that the drawdown at the well created by pumping at 2 gpm is 5 feet. When we combine the effects of pumping (Figure 4) with the vacuum effects (Figure 3), the results show that the cone of depression formed by the same extraction well pumping at the same 2 gpm rate is now only 3 feet, due to the applied vacuum (Figure 5). This is what you

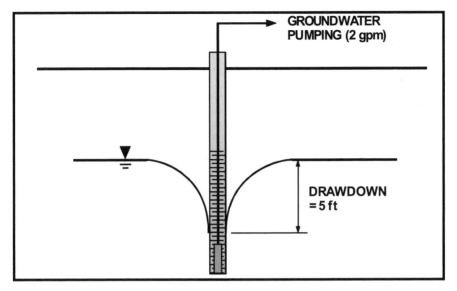

Figure 4   Effect of groundwater pumping on water levels.

would actually measure in the well; however, the effective drawdown, which is the combination of the pressure gradient (2 feet) and the liquid gradient (3 feet), is still 5 feet. And although water level measurements alone would suggest in the second case a smaller capture zone, in fact, the capture zone is the same in both cases because the pumping rate remained the same. The gradient required to produce 2 gpm did not change, just the method in which the gradient was created. The benefits

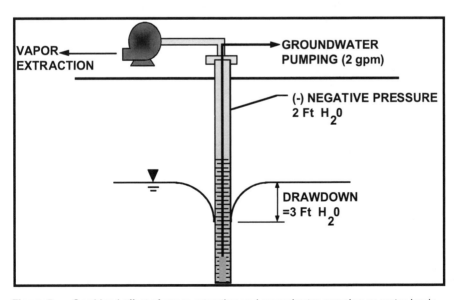

Figure 5   Combined effect of vapor extraction and groundwater pumping on water levels.

of using a vacuum-enhanced system are not simply to maintain the same pumping rate and capture zone, but to increase the yield of formations (pumping rate) beyond that which could be achieved by pumping alone. This increases your capture zone.

Now that we have introduced the concept of effective drawdowns, let us look in more depth at what we mean and how it relates to capture zones. The concept of effective drawdown is illustrated graphically in Figure 6 where Q_1 represents the maximum discharge rate that can be obtained by pumping alone due to limited saturated thickness. The figure shows the fluid surface and drawdown associated with Q_1. By applying a vacuum to the well, the discharge rate increases and the capture zone expands as depicted by the drawdown and fluid surface associated with Q_2 in the figure. The combination of the vacuum and the fluid drawdown represents the effective drawdown depicted here. The effective drawdown is greater than the saturated thickness due to the vacuum within the vacuum zone (area of measurable vacuum). The increased drawdown outside of the vacuum zone is a result of the increased discharge Q_2. The goal of a vacuum-enhanced recovery system under this scenario is to increase the capture zone; the increased available drawdown resulting from applying a vacuum is the means to achieve this goal. It is easy to see that vacuum-enhanced recovery systems can be advantageous since when properly used they can increase the capture zone associated with an individual extraction well, thus decreasing the number of extraction wells needed.

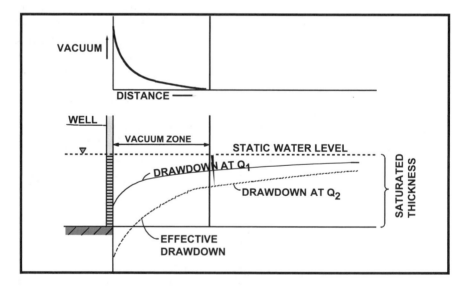

Figure 6   Vacuum effect on water levels.

APPLICABILITY

Vacuum-enhanced recovery is applicable to a limited range of hydrogeologic parameters and settings. This is important because if your site falls outside of these ranges, then the use of this technology should be critically evaluated and in many cases should not be applied. Generally low transmissivity is a requirement (<500 gpd/ft) in order to develop a vacuum of sufficient magnitude to have an effect at reasonable airflow rates. Low transmissivity is indicative of low permeability and small saturated thickness. The system generally needs to exhibits a permeability in the range of $1x10^{-3}$ to $1x10^{-6}$ cm/sec. If the permeability is too high then the system will not work because the high permeability prevents the formation of a significant cone of impression formed by the vacuum. In other words, the effective drawdown is only marginally greater than the available drawdown with pumping alone. If the permeability is too low, even vacuum-enhanced recovery systems cannot create a significant captive zone. In these cases, increasing the permeability of the formation, as discussed in Chapter 10 (Fracturing) would be needed to make these formations suitable for use for vacuum-enhanced recovery systems. Perched aquifers composed of interbedded sands and clays have been demonstrated to be particularly suitable for use of vacuum-enhanced systems. Vacuum-enhanced recovery systems may also be applicable to some fractured systems, those that fall within the range of permeability for this technology.

On the management side, there are a number of criteria that could dictate the use of a vacuum-enhanced system. Where suitable, it is generally considered to be a more effective cleanup program with shorter remediation times. In situations where LNAPLs are present, it can avoid source control programs such as excavation and it can be more cost effective from a lifecycle cost perspective than standard technologies due to reduction in capital costs and in O&M costs as a result of shorter remediation periods.

TYPES OF SYSTEMS

There are basically two types of systems: a single pump system, which uses a combined liquid/vapor pump, and a dual pump system, which uses separate liquid and vapor pumps. The single pump system (which is also referred to as bioslurping) uses a single pump with a drop pipe to withdraw both fluids and vapors (Figure 7). Normally with a suction lift system, we are limited to a maximum theoretical lift of about 34 feet, and practically to lifts of 15 feet or so. However, fluid recovery from deeper depths (50 feet or so) can be achieved by using an extraction tube with holes drill at strategic points along it to reduce the density of the fluid extracted. The benefits of a single pump system are lower capital and lower operational and maintenance costs. In addition, since all the equipment is above-ground, it is easier to install and maintain. Presented in Figure 8 is a photo showing a typical single pump system that is pneumatic and operates off of compressed air that was available at the facility. Disadvantages of the suction lift systems include limited depth, the problem of balancing vacuum at multiple wells, and the higher vacuum that is required to maintain the lift to extract fluids, which limits design flexibility.

Two pump systems can be used for greater depths and for these systems, a vacuum is applied at land surface and a second down hole pump is used to withdraw fluids (Figure 9). These systems are easier to balance and operate when multiple

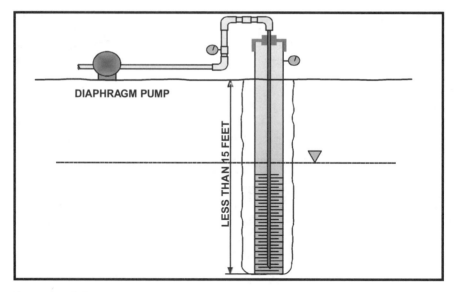

Figure 7   Schematic showing the configuration of a single pump vacuum-enhanced recovery
system.

Figure 8   Photograph of pneumatic single pump system.

extraction wells (greater than five) are involved and allow more design flexibility
after selecting the optimum vacuum pressure. Shown in Figure 10 is a photo of a
two pump mobile system. Disadvantages include the increased capital, operation
and maintenance costs of two pumping systems, and the care that must be taken in

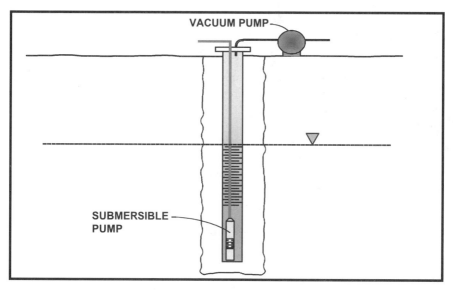

Figure 9   Schematic showing the configuration of a two pump vacuum-enhanced recovery system.

Figure 10   Photo of a two-pump, liquid ring mobile system.

selecting a down hole pump that will operate under vacuum. If the pump is not designed properly (available net positive suction head is not greater than that required by the pump) many serious problems can result. There will be a marked reduction in head and capacity, or even a complete failure to operate. Problems can also include excessive vibration or erosion of the pump parts resulting in reduced life.

These systems can be simple and inexpensive to install and operate, such as using a small pneumatic diaphragm pump to extract both fluids and gases in a low permeability setting. The pump may only be capable of moving about 1 scfm of air; however, that can be quite effective in low permeability systems, particularly if your goal is to maintain a good vacuum as opposed to maintaining high airflows. They also can be relatively complex, such as a liquid ring vacuum-enhanced system that can be modified for either single pump or dual pump applications. The liquid ring is capable of moving larger volumes of air at higher vacuums.

These two basic systems can be applied to the vacuum-enhanced applications to remove liquid and residual contaminants including:

- Liquid recovery—where the objective is simply to increase the rate, influence, and overall recovery of liquid contaminants
- Vacuum dewatering and vapor recovery—where the objective is to dewater a low permeability formation and then use the vacuum to move air through the formation and volatilize or biodegrade the residuals
- Combined liquid and vapor recovery—where the objective is to both remove liquid contaminants and volatilize or biodegrade residuals

PRELIMINARY EVALUATION OF SITE APPLICABILITY

Pilot studies are normally required to design vacuum-enhanced systems. It is important that the pilot study is planned and executed properly so that you collect all of the right data. Before you even consider a test, preliminary calculations should be performed to ensure that the equipment you select is properly sized for the field application. It is important to first make sure that the hydrogeology is suitable for the application of vacuum-enhanced recovery. The expected air and water flows at various pressures should be estimated to ensure, for instance, that the flow rates are not too high for vacuum-enhanced recovery. The basic equation for estimating water and airflow rates is the Cooper Jacob modification of the Theis equation shown below (Schafer 1995). It can be used to give a quick estimate of the water flow rates that can be expected at a pumping well using the estimated vacuum application and total effective drawdown at the well. In this application of the Cooper Jacob modification to the Theis equation, you substitute effective drawdown corrected for dewatering (s') for drawdown (s) and for the distance from the center of pumping of a pumped well to a point where the drawdown is measured (r) you use the well bore radius.

With these modifications, the Cooper Jacob equation permits calculating discharge as follows

$$Q = \frac{s' T}{264 \log \frac{0.3 T t}{r^2 S}} \tag{1}$$

where Q = discharge, in gpm; s' = effective drawdown corrected for dewatering, in ft; T = transmissivity, in gpd/ft; t = pumping time, in days; r = well bore radius, in ft; and S = storage coefficient.

To correct for dewatering

$$s' = s_a - \frac{s_a^2}{2 b_w} \tag{2}$$

where s_a = actual drawdown; and b_w = aquifer thickness.

This formula requires that you also have estimates of the site transmissivity, storage coefficient, and expected time of pumping.

Let us run through a typical example where the hydrogeology is suitable for consideration of a vacuum-enhanced recovery system.

Assumptions: Hydraulic conductivity, $K = 1 \times 10^{-4}$ cm/sec = 2.1 gpd/ft^2; b_w = 20 ft for saturated zone; b_g = 20 ft for vadose zone; 100 percent efficient well (both air and water); and $T = K b_w = 42$ gpd/ft.

At maximum drawdown

$$s' = 20 - \frac{(20)^2}{(2)(20)}$$
$$= 10 \text{ ft}$$

This is the effective available drawdown if no vacuum is applied.

Assume a liquid ring pump will be used to apply a vacuum of 20 inches of mercury. To get inches of water vacuum, multiply by 13.55, yielding 271 inches of water, or 22.6 feet.

This is the additional available drawdown applied to the well, bringing the total available drawdown to 22.6 plus 10, or 32.6 feet. At maximum drawdown, the water production with vacuum will be 3.26 times that without vacuum.

Using T = 42 gpd/ft, t = 30 days, S = 0.05, and r = 0.5 ft (bore hole radius), the pumping rate (Q) is 0.355 gpm at 10 feet of drawdown (no vacuum) and increases to 1.16 gpm at 32.6 feet of drawdown (with 20 inches Hg vacuum). This preliminary result suggests that vacuum-enhanced recovery is suitable for this application so the next step is to get a handle on expected airflow rates.

The airflow rates can be estimated in a similar manner using several formulas as follows:

Determine gas transmissivity, T_g. Gas conductivity is

$$K_g = K_w \frac{P_g \mu_w}{P_w \mu_g} \tag{3}$$

where K_g = gas conductivity; K_w = hydraulic conductivity; P_g = gas density (0.0013 g/cm^3 at 68°F); P_w = water density (1 g/cm^{3}at 68°F); μ_g = gas viscosity (183 micropoise at 68°F); and μ_w = water viscosity (10,000 micropoise at 68°F). Thus,

$$K_g = 2.1 \frac{(0.0013)(10,000)}{(1)(183)}$$
$$= 0.149 \text{ gpd/ft}^2 \text{ or } 0.02 \text{ ft/day}$$

Finally,

$$T_g = K_g b_g$$
$$T_g = (0.02)(20) \tag{4}$$
$$= 0.4 \text{ ft}^2/\text{day}$$

To estimate the airflow, we can use the Theis equation after correcting the drawdown (vacuum) for gas expansion:

1. The gas expansion correction has the same form as the dewatering correction

$$s_{eff} = s_a - \frac{s_a^2}{2P_{ATM}} \tag{5}$$

where s_{eff} = effective vacuum used in Theis equation; s_a = actual vacuum; and P_{ATM} = atmospheric pressure (405 inches of water).
Since the available well head vacuum is 271 inches of water

$$s_{eff} = 271 - \frac{(271)^2}{(2)(405)}$$
$$= 180 \text{ inches of water}$$

2. The Theis equation is

$$s = \frac{528Q}{T} \log \frac{R}{r} \tag{6}$$

where s = vacuum, in ft of air; Q = discharge, in cfm; T = transmissivity, in ft^2/day; R = radius of influence, in ft; and r = bore hole radius, in ft.

By assuming the expected vacuum at the well and estimating air permeability and the expected radius of influence; the extraction rate can be estimated. The radius of influence is obviously not known until the pilot studies are performed; however, the calculations are not highly sensitive to this parameter. Thus, using a value of 40 feet provides a good approximation. To express s in inches of water instead of feet of air, multiply by $12 \cdot P_g$.

$$s = \frac{8.23Q}{T}\log\frac{R}{r} \tag{7}$$

Solving for Q,

$$Q = \frac{sT}{8.23\log\frac{R}{r}}$$

Using s = 180 inches of water, T = 0.4 ft²/day, R = 40 ft, and r = 0.5 ft,

$$Q = 4.6 \text{ cfm}$$

The preliminary calculations, as demonstrated in this example, indicate that conditions appear favorable for applying vacuum enhancement. As a next step, a pilot study needs to be performed.

PILOT TEST PROCEDURES

When the results of the detailed evaluation and comparative analysis indicate vacuum-enhanced recovery may be applicable, then a pilot study should be conducted at the site to determine site specific parameters. Data from the pilot test program would be utilized to substantiate the preliminary calculations and to refine assumptions regarding such fundamental parameters as radius of influence, pumping rate, vacuum to be applied, and air and water quality. The collection and analysis of site-specific data would facilitate more accurate evaluation and final design of the remedial system.

The pilot test plan would include the following:

- Installation of test wells
- Test method and monitoring
- Mass removal estimation

Test and Monitoring Wells

The test recovery well and monitoring wells should be installed for conducting the pilot test. Existing monitor wells can be used if properly designed. The test recovery well should be installed in the vicinity of the impacted area of groundwater and soils. The test recovery well should be screened both in the saturated and unsaturated area of the subsurface. If the well has to be screen only in the saturated zone, then it is important to turn on the pump first and allow for some dewatering (drawdown) before adding the vacuum to the well. It is recommended that four monitoring wells be installed at different distances based on site conditions. Typical spacings would be 5, 15, 25, and 50 feet away from the test recovery well and at different compass directions. Final spacing would be dependent upon the local geology and surface structures. The wells should be designed and developed properly using a continuously wrapped well screen (No. 10 slot or smaller) and a fine grained sand pack.

Test Method

The test method described below is the methodology used in testing a system using a single pump system. Single pump systems are applicable to depths of 15 to 20 feet with no modifications to the drop tube. Single pump systems can be used to deeper depths (40 to 50 feet) if air inlet holes are drilled along the length of the drop tube which reduces the density of the water in the drop tube and allows water to be pulled from deeper depths.

A vacuum pressure is applied to the test recovery well to evacuate the well and surrounding soils of liquids and air/VOCs. The lower section of the test and monitoring well casings are screened and the test well is equipped with a drop tube which will extend below the well static liquid level to near the bottom of the well. The well casing head is sealed to withstand the applied vacuum pressure, and the vacuum is applied to the drop tube to evacuate the well.

A number of the monitoring wells are also equipped with drop tubes, which will extend below the well static liquid level to near the bottom of the well, and the well casing heads sealed. Well liquid levels are measured in the drop tube (with adjustments for casing side vacuum pressure). A pressure tap is provided on the casing to measure the induced soil vacuum pressure.

The portable vacuum pump is a centrifugal, liquid ring type and is capable of producing and sustaining vacuum pressures up to 24 inches of mercury (27 feet of water). The liquid ring pump is especially suited for this application because of its high vacuum pressure capability and because of the minimum risk for internal source of ignition while compressing potentially explosive mixtures of air/VOCs.

Monitoring

Prior to beginning the pilot test, the fluid levels in all of the on-site monitoring wells are measured using an electronic water level indicator or electronic product/water interface probe. The recovery well should be surveyed and tied into the elevation network for the existing monitoring wells. During the course of the design test, all parameter measurements are recorded. Measured parameters, monitoring techniques, and time schedules include the following:

- Monitoring well fluid levels with electric fluid level indicators inside a drop tube
- Monitoring well casing vacuum pressure at the well head with calibrated vacuum pressure gauge or well manometer each hour
- Measure recovery well applied vacuum pressure with calibrated vacuum pressure gauge or mercury manometer each hour
- Recovery well liquid production measured continually with totalizing type, turbine meter; accumulated total read and recorded hourly. Water tank will be gauged at the conclusion of the test to verify total production
- Recovery well VOC vapor volume continuously measured and recorded using an orifice meter and chart recorded

Figure 11 shows the typical layout of the pilot test for a vacuum-enhanced recovery system using a single pump system where LNAPLs are present. Note the placement of vacuum gauges and flowmeters. Two vacuum gauges are present, one at the recovery well and the other at the suction separator. For a two pump system a second pump would be installed within the well (Figure 9) and fluid recovery routed directly to the oil/water separator.

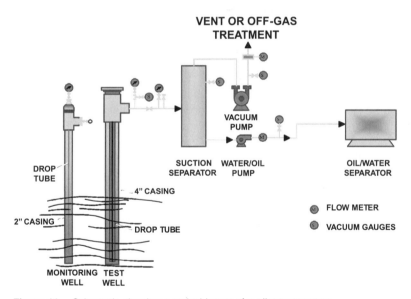

Figure 11 Schematic showing a typical layout of a pilot test system.

The most important aspect of the vacuum well head are tight seals (Figure 12). These well heads should be fabricated and tested prior to going into the field and should have a drop tube installed to measure fluid levels. The drop tube must be extended below the fluid level and cannot be used if the vacuum is expected to exceed the submergence of the drop tube. This is only a consideration in close proximity to the extraction well. The use of a drop tube will prevent the measurement of floating hydrocarbons, if present. However, this is not critical since it is the total head we are interested in for design calculations.

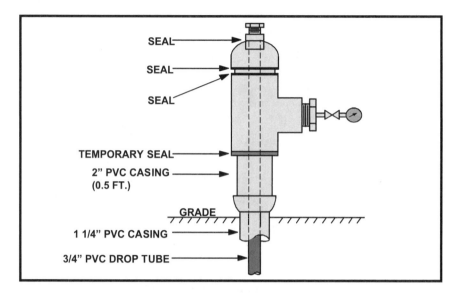

Figure 12   Schematic showing typical well head details for monitor wells.

Mass Removal Estimation

Vacuum-enhanced recovery will reduce the product mass by removing product in the free phase, dissolved phase, and vapor phase. The total mass removed during vacuum-enhanced recovery should be calculated based on the pilot test results. The free product mass is generally measured by calculating the amount removed from the oil/water separator. The dissolved groundwater portion is calculated by multiplying the total volume of groundwater pumped by the weighed average concentration of dissolved constituents measured in water samples collected during the test (three samples). Lastly, the mass in the air phase is calculated by multiplying the total volume of air discharged throughout the test by the weighed average concentration measured in air samples (three samples) collected during the test.

In addition, soil vapor concentrations should be monitored to assess the potential biological activity during the pilot test. The best way to estimate this removal process is to measure the production rate of CO_2. This can be measured at the same time that the organic compounds are tested in the air sample. Once again, the total mass is calculated by multiplying the weighted average concentration of CO_2 by the total volume of air discharged. This value may represent a high or low estimate of the total biological destruction depending upon site conditions. It may represent a high estimate if the pilot test is of short duration and the CO_2 represents biodegradation that occurred over a long period of time. On the other hand, for longer-term tests it may provide a low estimate, because bacteria usually have a lag time while they get adjusted to the increased supply of oxygen provided by the treatment process. The pilot test will not operate for a sufficient time for the bacteria to complete this adjustment. Therefore, judgement needs to be made by looking at the trend in CO_2 concentrations throughout the test to determine the anticipated amount of biological degradation of aerobically degradable organics expected to occur with the full-scale system. A more complete explanation of this process is provided in Chapter 7 (*In Situ* Bioremediation).

SYSTEM DESIGN

After the pilot test is completed, there are a number of engineering design parameters that must be determined to design a full-scale system as shown below:

- Determine groundwater influence
- Determine well spacing based on groundwater influence
- Determine design flow rate
- Select water treatment option (from groundwater concentrations)
- Determine required vacuum pressure
- Determine venting influence
- Determine airflow rate
- Select off-gas treatment (from vapor concentrations)
- Evaluate the presence of biological activity
- Select equipment
- Estimate cleanup time

The design is complicated by the need to evaluate both the groundwater and vacuum systems. As such, there is no cookbook approach that optimizes the design, but rather a detailed analysis of tradeoffs needs to be made to select the most suitable application for achieving your remedial goals. As a starting point, let us review the effect of vacuum application on liquid production, effective drawdown, and selection of an appropriate pumping level. The pumping level in the well is also a consideration in system design. Depending on the remedial objectives, you may want to maintain fluid levels in the well or maximize the drawdown, which will affect pump selection, system selection, and operation. Scenarios where you may want to maximize fluid levels include situations where you are trying to use water as the carrier and thus want to maintain high water movement rates (for instance when the contaminant is soluble in water and not volatile).

On the other hand, scenarios where you want to maximize drawdown would include situations where you would like to dewater the aquifer and rely on air to be the primary carrier to purge the dewatered zone of contaminants or to stimulate bioactivity in the dewatered zone. In this situation, the objective is to lower groundwater levels so that air can move through a larger portion of the contaminant zone. Thus, maintaining high airflow is more important than maintaining water movement.

As we indicated in the beginning, when you apply a vacuum of 10 feet of water, that amount is added to the available drawdown. If for example you have 9 feet of available drawdown without a vacuum, this converts to 19 feet of available drawdown when 10 feet of vacuum is applied. Consequently, the flow rate would essentially double using a vacuum-enhanced recovery system. During pump selection, you need to keep in mind that you have just added 10 feet to your total dynamic head requirements. Also, the available NPSH (net positive suction head) should be greater than the NPSH required by the pump.

Well Design

The use of high vacuums coupled with the low permeability formations can result in rapid well plugging and/or silting. The reduced pressures can result in more rapid precipitation of dissolved inorganic constituents on the well screen, gravel pack, or within the formation. The increased gradients can result in fines plugging the screens or silting of the wells in poorly designed wells. Thus, proper well design and wrapped screens are normally employed and periodic redevelopment of the wells may be necessary. Another tradeoff with the design is to lower the vacuum in order to reduce the above affects. While less flow per well would result in more wells being required, the savings in O&M could easily pay for the added capital costs. The pilot test can also be used to collect data for this analysis.

A thorough understanding of hydrogeologic conditions, aquifer conditions, and most effective well development procedures are required for the completion of high yielding efficient recovery wells.

Listed below are some of the parameters that should be considered in designing recovery wells:

- Proper design procedures for sizing gravel pack and screen to reduce plugging
- Sufficient well diameter for the recovery equipment
- Screen length selected to optimize well head vacuum pressure and airflow
- Depth to product/water from ground surface for selection of type of system (one or two pumps)
- Product characteristics to design oil/water separator and/or air treatment system

A good reference book on designing wells including slot size and gravel paths and on proper well development techniques is *Groundwater and Wells* (Driscoll 1986).

Well Spacing

The first step in system design is to determine well spacing. The optimal well spacings will vary depending on your remedial objective. For instance, if your objective is to contain groundwater and prevent further movement down-gradient, then the recovery wells would be spaced farther apart than if you were trying to achieve maximum dewatering so air can be the primary carrier. In the latter case, the system would be designed with recovery wells closer together and then applying principals set forth in Chapter 3 (Vapor Extraction and Bioventing) to achieve optimum airflow in the dewatered zone.

In either case, well spacings should first be determined based on groundwater influence. Information from the pilot test is used to determine the effective drawdown and flow rate. These parameters are affected by the vacuum level and associated pumping rate information such as that shown in the graph in Figure 13, which depicts

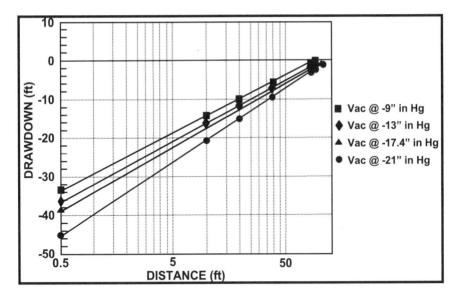

Figure 13   Distance versus drawdown from a vacuum-enhanced recovery pilot test.

the effective drawdown (s') vs. distance at various flow rates generated from a vacuum test. These can be used to estimate well spacing and influence of pumping on liquid levels. As a rule of thumb, do not consider the effective radius of influence beyond drawdown values of 0.10 feet. One important note concerning Figure 13 is that it shows that by increasing the vacuum, you are not necessarily increasing the capture zone significantly. You get diminishing returns as vacuum pressure continues to increase.

As a first cut in determining well spacings, use the effective radius of influence of the well at the expected vacuum pressure and plot overlapping cones to estimate the number and location of wells. This is generally a first cut evaluation and leads to system overdesign, where the remedial objective is groundwater containment, because the radius of influence during the short-term pilot studies are under non-equilibrium conditions. However, in certain cases (particularly in systems where you are trying to dewater and use air as a carrier) this approach may suffice.

For larger systems it is more appropriate to design the well spacings based on long-term operations when apparent equilibrium conditions are reached. To determine the capture zone under these conditions, the data collected during the pilot studies can be used with the approaches outlined below to evaluate capture zones under apparent equilibrium conditions and optimize well placement (Schafer 1995). It is important to note that use of conventional capture zone equations to calculate capture zones in tight sediments are not appropriate. Use of these equations results in prediction of capture zones that are unreasonably large. To demonstrate this let us look at an example of an over-predicted capture zone in a low-permeability formation. Let us examine a hypothetical 20 foot thick silt formation having a hydraulic conductivity of 2.1 gpd/ft^2 (1x10^{-4} cm/sec), a storage coefficient of 0.05, and a hydraulic gradient of 0.01 ft/ft. Assume a fully penetrating, 100 percent efficient extraction well operates for 30 consecutive days, resulting in a drawdown of 10 feet. Assume further that the bore hole radius is 0.5 ft (12 inch bore hole).

The discharge rate of the well may be calculated using the Cooper Jacob equation, but the observed drawdown must be corrected for dewatering first. As stated earlier in this chapter, the dewatering correction is as follows

$$s' = s_a - \frac{s_a^2}{2b_w} \tag{2}$$

With an observed drawdown of 10 feet

$$s' = 20 - \frac{(10)^2}{(2)(20)}$$
$$= 7.5 \text{ ft}$$

Similarly using the Cooper Jacob equation, the discharge is calculated as follows

$$Q = \frac{s'T}{264\log\frac{0.3Tt}{r^2 S}} \tag{1}$$

The resulting discharge rate is 0.266 gpm.

The standard capture zone equations calculate the distance to the stagnation point (x_0), the capture width at the well (w_0), and the up-gradient capture width (w) as follows (assuming consistent units)

$$x_0 = \frac{Q}{2\pi TI} \tag{8}$$

$$w_0 = \frac{Q}{2TI} \tag{9}$$

$$w = \frac{Q}{TI} \tag{10}$$

For the example presented here, a flow rate of 383 gpd (0.266 gpm) gives the following results

$$x_0 = 145\,ft$$
$$w_0 = 456\,ft$$
$$w = 912\,ft$$

Using vacuum-enhanced recovery techniques, the yield of this well would double or triple depending upon the amount of vacuum applied. Assuming we doubled the discharge rate to 766 (0.532 gpm), the capture zone dimensions compute to the following

$$x_0 = 290\,ft$$
$$w_0 = 912\,ft$$
$$w = 1824\,ft$$

Based on this analysis, it is easy to see that remediation design engineers should not rely on a 0.5 gpm well to provide more than one-third of a mile of capture width. In practice, intermittent recharge events swamp out the cone of depression periodically because an ordinary recharge event can overwhelm the small discharge rates generally associated with tight formations. With the cone of depression being flooded out periodically, its lateral extent is limited and the well is not able to influence gradients at the great distances predicted by conventional capture theory.

To overcome the problems associated with conventional capture zone equations, an algorithm has been prepared which should lead to a more conservative design. The essence of the procedure is to base capture on the configuration of the cone of depression after a fixed, limited pumping time—say, 30, 60, or 90 days. In other words, after the fixed time has passed, we assume we gain no further growth in the cone of depression. After selecting the arbitrary pumping time (perhaps 10 to 30 days for humid climates and 90 days for desert climates), drawdowns and gradients are calculated based upon the Theis equation (not the log equation) and capture analysis is based upon the resulting drawdown configuration. A description of this procedure follows, along with several required graphs and a few examples.

After the arbitrary pumping time has been chosen, the first step is to compute the so-called log-extrapolated radius of influence of the well, R. This is done primarily for mathematical convenience because R consolidates several other parameters. (R is commonly called the radius of influence because it is the distance to zero drawdown on an extrapolated semi-log distance-drawdown graph. However, it is not a true radius of influence because the Theis equation predicts some additional drawdown beyond this point.) R may be computed from the following equation

$$R = \sqrt{\frac{0.3Tt}{S}} \tag{11}$$

where, R = log-extrapolated radius of influence, in ft; T = transmissivity, in gpd/ft; t = pumping time, in days; and S = storage coefficient.

If you are using consistent units, such as transmissivity in ft²/day, the equation is as follows

$$R = \sqrt{\frac{2.246Tt}{S}}$$

The log-extrapolated radius of influence is significant in that we will want to express capture zone dimensions in terms of R.

Differentiating and manipulating the Theis equation gives rise to the following expression relating discharge (Q), and distance, to down-gradient stagnation point (x_0).

$$Q = 2\pi T I x_0 e^{\frac{x_0^2}{1.781 R^2}} \qquad (12)$$

This is the same as the conventional capture zone equation except for the exponential term. When the exponent is small (x_0 is small in relation to R), the exponential term is close to 1 and the standard capture zone equations work just fine. As x_0 increases, however, to a significant fraction of R or beyond, the exponential term is substantially greater than 1 and the extraction well must produce more water than would be determined by conventional analysis. In essence, the exponential term represents a multiplier that must be applied to the Q computed from the standard equations.

Figure 14 shows the magnitude of the exponential term as a function of the ratio x_0/R. For example, if x_0 is half the radius of influence, Q will be 1.15 times the conventional calculation. If x_0 equals R, the discharge must be 1.75 times that calculated from conventional analysis. And if x_0 equals 2 R, nearly a ten-fold increase in discharge rate is required over conventional theory.

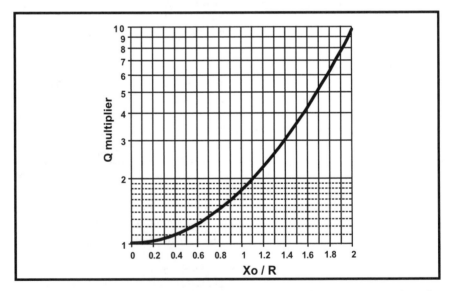

Figure 14   Ratio of required Q to conventional Q as a function of distance to stagnation point.

Conventional theory predicts that the width of the capture zone at the extraction well will be π times the distance to the stagnation point. That is

$$w_0 = \pi x_0 \tag{13}$$

For tight formations (time limited cones of depression), however, the ratio w_0/x_0 decreases as discharge rate and capture zone size increase. Figure 15 shows how this ratio decreases with increasing x_0/R. Figure 15 was developed empirically by using analytic element modeling and particle tracking to assess the relationship between capture width and distance to stagnation point.

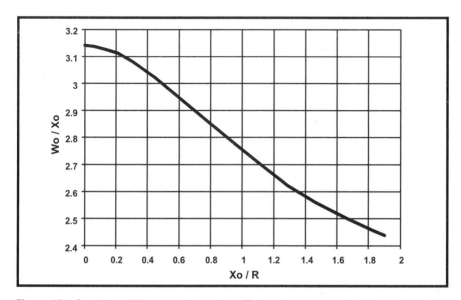

Figure 15 Decline in W_0/X_0 with increasing X_0 (R = log-extrapolated radius of influence).

Figure 14 was based on an exact equation, whereas Figure 15 was determined empirically. Combining these two graphs produces Figure 16, which shows the magnitude of the discharge increase required for capture compared to conventional theory, all as a function of desired capture width. For example, reading from the graph, when the capture width is twice the radius of influence, the discharge must be 1.4 times the value calculated using the conventional equations. For a capture width equal to the radius of influence, the graph shows that the required discharge is 1.09 times the conventional discharge.

Figure 16 shows that in tight formations we pay a penalty in terms of discharge rate, and that the penalty increases as the capture zone dimensions approach and exceed the log-extrapolated radius of influence. If the remediation design includes a large number of wells, each with a small capture zone, the total required discharge per well is minimized. As the number of wells is reduced and the individual capture zone size is increased, the required discharge rate increases by the multiplication factor shown on the vertical scale of the figures. Since the discharge rate is limited by the aquifer characteristics (hydraulic conductivity and saturated thickness) and applied vacuum (limited by equipment selection) this procedure can be used to estimate the minimum number of wells required for groundwater containment.

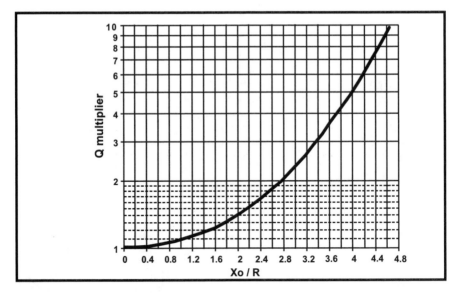

Figure 16   Ratio of required Q to conventional Q as a function of desired capture width.

An example will illustrate these procedures. Calculations will be made based upon capture width, w_0, because this is the parameter of greater importance.

Using the layer described above and a 30 day pumping time, the radius of influence is calculated as follows

$$R = \sqrt{\frac{0.3(42)(30)}{0.05}}$$ (11)

$$= 87 \text{ ft}$$

The required flow rate for a particular capture width may be computed as follows

1. Select the desired capture width. For example, a w_0 of 200 feet.
2. Compute discharge required using conventional equation.

$$Q = 2TIw_0$$
$$Q = 2(42)(0.01)(200)$$ (14)
$$= 168 \text{ gpd } (0.12\text{gpm})$$

3. Calculate w_0/R and use Figure 16 to determine the multiplier to apply to the conventional rate. The ratio w_0/R equals 2.3 and, reading from the graph, the multiplier is 1.6.

Thus, the required Q is

$$Q = (168)(1.6)$$
$$= 269 \text{ gpd } (0.19 \text{ gpm})$$

Repeating this process for another assumed value of w_0, for example, 300 feet, yields the following

$$Q = (2.42)(0.01)(300) \quad (\text{conventional equation})$$
$$= 252 \text{ gpd } (0.19 \text{ gpm})$$
$$w_0/R = 3.45$$
$$\text{Multiplier} = 3.2 \text{ (from Figure 16)}$$
$$Q \text{ actually required} = (252)(3.2)$$
$$= 806 \text{ gpd } (0.56 \text{ gpm})$$

These same calculations were performed for capture widths ranging from 50 to 400 feet in 50 foot increments. Table 1 shows the results, including the discharge associated with the conventional approach and the actual required discharge associated with this procedure. Recall that earlier analysis showed a required discharge rate of 0.266 to achieve a w_0 of 456 feet. According to Table 1, this discharge rate would result in a capture width between 200 and 250 feet. Similarly, earlier calculations showed that a discharge rate a little over 0.5 gpm resulted in more than 900 feet of capture, whereas Table 1 predicts less than 300 feet of capture.

Table 1   Comparison of Conventionally Calculated Discharge Rates and Actual Required Rates for 30 Days of Uninterrupted Pumping

w_0, in ft	w_0/R	Q from Conventional Equation ft³/day	gpm	Multiplier from Figure 16	Q actually Required ft³/day	gpm
50	0.58	42	0.029	1.02	43	0.030
100	1.15	84	0.058	1.1	92	0.064
150	1.73	126	0.088	1.29	163	0.113
200	2.30	168	0.117	1.6	269	0.187
250	2.88	210	0.146	2.2	462	0.321
300	3.45	252	0.175	3.2	806	0.560
350	4.03	294	0.204	5.1	1499	1.041
400	4.60	336	0.233	9.5	3192	2.217

The calculations described above may still over-predict the capture zone radius, particularly if the storage coefficient is low (10^{-3}). The calculations can more accurately reflect actual conditions by choosing a shorter pumping time. Typically, discharge rates decline with time so if declining discharge rates are expected, using a shorter pumping time in your calculations is more appropriate. For example, if a pumping time of 10 days is used in the above example, the radius of influence, R, is 50 feet. Recalculating the capture zone, then, produces the results shown in Table 2. Note that in this case, a discharge rate of more than 0.5 gpm would produce less

Table 2   Comparison of Conventionally Calculated Discharge Rates and Actual Required Rates for 10 Days of Uninterrupted Pumping

w_0, in ft	w_0/R	Q from Conventional Equation ft³/day	gpm	Multiplier from Figure 16	Q actually Required ft³/day	gpm
50	1.00	42	0.029	1.09	46	0.032
75	1.49	63	0.044	1.2	76	0.053
100	1.99	84	0.058	1.4	118	0.082
125	2.49	105	0.073	1.75	184	0.128
150	2.99	126	0.088	2.35	296	0.206
175	3.49	147	0.102	3.35	492	0.342
200	3.98	168	0.117	5	840	0.583
225	4.48	189	0.131	8.2	1550	1.076

than 200 feet of capture width and a discharge rate about 0.25 gpm would produce a little more than 150 feet of capture width.

When using the formulas presented previously, it is important to incorporate well efficiency into these formulas. In our example problem shown, the well efficiency was assumed to be 100 percent; however, under field conditions associated with the types of sediments where vacuum-enhanced recovery is applied, the well efficiency is usually substantially less than 100 percent. The pilot test will provide an insight into the expected well efficiency; however, it is important to keep in mind that well efficiency may vary in accordance with geologic changes across a site. Generally, as the grain size of the sediments opposite the well screen decrease, the well efficiency also decreases.

The well efficiency from a pilot test can be estimated by using the distance-drawdown graph (Figure 13). This is done by extending the straight line representing the profile of the cone of depression to show drawdown in the aquifer just outside the well. The theoretical drawdown represented by this point is for a 100 percent efficient well. (In a filter packed well, the radius is taken from the center of the well to the outside of the filter pack.) The well efficiency is calculated by dividing the theoretical drawdown in the well by the actual drawdown (Driscoll 1986).

These methodologies will only give an approximation; the size and complexity of the job will determine the extent of the hydrogeologic analysis that is required. In some instances, you are dealing with multiple tests of varying values as a result of variations in hydrogeology. If these variations in hydrogeology can be defined, they must be taken into account in determining well spacing and system influence. If not, a conservative estimate must be made.

Fluid Flow Rate

Based on well spacing and a hydrogeologic evaluation, the design fluid flow rate can be determined. As discussed previously, typically the fluid flow rate will decline with time, so this should be factored into the design flow rate. The design flow rate will be a summation of the flows from the individual extraction wells and is actually developed hand-in-hand with well spacing. If the capture zones from individual wells overlap one another, then fluid flow from individual extraction wells will be less than if the capture zones merely touch each other. Thus, if the system design is based on plotting capture zones from individual wells with data collected from pilot studies, then the calculated fluid recovery rate will be higher than will actually occur, so this should be kept in mind. Using the capture zone equations discussed in the previous section is generally the preferred method; however, often times data are not sufficient to get this sophisticated, so using simpler methodologies and good engineering judgment may suffice. Either way, a well-run pilot study is the key to collecting good site specific data. If you have limited funds, the money is better spent basing system design on this in lieu of sophisticated modeling approaches that are performed without the aid of site specific pilot tests.

Vacuum Pressure

Vacuum pressure selection can then be determined by a series of curves that are generated from the pilot testing. Ideally, your optimum vacuum is the lowest vacuum that will give you the water production and air influence you need. However, other criteria may control such as suction lift or water flow requirements. You must balance your desired water discharge, air influence, and airflow to arrive at the optimum vacuum. The data collected from the pilot studies should be evaluated in a manner similar to that shown in Figure 17. Typically, the vacuum decreases with time as the dewatering exposes more formation to airflow. You should consider this during the design of the vacuum pump. The optimum vacuum is defined as the point at which higher vacuums result in diminishing benefits for water discharge, air influence, and airflow. This may drive your pumping system selection, and the cost of increased vacuum would be weighed against the cost of a two pump system.

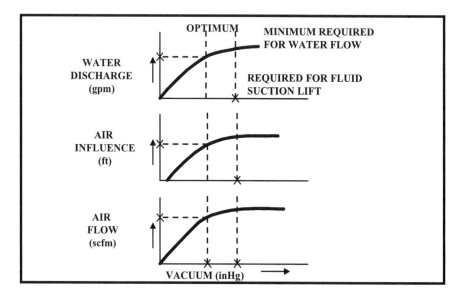

Figure 17   Selecting vacuum pressure based on pilot study results.

A higher vacuum will result in a higher airflow rate and venting influence, and generally, greater fluid recovery rates and groundwater influence. However, this is balanced by the increased cost and energy requirements of increasing the vacuum and the increase is not linear. As the vacuum continues to increase, the relative increase in water and airflow rates declines. This is depicted in Figure 17, which shows a comparison of the pilot test data at various vacuum pressures versus airflow rate and influence. Higher vacuums may also cause maintenance problems with the wells. The O&M costs also have to be included in full-scale design considerations.

The selection of the vacuum pumping system will also influence the design vacuum pressure. The design pressure may be reduced below optimum levels if the use of a liquid ring application can be avoided in favor of a less expensive alternative at a lower vacuum pressure, without significantly sacrificing effectiveness. This would be a function of a capital cost evaluation and O&M considerations. The final factor to evaluate is the type of pumping system selected. If a single pump vacuum system is used, the minimum vacuum will be defined by the suction lift required to produce fluids from the well. Evaluation of the other factors may indicate that a two pump system is more cost effective since a lower vacuum pressure can be selected. Once the vacuum pressure has been selected, venting influence must be determined. In some applications, the use of air as the carrier for removal of residuals through volatilization or enhancement of bioactivity will be one of the primary applications of the vacuum system. A general rule of thumb is using a value of 0.1 inches of water vacuum in the monitoring wells as the limiting effective vapor extraction influence. In most cases, the dominant issue for well spacing is the groundwater influence and not the venting influence. However, there may be some instances where the venting influence drives the selection of well spacing. This will be a function of the objectives of your remedial program.

Airflow Rate

The vacuum pressure selected above will set the airflow rate for the system. The application of the pilot test results to the remainder of the site requires a hydrogeo-logic evaluation as the consistency and/or variation of geologic materials at the site, similar to that done for the groundwater flow estimation. If there are well-defined areas or known variations in site permeability, it may be wise to run several short-term pilot tests (a few hours) with fewer observation wells to get a range of design parameters and flow rates. These can be used to estimate the variability in site-wide system production.

Off-Gas Treatment

An off-gas treatment system can be the most costly portion of the vacuum-enhanced treatment system. Care must be exercised to assure that the most cost effective approach is used, and that changes in off-gas concentrations versus time, which can occur quite rapidly, are understood. One of the primary concerns with the use of vacuum-enhanced systems at petroleum hydrocarbon sites is the high concentration of hydrocarbons in the off-gas during the early phases of the remediation. These concentrations in the several percent range must be factored into the design selection. The high concentrations usually require the use of some form of thermal destruction. Explosion-proof equipment is needed in many cases. If off-gas concentrations are expected to decrease rapidly, you may want to switch technologies to reduce lifecycle costs as discussed in Chapter 2 (Lifecycle Design).

Equipment Selection

All of the factors evaluated above will play a role in the selection of the equipment for a particular vacuum-enhanced option. Two of the factors will be discussed below, but a cost benefit analysis evaluating all of the parameters needs to be performed before a final selection is made.

A selection of the design vacuum level will dictate to some extent the type of vacuum system that must be employed at the site. At high vacuums (greater than 15 inches of mercury or 19 feet of water) liquid ring vacuum pumps (Figure 10) are normally employed. At lower vacuums (less than 8 inches of mercury or 10 feet of water) regenerative blowers or positive displacement blowers (Figure 18) are normally employed. In the mid-range (between 8 and 15 inches of mercury) lobe pumps (Figure 19) are used. These are strictly general ranges and they vary widely depending upon the site conditions and the needs at the site. For instance, regenerative blowers could be put in series to increase their vacuum production rather than going to a lobe pump.

Figure 18   Regenerative blower.

Figure 19   Lobe pump.

For petroleum hydrocarbon remedial programs, another factor to consider is the need for explosion-proof systems due the high hydrocarbon vapor concentrations in the off-gas. This should be one of the primary factors considered in the system design and will impact the pump selection and system configuration.

In a single pump system the vacuum blower is used to remove both the fluids and the vapors. Selection of the equipment is based on the expected air and water flow rates. Normally, as the depth to water increases, the vacuum required also increases. For two pump systems, a down hole pump is used to remove fluids. The down hole pump could be a skimmer pump that is used to extract hydrocarbons off of the top of the water table. In other cases the down hole pump is submerged such as a submersible pump. In two pump systems care must be exercised to ensure that the down hole pump remains submerged, or cavitation can occur. To avoid this the system can be designed with low level shutoff so if the groundwater in the well drops below a certain level the system automatically shuts off.

MASS REMOVAL AND REACHING CLEANUP GOALS

When applied to the right compounds and in a favorable geologic setting, vacuum-enhanced recovery can be very effective in both mass removal and achieving cleanup goals. Its ability to reach these goals varies depending on the application and contaminant involved. In all cases, if the permeability is too low throughout the saturated thickness, then vacuum-enhanced systems are not applicable unless the geology is manipulated (Chapter 10, Fracturing) to make it suitable for use of a vacuum-enhanced system. Vacuum-enhanced recovery has been shown to be cost effective for removing mass from a contaminated site. The main reason for this good performance is the reduced time requirements needed by the vacuum-enhanced recovery design compared to other mass removal techniques. Vacuum-enhanced recovery is usually able to enhance the removal of LNAPLs, dissolved phase, air phase, and DNAPLs from most sites.

LNAPLs

Vacuum-enhanced recovery systems can be effective in mass removal of LNAPLs since applying a vacuum adds a second force (pressure differential) to fluid removal and can significantly extend the capture zone. The vacuum force can become the major transfer force on the LNAPL when the hydrogeological conditions prevent or restrict water pumping. Vacuum-enhanced recovery is effective in fairly rapid recovery of LNAPLs and the use as air as a carrier is much more effective in removing volatile organics from the dewatered zone than using water as a carrier. Since 70 to 90 percent of mass at sites containing LNAPLs is often in the nonaqueous phase, use of a vacuum-enhanced system for mass removal can be beneficial. If the cleanup goal is removal of LNAPLs from the subsurface, this technology can be effective in achieving this goal in a reasonable time frame, less than one year to several years, depending on site conditions.

An example where the limited available drawdown was overcome by using a vacuum-enhanced recovery system was at a bulk storage facility (terminal) located on the west coast. At this facility, hydrocarbon products were found to be discharging along the bank adjacent to the terminal into a harbor. When the discharges were discovered, daily fines were assessed until the seepage ceased; therefore, the remedial objective was to install a system to prevent further seepage into surface water in the shortest possible time frame. Other remedial options such as trenches and barriers were readily evaluated, but were rejected due to space constraints, costs, soils disposal, and time required for construction. Since the aquifer was thin, with a total depth of 10 feet and a saturated thickness of 8 feet, a vacuum-enhanced system was deemed to be a good possibility.

A preliminary evaluation was undertaken to assess the suitability of using an enhanced vacuum system and the results looked favorable. Consequently, a recovery well was installed and a short pilot test was performed to determine the benefits and costs of using a vacuum-enhanced approach compared to a more conventional pumping approach. The pilot study showed that using a conventional pumping approach, the expected withdrawal rate from a single recovery well would be less than 0.1 gpm and its hydrocarbon recovery rate would be less than one barrel per day. Its capture zone in a direction parallel to the bank where seepage was occurring would be small. When a vacuum-enhanced recovery approach was tested on this same well, its withdrawal rate increased to 0.6 gpm, and its daily hydrocarbon production rate and its capture zone increased significantly. The pilot test results demonstrated the benefits of using a vacuum-enhanced approach and installation of a full-scale system was initiated. Not only did a vacuum-enhanced system allow the use of significantly fewer recovery wells (Figure 20), the capture zones from a

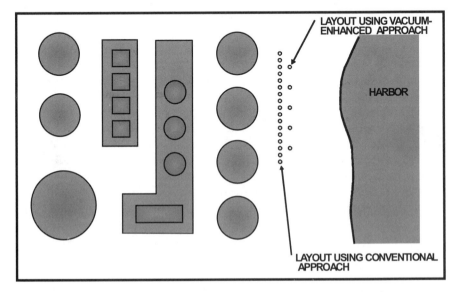

Figure 20   Site layout displaying recovery well layouts using conventional vs. vacuum-enhanced recovery approach.

vacuum-enhanced system extended much further down-gradient beyond what could be achieved using conventional pump and treat. Seepage to the harbor ceased within days of system installation as opposed to weeks using the conventional approach.

In situations where the geology is more permeable and the constituents less volatile, a more effective technique for mass removal of LNAPL may be the use of skimmer pumps combined with a vacuum system. Operation of this type of system enhances removal of the LNAPL by a moderate increase in the effective radius of the skimmer system, while at the same time removing mass from the vadose zone using the air phase. In this application the goal is to enhance the effective radius of the recovery well while at the same time minimizing the depth of the smear zone. For less volatile constituents such as diesel fuels, application of the vacuum-enhanced recovery in this fashion may be more appropriate.

Dissolved Phase

We have discussed how vacuum-enhanced systems are effective in increasing the size of capture zones and in increasing water production rates. However, if the remedial goal is to achieve groundwater cleanup to low constituent levels, such as drinking water standards, then vapor enhanced recovery suffers from the same limitations as pump and treat programs. Achieving cleanup at low levels is difficult to achieve because although air is a good carrier of volatile organic compounds, it can be applied only to the zone above-groundwater levels (under pumping conditions). Thus, it may reduce contaminants to acceptable levels in this zone within reasonable time frames, but in the zone below the water table, where water is the only carrier, you still have the limitations associated with conventional pumping systems. Moreover, the lower permeable deposits often contain highly adsorptive materials which makes reaching such cleanup goals even more difficult than if the LNAPLs were located in a sandy environment where there is less adsorption to the earth materials.

One thing to keep in mind in setting cleanup goals at sites suitable for vacuum-enhanced systems is that these are generally used in lower permeability environments which are generally not used for drinking water purposes. As such, the cleanup goals should reflect higher concentrations of contaminants than if it were a drinking water aquifer and should be set at levels that will protect the receiving body whether it is surface water or deeper aquifers. Realistic cleanup goals can be important in shaving years off operating vacuum-enhanced recovery systems and should not be overlooked.

In some situations, vacuum-enhanced recovery is used to control the migration of constituents into more permeable aquifer zones. An example of this is a terminal where dissolved constituents originating from releases from above-ground storage facilities were discovered. The terminal was located adjacent to a river, and a creek separated it from an adjacent industrial facility, which operated on-site production wells (see Figure 21). It was determined that the creek was receiving groundwater discharges from the uppermost sediments, which contained dissolved and liquid hydrocarbons. Discharges to the creek prompted regulatory action, although it was

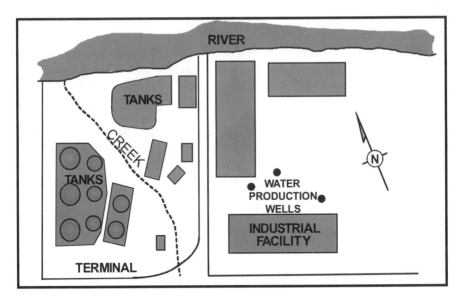

Figure 21   Map showing site layout.

later discovered that the production wells at the adjacent production facility were also impacted.

The cross section in Figure 22 shows the two major aquifers, fluid levels in wells, and general flow directions. The upper aquifer consisted of interbedded sands and clays with hydraulic conductivities of the clays ranging from 10^{-4} to 10^{-5} cm/sec. The lower aquifer was extremely productive with a hydraulic conductivity of 10^{-1} cm/sec and was used as a source of water supply in the area. Several off-site production wells were withdrawing millions of gallons of water a day, and the pumpage caused a strong vertical gradient in the over lying silty sand/clay aquifer. As a result, pockets of LNAPL within sand lenses were scattered throughout the upper zone.

The hydrogeologic investigation confirmed that dissolved hydrocarbons entering the lower aquifer were in the capture zone of the off-site production wells and consequently moved toward these wells. Although the off-site production wells were not used for drinking water purposes, the regulatory agency deemed the risks associated with releases from the site as unacceptable, requiring that an on-site remedy be developed. One approach would be to install a groundwater recovery system into the lower aquifer; however, this would have required excessive pumping rates (from 1500 to 2000 gpm) and the associated high costs to build and operate a large treatment system. In addition, this approach lacks any source removal, since the LNAPLs would still remain in the upper aquifer and remain as a continuing source of contamination to the lower aquifer.

Consequently, an alternative approach was considered, with the goal being to capture hydrocarbon liquids and dissolved hydrocarbons before they migrated into the lower aquifer by installing a recovery system to the base of the upper aquifer. This approach, as illustrated in Figure 23, would be designed to reverse the vertical

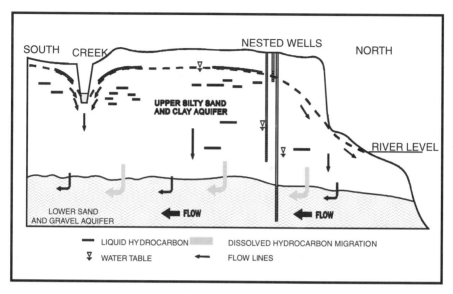

Figure 22   Hydrogeologic cross section.

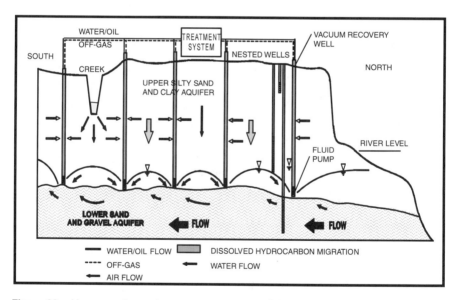

Figure 23   Vacuum-enhanced recovery system remedial approach.

gradient between the upper and lower aquifers, and would cause both the creek and the river to be recharge boundaries (as opposed to discharge boundaries) which would prevent discharges of hydrocarbons into these surface water bodies.

In order to evaluate whether or not a vacuum-enhanced approach was cost effective, a pilot study was initiated by pumping several recovery wells with and without the aid of a vacuum. The results of the pilot studies showed that under normal pumping conditions, without applying a vacuum, there was limited fluid production and limited influence. However, with the application of a vacuum, the flow rates and water level influence (capture zone) increased by a factor of five. These studies confirmed the benefits of a vacuum-enhanced recovery approach to intercepting the contamination in the upper aquifer and preventing its movement into the lower aquifer. The pilot studies along with computer modeling showed that a total flow rate of 30 gpm was required to manage the plume, and that could be achieved with significantly fewer wells using the vacuum-enhanced approach. Using air as a carrier to remove VOCs in the unsaturated zone also resulted in a much shorter cleanup time.

Air Phase

In both of the above examples, the goal of using vacuum-enhanced recovery was to supplement the force of gravity (by adding pressure differentials) to increase the capture zones and recovery rates of individuals recovery wells. Use of vacuum-enhanced systems for dewatering and mass removal via vapor extraction can also be effective, particularly in perched zones containing alternating lenses of sand and clays, and in zones that are relatively shallow and thin. Dewatering contaminant zones allows remediation to occur using air as the carrier which is more effective than using groundwater alone. This is true where LNAPLs are present or if the contaminants are in the dissolved phase. Thus, not only do you achieve good mass removal, but it also becomes easier to achieve cleanup goals in more reasonable time frames because air can more effectively remove VOCs.

In this example we will show how vacuum-enhanced recovery was used to incorporate air as the main carrier of contaminants. This example is typical of many sites were releases of LNAPLs from underground storage tanks result in the LNAPLs being trapped in the sand lenses of interbedded sands and clays, as depicted in Figure 24. The figure shows a release from a site located in the southeast where the LNAPL was confined primarily to a sand seam interbedded within clayey deposits with a permeability on the order of 1×10^{-4} cm/sec and a transmissivity of 40 gpd/ft. Previous attempts to recover the LNAPL using conventional pumping techniques, that is to say without vacuum enhancement, proved unsuccessful due to the low transmissivity and thin saturated thickness. Consequently, a decision was made to pursue a vacuum-enhanced recovery approach with the objective being to dewater the sand lense and to switch to air as the primary carrier to remove residuals LNAPLs that remained in the sand lense after dewatering. Since the LNAPLs were petroleum hydrocarbons, it was anticipated that using air as the carrier would also stimulate bioactivity by increasing the available oxygen to the sand seam.

A pilot test was performed to determine the effectiveness of the vacuum-enhanced recovery approach and to develop design data. The pilot test utilized an explosion-proof, liquid ring system capable of moving 50 scfm and producing a vacuum of 24 inches of mercury. The objective was to dewater the sand lense and rely on air to achieve the cleanup standards. This is important since the goal of this remedial program was to maximize airflow and volatilize or biodegrade *in situ* the hydrocarbons by the use of vacuum dewatering. In other words, the objective was to maintain good airflows, not to maintain a high vacuum.

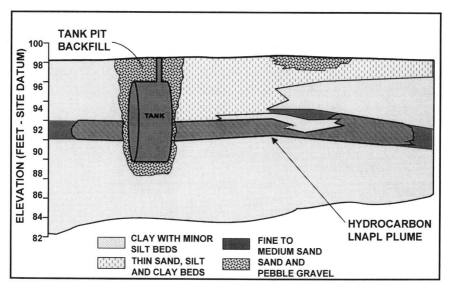

Figure 24   Geologic cross section at release from underground storage tank site.

The results of the pilot study showed that the groundwater influence (capture zone) increased from less than 15 feet using pumping alone to up to 100 feet with a vacuum of 12 inches of mercury applied. In addition, the groundwater yield increased from less than 0.1 gpm to 0.15 gpm. More importantly was the effect on airflow as the pilot study progressed, which is shown in Figure 25. As the sand lense dewatered, air could more easily flow through it, and the airflow increased with a

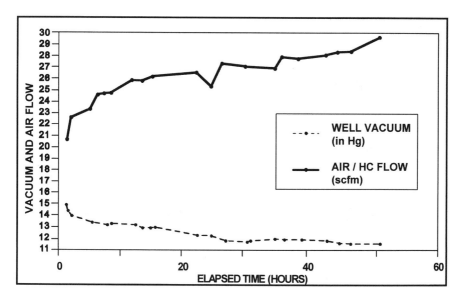

Figure 25   Airflow and vacuum vs. dewatering during vacuum-enhanced recovery pilot test.

resulting decrease in well head vacuum. The airflow increased 50 percent during the test and the vacuum decreased 25 percent.

The results of the pilot study also demonstrate the value of collecting concentrations of hydrocarbons in the off-gases, not only to show how effective the system is in volatilizing contaminants, but also to evaluate off-gas treatment requirements. Figure 26 shows the hydrocarbon concentration versus time in the off-gases. The concentration of about 9 percent total hydrocarbons declined to about 1 1/2 percent after 50 hours of operation. This information is needed during system design (the need for explosion-proof systems) and off-gas treatment evaluation as discussed in Chapter 6 (Air Treatment for *In Situ* Technologies).

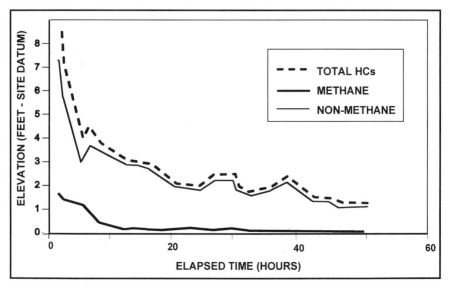

Figure 26   Hydrocarbon concentration in off-gases during vacuum enhancement pilot test.

The pilot study demonstrated the value of using vacuum-enhanced recovery in this type of application. A modular vacuum-enhanced recovery was installed and operated for approximately one and a half years at which time concentrations were significantly reduced and regulatory approval for site closure was secured.

The above is a good example of remediating LNAPL in a sand and clay environment. Other situations exist where there are dissolved constituents only, and where it is desirable to dewater the sediments and remove constituents absorbed to the aquifer matrix. Often the absorbed mass represents significantly higher mass than that found in the dissolved phase. In these situations, using a conventional pump and treat program relies on groundwater to desorb the constituents from the aquifer matrix, which can be considerably less efficient then dewatering the sediments and using air to remove the adsorbed phase. Two case studies are presented at the end of this chapter which demonstrate the application of vacuum-enhanced recovery using this remedial strategy.

DNAPLs

We all know the difficulties in trying to remediate groundwater impacted by DNAPLs. If the DNAPL is trapped in the pore spaces in the upper portion of a contaminated zone and the air can access these zones, then high contaminant recovery rates are achievable and removal of a considerable mass of DNAPL may occur. In situations where the DNAPL is at greater depth (below the dynamic water level) the primary benefit of vacuum-enhanced systems is the increased effective radius of the recovery well for groundwater, which can significantly decrease the number of recovery wells needed to achieve containment of the DNAPL hot spot.

To evaluate the effectiveness of vacuum-enhanced recovery versus conventional technologies to remediate DNAPLs, a technology demonstration project was performed by the Air Force Center for Environmental Excellence (AFCEE) at Hanscom Air Force Base. The technology demonstration was performed at a former burn pit area, which was reportedly used from the late 1960s through 1973 for training exercises. The area is underlain by approximately 20 to 25 feet of glacial till (overburden), that overlies granitic bedrock. The till typically consists of dense, coarse to fine sand with variable amounts of silt, fine to coarse gravel, cobbles, and boulders. *In situ* permeability tests of the till and nine packer tests of the bedrock indicated a hydraulic conductivity value for the till of 8.6×10^{-4} to 6.6×10^{-3} cm/sec, and for the bedrock of 0 to 1.3×10^{-3} cm/sec. Since groundwater flow in the bedrock occurs in fractures, the hydraulic conductivity of the bedrock is expected to vary considerably across the site.

Four extraction wells were installed in the area near existing wells RAP1-3R and RAP1-3S. DNAPL was detected previously in the existing bedrock well RAP-3R. The extraction wells were spaced 40 feet apart in a square-type orientation with RAP1-3R in the center of the square. The locations for the extraction wells and monitoring wells are shown in Figure 27.

The 4 inch diameter extraction wells were installed to a depth of approximately 50 feet, with 30 feet of 0.020 inch slot PVC screen opposite the bedrock zone. Vapor and liquid were drawn from each extraction well through a 1 inch diameter flexible polyethylene drop tube that extended to approximately 1 to 3 feet from the bottom of the well.

The main component of the extraction module was a 15 hp, 460 volt/3 phase explosion-proof, oil sealed, liquid-ring vacuum pump capable of generating a maximum of 28 inches Hg vacuum at the pump inlet.

The pilot study was performed in two phases. The first phase was performed between December 1997 and June 1998. The second phase was performed between October 1998 and March 1999. Prior to initiating the second phase of the vacuum-enhanced recovery pilot study, a 24 hour pump test was performed on the extraction well system using the liquid-ring vacuum pump and allowing air to move down the extraction wells. During the test, the groundwater extraction rate from the four extraction wells varied from 0.40 to 0.47 gpm, and the mass removal rate varied from 2.3 to 2.8 lbs per day. A soil vapor extraction test was also performed using a pump capable of producing several inches of vacuum. Because of strong capillary

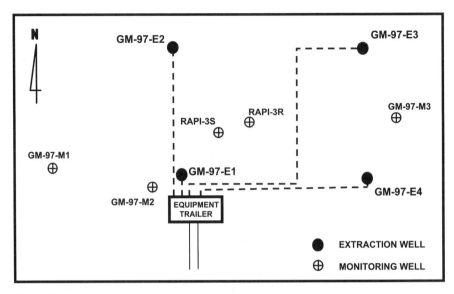

Figure 27   Location of monitoring and extraction wells.

forces associated with the fine grained sediments, the vacuum pump was incapable of producing formation airflow.

The vacuum provided by the liquid-ring pump was capable of inducing formation airflow. The formation airflow that occurred during vacuum-enhanced recovery varied throughout the test from about 15 to 32 scfm, primarily in response to fluctuations in groundwater levels. The groundwater extraction rate varied from about 1.3 gpm to 2.7 gpm with the highest flow rates occurring at the beginning of each of the two test phases.

CVOC (chlorinated volatile organic carbon) removal rates are shown in Figure 28. During the initial stages of Phase 1, high CVOC removal rates (approximately 5.5 lbs/day) were measured, but these rates declined to about 1.1 lbs/day after about five weeks. The removal rates gradually increased back up to about 6.5 lbs/day by the end of the first phase of the test. During the second phase, CVOC rates were varied between 3.6 and 6.2 lbs/day.

Throughout the test, data were also collected on the relative contributions to removal of CVOC mass by the water phase and by the air phase. These data are also shown on Figure 28. In general, the mass removed by the water phase is fairly steady and averages about 1 lb/day. However, the amount removed by the air phase varied considerably and averaged about three times that removed by the water phase.

The effects of the vacuum-enhanced recovery pilot test on TCE concentrations are shown in Figure 29. Significant fluctuations in constituent concentrations in the extraction wells occurred with no discernable trend due to the presence of the DNAPL. The technology was capable of increasing groundwater extraction rates by a factor of four, indicating that the capture zone of the vacuum-enhanced recovery

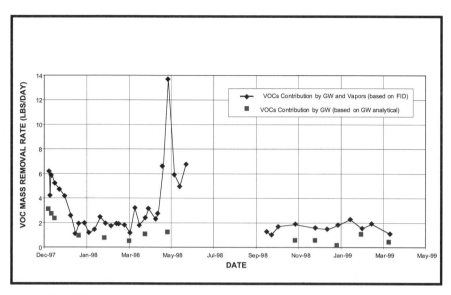

Figure 28   VOC mass removal rates.

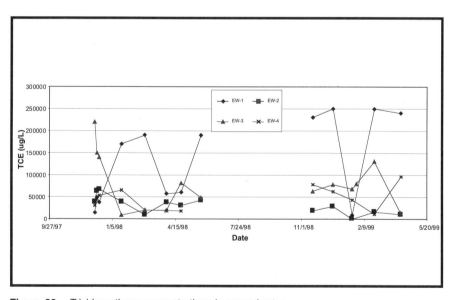

Figure 29   Trichloroethene concentrations in groundwater.

system is approximately four times larger than the one created using conventional pumping. In addition, air movement provided by vacuum-enhanced recovery accounted for removal of contaminant mass from the adsorbed phase in the vadose and in the dewatered zones.

CASE STUDY

Background

During 1995 and 1996, an aerospace research and development and manufacturing facility was demolished in preparation for residential redevelopment. The 100 acre property was located in southern California, and the regional board was responsible for authorizing cleanup requirements and approving closure.

Numerous soil and groundwater investigations had been completed prior to 1992. The results of the investigations identified an area of the property where the groundwater was impacted with CVOCs. The primary constituent was TCE.

The portion of the property with impacted groundwater was approximately 8 acres. This area was underlain by approximately 50 feet of silty sands and sands with silts, which in turn were underlain by bedrock deposits, predominantly semi-consolidated silts and clays. The primary source of groundwater was cultural, including leakage from subsurface utilities and irrigation. Up to 15 feet of groundwater accumulated on top of the interface between the sands and the underlying bedrock.

Strategy

A strategy was developed which expedited groundwater remediation and allowed the property to be redeveloped in 1997. The strategy was based on selecting the best available technology (BAT) to remediate groundwater where the total CVOC concentration exceeded 500 ug/l. The 500 ug/l figure was selected based upon risk assessment analysis that showed that the present concentrations posed no risk to future houses built over the area. However, the 8 acres of contamination were serving as a source area for a plume that had migrated off-site, and that, in the future, might reach an ocean bay. The regional board was most interested in removing the main mass in this area so that it would stop acting as a source area for the plume. The objective was to remove approximately six saturated pore volumes of groundwater and approximately 200 vadose zone pore volumes during six months. Using groundwater and air as carriers, the goal was to remove the majority of CVOCs, thereby minimizing further migration.

The BAT for this property was vacuum-enhanced recovery for groundwater dewatering and soil vapor recovery. This technology was selected because dewatering could rapidly be achieved. Vacuum-enhanced recovery could then rapidly remove CVOCs in the dewatered zone. Other technologies that were evaluated included pump and treat, air sparging, and enhanced biological degradation. None of these other technologies could achieve the mass removal in the required short period of time.

Project Design

Groundwater extraction and vacuum-enhanced recovery was performed using 38 recovery wells and a two pump system. The optimal spacing between wells was established to be approximately 80 feet during a two day pilot test and subsequent modeling of the pilot data. Individual groundwater delivery lines from extraction wells were connected to a common groundwater extraction header that delivered extracted water to the groundwater treatment system (granular-activated carbon). The treated water was then discharged to a National Pollutant Discharge Elimination System (NPDES) permitted outfall. Individual vacuum-enhanced recovery lines were connected to a common header that delivered extracted vapors to the treatment system (granular activated carbon).

Operations

Prior to pumping groundwater, the thickness of the saturated zone ranged from 8 to 13 feet. At the beginning of the program, the groundwater system extracted and treated approximately 25 gpm from the 38 wells. After the system was sealed and 60 inches (water) of vacuum was applied to the system, groundwater recovery rates increased almost three-fold to 70 gpm. At the end of the program, the average groundwater thickness was 3 feet (Figure 30) and the flow rate declined to approximately 14 gpm. The saturated thickness was 0.5 feet in the center of the dewatered area. This was also the area of highest concentration of CVOCs. Based on these measurements, the zone of impacted groundwater was approximately 70 percent dewatered, and the water pumping was able to exchange approximately 5.7 saturated pore volumes over the eight month operation period.

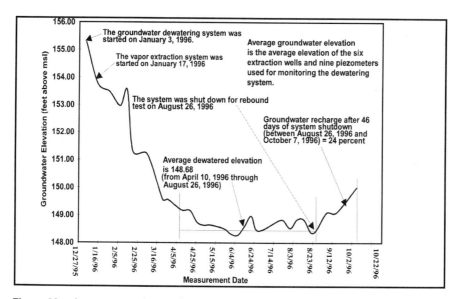

Figure 30   Average groundwater elevation vs. time.

Between start up of the remediation system and when the system was shut down, the combined groundwater CVOC concentrations declined from 685 ug/l to 248 ug/l (Figure 31). The initial rate of CVOC extraction from the saturated zone was approximately 0.5 pound per day. At the end of the program, the rate was 0.04

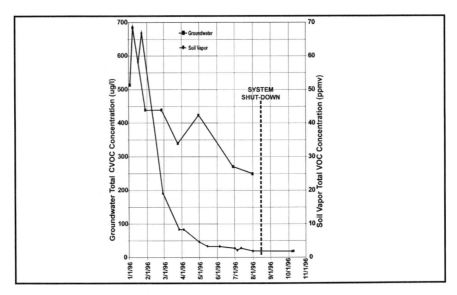

Figure 31   Total CVOC concentrations.

pounds per day. The cumulative amount of total CVOCs removed via groundwater extraction was 32 pounds (Figure 32). Confirmation data demonstrated that the CVOC concentrations did not rebound once the system was shut down; in fact, the concentrations continued to decrease.

The soil vapor combined flow rate ranged between 430 and 742 cfm. The vacuum at the wells ranged between 45 and 65 inches of water that declined as the saturated

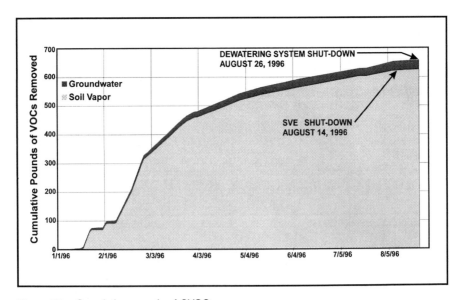

Figure 32   Cumulative pounds of CVOC.

zone was dewatered. Approximately 350 vadose zone pore volumes were extracted, which exceeded the 200 pore volume target.

Between start up of the vacuum-enhanced recovery system and shut down, the combined CVOC concentrations declined from 67 ppmv (based on volume) to 2 ppmv, Figure 31. The CVOC concentrations reached asymptotic conditions and did not increase once the system was shut down. The CVOC extraction rate ranged from 14.7 pounds per day to 0.8 pounds per day and the cumulative recovery was approximately 622 pounds (Figure 32).

The remediation program used vacuum-enhanced recovery to remove approximately six saturated pore volumes and 367 air filled pore volumes and the CVOCs were reduced to asymptotic concentrations in eight months. Because the technology-based goal was achieved and future groundwater contamination source was eliminated, the regulators granted project closure and property redevelopment was initiated.

CASE STUDY

Background

The site is a former residential property covering approximately two acres in southern New Hampshire. It is in a forested rural residential neighborhood. The site is underlain by 20 to 30 feet of relatively permeable glacial till (upper overburden), below which is a similar thickness of very dense lodgment till (lower overburden). The depth to the uppermost water-bearing zone is approximately 5 to 10 feet below land surface: shallower in the spring, deeper in the fall. The hydraulic conductivity of the upper till is on the order of 10^{-3} to 10^{-4} cm/sec. The hydraulic conductivity of the lower till is on the order of 10^{-6} cm/sec.

The site was discovered in mid 1980 with 400 drums of mismanaged chemicals. The groundwater contained CVOCs (TCE, 1,2-DCE, PCE), and VOCs (BTEX and ketones). These compounds were individually present up to single and double digit parts per million concentrations. Source removal of the shallow vadose zone contaminated soils was performed. Three separate plumes of compounds required remediation.

Strategy

The remedial strategy involved extracting groundwater and allowing air to be used as the primary carrier to maximize VOC removal and to minimize extracted groundwater during the short-term, active, mass removal, phase of the project. The site was capped to minimize infiltration of rainwater and to expand the influence of the air removal system. The active phase extraction was only to be performed while such removal was most cost effective, that is, until the mass removal versus time curves became asymptotic. At that point, natural attenuation would be relied upon to reach cleanup levels for the residuals.

Project Design and Installation

In spring 1995, an asphalt cap was installed over the entire two acres of the site. The vacuum-enhanced recovery system was installed soon thereafter. The vacuum-enhanced recovery system included two skid-mounted, 5 hp, liquid-ring pumps. The liquid-ring pumps remove both air and water from the subsurface. The extracted air and water from the wells were manifolded near the 10 foot x 10 foot treatment plant area. After the knock-out tank, the vapor and liquid phases were then passed through separate granular activated carbon units. The vacuum-enhanced recovery pump and treatment system was installed in a central location of the site so that it could affect all three treatment cells with the minimum of hardware. Due to the small size of the contaminated area, the pilot plant was expanded to the full-scale system.

Based on prior data, one upper overburden extraction well was installed in May 1995 at a location inside the fenced portion of the site, forming Treatment Cell 1 for the overburden aquifer pilot test. In addition, five piezometers and five vacuum probes were installed near the extraction well to monitor the impact of Treatment Cell 1 on groundwater and the vadose zone. In August 1995, two additional extraction wells were installed at locations northeast of the existing extraction well to expand the area of influence of Treatment Cell 1. Three vacuum probes were also installed near these new wells to monitor the effectiveness of the expanded treatment cell.

In August 1995, two new treatment cells were created at the site (see Figure 33). The locations of these cells were selected based on the results of a microwell investigation that was conducted at the site, which refined the delineation of the three areas of impacted groundwater.

In July and September 1996, two additional extraction wells were installed within Treatment Cell 1 to expedite the remediation of groundwater in the Treatment Cell

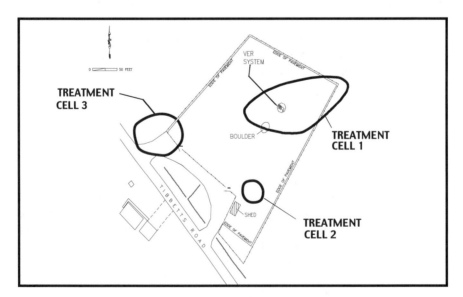

Figure 33   Treatment cell layout plan.

1. These wells were installed because the mass removal from one of the extraction wells was not declining as rapidly as was desired. Two vacuum probes were installed near these new wells to monitor the effectiveness of the expanded treatment cell.

Operations

Vacuum-enhanced recovery was initiated as a pilot test in spring 1995, and expanded and operated through 1997. Depending upon the maintenance being performed, up to seven extraction wells were operated at any given time. Groundwater was extracted from the individual extraction wells at an initial rate of about 2 gpm, declining to a sustained rate of about 0.5 gpm, under vacuum-enhanced conditions. Typical water level depressions produced by the system are shown in Figure 34.

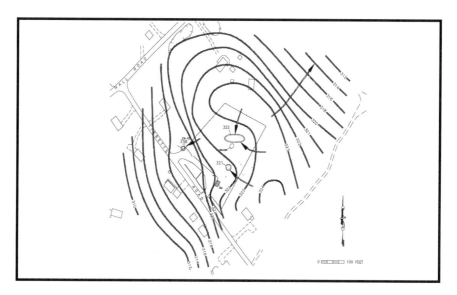

Figure 34   Overburden water levels, December 1996.

Vacuum applied at the wellheads was typically 10 to 15 inches Hg. Airflow was zero at spring start up when the well screens were filled with water, and then increased to about 50 scfm total (about 10 scfm from each well) as groundwater was removed and air was able to enter the wells. Air concentrations of the key compounds in the most contaminated single well initially totaled 30 ppm at start up, increasing to about 80 ppm, and declining again to about 20 ppm at the asymptote.

Figure 35 shows the removal of approximately 60 pounds of the key compounds in 1995. The mass removal rate peaked at 0.5 pound per day, dropping to about 0.1 pound per day late in the year. After a winter hiatus, vacuum-enhanced recovery was restarted in April 1996. During that treatment year, another 10 pounds of organic compounds were removed. The mass removal rate in 1996 began at 0.08 pounds per day, dropping to an asymptote of about 0.04 pounds per day late in the year.

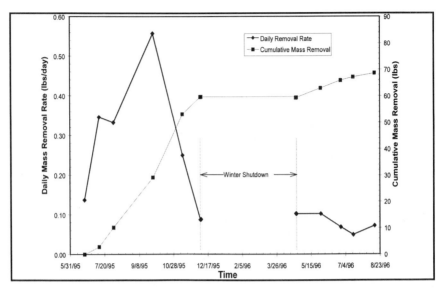

Figure 35   VOC mass removal: all cells combined.

The last year of operation of the vacuum-enhanced recovery was 1997. It was decided after 1997 that most of the mass had been removed. The system was shut down and the second phase of treatment begun.

DESIGN EXAMPLE

Problem

A former manufacturing facility has been impacted from past material handling and storage practices. Site investigation results indicate that both soil and groundwater have been impacted by petroleum hydrocarbon compounds and to a lesser extent by chlorinated VOCs. The constituents of interest identified in the shallow groundwater are benzene, toluene, ethyl-benzene, xylenes, and to a lesser extent tetrachloroethene and trichloroethene, as depicted on Figure 36. The site is approximately 1.5 acres in size. The groundwater plume is restricted to the site and underlies about a third of the site. There is no surface seal and the site is overgrown with weeds. The upper surficial sediments are approximately 20 feet thick that consist of silty sands grading to fine sands. The depth to water is 10 feet below land surface. From aquifer slug tests, the average hydraulic conductivity of these soils is approximately 5×10^{-4} cm/s. These sediments are underlain by a clay confining layer.

Solution

To evaluate the potential effectiveness of the application of vacuum-enhanced recovery, several steps are considered. The first step is to evaluate whether the

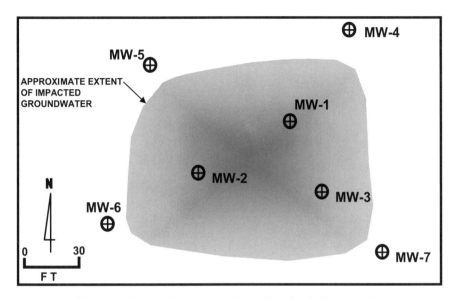

Figure 36   Site plan with groundwater contaminant plume for design example.

contaminant mass will be removed by the water phase only or whether the constituents are amenable to removal or biodegradation with the air phase. The contaminants identified are either volatile as indicated from Table 3 and/or biodegradable. These characteristics indicate that vacuum-enhanced recovery may be effective in not only increasing the well yield, but also capable of removing/degrading mass using the air phase.

The second step is to evaluate whether the geologic conditions are acceptable for vacuum-enhanced recovery. Generally, low transmissivity is a requirement (<500 gpd/ft) to develop a vacuum of sufficient magnitude to have an effect at reasonable airflow rates. The geological system generally needs to exhibit permeability in the range of 1×10^{-3} to 1×10^{-6} cm/sec. Preliminary data suggest that the permeability is within the acceptable range. Expected water and airflow rates are then calculated using Equation (7).

$$Q = \frac{s'T}{264}\log^{-1}\left[\frac{0.3tT}{r^2 S}\right] \tag{1}$$

where Q = discharge, in gpm; s' = effective drawdown corrected for dewatering, in ft; T = transmissivity, in gpd/ft; t = pumping time, in days; r = bore hole radius or 0.25 ft; and S = storage coefficient.

To correct for dewatering

$$s' = s_a - \frac{s_a^2}{2b_w} \tag{2}$$

Table 3   Henry's Law Constants

	Compound	Henry's Law Constant[a] atm	Reference
1	Acetone	0	1
2	Benzene	230	1
3	Bromodichloromethane	127	1
4	Bromoform	35	3
5	Carbon tetrachloride	1282	1
6	Chlorobenzene	145	2
7	Chloroform	171	1
8	2-Chlorophenol	0.93	2
9	p-Dichlorobenzene (1,4)	104	4
10	1,1-Dichloroethane	240	1
11	1,2-Dichloroethane	51	1
12	1,1-Dichloroethylene	1841	1
13	Cis-1,2-Dichloroethylene	160	1
14	Trans-1,2-Dichloroethylene	429	1
15	Ethylbenzene	359	1
16	Hexachlorobenzene	37.8	2
17	Methylene chloride	89	1
18	Methylethylketone	1.16	2
19	Methyl naphthalene	3.2	2
20	Methyl tert-butyl-ether	32.6	1
21	Naphthalene	20	4
22	Pentachlorophenol	0.15	2
23	Phenol	0.017	2
24	Tetrachloroethylene	1035	1
25	Toluene	217	1
26	1,1,1-Trichloroethane	390	1
27	1,1,2-Trichloroethane	41	2
28	Trichloroethylene	544	1
29	Vinyl chloride	355000	3
30	o-Xylene	266	1

[a] = at water temperature 68°F

1. Per Hydro Group, Inc., 1990.

2. Solubility and vapor phase pressure data from *Handbook of Environmental Data on Organic Chemicals.* 2nd edition, by Karel Verschueren, Van Nostrand Reinhold, New York 1983.

3. Michael C. Kavanaugh and R. Rhodes Trussel, "Design of aeration towers to strip volatile contaminants from drinking water"" *Journal AWWA*, December 1980. p. 685.

4. Coskun Yurteri, David F. Ryan, John J. Callow, Mirat D. Gurol, "The effect of chemical composition of water on Henry's law constant", *Journal WPCE*, Volume 59, Number 11, p. 954, November 1987.

where s_a = actual drawdown; and b_w = saturated zone thickness.

Assumptions made based on our hydrogeologic setting are as follows:

Hydraulic conductivity, $K = 5 \times 10^{-4}$ cm/sec = 8.0 gpd/ft^2

b_w = 10 ft for saturated zone

100 percent efficient well (both air and water)

$T = Kb_w = 80$ gpd/ft

$$s' = (10 \text{ ft}) - \frac{(10 \text{ ft})^2}{2(10 \text{ ft})}$$

$$s' = 5 \text{ feet}$$

This is the effective available drawdown if no vacuum is applied.

Now, assume a liquid-ring pump will be used to apply a vacuum of 18 inches of mercury at the extraction well head. Converting vacuum in inches of mercury to inches of water

$$H_{water} = H_{mercury} \frac{SG_{mercury}}{SG_{water}} \qquad (15)$$

where H_{water} = applied vacuum in inches of water; $H_{mercury}$ = applied vacuum in inches of mercury; $SG_{mercury}$ = specific gravity of mercury = 13.55; and SG_{water} = specific gravity of water = 1.

$$H_{water} = (18 \text{ in}) \frac{(13.55)}{(1)}$$

$$H_{water} = 243 \text{ inches of } H_2O$$

The total available drawdown for the well with this vacuum applied is calculated as follows

$$S_{available} = s' + H\left(\frac{1 \text{ ft}}{12 \text{ in}}\right)$$

$$S_{available} = 5 \text{ ft} + (243 \text{ in})\left(\frac{1 \text{ ft}}{12 \text{ in}}\right)$$

$$S_{available} = 25.3 \text{ ft}$$

Calculating a pumping rate (Q) using vacuum-enhanced extraction using $T = 80$ gpd/ft, $t = 30$ days, $S = 0.005$, $r = 0.25$ ft and using $s_{available}$ for s'

$$Q = \frac{S_{available}T}{264} \log^{-1}\left[\frac{0.3tT}{r^2S}\right]$$

$$Q = \frac{(25.3 \text{ ft})(80 \text{ gpd/ft})}{264} \log^{-1}\left[\frac{0.3(30 \text{ days})(80 \text{ gpd/ft})}{(0.25 \text{ ft})^2(0.005)}\right] \qquad (1)$$

$$Q = 1.2 \text{ gpm}$$

This preliminary result suggests that vacuum-enhanced recovery is suitable for this application.

The next step is to estimate the expected airflow rates. The airflow rates can be estimated in a similar manner using several formulas.

Gas conductivity is calculated as follows

$$K_g = K_w \frac{P_g \mu_w}{P_w \mu_g} \tag{3}$$

where K_g = gas conductivity (ft/day); K_w = hydraulic conductivity (ft/day); P_g = gas density (0.0013 g/cm³ at 68°F); P_w = water density (1 g/cm³ at 68°F); μ_g = gas viscosity (183 micropoise at 68°F); and μ_w = water viscosity (10,000 micropoise at 68°F).

Thus,

$$K_g = (8 \text{ gpd/ft}^2)\frac{(0.0013 \text{ g/cm}^3)(10,000 \mu p)}{(1 \text{ g/cm}^3)(183 \mu p)}$$

$$K_g = 0.57 \text{ gpd/ft}^2 \text{ or } 0.08 \text{ ft/day}$$

Finally,

$$T_g = K_g b_g \tag{4}$$

where b_g = 10 feet, the unsaturated zone thickness; T_g = (0.08 ft/day)(10 ft.); and T_g = 0.8 ft²/day.

To estimate the airflow, we can use the Theis equation after correcting the drawdown (vacuum) for gas expansion:

1. The gas expansion correction has the same form as the dewatering correction

$$s_{eff} = s_a - \frac{s_a^2}{2P_{atm}} \tag{5}$$

where s_{eff} = effective vacuum used in Theis equation; s_a = actual vacuum; and P_{atm} = atmospheric pressure (405 inches of water).

Since the available well head vacuum is 243 inches of water, the effective vacuum is

$$s_{eff} = (243 \text{ in}) - \frac{(243 \text{ in})^2}{2(405 \text{ in})}$$

$$s_{eff} = 170 \text{ inches of water}$$

2. The Theis equation is

$$s = \frac{528Q}{T}\log\frac{R}{r} \qquad (6)$$

where s = vacuum, in ft of air; Q = discharge, in cfm; T = transmissivity, in ft²/day; R = radius of influence, in ft; and r = bore hole radius, in ft.

By assuming the expected vacuum at the well and estimating air permeability and the expected radius of influence, the extraction rate is calculated. The radius of influence obviously is not known until the pilot studies are performed; however, the calculations are not highly sensitive to this parameter. Thus, using a value of 40 feet provides a good approximation. To express s in inches of water instead of feet of air, multiply the Theis equation by $12*P_g$

$$s = \frac{8.23Q}{T}\log\frac{R}{r}$$

Solving for Q,

$$Q = \frac{sT}{8.23\,\log\dfrac{R}{r}}$$

Using s = 170 inches of water, T = 0.8 ft²/day, R = 40 ft, and r = 0.25ft,

$$Q = \frac{(170\ \text{in})(0.8\ \text{ft}^2/\text{day})}{8.23\,\log\dfrac{(40\ \text{ft})}{(0.25\ \text{ft})}}$$

$$Q = 7.5\ \text{cfm}$$

Pilot Test Planning

The preliminary calculations indicate a single extraction point under an applied vacuum of 18 inches of mercury will produce approximately 1.2 gpm of groundwater and 7.5 cfm of airflow. As demonstrated in this example, the conditions appear favorable for applying vacuum-enhanced recovery. As a next step, a pilot study needs to be performed.

A review of the well construction details indicates that only one of the existing monitor wells can be used as an observation well for the pilot test. To have a good distribution of data collection points, more wells will be required. For this example, three additional observation wells will be installed at different distances and compass directions based on site conditions. Radial distribution should account for site heterogeneities and obtain the data necessary for full-scale design. Spacings of 5, 10, 20, and 40 feet away from the test recovery well are selected. The test extraction well will be 20 feet deep, extending to the top of the confining layer, with the lower 15 feet composed of slotted well screen. The well head assembly at the extraction well will include a drop tube for extracting both liquid and vapor, a vacuum gauge, and an air bleed valve. Each observation well will be constructed with well screen sections extending from 5 feet above the water table down to the bottom of the aquifer at 20 feet bls with a drop tube inside extending below the water table. This will allow monitoring of the water table drawdown (via the drop tube) and the induced vacuum in the vadose and dewatered zones. Figure 37 is a site plan depicting the extraction point and observation well locations.

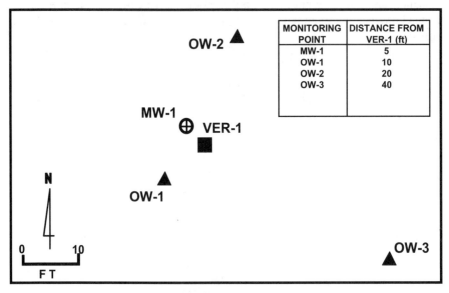

MONITORING POINT	DISTANCE FROM VER-1 (ft)
MW-1	5
OW-1	10
OW-2	20
OW-3	40

Figure 37   Pilot test layout for design example.

Finally, we need to select the type of extraction equipment that will be used. Because both soil and groundwater are impacted, combined liquid and vapor recovery is desired. Also, since the depth of groundwater to be recovered is less than 20 feet, a single vacuum pump will be used to withdraw both liquid and vapors from the extraction well drop tube. Based on the preliminary calculations, a vacuum pump capable of delivering 15 cfm at 20 inches of mercury vacuum will be used. A wide variety of portable pilot test units are readily available for this application. The selected pilot test unit was a portable two hp liquid-ring pump capable of delivering 22 cfm at 20 inches of mercury equipped with an air/water separator and a groundwater transfer pump. The air phase emissions were treated using a 1,000 pound granular activated carbon canister prior to atmospheric discharge.

Conducting the Pilot Test

The vacuum-enhanced recovery pilot test included groundwater withdrawals in two phases. First, a standard pump test was conducted for one hour to determine the quantity of groundwater that can be recovered without the influence of the vacuum system. This was accomplished by operating the vacuum pump with the extraction well bleed valve completely open. A groundwater extraction flow rate of 0.1 gpm was observed. The remainder of the pilot study was conducted over a nine-hour period. The radius of influence was evaluated by recording the change in vacuum readings and water levels at the observation wells. These monitoring parameters along with the pump inlet vacuum, extraction wellhead vacuum, and extraction flow rates (both air and liquid) were recorded at 5-minute intervals during the first 30 minutes of the test then at approximately 15 to 30 minute intervals thereafter. Air samples were collected from the vacuum pump discharge for analysis to estimate the mass of contaminant emissions during the pilot test for use in sizing the off-gas treatment system.

Evaluating the Data

The pilot test data indicated that vacuum-enhanced recovery created a ground-water capture zone and induced airflow through the unsaturated zone. Results of the pilot test are summarized below.

1. *Water Table Drawdown.* As shown on Figure 38, vacuum-enhanced pumping resulted in a maximum drawdown ranging from 0.3 to 2.25 feet bls in the four observation/monitoring wells over the nine hour test. The greatest drawdown was observed at MW-1 located within 5 feet of the extraction well. A steady drawdown was observed at MW-1.

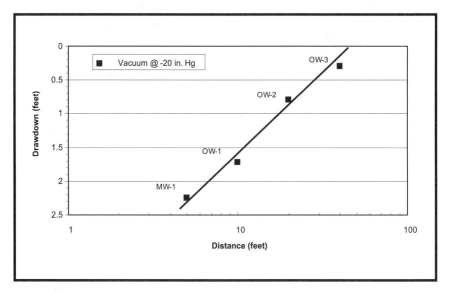

Figure 38 Distance versus drawdown from a vacuum-enhanced recovery pilot test.

2. *Airflow Rates.* The relationship between the well head vacuum and the observed formation airflow rates, which is the total airflow rate less the air bleed valve contribution, is illustrated on Figure 39. These data show that as the well head vacuum increased, the subsurface airflow increased from approximately 8 to 12

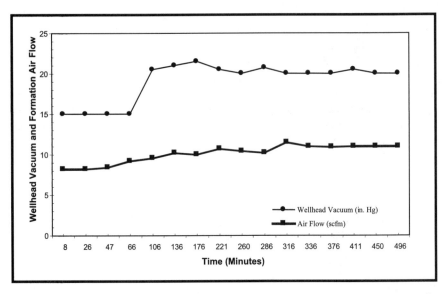

Figure 39 Observed vacuum influence and airflow during vacuum-enhanced recovery pilot test.

cfm. This increase is expected as the subsurface slowly becomes dewatered, creating a greater area for formation airflow.

3. *Groundwater Recovery Rates.* Under vacuum influence, the observed groundwater extraction rate remained relatively constant at 0.36 gpm throughout the test. As expected, this is greater than three times the recovery rate observed during the one-hour pump test under no induced vacuum.

System Design

The pilot test results will be used in designing the full-scale vacuum-enhanced recovery system. Because of the relatively narrow saturated zone of the surficial sediments, the intent of the vacuum extraction system is to dewater the saturated impacted zone using air as the primary mass removal mechanism. This will expedite the removal of adsorbed and dissolved phase contaminants from the soil pore spaces of the once saturated zone. The first step will be to calculate the influence area of the extraction well to estimate the appropriate quantity and spacing for the network of extraction wells. A plot of water table drawdown versus distance as shown in Figure 38 is used to calculate the actual transmissivity (T) and storage coefficient (S) of the aquifer using the Theis equation. It should be cautioned here that the transmissivity value that is calculated will be the maximum value that the system sees. With continued dewatering the saturated zone will decrease with an attendant reduction in transmissivity.

Once T and S of the aquifer are known, the radius of influence can be calculated. From the semi-logarithmic distance-drawdown plot presented in Figure 38, the slope of the best fit line represents a Δs, which was used to calculate the T and S values based on the pilot test data. The equations used below to calculated T and S are derived from the Theis modified nonequilibrium equation

$$T = \frac{528Q}{\Delta s} \tag{16}$$

where T = calculated transmissivity in gpd/ft; Q = pumping rate observed during the pilot test, in gpm; and Δs = the slope of the distance-drawdown curve, Figure 38.

$$S = \frac{0.3Tt}{r_0^2} \tag{17}$$

where T = calculated transmissivity in gpd/ft; t = time since pumping started, in days; and r_0 = intercept of straight line at zero drawdown, in feet.

To calculate Δs, the observed drawdown observed at distances of d_1=30 feet and d_2=7 feet from the vacuum extraction well was used as demonstrated below.

$$\Delta s = \frac{s_2 - s_1}{\log \dfrac{d_1}{d_2}} \tag{18}$$

where s_1 = measured drawdown of 0.40 feet at 30 feet from VER-1; and s_2 = measured drawdown of 2.0 feet at 7 feet from VER-1.

$$\Delta s = \frac{2.0 - 0.40}{\log \dfrac{30}{7}}$$

$$\Delta s = 2.22$$

Using this Δs value, transmissivity (T) coefficient is calculated below

$$T = \frac{528Q}{\Delta s} = \frac{528(0.36 \text{ gpm})}{2.22}$$

$$T = 86 \text{ gpd/ft}$$

Using the calculated T value, the storage coefficient S is calculated below

$$S = \frac{0.3Tt}{r_0^2} = \frac{0.3(86 \text{ gpd/ft}^2)(0.375 \text{ days})}{(45 \text{ ft})^2}$$

$$S = 0.0048$$

where t = 9 hours of pumping or 0.375 days; r_0 = the zero drawdown occurs at 45 feet as shown in Figure 38; and Q = the average flow rate measured during the pilot test was 0.36 gpm.

A comparison of the calculated T and S coefficients with those estimated during the pilot test planning revealed similar results. However, the anticipated flow rate of 1.2 gpm exceeded the actual observed flow rate of 0.36 gpm during the test. The extraction well was assumed to be 100 percent efficient to be conservative during pilot test planning. A comparison of the actual recovery rate to that calculated before the pilot test suggests that the extraction well is approximately 30 percent efficient, which is not unexpected for silty sand sediments present.

As shown on Figure 38, the distance drawdown plot suggests that the radius of influence of the vacuum-enhanced recovery system is approximately 40 feet based on a minimum drawdown of 0.1 feet. The radius of influence can also be calculated as shown below using the pump test data

$$R = \sqrt{\frac{0.3Tt}{S}} \tag{11}$$

where R = log-extrapolated radius of influence, in ft; T = transmissivity, in gpd/ft; t = pumping time, in days (the test duration of 9 hours or 0.375 days was used); and S = storage coefficient.

$$R = \sqrt{\frac{0.3(86 \text{ gpd/ft})(0.375)}{0.0048}}$$

$$R = 45 \text{ feet}$$

The calculated radius of influence is similar to that noted during the pilot test.

As previously mentioned in this chapter, the observed radius of influence from the pilot test can be used to determine the number and spacing of the extraction well network by placing the wells with overlapping influences areas. This extraction well configuration would provide a high degree of dewatering and capture of the contaminant plume. In many instances, however, this has lead to a more conservative design requiring more extraction wells and a larger capacity vacuum pump than needed. In addition, the increased vacuum pump capacity may result in the need for larger groundwater and air phase treatment equipment. The remediation professional must weigh these cost considerations with the overall remedial benefit.

Because of the complex hydrogeologic calculations involved and numerous variables, a computer model such as MODFLOW or QUICKFLOW was used to simulate the optimal capture zones and dewatering area of an extraction well network using the pilot test results. The layout for the full-scale vacuum-enhanced recovery system was determined using a simplified model based on a 2D, uniform thickness, flat lying, homogeneous, and confined aquifer. As shown on Figure 40, the modeling results predicted that five extraction wells, operating at a total extraction flow rate of 1.2 gpm (0.24 gpm per well), should result in dewatering and capture of the half

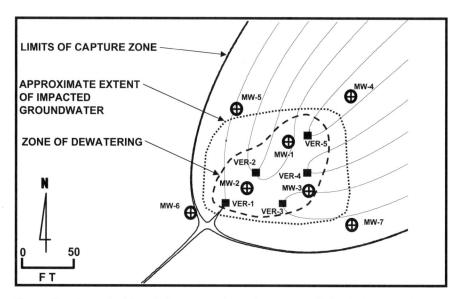

Figure 40   Conceptual layout of vacuum-enhanced recovery wells for design example.

acre groundwater contaminant plume. In addition, the modeled capture zone based on the contribution of the five extraction wells extended laterally a distance of approximately 200 feet from the center of the plume.

Based on a design well head vacuum of 20 inches of mercury and a total extraction flow rate of 60 cfm (five wells at 12 cfm per well), a 7.5 hp liquid-ring vacuum pump would be suitable for this scenario. A 0.5 hp centrifugal pump would be suitable for transferring the recovered groundwater from the air/water separator to the groundwater treatment system. Because off-gas treatment for vacuum-enhanced recovery systems can be costly, the design professional should evaluate the air emissions treatment requirements prior to selecting the full-scale vacuum pump. Discussion of off-gas treatment technologies is provided in Chapter 6.

The design professional should conduct a lifecycle cost analysis prior to system design. Trade offs between the quantity of extraction wells and extraction flow rates can be evaluated to adjust the project costs and cleanup duration to meet the project's goals and objectives.

REFERENCES

Blake, S. B. and Gates, M.M., "Vacuum-Enhanced Recovery: A Case Study," Proceedings of Petroleum Hydrocarbons and Organic Chemicals in Groundwater: Prevention, Detection and Restoration. NWWA/AP I., Houston, Texas, 1986.

Blake, S.B. and Hall, R. A., "Monitoring Petroleum Spills with Wells: Some Problems and Solutions," Proceedings of Fourth National Symposium on Aquifer Restoration and Ground Water Monitoring, NWWA, 305-310, 1984.

Cooper, H.H., Jr. and Jacob, C.E., "A Generalized Graphical Method for Evaluating Formation Constants and Summarizing Well Field History," Trans. Am. Geophysical Union, Vol. 27, No. 4, Washington, 1946.

Driscoll, F. G., *Groundwater and Wells*, 2nd Edition, Johnson Div., 1986.

Powers, J.P., "Construction Dewatering" A guide to theory and practice. Wiley and Sons, Inc., 1981.

Schafer, D., Internal Memorandum on Capture Zones in Low-Permeability Formations, Geraghty & Miller Inc., 1995.

CHAPTER 5

In Situ Air Sparging

James M. Bedessem

CONTENTS

1-56670-528-2/01/$0.00+$.50

INTRODUCTION

In situ air sparging is a remediation technique which has been used since about 1985, with varying success, for the remediation of volatile organic compounds (VOCs) dissolved in the groundwater, sorbed to the saturated zone soils, and trapped in soil pores of the saturated zone. This technology is often used in conjunction with vacuum extraction systems (Figure 1) to remove the stripped contaminants, and has broad appeal due to its projected low costs relative to conventional approaches.

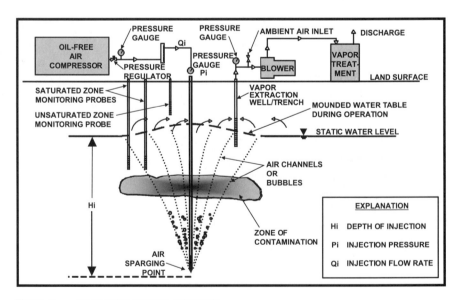

Figure 1   Air sparging process schematic.

The difficulties encountered in modeling and monitoring the multiphase air sparging process (i.e., air injection into water saturated conditions) have contributed to the current uncertainties regarding the processes responsible for removing the contaminants from the saturated zone. Engineering design of these systems, even today, is largely dependent on empirical knowledge. At this point, the air sparging process should be treated as a rapidly evolving technology with a need for continuous refinement of optimal system design and mass transfer efficiencies. The mass transfer mechanisms during *in situ* air sparging relies on the interactions between complex physical, chemical, and microbial processes, many of which are not well understood.

A typical air sparging system has one or more subsurface points through which air is injected into the saturated zone. When this technology was first emerging, it was commonly perceived that the injected air traveled up through the saturated zone in the form of air bubbles (Angell 1992, Brown 1992a, Brown 1992b, and Sellers and Schreiber 1992); however, it is more realistic that the air travels in the form of continuous air channels (Johnson et al. 1993, Wei et al. 1993, and Ardito and Billings 1990). While the airflow path will be influenced by the pressure and flow rate of injected air and depth of injection, the structuring and stratification of the saturated zone soils appear to be the predominant factors (Johnson et al. 1993, Wei et al. 1993, and Ardito and Billings 1990). Significant channeling may result from relatively subtle permeability changes, and the degree of channeling will increase as the size of the pore throats get smaller. Research (Wei et al. 1993) shows that even minor differences in permeability due to stratification can impact the sparging effectiveness.

In addition to conventional air sparging where air is injected as shown in Figure 1, many modifications of the technique to overcome geologic/hydrogeologic limitations will also be discussed in this chapter.

GOVERNING PHENOMENA

In situ air sparging is potentially applicable when volatile and/or easily aerobically biodegradable organic contaminants are present in water-saturated zones, under relatively permeable conditions. The *in situ* air sparging process can be defined as injection of compressed air at controlled pressures and volumes into water-saturated soils. The three primary contaminant mass removal mechanisms that occur during the operation of air sparging systems include, (1) *in situ* stripping of dissolved VOCs; (2) volatilization of trapped and adsorbed phase contamination present below the water table and in the capillary fringe; and (3) aerobic biodegradation of both dissolved and adsorbed phase contaminants resulting from delivery of oxygen.

It was determined during *in situ* air sparging of petroleum hydrocarbon sites that stripping and volatilization account for much more of the hydrocarbon mass removal in the initial weeks/months of operation than does biodegradation. Biodegradation becomes more significant for mass removal only during long-term system operation.

In Situ Air Stripping

Among the three contaminant mass removal mechanisms mentioned above, *in situ* air stripping may be the dominant process for some dissolved contaminants. The ability of a dissolved contaminant to be removed by air sparging through a stripping mechanism is a function of its Henry's Law constant (vapor pressure/solubility). Compounds such as benzene, toluene, xylene, ethylbenzene, trichloroethylene, and tetrachloroethylene are considered to be very easily strippable (see Chapter 3 for a discussion of Henry's Law constants). However, a basic assumption made in analyzing the air stripping phenomenon during air sparging is that Henry's Law applies to the volatile contaminants, and that all the contaminated water is in close communication with the injected air. In depth evaluation of these assumptions exposes the shortcomings and complexities of interphase mass transfer during air sparging.

First of all, Henry's Law is valid only when partitioning of dissolved contaminant mass has reached equilibrium at the air/water interface. However, the residence time of air, traveling in discrete channels, may be too short to achieve the equilibrium due to the high air velocities and short travel paths. Another issue is the validity of the assumption that the contaminant concentration at the air/water interface is the same as in the bulk water mass. Due to the removal of contaminants in the immediate vicinity of the air channels, it is safer to assume that contaminant concentrations are going to be lower immediately around the channels than away from the channels. To replenish the mass lost from the water around the air channels, mass transfer by diffusion and convection must occur from water away from the air channels. Therefore, it is likely that the density of air channels plays a significant role in mass removal and that mass transfer efficiencies increase as the distance between air channels decreases. In addition, the density of air channels will also influence the interfacial surface area available for mass transfer. This mass transfer limitation may also prevent this technology from reaching final cleanup criteria as discussed in Chapters 1 and 2.

The literature suggests that the air channels formed during air sparging mimic a viscous fingering effect, and that two types of air channels are formed: large-scale channels and pore-scale channels (Clayton, Brown, and Bass 1995). The formation of both types of channels enhances the channel density and the available interfacial surface area.

It has been proposed that *in situ* air sparging also helps to increase the rate of dissolution of the sorbed phase contamination and eventual stripping below the water table. This is due to the enhanced dissolution caused by increased mixing, and the higher concentration gradient between the sorbed and dissolved phases under sparging conditions.

Direct Volatilization

The primary mass removal mechanism for VOCs present in the saturated zone during pump and treat operations is resolubilization into the aqueous phase and the eventual removal with the extracted groundwater. During *in situ* air sparging, direct volatilization of the sorbed and trapped contaminants is enhanced in the zones where airflow takes place. The volatile compounds do not have to transfer through the water to reach the air. If an air channel intersects pure compound, direct volatilization can occur. Direct volatilization of any compound is governed by its vapor pressure. Most volatile organic compounds are easily removed through volatilization. The schematic presented in Figure 1 includes air channels or bubbles moving through an aquifer containing sorbed or trapped NAPL contamination. In the regions where the soil is predominantly air saturated or the air channel is next to the zone of trapped contamination, the process is similar to soil vapor extraction or bioventing, albeit on a microscopic scale.

Where significant levels of residual contamination of VOCs or NAPLs are present in the saturated zone, direct volatilization into the vapor phase may become the dominant mechanism for mass removal where air is flowing. The high level of mass that the air can carry, combined with the rapid exchange of pore volumes, results in a process that can remove significant pounds of contaminants in a relatively short period of time. This may explain the significant increase in VOC concentrations typically observed in the soil vapor extraction effluents at many sites (Geraghty & Miller 1995).

Biodegradation

In most natural situations, aerobic biodegradation of biodegradable compounds in the saturated zone is rate limited by the availability of oxygen. Biodegradability of any compound under aerobic conditions is dependent on its chemical structure and environmental parameters such as pH and temperature. Some VOCs are considered to be easily biodegradable under aerobic conditions (e.g., benzene, toluene, acetone, etc.) and some of them are not (e.g., trichloroethylene and tetrachloroethylene).

Typical dissolved oxygen (DO) concentrations in uncontaminated groundwater are less than 4.0 mg/l, and under anaerobic conditions induced by the natural degradation of the contaminants, are often less than 0.5 mg/l. DO levels can be raised by air sparging up to 6 to 10 mg/l under equilibrium conditions (Brown 1992a, Geraghty & Miller 1995, and Brown, Herman, and Henry 1991). An increase in the DO level will contribute to enhanced rates of aerobic biodegradation in the saturated zone. This method of introducing oxygen to increase the DO level is one of the inherent advantages of *in situ* air sparging. However, the oxygen transfer into the bulk water is a diffusion limited process. The diffusion path lengths for transport of oxygen through the groundwater are defined by the distances between air channels. Where channel spacing is large, diffusion alone is not sufficient to transport oxygen into all areas of the aquifer for enhanced biodegradation. The pore-scale channels formed and the induced mixing during air sparging enhance the rate of oxygen transfer (Clayton, Brown, and Bass 1995). The specific costs and methodology of enhanced biodegradation will be discussed in Chapters 7 and 8.

APPLICABILITY

Examples of Contaminant Applicability

Based on the discussion in the previous section, Table 1 describes the applicability of air sparging for a few selected contaminants in terms of the contaminant properties of strippability, volatility, and aerobic biodegradability. For air sparging to be effective, the VOCs must transfer from the groundwater or from the saturated zone into the injected air, and oxygen present in the injected air must transfer into the groundwater to stimulate biodegradation.

Table 1   A Few Examples of Contaminant Applicability for In Situ Air Sparging

Contaminant	Strippability	Volatility	Aerobic* Biodegradability
Benzene	High (H = 5.5 x 10^{-3})	High (V$_P$ = 95.2)	High (t$_{1/2}$ = 240)
Toluene	High (H = 6.6 x 10^{-3})	High (V$_P$ = 28.4)	High (t$_{1/2}$ = 168)
Xylenes	High (H = 5.1 x 10^{-3})	High (V$_P$ = 6.6)	High (t$_{1/2}$ = 336)
Ethylbenzene	High (H = 8.7 x 10^{-3})	High (V$_P$ = 9.5)	High (t$_{1/2}$ = 144)
TCE	High (H = 10.0 x 10^{-3})	High (V$_P$ = 60)	Very low (t$_{1/2}$ = 7,704)
PCE	High (H = 8.3 x 10^{-3})	High (V$_P$ = 14.3)	Very low (t$_{1/2}$ = 8,640)
Gasoline Constituents	High	High	High
Fuel Oil Constituents	Low	Very low	Moderate

where H = Henry's Law constant (atm-m^3/mol); V$_P$ = Vapor pressure (mm Hg) at 20°C; t$_{1/2}$ = Half life during aerobic biodegradation, hours; and * = It should be noted that the half lives can be very dependent on the site specific subsurface environmental conditions.

In practice, the criterion for defining strippability is based on the Henry's Law constant being greater than 1×10^{-5} atm-m^3/mole. In general, compounds with a vapor pressure greater than 0.5 to 1.0 mm Hg can be volatilized easily; however, the degree of volatilization is limited by the flow rate of air. The half lives presented in Table 1 are estimates in groundwater under natural conditions without any enhancements to improve the rate of degradation (enhancements are discussed in Chapter 8, Reactive Zone Remediation).

Many constituents present in heavier petroleum products such as No. 6 fuel oil will not be amenable to either stripping or volatilization (Figure 2). Hence, the primary mode of remediation, if successful, will be due to aerobic biodegradation. Required air injection rates under such conditions will be influenced only by the requirement to introduce sufficient oxygen into the saturated zone.

Figure 2 qualitatively describes different mass removal phenomena in a simplified version under optimum field conditions. The amounts of mass removed by stripping and volatilization have been grouped together, due to the difficulty in separating them in a meaningful manner. However, the emphasis should be placed on total mass removal, particularly of mobile volatile constituents, and closure of the site regardless of the mass transfer mechanisms.

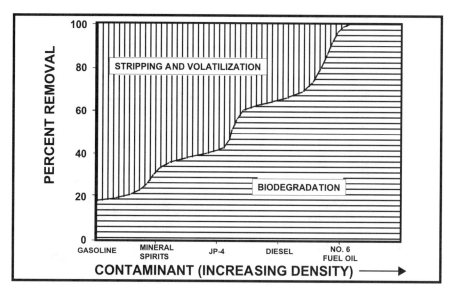

Figure 2   Qualitative presentation of potential air sparging mass removal for petroleum compounds.

Geological Considerations

Successful implementation of *in situ* air sparging is greatly influenced by the ability to achieve significant air distribution within the target zone. Good vertical pneumatic conductivity is essential to avoid bypassing or channeling of injected air horizontally, away from the sparge point. It is not an easy task to evaluate the pneumatic conductivities in the horizontal and vertical direction for every site considered for *in situ* air sparging.

Geologic characteristics of a site are important when considering the applicability of *in situ* air sparging. The most important geologic characteristic is stratiographic homogeneity or heterogeneity. The presence of lower permeability layers under stratified geologic conditions will impede the vertical passage of injected air. Laboratory-scale studies have illustrated the impact of geologic characteristics on air channel distribution (Wei et al. 1993). Under laboratory conditions, injected air was shown to accumulate below the lower permeability layers and travel in a horizontal direction. In field application, this condition may have the potential to enlarge the contaminant plume (Figure 3). High permeability layers may also cause the air to preferentially travel laterally, again potentially causing an enlargement of the plume (Figure 3). Horizontal migration of injected air limits the volume of soils that can be treated by direct volatilization due to the inability to capture the stripped contaminants. Horizontal migration can also cause safety hazards if hydrocarbon vapors migrate into confined spaces such as basements and utilities. Hence, homogeneous geologic conditions are essential for the success and safety of *in situ* air sparging.

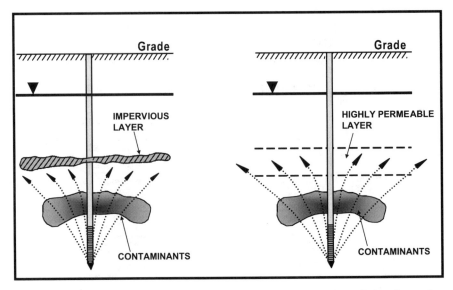

Figure 3   Potential situations for the enlargement of a containment plume during air sparging.

Both vertical pneumatic conductivity and the ratio of vertical to horizontal permeability decrease with decreasing average particle size of the sediments in the saturated zone. The reduction of vertical permeability is directly proportional to the effective porosity and average grain size of the sediments (Bohler, Hotzl, and Nahold 1990). Hence, based on the empirical information available, it is recommended that application of *in situ* air sparging be limited to saturated zone conditions where the hydraulic conductivities are greater than 10^{-3} cm/sec (Johnson et al. 1993, and USEPA 1993).

It is unlikely that homogeneous geologic conditions across the entire cross section will be encountered at most sites. The optimum geologic conditions for air sparging may be where the permeability increases with increasing elevation above the point of air injection. Decreasing permeabilities with elevation above the point of air injection will have the potential to enlarge the plume due to lateral movement of injected air.

DESCRIPTION OF THE PROCESS

Air Injection into Water-Saturated Soils

The ability to predict the performance of air sparging systems is limited by the current understanding of airflow in the water-saturated zone and limited performance data. There are two schools of thought in the literature describing this phenomenon. The first, and the widely accepted one, describes that the injected air travels in the vertical direction in the form of discrete air channels. The second describes that the injected air travels in the form of air bubbles. Airflow mechanisms cannot be directly observed in the field; however, conclusions can be reached by circumstantial evidence collected at various sites and from laboratory-scale visualization studies.

Sandbox model studies performed (Wei et al. 1993 and Johnson 1995) tend to favor the air channels concept over the air bubbles concept. In laboratory studies simulating sandy aquifers (grain sizes of 0.075 to 2 mm) stable air channels were established in the medium at low injection rates, whereas, under conditions simulating coarse gravel (grain sizes of 2 mm or larger), the injected air rose in the form of bubbles. At high air injection rates in sandy, shallow, water table aquifers, the possibility for fluidization (loss of soil cohesion) around the point of injection exists (Johnson et al. 1993, and Johnson 1995), and thus the loss of control of the injected air may occur.

Mounding of Water Table

When air is injected into the saturated zone, groundwater must necessarily be displaced. The displacement of groundwater will have both a vertical and lateral component. The vertical component will cause a local rise in the water table, sometimes called water table mounding. Mounding has been used by some as an indicator of the radius of influence of the sparge well during the early stages of development of this technology (Brown 1992a, Brown 1992b, Brown, Herman, and Henry 1991, Kresge and Dacey 1991, and Boersma, Diontek, and Newman 1995). Mounding is also considered to be a design concern because it represents a driving force for lateral movement of groundwater and dissolved contaminants and can therefore lead to spreading of the plume. The magnitude of mounding depends on the site conditions and the location of the observation wells relative to the sparge well. Mounding can vary from a negligible amount to several feet in magnitude.

Simulations of the flow of air and water around an air sparging well were performed with a multiphase, multicomponent simulator (TETRAD) originally developed for the study of problems encountered during exploration of petroleum and geothermal resources (Lundegard and Anderson 1993, and Lundegard 1995). The simulations were performed by defining two primary phases of transient behavior that lead to a steady state flow pattern (Figures 4 and 5). The first phase is characterized by an expansion in the region of airflow (Figure 4). During this phase, the rate of air injection into the saturated zone exceeds the rate of airflow through the saturated zone into the vadose zone. It is during this transient expansion phase that groundwater mounding first develops and reaches its highest level. The groundwater mound during this phase extends from near the injection well to beyond the region of airflow in the saturated zone. When injected air breaks through to the vadose zone, the region of airflow in the saturated zone begins to collapse or shrink (Figure 5). During this second transient phase of behavior, the preferred pathways of higher air permeability from the point of injection to the vadose zone are established. The air distribution zone shrinks until the rate of air leakage to the vadose zone equals the rate of air injection. During this collapse phase, mounding near the sparge well dissipates. When steady state conditions are reached, little or no mounding exists. This behavioral pattern has also been observed in the field (Johnson et al. 1993, Boersma, Diontek, and Newman 1995, and Lundegard 1995). The timeframe over which these phases occur is dependent on geology. At Port Hueneme in

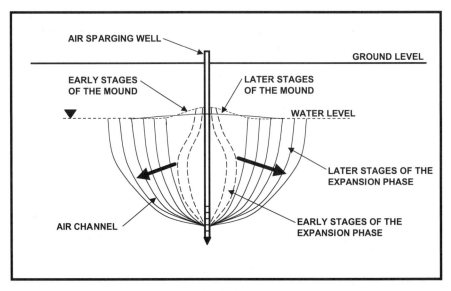

Figure 4   The first transient behavior after initiating injection into the saturated zone.

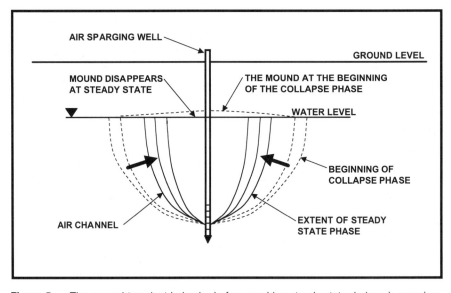

Figure 5   The second transient behavior before reaching steady state during air sparging.

a generally uniform, permeable aquifer without any apparent lenses between the injection point and the vadose zone, the injected air is able to travel to the vadose zone relatively rapidly and the groundwater mounding effects reach a steady state condition in a matter of a few hours as shown in Figure 6a (Leeson et al. 1999). Conversely, it is hypothesized that in conditions where a finer grained lens may occur between the sparge point and the vadose zone, causing more lateral movement of the injected air prior reaching the vadose zone, steady state conditions would

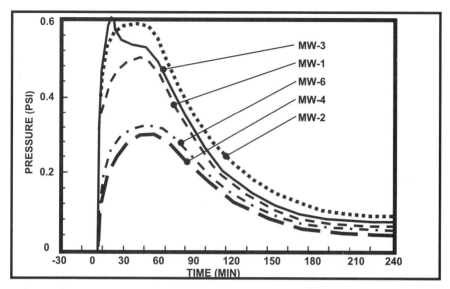

Figure 6a   Appearance and disappearance of groundwater mound during *in situ* air sparging as indicated by pressure transducer response at port hueneme site: indicative of uniform permeable conditions between sparge point and vadose zone.

require several days to achieve (Leeson 1999). An example of this type of behavior was observed during a pilot test at Hill Air Force Base and is portraited in Figure 6b (Leeson 1999). In both cases, the shape of groundwater elevation response curve generated to reflect the building and decay of the groundwater mound at the sparging location was similar.

The transience of groundwater mounding at most sites has important implications for the risk of lateral movement of the contaminant plume. Because the water table returns close to its presparging position during continuous air injection, the driving force for lateral movement of groundwater caused by air injection becomes very small.

An important aspect of groundwater mounding is that it is not a direct indicator of the physical presence of air in the saturated zone. Water table mounding at a given place and time may or may not be associated with the movement of air in the saturated zone at the same location. Some mounding will occur beyond the region of airflow in the saturated zone. Additionally, a transient pressure increase without water table mounding commonly occurs beyond the limits of airflow, especially where airflow is partially confined. Because of its transient nature and the fact that the water table is displaced ahead of injected air, water table mounding can be a misleading and overly optimistic indicator of the distribution of airflow within the saturated zone.

Distribution of Airflow Pathways

It is often envisioned that airflow pathways developed during air sparging form an inverted cone with the point of injection being the apex. This would be more

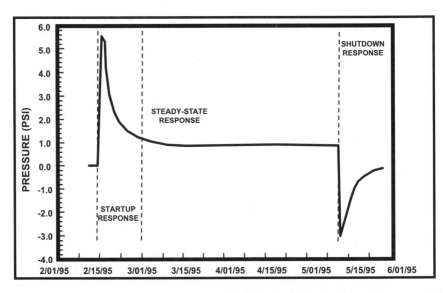

Figure 6b   Pressure transducer response at Hill AFB site: indicative of lateral movement of
air under a fine-grained layer between sparge point and vadose zone.

true if soils were perfectly homogeneous or composed of coarse grained sediments,
and injected airflow rate was low. During laboratory experiments using homogeneous media with uniform grain sizes, symmetrical airflow patterns about the
vertical axis were observed (Wei et al. 1993). However, media simulating mesos-
cale heterogeneities resulted in unsymmetrical airflow patterns (Wei et al. 1993).
The asymmetry apparently resulted from minor variations in the permeability and
capillary air entry resistance which resulted from pore scale heterogeneity. Hence,
under natural conditions, it is realistic to expect that symmetric air distribution
will never occur.

These same experiments also indicated that the channel density, and thus the
interfacial surface area, increased with increased airflow rates, since higher volumes
of air occupy an increased number of air channels. Assuming that the air channels
are cylindrical in shape and that the number of channels and air velocity in the
channel remains the same even for a change in airflow rate, the interfacial surface
area will change by a ratio $(Q_{final}/Q_{initial})^{0.5}$, where Q is the airflow rate.

It is reported in some literature (Brown 1992a) that at low sparge pressures, air
travels 1 to 2 feet horizontally for every foot of vertical travel. However, it has to
be noted that this correlation was not widely observed. It was also reported that as
the sparge pressure is increased, the degree of horizontal travel increases (Johnson
1995, Lundegard and Andersen 1993, and Brown 1992b). Field observations have
indicated that airflow channels extend 10 to 40 feet away from the air injection point,
independent of flow rate and depth of sparge point (Bohler, Holtz, and Nahold 1990,
Geraghty & Miller 1995, and Lundegard and Andersen 1993).

Groundwater Mixing

Mixing of groundwater during air sparging is an important mechanism to overcome the diffusion limitations of contaminant mass transfer out of and oxygen transport into the aquifer. Groundwater mixing during air sparging may significantly reduce the diffusion limitations of mass transfer without generating any changes in the bulk groundwater flow. It has been shown that nonsteady state mixing mechanisms induced in opposite directions at different times as a result of pulsed sparging operations will enhance mass removal efficiencies (Clayton, Brown, and Bass 1995, and Boersma, Diontek, and Newman 1995).

There are many possible mechanisms for groundwater mixing during air sparging (Clayton, Brown, and Bass 1995). Several possible mechanisms are listed below:

- Physical displacement by injected air
- Capillary interaction of air and water
- Frictional drag by flowing air
- Water flow in response to evaporative loss
- Thermal convection
- Migration of fines

Groundwater is physically displaced by air as it moves through the saturated zone soils during air sparging. This process occurs during nonsteady state airflow conditions, where the percentage of air saturation changes with time until the formation of spatially fixed air channels. The amount of mixing due to this physical displacement is dependent on the amount of groundwater displaced and the duration of nonsteady state flow conditions. The rate of water displacement is permeability limited and, therefore, the duration of these effects is generally greater in low permeability soils. The process will take place over both microscopic distances (inches) and site scale distances. Pulsed sparging will frequently create nonsteady state conditions and enhance groundwater mixing (Clayton, Brown, and Bass 1995).

While physical displacement of water by air involves changes in fluid saturation, capillary fluid interactions during sparging can cause groundwater movement without a change in air saturation. This process can be expected to be more pronounced during nonsteady state conditions, when higher air injection pressures can be maintained (Clayton, Brown, and Bass 1995, Boersma, Diontek, and Newman 1995, and Lundegard 1995). Pulsed sparging may enhance this mixing mechanism by increasing the time during which conditions are in a nonsteady state.

Frictional drag on groundwater can be induced by transfer of shear stresses from flowing air to pore water during nonDarcy airflow conditions (Clayton, Brown, and Bass 1995). For fluid flow in a porous medium, a critical value of Reynolds number (Re) for nonDarcy flow is 1 (Clayton, Brown, and Bass 1995), which corresponds to an air velocity of 0.015 to 0.15 m/s for fine sands to coarse sands.

Evaporative loss of water to the injected air can result in water inflow to the sparge zone to maintain volume balance. This volume balance approach must consider changing air saturations and is sensitive to the degree of air saturation and relative humidity of the injected air and their effects on the rate of evaporation. This is also a thermodynamic process, where heat lost to evaporation cools the groundwater, leading to downward density driven flow. This flow would be opposite to that induced by frictional drag (for upward airflow) (Clayton, Brown, and Bass 1995).

Thermal convection can occur through density driven flow of cooled groundwater as indicated above, or through heating of groundwater by injecting heated air. This process is sensitive to the air saturation developed by its effect on heat transfer. The heat capacity of air is much less than that of water, potentially limiting the warming of groundwater (Clayton, Brown, and Bass 1995).

The migration of fine sediments has been shown to significantly reduce the permeability of petroleum reservoirs by sealing pore throats (Clayton, Brown, and Bass 1995). Fines migration also has been observed during sparging in both laboratory sand tank studies and in field studies. Airflow paths may be destabilized by changes in air permeability caused by fines migration, and the resulting redirection of airflow may cause groundwater mixing as the water is displaced by or displaces air.

Based on the above discussion, physical displacement of water and capillary interactions seem to be relevant primarily during nonsteady state conditions. Frictional drag, evaporative loss, thermal convection, and fines migration may also cause groundwater mixing after steady state conditions are reached, but the magnitude of mixing resulting from these processes may be less than that which occurs during the nonsteady state (Clayton, Brown, and Bass 1995).

Groundwater mixing is important during air sparging to effectively transport dissolved oxygen for *in situ* bioremediation. Groundwater mixing can be effective if it occurs at the pore scale as well as over site-scale distances, since either process can reduce the diffusion limitations of sparging. This mixing is commonly bidirectional, which may prevent development of a discernible site-scale flow pattern. Because sparging without groundwater mixing will be of limited effectiveness, the increased VOC removal and DO addition that occurs during sparging and is enhanced by pulsing provides strong indirect evidence that mixing does occur (Clayton, Brown, and Bass 1995).

SYSTEM DESIGN PARAMETERS

In the absence of any reliable models for the *in situ* air sparging process, empirical approaches are used in the system design process. The parameters that are of significant importance in designing an *in situ* air sparging system are listed below:

- Air distribution (zone of influence)
- Depth of air injection
- Air injection pressure and flow rate
- Injection mode (pulsing or continuous)
- Injection well construction
- Contaminant type and distribution

Air Distribution (Zone of Influence)

During the design of air sparging systems, it may be difficult to define a radius of influence the way it is used in pump and treat and/or soil venting systems. Due to the asymmetric nature of the air channel distribution and the variability in air channel density, it is safer to assume a zone of influence than a radius of influence (Johnson et al. 1993, Johnson 1995, and Ahlfeld, Dahmani, and Wei 1994).

It becomes necessary to estimate the zone of influence of an air sparging point, similar to any other subsurface remediation technique, to design a full-scale air sparging system consisting of multiple points. This estimation becomes an important parameter for the design engineer to determine the number of required sparge points. The zone of influence should be limited to describing an approximate indication of the average distance traveled by air channels from the sparge point in the radial directions, under controlled conditions.

The zone of influence of an air sparging point is assumed to be an inverted cone; however, it should be noted that this assumption implies homogeneous soils of moderate to high permeability, which is rarely observed in the field. As noted earlier, during a numerical simulation study on air sparging (Lundegard and Andersen 1993), three phases of behavior were predicted following initiation of air injection (Figures 4 and 5). These are, (1) an expansion phase in which the vertical and lateral limits of airflow grow in a transient manner; (2) a second transient period of reduction in the lateral limits (collapse phase); and (3) a steady state phase, during which the system remains static as long as injection parameters do not change. The zone of influence of air sparging was found to reach a roughly conical shape during the steady state phase.

Based on the inverted cone airflow distribution model, many air sparging system designs are performed based on the zone of influence measured by conducting a field design test. Many applications require multiple zones to cover an entire area. When a hot spot or source area is under consideration for cleanup, it is prudent to design the air sparging system in a grid fashion, Figure 7. The grid should be designed with overlapping zones of influence providing complete coverage of the area under consideration for remediation. If an air sparging curtain is designed to contain the migration of dissolved contaminants, the curtain should be designed with overlapping zones of influence in a direction perpendicular to the direction of groundwater flow.

A properly designed test can provide valuable information. The limitations of time and money often restrict field evaluations to short duration single well tests. Potential measuring techniques (Figure 8) of the zone of influence have evolved with this technology during the last few years.

1. Measurement of the lateral extent of groundwater mounding in adjacent monitoring wells (Brown 1992a, Brown 1992b, Boersma, Diontek, and Newman 1995, and Brown, Herman, and Henry 1991).
   This was the earliest technique used during the early days of implementation of this technology. It did not take long to conclude that the lateral

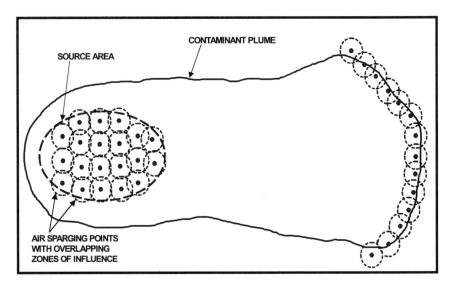

Figure 7   Air sparging points location in a source area and in a curtain configuration.

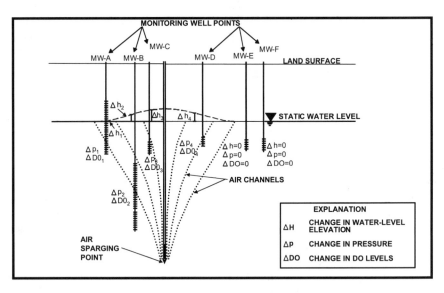

Figure 8   Air sparging test measurements.

   extent of the mound is only a reflection of the amount of water displaced
   and does not correspond to the zone of air distribution.

2. Measurement of the increase in DO levels and redox potentials in comparison to pre sparging conditions (Brown 1992b, Sellers and Schreiber
   1992, Kresge and Dacey 1991, Marley, Li, and Magee 1992, Marley,
   Walsh, and Nangeroni 1990, and Ardito and Billings 1990).
   These parameters should be measured in field only using downhole probes
   or flow through cells. Oxygen transfer could take place during other forms

of measurement when collecting and handling the sample, and may bias the results of the analysis. This concept lost its value when it was realized that the injected air travels in the form of channels more than as bubbles. Increases in DO levels in the bulk water due to diffusion limited transport of oxygen will be noticeable only during a long-term pilot study. In most cases, the increased DO levels observed during short duration pilot tests were due to the air channels directly entering the monitoring wells and not due to overall changes in dissolved oxygen levels in the aquifer.

3. Measurement of soil gas pressures.

   This technique involves the measurement of an increase in the soil gas pressure above the water table due to the escape of injected air into the vadose zone. The escaped air will quickly equilibrate in the vadose zone, and may spread over a larger area than the zone of air distribution into the vadose zone. As a result, during the combined operation of soil venting and air sparging, measurement of this parameter may be totally misleading.

4. Increase in head space pressure within sealed saturated zone monitoring probes which are perforated below the water table only.

   This technique is the most widely used and is currently considered to be the most reliable in terms of detecting the presence of air pathways at a specific distance in the saturated zone. When an air channel enters a monitoring probe via the submerged screen, the head space pressure could increase up to the hydrostatic pressure at the point of entry. However, the actual distribution of air channels may extend beyond the farthest monitoring probe.

5. The use and detection of insoluble tracer gases, such as helium and sulfur hexafluoride (Johnson et al. 1993, Ahlfeld, Dahmani, and Wei 1994, Marley, Bruell, and Hopkins 1995, Sellers and Schreiber 1992, and Leeson et al. 1999).

   Initial monitoring of the tracer gas in the vadose zone is typically performed while the soil vapor extraction system is off. The potential to balance the mass of the injected tracer and the amount of recovered tracer raises the level of confidence in the estimation of the capture rate of injected air. The use of sulfur hexafluoride as a tracer has the advantage that its solubility is similar to that of oxygen. Hence, the detection of sulfur hexafluoride in bulk water will be an indicator for the diffusional transport of oxygen. This technique will also provide information on vapor flow paths and vapor recovery efficiencies during air sparging.

6. Measurement of the electrical resistivity changes in the target zone of influence as a result of the changes in water saturation due to the injection of air (electrical resistivity tomography (ERT) method).

   Tomography is a method of compiling large amounts of one dimensional information in such a way as to produce a three dimensional image (CAT scans, MRIs, and holograms make use of tomography). ERT is a process in which a three dimensional depiction of air saturation within the saturated zone is generated by measuring the electrical resistance of the soil between electrodes placed at various locations on wells during installation. Other tomographic methods include vertical induction profiling (VIP) and geo-physical diffraction tomography (GDT).

VIP is similar to ERT, except that ERT uses a direct current (DC) potential while VIP uses a 500 Hz alternating current (AC). The use of AC makes it possible to detect electrical field strength by induction, so existing PVC wells can be used without the requirement for subsurface electrode installation.

GDT is a high resolution acoustic technique that provides quantitative subsurface imaging by measuring variations in acoustic velocity between various locations on the ground surface and depths in monitoring wells.

ERT may be the most reliable method among all techniques discussed in this section. The high drilling costs associated with installing large numbers of electrodes in the subsurface, preclude it from being used widely.

7. Measurement of moisture content changes within the target zone of influence using time domain reflectometry (TDR) technique.

TDR is a well established and accurate means to measure the moisture content of soils and has been widely used in the agricultural industry. When injected air travels within the zone of influence, moisture content will decrease due to the displacement of water. TDR data, collected from probes placed in the aquifer, can accurately reflect the changes in moisture content

8. Neutron probe technique to measure the changes in water saturation (Acomb 1995).

This technique uses a neutron probe to measure changes in water saturation (thus air saturation) below the water table during air sparging. The neutron probe detects hydrogen in water and thus translates into a water saturation value. The water saturation values can be converted to air saturation values. The neutron probe can also detect the hydrogen in petroleum hydrocarbons and hence can bias the fluid saturation values

9. The actual reduction in contaminant levels due to sparging.

This evaluation gives an indication of the extent of the zone of influence in terms of contaminant mass removal, but the test has to be run long enough to collect reliable data

Since cost and budgetary limitations influence how a field design test is performed, availability of resources will determine the type of method that is used. The most reliable method is the one that measures the changes in electrical resistivity due to changes in air/water saturation. The most cost effective method is the one that determines the head space pressure within the saturated zone probes.

Depth of Air Injection

Among all the design parameters, depth of air injection may be the easiest to determine since the choice is influenced by the contaminant distribution. It is prudent to choose the depth of injection at least a foot or two deeper than the deepest known point of contamination. However, in reality, the depth determination is influenced by soil structuring and extent of layering since injection below any impermeable or very permeable zones should be avoided. The current experience in the industry is mostly based on depths less than 30 to 60 feet below the water table (Brown 1992a, Geraghty & Miller 1995, and Marley, Bruell, and Hopkins 1995).

The depth of injection will influence the injection pressure and the flow rate. The deeper the injection point is located, the greater the zone of influence will be expanded, and thus the more air will be required to provide a reasonable percentage of air saturation within the zone of influence.

Air Injection Pressure and Flow Rate

The injected air will penetrate the aquifer only when the air pressure exceeds the sum of the water column's hydrostatic pressure and the threshold capillary pressure, or the air entry pressure. The air entry pressure is equal to the minimum capillary entry resistance for the air to flow into the porous medium. Capillary entry resistance is inversely proportional to the average diameter of the grains and porosity (Bohler, Holtz, and Nahold 1990 and Lundegard and Andersen 1993).

The injection pressure necessary to initiate *in situ* air sparging should be able to overcome the following:

1. The hydrostatic pressure of the overlying water column at the point of injection.
2. The capillary entry resistance to displace the pore water; this depends on the type of sediments in the subsurface.
3. The resistance of the well, screen, and packing material.

Hence, the pressure of injection (P_i) could be defined as

$$P_i = P_h + P_a + P_d \tag{1}$$

where P_h = hydrostatic pressure; P_a = air entry pressure of formation; and P_d = air entry pressure for the well screen and packing.

The hydrostatic pressure of the overlying water column at the point of injection can be described by the following equation

$$P_h = \delta H_I \tag{2}$$

where P_h = hydrostatic pressure; δ = density of water; and H_i = hydrostatic head (distance from the top of the well screen to the water table).

Therefore, for each foot of water column above the top of the well screen, the injection pressure required to overcome hydrostatic pressure is as follows

$$P_h = (62.4 \text{ pounds/ft}^3)(1 \text{ ft})$$

$$P_h = (62.4 \text{ pounds/ft}^2)$$

Converting to pounds psi

$$P_h = (62.4 \text{ pounds/ft}^2)(144 \text{ in}^2/\text{ft}^2)$$
$$P_h = (0.43 \text{ psi})$$

The air entry pressure for a formation is heavily dependent on the type of geology and includes the capillary pressure to displace the pore water in the formation. The capillary pressure can be quantitatively described (Ardito and Billings 1990), under idealized conditions, by the following equation

$$P_c = 2 \text{ s/r} \qquad\qquad (3)$$

where P_c = capillary pressure; s = the surface tension between air and water; and r = the mean radius of curvature of the interface between fluids.

This equation reveals that as r decreases, the capillary pressure increases. Generally, r will decrease as grain size decreases. Therefore, the required pressure to overcome capillary resistance increases with decreasing sediment size.

In reality, the air entry pressure of the formation will be higher for fine grained sediments (0.43 to 4.3 psi) than for coarse grained sediments (0.04 to 0.4 psi).

When P_h is significantly greater than P_a and P_d combined, it is likely that air will enter the formation primarily near the top of the injection screen.

The notion that higher pressures and flow rates correspond to better air sparging performance is not true. Increasing the injection rate to achieve a greater flow and wider zone of influence must be implemented with caution (Johnson et al. 1993 and Lundegard and Andersen 1993). This is especially true during the start up phase due to the low relative permeability to air attributable to low initial air saturation. The danger of pneumatically fracturing and thus creating secondary permeability in the formation under excessive pressures should also be taken into consideration in determining injection pressures. As such, it is important to gradually increase the pressure during system start up.

While a more detailed discussion of pneumatic fracturing is provided in Chapter 10, a conservative approximation of the upper bound of air sparging pressures that, if exceeded, could cause fracturing is 0.75 psi per foot of soil overburden (Leeson et al. 1999). Thus, at most sites, the air sparging injection pressures will range from a low of 0.43 psi per foot of water column above the top of the sparge screen to a high of 0.75 psi per foot of soil overburden above the top of the sparge screen or the limit of the materials used (such as piping), whichever is less.

The typical values of injected airflow rates reported in the literature (Lundegard and Andersen 1993, Johnson et al. 1993, and Leeson et al. 1999) range from 1 cfm to 20 cfm. Injection airflow determinations are influenced more by the ability to recover the stripped contaminant vapors through a vapor extraction system, thus containing the injected air within a controlled air distribution zone.

Injection Mode (Pulsing and Continuous)

Direct and speculative information available in the literature indicates that the presence of air channels impedes, but does not stop, the flow of water across the sparging zone of influence. The natural groundwater flow through a sparged zone of an aquifer will be slowed and diverted by the air channels due to changes in water saturation and thus relative hydraulic permeability. This potentially negative factor could be overcome by pulsing the air injection and thus minimizing the decrease in relative permeability due to changes in water saturation.

An additional benefit of pulsing will likely be due to the increased mixing of groundwater resulting from air channel formation and collapse during each pulse cycle. This should also help to reduce the diffusional rate limitation for the transport of contaminants in the bulk water phase towards the air channels, due to the cyclical displacement of water during pulsed air injection. As noted earlier, the expansion phase during air sparging (Figures 4 and 5) appears to have a greater zone of influence than under the steady state conditions; therefore, pulsing may improve the efficiency of air sparging by creating cyclical expansion and collapse of the zone of influence.

Injection Well Construction

Injection wells must be designed in such a way as to accomplish the desired distribution of airflow in the formation. Conventional design of an air sparging well under shallow sparge depth conditions (less than 20 feet) and deeper sparge depth conditions (greater than 20 feet) are shown in Figures 9 and 10. Schedule 40 or 80 PVC piping and screens in various diameters can be used for the well construction. In both configurations, the sparge point should be installed by drilling a well to ensure an adequate seal to prevent short circuiting of the injected air up the well bore. At large sites where many wells are required, the cost of installing multiple sparge points may prohibit the consideration of air sparging as a potential technology.

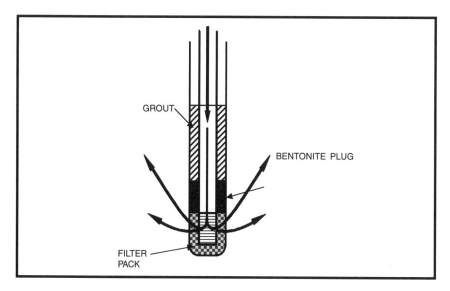

Figure 9   Schematic showing conventional design of an air sparging point for shallower applications.

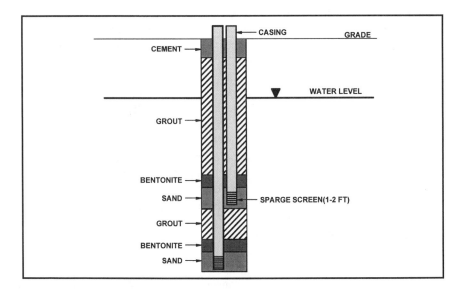

Figure 10   Diagram of nested sparge well for deeper applications.

Injection well diameters range from 1 to 4 inches. The performance is not expected to be affected significantly by changes in well diameter. Economic considerations favor smaller diameter wells (1 to 2 inch), since they are less expensive to install. However, as the diameter of the well is reduced, the pressure drop due to the flow through piping increases and may become significant, especially at deeper injection depths.

Driven air sparge well points made out of small diameter (3/4 inch to 1 1/2 inch) 8 to 10 feet cast iron, flush jointed sections (Figure 11) will help in making this technology more cost effective under some conditions. However, the absence of a sand pack ground the sparge points may allow clogging of the sparge points to develop over a long period, particularly under pulsing conditions. Specifically, continuous expansion and collapse of the soils around the sparge point during the pulse cycles will act like a sieving action, thus allowing finer sediments to accumulate around the sparge points and eventually clog them.

The well screen location and length should be chosen to maximize the flow of injected air through the zone of contamination. At typical injection flow rates, most of the air will escape through the top 12 inches of the screen. A 10-slot PVC screen is normally used for air sparging applications.

Contaminant Type and Distribution

Volatile and strippable compounds will be most amenable to air sparging. Nonvolatile, but aerobically biodegradable compounds can also be addressed by this technique. This means that most petroleum hydrocarbons and chlorinated solvents can be treated with air sparging. Even compounds like acetone and other ketones that cannot be treated with an air stripper above-ground can be affected by air

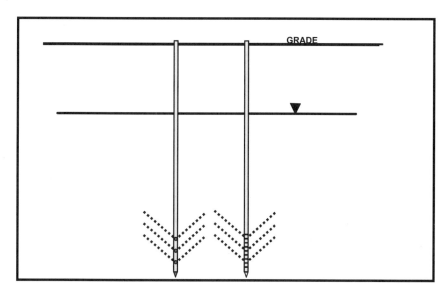

Figure 11   Small diameter air sparging well configuration.

sparging due to the biological activity. A compound like 1,4 dioxane which is very soluble and nondegradable will not be remediated.

There is no limit on the dissolved concentrations of contaminants treatable by air sparging. For air sparging to be effective, the air saturation percentage and the radius and density of air channels are an important factor for mass transfer efficiencies of both contaminants and oxygen. The rates of stripping and biodegradation are both limited by diffusion through water. It is not possible to optimize them separately (Mohr 1995).

PILOT TESTING

Since the state-of-the-practice in designing an *in situ* air sparging system has not progressed beyond the empirical stage, a pilot study should be considered to prove its effectiveness as well as to gather the data necessary for full-scale design. The pilot study could be more appropriately defined as a field design study, since the primary objective would be to obtain site-specific design information. However, due to the still unknown nature of the mechanics of the process, the data collected from a pilot test should be treated with caution. The collected data should be valued as a means of overcoming prior concerns, if any, regarding the implementation of this technology. Since vapor extraction is a complimentary technology to *in situ* air sparging, pilot testing of the integrated system is highly recommended.

Short-term pilot tests play a key role in the selection and design of *in situ* air sparging systems. Most conventional pilot tests are less than 24 to 48 hours in duration and consist of monitoring changes in:

- Pressure buildup in sealed piezometers screened below the water table
- Dissolved oxygen levels
- Water levels in wells
- Soil gas pressures
- Contaminant concentrations in soil gas
- Presence and capture of tracer gases

These parameters are assumed to be indicators of air sparging feasibility and performance and are also used in the design of full-scale systems. As noted earlier, the understanding regarding the value and usefulness of each of the parameters listed above is improving. However, it is suggested that collection and comparison of as many parameters as possible will provide valuable insight on site-specific applicability of air sparging as a remediation technique.

It is important to perform a preliminary evaluation of the geologic and hydrogeologic conditions for the applicability of *in situ* air sparging prior to the pilot study. In addition, a thorough examination of the degree and extent of contamination should be performed.

Evaluating the site-specific parameters listed in Table 2 prior to designing a pilot test will enhance the quality of data that would be collected.

Table 2   Site-Specific Parameters

Condition	Favorable Condition	Impact
Saturated Zone Soil Permeability (horizontal)	10^{-3} cm/s	Applicability, Flow Rate vs. Pressure, Mass Removal Efficiencies vs. Transport Rate
Geologic Stratification and Anisotropy	Sandy, Gravelly Soils, Homogeneous	Applicability, Air Distribution and Flow Pattern
Aquifer Type	Unconfined	Recovery of Injected Air
Depth of Contamination Below the Water Table	Less than 40 to 50 feet	Sparging Depth (Injection Pressure)
Type of Contaminant	High Volatility, High Strippability, High Aerobic Degradability	Applicability - Volatility/Strippability/ Biodegradability
Extent of Contamination	No Separate Phase Contamination	Applicability and Mass Removal Efficiency (Multiple Sparge Points)
Soil Conditions Above the Water Table	Vadose Zone >5 Feet Thick Permeable Soils	Ability to Capture the Stripped Contamination by Vapor Extraction, Vapor Flow Paths

The typical equipment setup used for an air sparging pilot test is similar to that shown in Figure 1. The data that should be collected during the pilot study include the following engineering parameters:

1. *Zone of Air Distribution:* For any subsurface remediation system, this is a key design parameter since this would determine the required number of injection points. The zone of influence under various pressure and flow combinations should be

measured. The methods to estimate the zone of influence have been described in the air distribution section of this chapter and Figure 8.

2. *Injection Air Pressure:* This parameter is significantly influenced by the depth of injection and subsurface geology. The required baseline pressure during the test should be equal to or just above the value necessary to overcome the sparging depth. The impact of any additional required pressure should be evaluated carefully in incremental steps because excessive pressures may fracture the soils around the point of injection.

3. *Injection Flow Rate:* The injection airflow rate should provide an adequate percentage of air saturation within the zone of air distribution. The greater the sparging depth, the higher the flow rate required to achieve a percentage of air saturation will be. Evaluation of the injection flow rate should be governed more by the ability to capture the stripped contaminant vapors and the net pressure gradient in the vadose zone. At a minimum, the airflow rate should be sufficient to promote significant volatilization rates and/or maintain dissolved oxygen levels greater than 2 mg/l. Typical injection flow rates are in the range of 1 to 20 cfm per injection point, depending on the type of geology and the sparging depth.

4. *Mass Removal Efficiency:* Another key objective during the field study should be to demonstrate the mass removal efficiency of the *in situ* air sparging process. This can be determined by measuring the net increase in contaminant levels in the vapor extraction system after the initiation of the air sparging system. To evaluate the net increase in contaminant levels in the effluent, the field test should be conducted as a sequential test in two phases.

The first phase should be to perform the vapor extraction test and monitor the effluent air levels under pseudo-steady state conditions, which can generally be accomplished after removing 1.5 to 2 pore volumes from the unsaturated zone. The air sparging is initiated during the second phase with continued monitoring of the contaminant levels in the vapor extraction system air stream. An increase in the contaminant level and the duration of increase would indicate the short-term mass removal efficiency due to air sparging (Figure 12). The second phase of the test should be continued until a decline in concentrations in the effluent air stream is observed, or until 2 to 3 pore volumes of air have been injected into the affected saturated zone. NAPLs and geologic heterogenaities may prevent the concentrations during the second phase from declining. Extended pilot plants may be required if it is important to determine the total time required for full-scale sparging operations. Conversely, if the dissolved contaminant concentrations within the zone of air distribution are low or the volume of soil vapor extracted is much greater than the volume of air being injected, the vapor phase contaminant increase described in Figure 12 may go unnoticed, especially without frequent sampling.

Determination of the increase in contaminant levels due to air sparging is important to evaluate the safety considerations associated with implementing this technology. Continuous removal of the contaminants transferred into the vadose zone is very important. Buildup of these contaminants to explosive levels should be avoided at any cost. The air injection and air extraction rates should be controlled to maintain a net negative pressure within the target area.

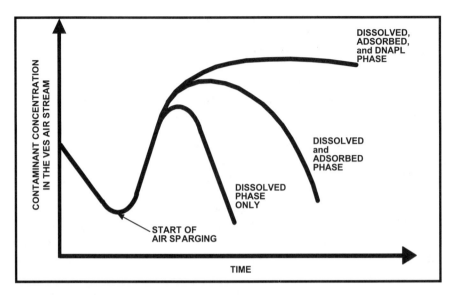

Figure 12   Contaminant removal efficiencies during a pilot test.

MONITORING CONSIDERATIONS

In situ and above-ground monitoring data should be used to assess the performance of operating conditions to determine whether system adjustments or expansions are necessary. Table 3 lists the various parameters that can be used to monitor system performance.

In situ field parameters obtained during air sparging are often subject to a wide range of interpretations. Data obtained from groundwater monitoring wells while air is being injected will always be questioned with respect to the potential of an air channel bubbling up through the water in the well. Therefore, it is recommended that groundwater quality samples be collected after air injection is shut down and sufficient time is provided for equilibrium to be reached. Dissolved oxygen, redox, and carbon dioxide levels are recommended to be measured with continuous flow through cells due to the potential for these parameters to change during conventional groundwater sampling and handling procedures.

PROCESS EQUIPMENT

Successful implementation of an air sparging system is dependent on the proper selection of the process equipment. Primary components of an *in situ* air sparging system are:

- Air compressor or blower
- Vacuum blower
- Fittings and tubing to connect the compressor to the well(s)

Table 3   *In Situ* Air Sparging System Monitoring Parameters

In Situ Parameters	Measurement
Groundwater quality improvement	Obtaining periodic groundwater samples from monitoring wells after shutting down air injection.
Dissolved oxygen levels/temperature	Field probes in the monitoring wells after shutting down air injection
Redox potential/pH	Field probes in the monitoring wells after shutting down air injection
Biodegradation byproducts such as CO_2	Groundwater samples obtained with a flow-through cell
Soil gas concentrations	FID, PID, explosimeter or field gas chromatograph or laboratory air samples
Soil gas pressure/vacuum	Pressure/vacuum gauge or manometer
Groundwater level	Water level meter
System Operating Parameters	**Measurement**
Injection well pressure	Pressure gauge or manometer
Soil vapor extraction well vacuum	Vacuum gauge or manometer
Injection well flow rate	Airflow meters
Soil vapor extraction flow rate	Airflow meters
Extraction vapor concentrations	FID, PID, explosimeter, field GC, or laboratory air samples
O_2, CO_2, N_2, CH_4	Laboratory analysis
Pulsing frequency	Timer

- Air filters
- Pressure regulator
- Flow meters
- Pressure gauges
- Air drying unit

Air Compressor or Air Blower

Selection of an air compressor or air blower will depend on the required pressure rating at which air has to be injected (obtained from the pilot test). Air blowers (positive displacement) can be used only when required pressure rating is less than 12 to 15 psi. There are various types of air compressors available, including the following.

1. Positive displacement compressors
   a. Reciprocating piston
   b. Diaphragm
   c. Rocking piston
   d. Rotary vane
   e. Rotary screw and lobed rotor
2. Non positive displacement compressors
   a. Centrifugal
   b. Axial flow
   c. Regenerative

Positive displacement compressors generally provide the most economic solution for systems that require relatively high pressures. Disadvantages include low flow rates, oil removal in some cases, and the heat generated during operation.

Unlike the positive displacement compressor, a nonpositive displacement compressor does not provide a constant volumetric flow rate over a range of discharge pressures. The most important advantage of nonpositive displacement compressors is their ability to provide high flow rates. Table 4 and Figure 13 provide a summary of air compressor characteristics. It should be noted that the size and capacity requirements of an air compressor required for an air sparging site will typically be below 100 hp and 150 psi.

Oil contamination in the injected air can affect the *in situ* air sparging system performance. A variety of filters have been developed to filter out the contained oil. An alternative is to use an oilless compressor. Higher capital and maintenance costs are typical of oilless equipment as compared to their oil-lubricated counterparts.

Some pneumatic systems cannot tolerate moisture formed by the cooling of air caused by compression. While a mechanical filter removes most of the solid and liquid particulates from the air, it is not effective for removing water and oil vapors. The moisture may later condense and freeze in the pipes downstream when low temperatures are encountered. An air dryer prevents condensation by reducing the humidity of the air stream. A practical type of dryer is a desiccant unit, which uses a moisture absorbing chemical, usually in palletized form. Water vapor can also be removed through condensation by passing the air through a chilling unit. Coalescing filters are also effective in removing mists of tiny water droplets. Less costly options include heat tracing of piping/manifold or the use of a receiver tank with a manual or automatic drain to remove the condensation.

Other Equipment

Selection of the vacuum blower will depend on the required airflow rate and vacuum levels necessary for efficient subsurface vapor recovery. Depending on the geologic conditions encountered, high vacuum/low flow versus low vacuum/high flow combination has to be evaluated. For further discussion of SVE equipment, see Chapter 3.

MODIFICATIONS TO CONVENTIONAL AIR SPARGING APPLICATION

For the purposes of discussion in this book, conventional application of air sparging is defined as shown in Figure 1. Due to the geologic and hydrogeologic conditions encountered at many sites across the country, this form of application may have to be limited to only 25 percent of the remediation sites (Geraghty & Miller 1995). However, the concept of using air as a carrier for removing contaminant mass still remains attractive and cost effective compared to currently available options. There are several modifications to the conventional air sparging design that can be used to extend its application.

Table 4   Summary of Compressor Characteristics

Class	Category	Type	Power Range (hp)	Pressure Range (psi)	Advantages
Positive displacment compressors	Reciprocating	Piston air-cooled	1/2 - 500	10 - 250	Efficient, light-weight
		Piston water-cooled	10 - 500	10 - 250	Efficient, heavy-duty
		Diaphragm	10 - 200	10 - 250	No seal, contamination-free
	Rotary	Sliding vane	10 - 500	10 - 150	Compact, high-speed
		Screw (helix)	10 - 500	10 - 150	Pulseless delivery
		Lobe, low-pressure	15 - 200	5 - 40	Compact, oil-free
		Lobe, high-pressure	7 1/2- 200	20 - 250	Compact, high-speed
Nonpositive displacement compressors	Rotary	Centrifugal	50 - 500	40 - 250	Compact, oil-free, high-speed
		Axial flow	1,000 - 10,000	400 - 500	High volume, high speed
		Regenerative peripheral blower	1/4 - 20	1 - 5	Compact, oil-free, high volume

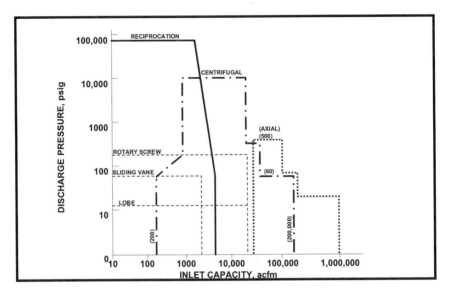

Figure 13   Air compressor characteristics.

Horizontal Trench Sparging

Trench sparging was developed to apply air sparging under less permeable geologic conditions when the depth of contamination is less than 30 feet. When the hydraulic conductivities (in the horizontal direction) are less than 10^{-3} cm/sec, it is prudent to be cautious when injecting air directly into the water-saturated formations. This technique is generally applicable where there is a shallow depth to groundwater and the formation is fine grained.

Trench sparging includes, (1) placement of a single or parallel trenches perpendicular to groundwater flow; (2) injection of air through lateral or vertical pipes at the bottom of the trench; and (3) extraction of air from lateral pipes in the trench above the water table. Figure 14 shows a typical trench sparging system.

The primary focus in this modified approach is to create an artificially permeable environment in the trenches where the distribution of injected air can be controlled. As the contaminated water travels through the trench, the strippable VOCs will be removed from the groundwater and captured by the vapor extraction pipe placed above the water table.

Due to the extremely slow groundwater velocities under conditions where this technique may be preferred, the residence time of the moving groundwater in the trench will be high. Hence, the injection of the air does not have to be continuous and a pulsed mode of injection can be instituted. When biodegradable contaminants are present, the trench can be designed to act like an *in situ* fixed film bioreactor and the rate of injection of air can be further decreased. The need for effluent air treatment may be avoided under these circumstances. Injection of nutrients, such as nitrogen and phosphorus, may enhance the rate of biodegradation in the trench.

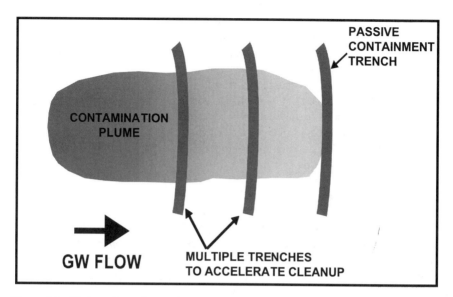

Figure 14a Horizontal trench sparging (section view).

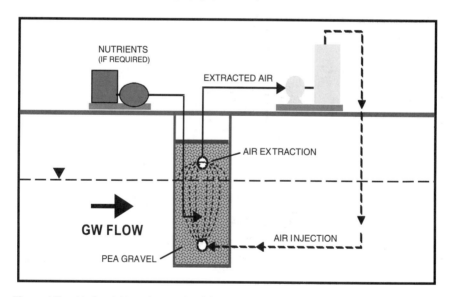

Figure 14b   Horizontal trench sparging (plan view).

The treated groundwater leaving the trench will be saturated with dissolved oxygen and nutrients (if added) and this may enhance the degradation of dissolved and residual contaminants down-gradient of the trench also. If the primary focus of remediation is containment only, this concept can be implemented as a passive containment technique using a single down-gradient trench (Figure 14). Depending on the need to cleanup the site faster, this variation on air sparging can be implemented with multiple trenches (Figure 14).

The biggest limitation of this technique will be the total depth of the trench. Total depths beyond 30 to 35 feet may preclude the implementation of this technique due to shoring costs, site accessibility, and the potential need to deal with a large volume of contaminated soils.

When the depths of the sparge trench will be limited to less than 35 feet, air injection into the trench can be accomplished with a blower instead of a compressor. The extracted air can be treated with a vapor treatment unit (probably vapor phase GAC due to the low levels of mass expected) and reinjected back into the trench as shown in Figure 14. This configuration may eliminate the need to take regular air samples for regulatory purposes.

In Well Air Sparging

This modification was developed as a means to use air as the carrier of contaminants and to overcome the difficulties of injecting air into nonoptimum geologic formations. In well air sparging, shown in Figure 15, can also overcome the difficulties of installing trenches, described in the previous section.

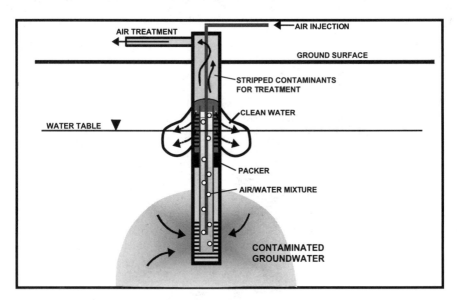

Figure 15   In well air sparging.

The injection of air into the inner casing (Figure 15) induces an air lifting effect that is limited only to the inner casing. The water column inside the inner casing will be lifted upwards (in other words water present inside the inner casing will be pumped) and will flow over the top of the inner casing as shown in Figure 15. As a result, contaminated water will be drawn into the lower screen from the surrounding formation and will be continuously air lifted in the inner tube.

Due to the mixing of air and contaminated water, as the air/water mixture rises inside the inner tube, strippable VOCs will be air stripped and captured for treatment as shown in Figure 15. Treated, clean water that spills over the top of the inner casing will be released back into the formation via the top outer screen. This approach can completely eliminate the need for extracting the water for above-ground treatment under some conditions.

An added benefit of in well air sparging is that the reinjected water, saturated with dissolved oxygen, can enhance the biodegradation of aerobically biodegradable contaminants present in the saturated zone. The need to inject nutrients inside the well can be evaluated on a site-specific basis.

Bio-Sparging

As discussed in previous sections, injection of air into water saturated formations has a significant benefit in terms of delivery of oxygen to microorganisms for *in situ* bioremediation. If delivery of oxygen for the biota is the primary objective for air injection, the volume of airflow does not have to be at the same level required to achieve stripping and volatilization. Control of air channel formation and distribution and capturing of the stripped contaminants also become less significant under these circumstances. Application of this technique to remediate a dissolved plume of acetone, for example, which is a nonstrippable but extremely biodegradable compound, will be appropriate.

Injection of air at very low flow rates (0.5 cfm to less than 2 to 3 cfm per injection point) into water-saturated formation to enhance biodegradation is defined as bio-sparging. Limitations caused by geological formations also become less significant, since the path of air channels can be allowed to follow the path of least resistance. However, it has to be noted that the time required to increase the DO levels in the bulk water depends on the time required for the diffusion of O_2 from the air channels into the water surrounding the channels. It is estimated that only 0.5 percent of the oxygen present in the injected air will be transferred into the dissolved phase during air sparging (Johnson 1995, Geraghty & Miller 1995, and Boersma, Diontek, and Newman 1995). Therefore, caution must be exercised in terms of evaluating the changes in DO levels after the initiation of biosparging. It is common practice to assume that the observed increase in DO levels in monitoring wells is due to the changes in the bulk water. Direct introduction of air into the monitoring wells due to an air channel being intercepted could also be a reason for increased DO levels in monitoring wells. Chapters 7 and 8, *In Situ* Bioremediation and Reactive Zone Remediation, provide more details of the designs, operations, and associated costs for biosparging.

Vapor Recovery via Trenches

Trench vapor recovery is a minor modification to conventional air sparging that involves the recovery of stripped vapors from fine grained formations (Figure 16). Trench vapor recovery can be used when there is a shallow depth to groundwater and overlying fine grained formations extend from the surface to below the water table. This geologic situation would normally inhibit extraction of stripped vapors by vapor extraction wells. Saturated zone mass transport, removal rates, and removal

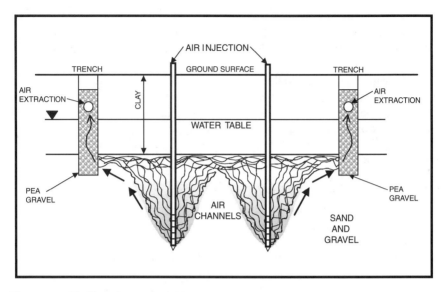

Figure 16   Modified air sparging with vapor recovery through trenches.

mechanisms are very similar to those of conventional air sparging, with the exception of the capillary fringe area. If contaminants are adsorbed to the fine grained formation matrix in the capillary zone, air sparging with trench vapor recovery may be ineffective in transferring and removing the contaminant mass from these areas. Trench vapor recovery systems are most effective when only dissolved phase contaminants need to be addressed.

Pneumatic Fracturing for Vapor Recovery

The application of *in situ* air sparging using pneumatic fracturing to enhance vapor recovery is applied to sites with fine grained formations that extend below the water table and depths to water that prohibit trenching (Figure 17). Pneumatic fracturing increases hydraulic and vapor flow conductivity near the top of the water table and in the overlying unsaturated zone, while allowing stripped contaminants to be collected without spreading out laterally. Balancing injection flow rates is critical when using this method of vapor recovery. Mass transport and recovery have limitations similar to trench vapor recovery. Mass transfer and removal of adsorbed contaminants near the top of the water table are limited in areas between fractures.

CLEANUP RATES

To date, there are no reliable methods for estimating groundwater cleanup rates. A mass removal model for *in situ* air sparging has been reported (Marley, Li, and Magee 1992) using air stripping as the only mass transfer mechanism. However, this model was based on the premise that injected air travels in the form of bubbles.

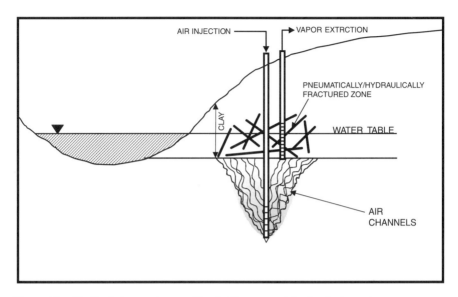

Figure 17   Modified air sparging combined with pneumatic/hydraulic fracturing.

Because it has been established that the primary mode of travel of injected air at most sites is in the form of air channels, the reliability of this model may be questionable.

In the presence of air channels, the rate of mass transfer will be limited either by kinetics of the mass transfer at the interface, or by the rate of transport of the contaminant through the bulk water phase to the air/water interface. Based on these assumptions, reaching nondetectable levels may be possible only with biodegradable contaminants (Figure 18). (See Chapter 1 for a full discussion on the geologic limitations of air as a carrier.)

Cleanup times of less than 12 months to three years have been achieved in many instances. Reports in the literature indicate that sites that have implemented air sparging have often met groundwater cleanup goals in less than one year (USEPA 1993, Brown 1992b, Brown, Herman, and Henry 1991, Marley, Li, and Magee 1992, and Ardito and Billings 1990). However, it should be noted that at most of these sites, the cleanup goal was around 1 mg/l for total BTEX and that BTEX compounds are very biodegradable.

The air saturation and the size of the air filled regions have the greatest effect on the mass transfer rates. Specifically, a high volume of air filled spaces that are finely distributed will promote faster mass transfer. Numerical analyses show that the air saturation must be greater than 0.1 percent and the size of air channels must be on the order of 0.001 meters for air sparging to be successful (Mohr 1995). At sites with optimum geologic conditions, the above criteria will be met easily.

The required cleanup times for a site will depend on the following

1. Target cleanup levels
2. Extent and phases of contamination

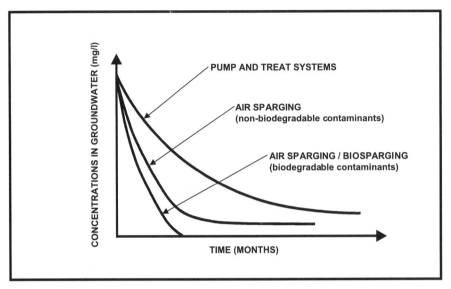

Figure 18   Cleanup rates for various contaminants during *in situ* air sparging.

    a. Contaminant mass present in the saturated zone and the capillary fringe
    b. Extent of dissolved and sorbed phase contamination
    c. The presence and absence of a DNAPL
3. Strippability, volatility, and biodegradability of contaminants present
4. Solubility, partitioning of the contaminants
5. Geologic conditions
    a. Percentage of air saturation
    b. Density of air channels
    c. Size of the air channels
    d. Heterogeneity of the aquifer

Figure 18 shows the effectiveness of air sparging on decreasing the dissolved groundwater contaminant levels in comparison to a pump and treat system. These data were summarized from operational history obtained from approximately 40 *in situ* air sparging systems (Geraghty & Miller 1995). Implementation of air sparging at sites with aerobically biodegradable contaminants led to more rapid attainment of cleanup standards. At sites with aerobically nonbiodegradable contaminants, specifically at sites with chlorinated organic compounds, an asymptotic concentration level was reached. However, this asymptote was at a lower concentration, and was reached in less time than what could have been accomplished with a pump and treat system. See Chapter 2 for a full discussion on the lifecycle of remediations.

LIMITATIONS

At first glance, *in situ* air sparging appears to be a simple process: injection of air into a contaminated aquifer below the water table with the intent of volatilizing

VOCs and providing oxygen to enhance biodegradation. Previous discussions in this chapter have included the applicablity of the *in situ* air sparging process. It is also important to know when not to apply this technology. The following provides a brief summary of the conditions under which conventional application of this technology is not recommended

1. Tight geologic conditions with hydraulic conductivities less than 10^{-3} cm/sec. The vertical passage of the air may be hampered, the potential for the lateral movement of contaminants will be increased, and there is the potential for inefficient removal of contaminants. Conventional form of air sparging should be evaluated with extreme caution under these conditions.
2. Heterogeneous geologic conditions with the presence of low permeability layers overlying zones with higher permeabilities. The potential for the enlargement of the plume exists, again due to the inability of the injected air reaching the soil gas above the water table.
3. Contaminants present are nonstrippable and nonbiodegradable.
4. Mobile free product has not been removed or completely controlled. Air injection may enhance the uncontrolled movement of this liquid away from the air injection area.
5. Air sparging systems that cannot be integrated with a vapor extraction system to capture all the stripped contaminants. In some instances, the stripped contaminants can be biodegraded in the vadose zone if optimum conditions are available. Thicker vadose zones and low injection rates are more appropriate to implement this than shallower depths.
6. The structural stability of nearby foundations and buildings may be in jeopardy due to the potential of soil fluidization or fracturing.
7. Potential for uncontrolled migration of vapor contaminants into nearby basements, buildings, or other conduits.

DESIGN EXAMPLE

Problem

A former manufacturing facility has been impacted from past material handling and storage practices. Site investigation results indicate that both soil and groundwater have been impacted by hydrocarbon compounds and to a lesser extent chlorinated VOCs. The site is approximately 1.5 acres in size. The groundwater plume is restricted to the site and underlies about a third of the site. There is no surface seal; the site is overgrown with weeds. Groundwater monitoring data indicate that the groundwater is impacted primarily by benzene, toluene, ethyl-benzene, xylenes, and to a lesser extent tetrachloroethene and trichloroethene, as depicted on Figure 19. The aquifer is approximately 20 feet thick and consists of fine to medium grain sands and is overlain by 10 feet of vadose zone soils of similar geology. The groundwater impacts are limited to the upper 10 feet of the aquifer.

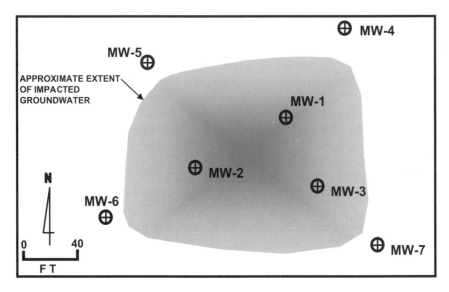

Figure 19   Site plan with groundwater contaminant plume for design example.

Solution

Is air sparging a viable technology for the site? First, we need to determine if the contaminants are amenable to air sparging. The contaminants identified are either volatile and/or biodegradable as indicated by Table 1, and air sparging should be considered. Secondly, we need to determine if the geologic and site surface features are acceptable for air sparging. The geology consists of fine to medium grained sand without restrictive lenses of silt or clay in either the saturated or vadose zone in the area of the plume. Therefore, air sparging should be viable.

The necessity for and viability of SVE technology for capturing the volatilized contaminants is often performed concurrently with an air sparging analysis, but that will not be examined here. The applicability of SVE and a design example under similar site conditions was presented in Chapter 3.

Pilot Test Planning

For sites with relatively large groundwater contaminant plumes, a pilot test is recommended to evaluate the effectiveness of air sparging and also to obtain site-specific data for the full-scale design. A pilot test requires detailed planning and typically includes three major components, (1) development of a monitoring program; (2) estimation of the minimum air injection pressure; and (3) selection of an appropriate range of air injection flow rates.

Development of the monitoring program includes the selection of the monitoring locations, parameters of interest, and frequency of data collection. A review of the existing monitor wells and construction details indicates that each of the existing shallow groundwater monitoring wells have screen sections that extend above the

top of the water table. To obtain more reliable pressure gradient data, four piezometers with 10 foot screen sections completed below the water table will be installed
for the pilot test. To have a good distribution of data collection points, these additional
monitoring points will be installed in a radial distribution at 5-foot intervals around
the sparge point to minimize potential interference from site heterogeneities and to
obtain the data necessary for full-scale design. The air sparging test will also be
performed in the impacted area to collect data as representative as possible, including
air emissions data. Figure 20 is a site plan depicting the sparge point, SVE well,
and piezometer locations.

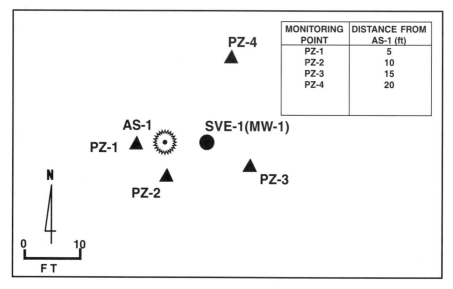

Figure 20   Pilot test layout for design example.

Various ways of determining the radius of influence from a pilot test have been
presented in the literature. The majority of methods base radius of influence as a
function of the observed increase in water level, formation pressure, and dissolved
oxygen. In addition, the physical observation of bubbles with the monitoring points
and an increase in well head vapor concentrations are common indicators of positive
influences induced by an air sparging system. Each of these parameters can be easily
measured using common portable meters. The monitoring frequency for each of
these parameters varies based on the pilot test length. The monitoring frequency
should include pretest sampling to establish a baseline for comparison and a 30
minute sampling interval throughout the test to ensure that an adequate number of
data points are collected.

The second component is to select an appropriate range of air injection flow
rates. Typical air sparge injection rates for shallow contaminant plumes typically
range from 3 to 10 cfm. Further, it is desirable to run a pilot test until at least until
two to three pore volumes of air are injected into the saturated zone. Assuming a
conical zone of influence, an injection depth of 12 ft below the water table and a

porosity of 0.3 for fine sands, injection flow rates ranging from 5 to 10 cfm should provide the desired two to three pore volume of air within 2 hours of initiating the test as shown below:

Estimate one pore volume (V_{pore}) based on a 30° angle of influence

$$V_{pore} = D^2hn/12 \tag{4}$$

where D = the diameter of cone of influence, which = $2h[\tan(30°)]$ when assuming a 30° angle of influence, or 14 feet; h = the height of the cone of influence, or in this case the depth of injection, 12 feet; and n = the porosity of the formation, 0.3.
$V_{pore} = (14 \text{ feet})^2(12 \text{ feet})(0.3)/12$
$V_{pore} = 185$ cubic feet

Estimate time (T) to inject three pore volumes of air at a flow rate (Q) of 5 cfm

$$T = 3V_{pore}/Q \tag{5}$$

T = 3(185 cubic feet)/5 cfm
T = 111 minutes

   The third component is to estimate the minimum air injection pressure for the pilot test. As indicated previously in the literature, the minimum air injection pressure required to begin distributing air into the aquifer is the sum of the hydrostatic pressure of the overlying water table and capillary pressure of the aquifer matrix. Because the vertical extent of groundwater impacts is limited to the upper 10 feet of the aquifer, the air sparge point will be completed 2 feet below the bottom of the impacted aquifer or 22 feet below land surface. The resulting hydrostatic pressure of the aquifer will be based on a water column height of 12 feet. The capillary pressure will be evaluated for an aquifer matrix comprised of fine grained sands. The minimum required air injection pressure will simply be the sum of the hydro-static pressure and the formation entry pressure as calculated below.
   The hydrostatic pressure of the overlying water table is the product of the specific weight of water and the water column height

$$P_{hydrostatic} = \delta_{water}H_{hydrostatic} \tag{2}$$

$P_{hydrostatic} = (62.4 \text{ lb/ft}^3)(12 \text{ ft})(\text{ft}^2/144 \text{ in}^2)$
$P_{hydrostatic} = 5.2$ psi

   The air entry pressure for a fine grained sand aquifer matrix can be estimated by calculating the height of capillary rise using the following equation (Kuo 1999)

$$h_{capillary} = 0.153/r \tag{6}$$

where r = pore radius of aquifer media; approximately 0.02 cm (Kuo 1999).
$h_{capillary}$ = (0.153)/(0.02cm)
$h_{capillary}$ = 7.65 cm or 0.23 ft

Convert the capillary rise to pressure knowing 14.7 psi (1 atm) is equivalent to 33.9 ft.

$P_{capillary}$ = (0.23 ft)(14.7 psi/33.9 ft)
$P_{capillary}$ = 0.10 psi

The total anticipated air injection pressure for the pilot test is calculated below

$$P_{injection} = P_{hydrostatic} + P_{capillary} \qquad (7)$$

$P_{injection}$ = 5.2 psi + 0.10 psi
$P_{injection}$ = 5.30 psi; use 6.0 psi

To prevent fracturing of the aquifer formation and creation of unwanted preferential flow paths, the estimated injection pressure should be compared to the formation fracture pressure. As mentioned previously in the literature, the fracture pressure is approximately 0.75 psi per foot of overlying soil above the air sparge screen section. The screen section of the sparge point will be set 22 feet below land surface. The resultant fracture pressure for this scenario is 16.5 psi, which is significantly greater than the proposed injection pressure.

The minimum injection pressure at the sparge point well head calculated above ignored pressure losses due to friction from airflow through the piping and well screen. For the pilot test, these pressure losses will typically be negligible because of the short lengths of piping as shown below.

Determine the pressure losses using an anticipated maximum injection flow rate range 10 cfm, a maximum injection pressure of 16.5 psi, a sparge point piping diameter of 1 inch, and a slotted well screen size of 0.010 inches:

From pressure loss tables presented in Ingersoll-Rand (1988), the pressure loss in a 1 inch pipe with a free airflow of 20 cfm and a line pressure of 15 psi (the conditions closest to this design example) is 0.38 psi per 100 feet of pipe. Thus for a 12 foot well, the pressure loss will be approximately

P_{pipe} = (0.38 psi/100 feet)(12 feet)
P_{pipe} = 0.05 psi

From pressure loss calculations provided by US Filter Johnson Screens, a manufacturer of well screens, the pressure loss through a 1 foot long, 0.010 inch slotted well screen is less than 0.001 psi (Seymour 2000).

Thus, the pressure loss through the well piping and screen for the pilot test are approximately 0.05 psi, which is neglible.

While the pressure losses from the piping and well screen were shown to be negligible for the pilot test, the piping losses cannot be ignored during design of the full-scale system. The losses through the well screens will still be negligible during full-scale design.

Conducting the Pilot Test

Figure 20 is a depiction of the pilot study layout. Based on the preliminary calculations, an air compressor capable of delivering at least 10 cfm at 6 psi will be used. The injection pressure will be kept below 16.5 psi to prevent fracturing of the aquifer matrix. The pilot study will be conducted over a period of at least 6 hours with air injection flow rates increasing incrementally every 2 hours from 6 cfm to 8 cfm then to 10 cfm. In addition to baseline monitoring, the specified monitoring parameters will be collected every 30 minutes providing at least four data points for each 2 hour step test.

Evaluating the Data

The pilot test data indicates that an obvious radius of influence was created at air injection flow rates of 8 cfm and 10 cfm. The initial test flow rate of 6 scfm did create observed changes to the groundwater conditions. Discussed below are the field measurements and physical observations that were noted during the test signifying an increase in subsurface airflow.

1. Water Table Mounding: As shown in Figure 21, a significant rise in water levels occurred after increasing the injection flow rate from 6 cfm to 8 cfm. The water table mound began to subside after 60 minutes of operation at 8 cfm and again increased upon increasing the air injection flow rate to 10 cfm. The water table mound was observed at PZ-4 located 20 feet from the air sparge point.

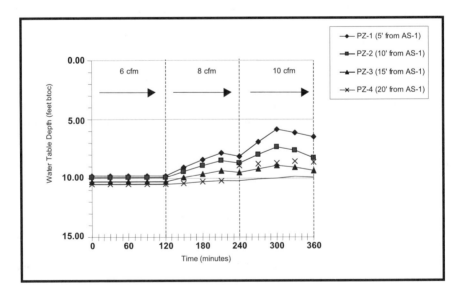

Figure 21   Air sparging water table mounding vs. time.

2. *Increased Dissolved Oxygen:* Dissolved oxygen increases were observed approximately 30 minutes after increasing the injection flow rate to 8 cfm. As shown in Figure 22, the dissolved oxygen content continued to increase during the pilot test and eventually reached asymptotic levels at the end of the test. The highest dissolved oxygen increases occurred at the two interior piezometers, PZ-1 and PZ-2, with lesser increases noted at PZ-3, located 15 feet from the air sparge point. Dissolved oxygen levels remained unchanged at PZ-4, located 20 feet away.

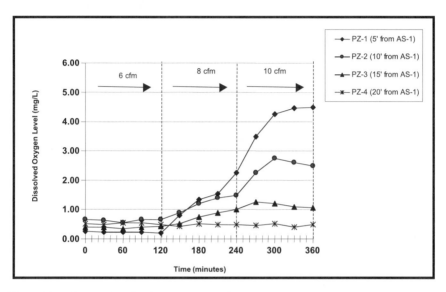

Figure 22   Air sparging dissolved oxygen vs. time.

3. *Increased Pressure Gradient:* As shown on Figure 23, notable pressure increases ranging from 1 to 30 inches of H_2O were observed in the three interior piezometers located within 15 feet of the air sparge point. These observed pressure increases began 30 minutes after increasing the injection flow rate to 8 cfm. The induced pressure reached maximum values and then began decreasing at the innermost piezometer PZ-1 approximately 30 minutes before terminating the pilot test. This decrease in pressure is normal and attributable to the formation of preferential airflow pathways within the aquifer matrix.

4. *Other Physical Observations:* At sites with relatively shallow water tables such as the test site, the presence of steady air bubbles in the test wells provides another indicator of positive air sparging influences. During the pilot test, air bubbles were immediately observed at injection flow rates of 8 cfm and 10 cfm. Steady air bubbles were noted in piezometers PZ-1 and PZ-2 and moderate, intermittent air bubbles were observed at PZ-3. The intensity of the air bubbles remained constant during both the 8 cfm and 10 cfm tests. No air bubbles were noted in PZ-4. Measurable organic vapor levels and hydrocarbon odors were noted at piezometers exhibiting air bubbles. These observations indicate significant volatilization of contaminants will occur within 15 feet of an air sparge point.

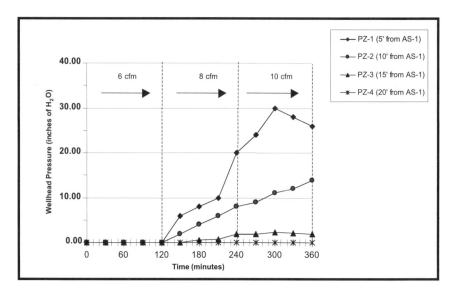

Figure 23   Air sparging pressure vs. time.

The pilot test results indicates that a maximum radius of influence of 15 feet can be created around an air sparge point installed 12 feet below the water table at a minimum air injection rate of 8 cfm. A design injection flow rate will be established at 8 cfm because the higher flow rate of 10 cfm did not increase the sparging influence area. Utilizing an approximate radius of influence of 15 feet, a layout of 34 air sparge wells were distributed within the impacted groundwater zone, as shown on Figure 24.

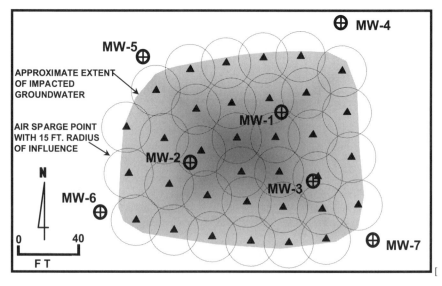

Figure 24   Conceptual layout of air sparge wells for design example.

Before we select a compressed air device, the injection flow rate must be corrected for pressure as described below

$$Q_{scfm} = Q_{acfm} \frac{(P_a)(T_s)}{(P_s)(T_a)} \tag{8}$$

where Q_{scfm} = vapor flowrate in standard cubic feet per minute (scfm); Q_{acfm} = vapor flow rate in ambient cubic feet per minute (conditions within piping); P_a = ambient absolute pressure (within piping); P_s = standard absolute pressure (14.7 psi at 60 °F); T_a = ambient absolute pressure (within piping)°; and T_s = standard absolute temperature (60 °F or 519 °R).

Based on a pilot test injection pressure of 6 psi and an air stream temperature of 125 °F (585 °R), the total injection flow rate corrected for pressure and temperature is calculated below

Q_{scfm} = (8 cfm)(6 psi + 14.7 psi)(519°R)/(14.7)(585°R)
Q_{scfm} = 10.0 scfm per sparge point
Q_{design}= ($Q_{injection}$)(Number of Sparge Points)
Q_{design}= (10 scfm)(34)
Q_{design}= 340 scfm

Based on a total design injection rate of 340 scfm and pressure of 7.5 psi (injection pressure plus 25 percent to account for system pressure drop), a rotary screw compressor would be suitable for this high flow rate but low pressure scenario. It should be noted that a detailed evaluation of the system pressure drop may be necessary so that the compressor is not undersized; however, those calculations are beyond the scope of this book. Detailed discussion of pressure losses in compressed air systems can be found in fluid mechanics texts (Prasuhn 1980) and from vendor literature (Ingersoll-Rand 1988).

To minimize capital costs, the sparge system could be operated in three zones (two zones with 11 sparge points and one zone with 12 sparge points) thus reducing the system operation requirements to 120 scfm at 7.5 psi. This would expand the compressed air system options to include rotary vane compressors and high pressure regenerative blowers. The smaller compressed air unit will result in lower capital costs and power usage. While a pulsed system may remove contaminant mass more efficiently and cost less to install and operate, these benefits must be weighed against a longer time to achieve the cleanup goals because less air is being injected into each zone for a given unit of time. Because there are many possible combinations of equipment and operations scenerios, experience is valuable in the final selection of the sparging equipment and development of the operation plan.

REFERENCES

Acomb, L.J., et al., Neutron Probe Measurements of Air Saturation Near an Air Sparging Well, Third International Symposium *In Situ* On-Site Bioreclamation, San Diego, April 1995.

Ahlfeld, D.P., Dahmani, A., and Wei, Ji, "A Conceptual Model of Field Behavior of Air Sparging and Its Implications for Application," *Groundwater Monitoring and Remediation*, Fall 1994.

Angell, K.G., "*In Situ* Remediation Methods: Air Sparging," *The National Environmental Journal*, pp. 20-23, January/February 1992.

Ardito, C.P. and Billings, J.F., "Alternative Remediation Strategies: The Subsurface Volatilization and Ventilation System," Proceedings of the Petroleum Hydrocarbons and Organic Chemicals in Groundwater: Prevention, Detection, and Restoration, Houston, 1990.

Boersma, P.M., Diontek, K.R., and Newman, P.A.B., "Sparging Effectiveness for Groundwater Restoration", Third International Symposium, *In Situ* and On-Site Bioreclamation, San Diego, April 1995.

Bohler, J. B., Hotzl, H., and Nahold, M., "Air Injection and Soil Air Extraction as a Combined Method for Cleaning Contaminated Sites-Observations from Test Sites in Sediments and Solid Rocks," in *Contaminated Soil,* Eds. Arendt, F., Hinsevelt, M., and van der Brink, W.J., 1990.

Brown, R., Herman, C., and Henry, E., "The Use of Aeration in Environmental Cleanups," Presented at HAZTECH International Pittsburgh Waste Conference, Pittsburgh, Pennsylvania, 1991.

Brown, R.A., (1992b), "Treatment of Petroleum Hydrocarbons in Groundwater by Air Sparging," Section 4, Research and Development, RSKERL - Ada, USEPA, Eds. Wilson, B., Keeley, J., and Rumery, J. K., Nov. 1992.

Brown, Richard A., (1992a), "Air Sparging: A Primer for Application and Design," Subsurface Restoration Conference, USEPA 1992.

Clayton, W.S., Brown, R.A., and Bass, D.H., "Air Sparging and Bioremediation: The Case for *In Situ* Mixing," Third International Symposium, *In Situ* and On-site Bioreclamation, San Diego, April 1995.

Geraghty & Miller, Inc., "Air Sparging Projects Data Summary," 1995.

Howard, P. H., et al., *Handbook of Environmental Degradation Rates,* Lewis Publishers, Inc. 1991.

Ingersoll-Rand, Cameron Hydraulic Data, 1988.

Johnson, R. L., Center for Groundwater Research, Oregon Graduate Institute, Beaverton, Oregon, Personal Communication, April 1995.

Johnson, R.L., Johnson, P.C., McWharter, D.B., Hinchee, R.E., and Goodman, I., "An Overview of *In Situ* Air Sparging, Groundwater: Monitoring and Remediation," Fall 1993.

Kresge, M.W., and Dacey, M.F., "An Evaluation of *In Situ* Groundwater Aeration," Proceedings of the Ninth Annual Hazardous Waste Materials Management Conference/International, Atlantic City, New Jersey, 1991.

Kuo, J., *Practical Design Calculations for Groundwater and Soil Remediation,* Lewis Publishers, 1999.

Leeson, A., Johnson, P. C., Johnson, R.L., Hinchee, R.E., and McWhorter, D.B., "Air Sparging Design Paradigm" - Draft, Battelle, June 14, 1999.

Lundegard, P.D., and Andersen, G., "Numerical Simulation of Air Sparging Performance," Proceedings of the Petroleum Hydrocarbons and Organic Chemicals in Groundwater: Prevention, Detection, and Restoration, Houston, Texas, 1993.

Lundegard, D., "Much Ado About Mounding," Third International Symposium, *In Situ* and On-site Bioreclamation, San Diego, April 1995.

Lyman, W. J., Reehl, W.F., Rosenblatt, D.H., *Handbook of Chemical Property Estimation Methods,* McGraw-Hill Book Company, New York. 1992.

Marley, M.C., Li, F., and Magee, S., "The Application of a 3-D Model in the Design of Air Sparging Systems," Proceedings of the Petroleum Hydrocarbons and Organic Chemicals Groundwater: Prevention, Detection, and Restoration, Houston, Texas, 1992.

Marley, M.C., Walsh, M.T., and Nangeroni, P.E., "Case Study on the Application of Air Sparging as a Complimentary Technology to Vapor Extraction at a Gasoline Spill Site in Rhode Island," Proceeding of HMCRIS 11th Annual National Conference, Washington, D.C. 1990.

Marley, M.C., Bruell, C.J., and Hopkins, H.H., "Air Sparging Technology: A Practice Update," Third International Symposium, *In Situ* and On-site Bioreclamation, San Diego, April 1995.

Mohr, D.H., Mass Transfer Concepts Applied to *In Situ* Air Sparging, Third International Symposium on *In Situ* and On-Site Bioreclamation, San Diego, April 1995.

Prasuhn, A. L., *Fundamentals of Fluid Mechanics,* Prentice-Hall, Inc. 1980.

Sellers, K. and Schreiber, R., "Air Sparging Model for Predicting Groundwater Cleanup Rate," Proceedings of Petroleum Hydrocarbons and Organic Chemicals in Groundwater: Prevention, Detection, and Restoration, Houston, Texas, 1992.

Seymour, USFilter Johnson Screens, St. Paul, MN, personal communication, 2000.

Trotta, L., USF Johnson Screens, St. Paul, Minnesota, Personal Communication, June 2000.

U.S. Environmental Protection Agency, "Evaluation of the State of Air Sparging Technology," Report 68-03-3409, Risk Reduction Engineering Laboratory, Cincinnati, Ohio 45286, USEPA, 1993.

Wei, J., Dahmani, A., Ahlfeld, D.P., Lin, J.D., and Hill III, E., "Laboratory Study of Air Sparging: Airflow Visualization," Groundwater: Monitoring and Remediation, Fall 1993.

CHAPTER **6**

Air Treatment for *In Situ* Technologies

Donald F. Kidd and Evan K. Nyer

CONTENTS

INTRODUCTION

In Chapter 1 we introduced the concept that most *in situ* treatment processes were simply a switch from water to air as the carrier. This chapter will look at treating the air carrier as it is brought above-ground.

Above-ground vapor treatment of emissions from soil vapor extraction, air sparging, and air stripping applications often represents the largest portion of the overall cost of implementing these technologies. Figure 1 represents a pie chart of overall project costs associated with a vapor extraction system which operates for 3 years at a vapor recovery rate of 300 cfm and a declining influent concentration from 2,000 ppm (hydrocarbon vapors) to 5 ppm over the project lifetime. It is assumed that vapors are treated using catalytic oxidation for the first two years, and that granular activated carbon (GAC) is used for treatment during the last year of treatment. Figure 1 shows that air emission control and O&M related costs might be over 50 percent of overall project costs.

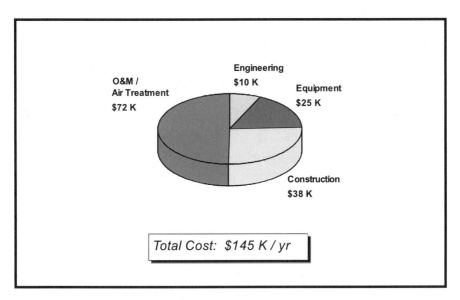

Figure 1   Pie chart of overall VES project costs.

Due to the magnitude of air emission control costs, the design engineer must carefully evaluate and select the most appropriate technology. Control technology selection must consider several criteria that will be introduced in this chapter:

- Regulatory requirements
- Overall mass of VOCs to be treated
- Anticipated decline rate of VOC concentrations over project lifetime (life-cycle design)
- Citing considerations
- Utility availability

- Organic and inorganic composition of process vapor stream
- Other project specific considerations

The most common air emission control technologies can be classified as adsorptive, oxidative, or biological. The adsorption based technologies include off-site regenerable/disposable vapor phase GAC, on-site steam regenerable GAC, and on-site regenerable macroreticular resin systems. Oxidative technologies include catalytic oxidation, and thermal oxidation. Biological based systems have gained attention in the last few years and have become commercially available. Less commonly utilized technologies include scrubbing, vapor compression, UV/ozone oxidation, and refrigeration. The most commonly utilized technologies will be introduced in this chapter.

DESIGN CRITERIA

Regulatory Requirements

Vapor emissions from site remediation activities are generally not permanent sources of discharge. The short duration of the emission may exempt its permitting and control in some states. Often however, in cities or states where overall air quality standards are not met (nonattainment areas), or in states with strict emission control standards, permitting and vapor treatment is required.

Emission requirements are quite variable within the different states. Emission requirements may be based on total mass emissions per hour or per day. Mass based emission criteria may be for total VOCs and/or may be compound specific. The design engineer must select the air emission control system based upon the most limiting criteria. For example, VES emissions for a gasoline release contain a variety of compounds. If air emission standards require a maximum 15 pounds/day total VOC limit, and an overall benzene limit of one lb/hr, then system design must be based upon the limiting regulatory requirement. In this example, the limiting criterion is likely based on total VOC emissions, since benzene usually constitutes only a small fraction of the total mass of gasoline. Because of the variable composition of gasoline and variations in constituent volatility, vapor samples collected during pilot testing are often used to determine the factors dictating emission control design.

Alternately, emission control criteria may simply require that the emission rates not cause an exceedance in ambient air quality or other risk based criteria. This generally requires that the point source of discharge be modeled using air dispersion techniques (Bethea 1978). Several states impose both mass emission and concentration criteria. Some states require that all emissions be treated using Best Available Control Technology (BACT) regardless of the magnitude of the emissions (this requirement calls for emission control even if the process stream already complies with mass emission requirements). Therefore, the first thing that the design engineer must do is to acquire the local and state regulations before trying to design a vapor treatment system. Based upon lifecycle design (Chapter 2), emission rates decline during site remediation. Permit preparation should account for this temporal change.

It is plausible to limit the site operation (hours/day or number of extraction wells) to stay within permitting limits (without treatment) until emission rates drop as the cleanup progresses. This approach will likely increase site remediation time frame and the design engineer must conduct a cost-benefit analysis in order to justify merits of this phased start up method.

Mass of Contaminants

An estimation of the total mass of VOCs that may be recovered by the remediation system is often a requirement prior to determining the appropriate treatment technology. This is particularly true for adsorbent based treatment systems such as carbon or resin-based controls. For example, if 1,000 pounds of gasoline are known to be in the subsurface, and one expects that 65 percent of the mass will be recovered by vapor extraction, 30 percent will be biodegraded, and 5 percent will not be recovered (ratios are based upon empirical projection), an estimate can be formulated for expected adsorbent consumption. Assuming a 7 percent by weight adsorption capacity for GAC, approximately 9,300 lbs of GAC will be required. The cost of other technologies can also be estimated based upon the mass of VOCs and expected flow rates.

There are several simple methods to estimate the mass of VOCs in the subsurface. Once this is calculated, an estimate can be made of the amount (percent) that is expected to be extracted for above-ground treatment. Often the final estimate is based on an average of the various estimation methods. An excellent starting point is direct knowledge of the amount of contaminants released. Time is also an important factor in that spills will weather and naturally degrade (see Chapter 7 on *In Situ* Bioremediation).

Second, the total mass may be estimated by using soil contaminant concentrations, groundwater concentrations, soil gas concentrations, and NAPL thickness at the various locations across the site (see Equation 1 in Chapter 3). The use of weighted average methods (concentration and expected flow from each zone) and subdivision of the site into small quadrants (based on the available data) will yield more accurate mass estimates.

It should be noted that soil analytical methods often underestimate the amount of adsorbed VOCs due to significant losses during the sampling procedures (USEPA 1991). Use of methanol extraction/preservation methods can often lead to soil contaminant levels that may be one to two orders of magnitude higher than conventional methods. It should also be noted that core samples represent a small statistical percentage of the sampled media, and therefore, are inherently inaccurate.

Third, in instances where limited information is available, gross estimates of the total mass of contamination in the subsurface may be evaluated using partitioning coefficients. For example, if no soil contamination data is available, groundwater data, knowledge of the compound octanol-water partition coefficient, and soil organic content can be used to estimate the amount of VOCs adsorbed to the soil (Equation 3 in Chapter 3). It should be stated that these equations assume equilibrium conditions persist in the subsurface. Nonequilibrium conditions generally dominate in the subsurface, and partitioning based calculations underestimate the adsorbed mass. This is because a large portion of the mass may be restricted from being in equilibrium with the surrounding soil vapor/groundwater due to nonequilibrium type adsorptive or mass transfer limitations (Brusseau 1991).

Lifecycle Emission Concentration

Design of cleanup strategies to accommodate the lifecycle of the project has been emphasized several times in this book. This is particularly true for the treatment of vapor emissions from VES, where concentrations may drop four orders of magnitude over a project lifetime (Figure 2). Emission control technology selection is more significantly affected by concentration than volume through-put for vapor phase treatment than for liquid phase treatment. For example, an air stripper will generally be chosen for the treatment of 100 ppm or 100 ppb of groundwater contaminated by BTEX compounds. This choice will be made for almost any flow rate. On the other hand, an emission stream of 10,000 ppm vapors from a VES stack (300 cfm), is best treated by a thermal oxidizer. As the concentrations drop to 1,000 ppm, the vapors are best treated by catalytic oxidation. At influent concentrations of 20 ppm, the optimal choice may be GAC. The original design must encompass all of these criteria, not just the initial influent concentration.

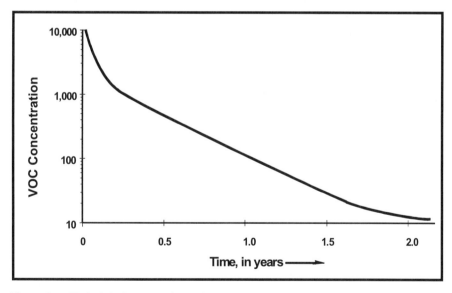

Figure 2   Typical decline curve for VES emissions.

This dynamic need to modify treatment technologies necessitates foresight from the design engineer for vapor emission system design. The systems must be designed and installed with sufficient flexibility to allow for future modifications. For example, at a site where VES emissions are expected to be above 300 ppm for 6 months and then drop off rapidly (typical small service station, limited spill situation), a catalytic oxidizer may be rented for the first six months and subsequently GAC may be installed at the site. Many localities have recognized the benefits and needs for a dynamic emission control scheme for these typically short duration emission sources. In efforts to streamline the local regulatory approval process, often permits are issued for various locations, allowing one system to be placed at several sites within the agency's jurisdiction. Such permits allow for more rapid deployment of the appropriate emission control system, while reducing the burden for detailed evaluation of an often lengthy permit application. When considering the merits of an adaptive emission control scheme, the treatment system citing must accommodate future needs of the contingent system allowing adequate space and utility connections for each component of the treatment process equipment.

Typically, the engineer needs to predict the decline curve for the emissions from the air treatment system and subsequently prepare a cost analysis for the various options at varying concentrations. A typical cost analysis table is shown in Table 1. Modeling of the remedial system performance to predict the decline curve may be conducted. In many instances, this modeling is not performed and empirical methods (fitting the concentration decay to an empirical logarithmic decay equation over a time period based upon past experience) are used for its prediction. The use of empirical methods is generally acceptable in the consulting industry for purposes of air emission selection due to the costs of modeling and its inherent uncertainties. For example, it is not critical to know whether a catalytic oxidizer will run for six or seven months before switching to GAC, what is important is the ability to plan for and switch to GAC.

Table 1   Cost Analysis Spreadsheet for Vapor Treatment Costs

	Influent Concentration	VOC/day	GAC cost/day	Catalytic Oxidation cost/day
1	50 ppm	1.47	$59	$136
2	100 ppm	2.93	$117	$128
3	200 ppm	5.86	$234	$120
4	500 ppm	14.65	$586	$112
5	1,000 ppm	29.31	$1,170	$110
6	2,000 ppm	58.62	$2,344	$100
7	3,000 ppm	87.93	$3,516	$100

Cost Assumptions:
1 ppm = 3.26 mg/cu. meter (benzene)
100 cfm operation for 24 hour per day
GAC adsorption capacity = 15% by weight
Carbon Costs = $6/lb. (new plus regeneration and changeout cost)
Catalytic oxidation unit rental is $3,000/month
Catalytic oxidation power consumtion is $350/month (at 500 ppm influent); assume costs are slightly higher at lower concentrations; slightly lower at higher concentrations assume costs are slightly higher at lower concentrations; slightly lower at high concentrations.
Cost per day of catalytic oxidizer is (3,000+350)/12= 114

Citing and Utility Considerations

There are several citing considerations that need to be evaluated prior to treatment technology selection. Some of these constraints are presented below:

- Availability of utilities
- Utility cost analysis
- Access issues relating to O&M
- Aesthetic issues
- Proximity to homes and buildings
- Winterization
- Other site specific considerations

The availability of utilities and their ability to accommodate the treatment equipment must be carefully evaluated. Most utility evaluations are for thermal and catalytic oxidation. For example, if natural gas is to be selected, it must be available in sufficient pressure to be utilized by the treatment equipment. In residential neighborhoods, natural gas lines may not have sufficient pressure for adequate operation of some thermal oxidation units. Even where available, high pressure gas line connections to the oxidizer typically require additional lead time for permitting and installation. Electrical power must be available in the appropriate phase and voltage to power the equipment. At remote sites, where utility availability is limited, propane tanks can be utilized. The design engineer needs to conduct a cost analysis in order to choose the most appropriate power source (natural gas, electrical, propane, oil, etc.) for powering the treatment unit. When available, natural gas tends to be the lowest cost option in many locations. The use of propane, in addition to increased cost per BTU (generally 1.5 to 2 times higher than natural gas), also presents other operational problems such as increased fouling of burner components, as well as logistical problems created from scheduling fuel deliveries.

Remedial systems are unplanned installations. Sites and neighborhoods are obviously developed without planning for a potential remedial system installation. This unplanned remedial system, therefore, needs to be located to accommodate several factors that may sometimes be conflicting. It must be located to attain permitting, meet regulatory stack height and air dispersion requirements, fit in with the natural setting, and be accessible for routine operation and maintenance. Concurrently, the system must not be offensive to neighbors and have a stack height that meets local zoning laws.

TREATMENT TECHNOLOGIES

Adsorption-Based Treatment Technologies

Adsorption is a process by which material accumulates on the interface between two phases. In the case of vapor phase adsorption, the accumulation occurs at the air/solid interface. The adsorbing phase is called the adsorbent and the substance being adsorbed is termed an adsorbate. It is useful to distinguish between physical adsorption, which involves only relatively weak intramolecular bonds, and chemisorption, which involves essentially the formation of a chemical bond between the sorbate molecule and the surface of the adsorbent. Physical adsorption requires less heat of activation than chemisorption and tends to be more reversible (easier regeneration).

GAC is the most popular vapor phase adsorbent in the site remediation industry. A number of new synthetic resins, however, have shown increased reversibility and have higher adsorption capacities for certain compounds.

The most efficient arrangement for conducting adsorption operations is the columnar continuous plug flow configuration known as a fixed bed. In this mode, the reactor consists of a packed bed of adsorbent through which the stream under treatment is passed. As the air stream travels through the bed, adsorption takes place and the effluent is purified (Figure 3).

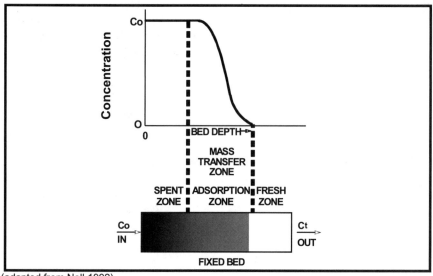

(adapted from Noll 1992)

Figure 3   Concentration profile along an adsorbent column.

The part of the adsorption bed that displays the gradient of concentration is termed the mass transfer zone (MTZ). The amount of adsorbate within the bed changes with time as more mass is introduced to the adsorbent bed. As the saturated (spent or used) zone of the bed increases, the MTZ travels downward and eventually exits the bed. This gives rise to the typical effluent concentration versus time profile, called the breakthrough curve (Figure 4). The reader is referenced to several textbooks for adsorption theory, multicomponent effects, isotherm description, and modeling (Noll, Vassilios, and Hou 1992 and Faust and Aly 1987). This basic knowledge of adsorption theory is critical to proper understanding and selection of the various adsorbents.

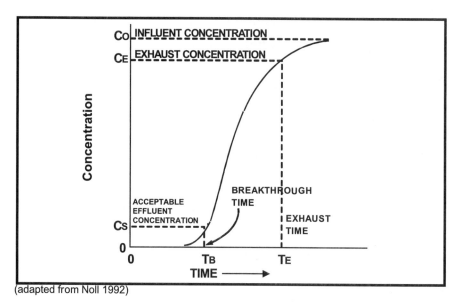

(adapted from Noll 1992)

Figure 4   Breakthrough curve for a typical adsorber column.

Off-Site Regenerable/Disposable Gas Phase GAC

Gas phase GAC is an excellent adsorbent for many VOCs commonly encountered in vapor extraction, air sparging, vacuum-enhanced recovery, and conventional groundwater extraction. The adsorption capacity of GAC is often quantified as the mass of contaminant that is adsorbed per pound of GAC. This nominal adsorption capacity is a useful guide for pure compound adsorption but can be misleading when complex mixtures of VOCs are treated. Breakthrough, or GAC bed life, is defined when breakthrough occurs for the compound most difficult to adsorb. In instances where multicomponent mixtures are present, the adsorption capacities for each compound are generally lower than for pure compounds. Isotherm data (mg adsorbate removed per g adsorbent at a constant temperature) and other product specific data are generally available for contaminants of interest from carbon vendors as well as in the published literature. Pilot testing for GAC feasibility is rarely conducted, except in instances where complex mixtures of VOCs are encountered.

GAC is generally a good adsorbent for hydrocarbon origin VOCs, and some chlorinated VOCs. GAC has limited adsorption capacity for ketones and generally poor adsorption of volatile alcohols. Table 2 shows typical adsorption removal efficiencies for a variety of VOCs by GAC under constant temperature and moisture conditions (as stated in the table).

GAC adsorption capacity is significantly enhanced if the vapor stream's relative humidity is kept low. The use of water knock outs, demisting, desiccants, and air stream temperature adjustments are therefore common pretreatment steps to enhance GAC performance. Adsorption capacity may be as much as 10 times higher for a low humidity stream than for a humid air stream. This is particularly true for lower

Table 2   Adsorption Capacity of GAC for Some Common VOCs

	Carbon Capacity* (lb org/100 lbs GAC)	
	@ 10 ppmv	@ 100 ppmv
Benzene	13	19
Carbon Tetrachloride	20	33
Methylene Chloride	1.3	2.7
Toluene	21	27
TCE	19	33

Note: Carbon capacities are based on Calgon BPL carbon at 70 F and
1 atm. Values adjusted to reflect usage per 100 lbs GAC.

Source: Calgon Bulletin #23-77a (2/86).

concentrations of VOCs. The humidity effects are less pronounced at higher VOC concentrations (Figure 5). Temperature elevation after water demisting/desiccation increases the amount of moisture that the air stream can sustain thus reducing the relative humidity. It should be noted that the adsorption vessels must also be kept warm so as to avoid water condensation on the GAC and for the air stream to maintain the reduced relative humidity.

The use of off-site regenerable/disposable GAC for emissions control is often limited to instances where the mass loading is low and therefore GAC consumption is low. As a general guideline adsorption capacities for adsorbable VOCs are in the range of 2 to 20 percent range by weight. GAC costs are in the range of $3/lb (if purchased in canisters; $1/lb if purchased as carbon only) plus an additional $0.5 to $1/lb for regeneration/disposal. GAC vessels may be purchased as canisters (200 lb size, where the vessel and GAC are replaced after consumption), larger replaceable plastic/fiberglass canisters, or as conventional steel vessels wherein only the GAC

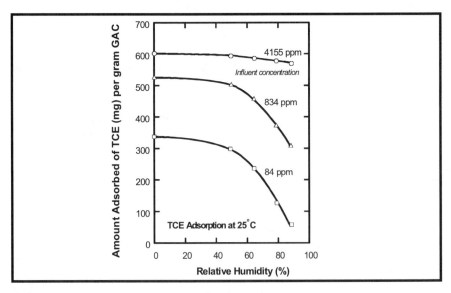

Figure 5   Effects of relative humidity on TCE adsorption by GAC.

is replaced. Vessels are most commonly used in series in order to allow for effluent stream sampling between vessels in order to more accurately predict breakthrough times (Figure 6). Single vessels may also be used in conjunction with vapor effluent detectors that can shut down the system or switch spent vessels to stand by unused vessels. GAC vessel/blower selection must account for head losses through the carbon system to ensure maintenance of the desired airflow.

Figure 6   Photo of typical GAC system in series.

On-Site Regenerable GAC

Liquid phase GAC is difficult to regenerate at low temperatures due to adsorption of background metals and TOC (typically naturally occurring humic and fulvic acids). In the vapor phase, the GAC does not see the metals nor the nonvolatile TOC and is therefore amenable to on-site, low temperature regeneration. Air emission control using steam regenerable GAC generally utilizes a two bed system whereby one bed is being utilized, while the second bed is either being regenerated or in a stand by mode. Regeneration is accomplished by passing low-pressure steam through the carbon vessel, which desorbs the contaminants (due to the raised temperature). The contaminated steam is subsequently cooled and separated from the nonaqueous organics by gravity. The condensed organics generally require disposal, whereas the contaminated steam may undergo water treatment (especially if a groundwater treatment system exists on-site) or may also require off-site disposal. If the condensed steam is treated on-site, it should be metered into the groundwater treatment system since it is generally much more contaminated than the groundwater.

If regenerable systems are used for adsorption of chlorinated VOCs, the vessels should be lined or made of an acid-resistant material. The adsorption/regeneration cycle results in formation of some hydrochloric acid (HCl) within the vessels. The HCl formation will reduce the pH of the condensed steam.

Regenerable GAC units are also available with nitrogen regeneration. This is particularly useful to minimize steam disposal and eliminate problems with HCl formation during steam regeneration. Regenerable beds will usually have a performance lifetime, since adsorption capacity tends to diminish with continued regeneration (typically decay to 70 percent of original capacity). System lifetime ranges are dependent on frequency of regeneration but are typically in the 3- to 7-year range.

Regeneration can be, (1) manual during a site operation and maintenance visits; (2) by a timer system that starts up the boiler/regeneration prior to the time of expected breakthrough; or (3) using an effluent detector system (typically either a flame ionization or photoionization detector) which initiates vessel alternation and regeneration based upon breakthrough. After completion of the regeneration, the vessel will usually undergo a drying cycle in order to prepare the vessel for the next adsorption cycle. Figure 7 provides a schematic of a regenerable GAC system. Figure 8 is photo of a commercially available system.

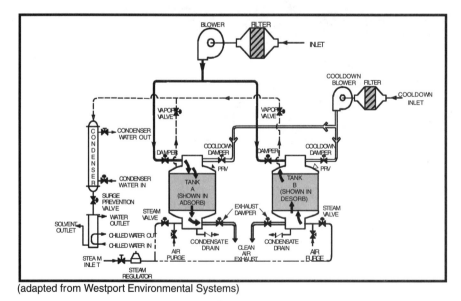

(adapted from Westport Environmental Systems)

Figure 7   Regenerable GAC system schematic.

Figure 8   Regenerable GAC system photo.

The regeneration capability of the GAC allows for treatment of more contaminated air streams than is possible with off-site regenerable GAC. In VES applications, regenerable GAC systems are often thought of as best suited for high flow applications with moderate VOC loadings. It can be used for the treatment of halogenated and nonhalogenated air streams. During adsorption of highly oxidizable VOCs (ketones) in the presence of ozone or other oxidizing agents, the GAC bed may be prone to bed fires. Fire detection and suppression systems may be considered under these circumstances.

As a general guideline, fully manual regenerable systems with a boiler may be purchased for roughly $60,000 for a 500 cfm dual bed application. System automation with effluent detector will generally add approximately $40,000.

Resin Adsorption Systems

The use of adsorbent resins for water treatment began in the 1960s with Rohm and Hass' introduction of their Macroreticular Ambersorb adsorbent (Faust and Aly

1987). These synthetic resins were designed to have low adsorption for background TOC and metals and thereby enable the liquid phase adsorbent to better adsorb the VOCs. Elimination of TOC and metal adsorption would also reduce biofouling (less TOC for microbial growth). The liquid phase resins were also designed to be steam regenerable since metals and TOC fouling would not impact regeneration. For a variety of economic and performance reasons, liquid phase adsorbent resins have gained limited acceptance. New resins by both Rohm and Hass and Dow within the past five years, however, may make liquid phase resin adsorption more economically competitive.

It had been commonly prescribed that these resins be prewet prior to their use. In the 1980s, it was observed that the resins have an almost similar adsorption capacity in their prewet condition (Yao and Tien 1990 and Rixey and King 1989). These resins tend to adsorb VOCs in a similar fashion to GAC, but also tend to demonstrate some absorption within the resin itself (Gusler, Brown, and Cohen 1993). The regeneration capability of absorbed VOCs is not currently well quantified. Despite this uncertainty, the resins appear to have a few advantages over vapor phase GAC. The resins have a higher adsorption capacity for some VOCs than GAC, are less influenced by relative humidity than GAC, and their on-site regeneration tends to produce less acid when adsorbing chlorinated VOCs than GAC. The resins, however, appear to have some catalytic ability, and therefore break down some of the halogenated VOCs to form the hydrochloric acid (but still less than GAC). The resins also have less oxidative capacity and are less prone to bed fires than GAC for adsorption of readily oxidizable compounds.

In particular, the resins appear to better adsorb vinyl chloride than conventional GAC systems. When present, vinyl chloride can effectively preclude GAC adsorption as the sole emission control process due to it low adsorptive capacity and typically low emission requirements. Chemical oxidation using potassium-permanganate impregnated activated alumina has been found an effective alternative. In this application, the activated alumina is placed at the end of the treatment train where vinyl chloride is essentially the only organic compound reaching the oxidizing agent. By this configuration, oxidation of other organic compounds is minimized, similarly minimizing the consumption of the oxidizer.

An additional advantage of the resin system over steam regenerable GAC, is the elimination of steam disposal/treatment. Although prices have been declining for these systems, they are still higher than conventional regenerable GAC systems. The average price of the resin is $70/lb in comparison to $1/lb for GAC (if purchased in bulk; $3/lb in canister form). The resin's superior performance for certain compounds and its minimal disposal and O&M costs, however, may make it more cost effective than regenerable GAC for certain installations. In general, the resin adsorption technology is not considered competitive for hydrocarbons, since these compounds can be more economically destroyed by catalytic or thermal oxidation. The resin technology is most applicable for moderately contaminated halogenated VOC vapor streams. However, even these economics are changing as better catalysts are developed for thermal destruction of chlorinated hydrocarbons.

Oxidation-Based Technologies

Oxidation-based technologies react to the VOCs at elevated temperatures with oxygen for sufficient time to initiate the oxidation reactions. The ultimate goal of any combustion/oxidation reaction is the conversion of the VOCs to carbon dioxide, water, sulfur dioxide, and nitrogen dioxide (end products of combustion). The mechanism involves rapid chain reactions that vary with the different VOCs being oxidized. The sequential reactions leading to combustion are generally not detrimental to the environment or to the oxidative equipment unless they are interrupted (possibly insufficient residence time or decayed catalysts). If the oxidation reactions are incomplete, partial oxidation byproducts will be released in the stack effluent. Sometimes these compounds may be more noxious than the parent compound. For example, oxidation of oxygenated organics may form carbon monoxide under unsatisfactory combustion conditions. To minimize such occurrences, excess air (above stoichiometric conditions) is used during oxidation reactions. The continuity of any oxidation reaction depends on maintaining the reaction mixture (air, VOCs, temperature, and catalyst, if used) in the optimal mixture range (Bethea 1978). The two most common oxidation methods in the remediation industry are thermal oxidation and catalytic oxidation. Figure 9 shows a schematic of the two methods.

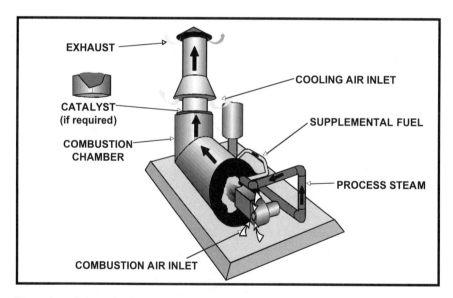

Figure 9   Schematic of thermal oxidation and catalytic oxidation.

Thermal Oxidation/Incineration

This technology burns the VOCs at an elevated temperature with resultant production of burn gases such as carbon dioxide and water vapor. Thermal oxidizers typically operate at 1400-1500°F with typical residence times of 0.75 seconds to 2.0

seconds. The burner chamber must be designed so that the VOCs pass directly through its flame, thus minimizing the possibility of incomplete combustion. Although thermal oxidation is often the most expensive combustion based control process, it is well suited for high concentration VOC air streams (2000 to 10000 ppm range). Operating costs can be offset to some extent if the gas stream is already considerably above ambient temperature by utilizing the combustion heat for pre-heating the air stream (heat recovery). The use of this technology for air streams that have significant BTU value also reduces fuel consumption. VOC destruction efficiencies of well operated thermal oxidizers are generally in the range of 95 to 99 percent. Thermal oxidizers generally have a maximum allowable concentration of influent VOCs (50 percent of LEL or in the range of 7000 ppm (by weight) for most hydrocarbon applications. While technically possible to operate thermal oxidizers above the 50 percent LEL range, the increased temperature created by oxidation of process streams with concentrations higher than this will often exceed the temperature design criteria for most commercially available oxidizers (typically 1600 to 1700°F maximum.) During setup and operation, care must be taken to ensure that the free oxygen in the exhaust stream maintains at a minimum of 10 percent. Situations such as reducing conditions in the subsurface being vented, or by high influent concentrations can lead to oxygen starvation by the oxidizer, resulting in less than optimal destruction efficiency. Figure 10 shows a photo of a commercially available thermal oxidizer.

Figure 10   Photo of a thermal oxidizer.

Thermal oxidizers have generally been used for the treatment of high concentrations of hydrocarbon VOCs. If the influent concentrations exceed 13,000 ppm, dilution air is added. The oxidation of halogenated hydrocarbons may result in formation of byproducts and hydrochloric acid that may necessitate the treatment of the combustion gases. This limitation generally leads many practitioners to use adsorption based technologies rather than burn based technologies for the treatment of halogenated VOCs.

This technology is best suited for VES emission control at heavily contaminated sites. The most frequent application has been in the venting of free phase hydrocarbons atop the water table. Often times, thermal oxidizers are trailer mounted and are utilized during the initial phase of site remediation when VOC levels are high. As concentrations drop, power/fuel consumption generally rises dramatically, and the trailer mounted unit is replaced by a more economical treatment method. As a general guideline, a 500 cfm thermal oxidation unit with heat recovery capabilities can be purchased for $60,000. Typical monthly fuel consumption for a 500 scfm, 5,000 ppm VOC air stream is in the range of $1500 to 1800 based on a natural gas cost of $0.40/therm (the standard unit of sale for natural gas, representing 100,000 BTU). As the concentration of this vapor stream decreases, the fuel consumption cost required to treat the process stream described above can increase to more than $2500/month.

Catalytic Oxidation

Catalytic oxidation occurs when the contaminant laden air stream is passed through a catalyst bed that promotes the oxidative destruction of the VOCs to combustion gases. The presence of the catalyst bed allows for the oxidation to occur at a lower temperature than would be required for direct thermal oxidation. The primary advantage of catalytic oxidation is the decrease in supplemental fuel requirement (BTU not provided by the VOCs).

The catalyst metal surface must be large to provide sufficient active sites on which the reactions occur. Its surface must be kept free from dust or other noncombustible materials. Catalysts are subject to both physical and chemical deterioration. Physical deterioration results from mechanical attrition or overheating of the catalyst. Chemical deterioration most frequently is due to the presence of impurities in the VOC stream or from byproduct formation. For example, in VES applications of leaded gasoline, catalyst poisoning from the tetraethyl and tetramethyl lead in the gasoline vapors will likely occur. Another form of catalyst deterioration is caused by exposure to halogens or sulfur containing compounds. Halogen poisoning may occur from entrained water particles that contain chloride (especially in remedial applications involving salt water) or from chlorinated VOCs. Metals in entrained water particles may also act as poisons. Mercury, arsenic, bismuth, antimony, phosphorous, lead, zinc, and other heavy metals are common poisons. Lastly, the presence of high methane levels (either naturally occurring or escaping from pipelines) may cause catalyst damage.

Catalytic oxidation burns the VOCs at approximately 600°F for most hydrocarbon remediation applications utilizing a platinum or palladium catalyst. The technology is best suited for treatment of nonhalogenated hydrocarbons to avoid catalyst poisoning. Catalytic oxidizers are generally limited to maximum influent VOC levels in the range of 3500 ppm (since VOCs are a BTU source and the catalyst has an upper temperature limit). Higher influent concentrations will require dilution in order to reduce the influent concentration. Dilution, however, can reduce the extracted volume from the subsurface. A cost analysis is frequently used to justify operation in the catalytic mode using dilution versus purchase of a combined catalytic/thermal oxidizer. As a general guideline, a 500 cfm catalytic oxidation unit, with heat recovery capabilities, can be purchased for $75,000 or $50,000 without heat recovery. Typical fuel consumption per month for a 2,000 ppm VOC, 500 cfm air stream is in the range of $900 to $1200 based on a natural gas cost of $0.40/therm.

Catalytic oxidizers are available as stand alone units with auxiliary heat sources as shown in Figure 11 or can be purchased as an internal combustion engine as shown in Figure 12. The combustion engine unit is attractive for pilot testing applications since it is a self-contained unit with its own fuel source. Internal combustion engine units, however, tend to be less reliable and require more maintenance than conventional units, and are limited in capacity.

Figure 11   Photo of conventional catalytic oxidizer.

Catalytic oxidizers that can treat chlorinated VOCs are also available. While treatment units for site remediation have been available commercially since the early 1990s, only in the last four to five years have they become commonplace. Often, these systems are fitted with a scrubber system to remove waste HCl gas from the effluent stream prior to atmospheric discharge. The addition of a scrubber, however, can significantly increase the O&M complexity of the system. The incremental cost of the scrubber maintenance should be factored into a cost analysis when considering competitive alternatives. Scrubbers are presented in more detail later in this chapter.

Figure 12   Photo of internal combustion engine catalytic oxidizer.

Biological Technologies

Biological treatment of air emissions has gained significant interest in Europe in the past several years. In the same fashion that vadose zone bioventing can be accomplished, it is envisioned that an above-ground reactor can be configured to degrade the VOCs. The above-ground reactor is commonly referred to as a biofilter. These units are now commercially available from several vendors. Figure 13 shows a photo of a biofilter.

This technology involves the vapor phase biotreatment of the air emissions by bacterial populations growing on fixed media. The fixed media (various combinations of plastic support media, compost, wood chips, and other) and water in the biofilter assist in the adsorbing/solubilizing the VOCs for subsequent breakdown by the bacteria. Use of biofilters is relatively new in the United States but has gained widespread attention in Europe. Treatment efficiency is generally around 40 to 95 percent (compound and biofilter system specific). The more water soluble compounds are better treated than the nonsoluble compounds, since all of the degradation occurs in the water phase which is adsorbed to the support media.

Scrubbers

While most of this chapter has been devoted to organic compounds, some VES remediations may have inorganic compounds in the contaminated air stream. Hydrogen sulfide is probably the most widely found inorganic contaminant at impacted sites. Some sites have been known to have up to 30,000 ppm of hydrogen sulfide in the extracted air stream.

Chemical scrubbers can be used to remove the H_2S from the air stream. The scrubber looks very similar to an air stripper but operates in reverse. The air/water contact serves to remove the contaminants from the air and transfer them to the

Figure 13   Photo of a commercially available biofilter from EG&G Rotron.

water stream. The water can then be treated with caustic and an oxidizing agent (hydrogen peroxide, ozone, supersaturated oxygen) to destroy the hydrogen sulfide. The high pH increases the rate of H_2S transfer and oxidizing agent hydrolyzes the H_2S. This rapid reaction tends to maximize the driving gradient from the air to the water. In some instances (low concentrations: 50 ppm range) the H_2S in air may be treated biologically rather than oxidatively.

Technology Selection Summary

Vapor emissions from site remediation should be evaluated on a case by case basis, based upon the available site characterization data, regulatory requirements, lifecycle considerations and other project specific data as previously outlined. Some generalizations, however, can be formulated regarding technology selection. Thermal oxidation is generally best suited for low airflow (air is expensive to heat) and high hydrocarbon VOC situations, such as VES emissions at sites impacted by NAPL. Catalytic oxidation is best utilized at hydrocarbon sites with moderate VOC impacts and low to moderate airflows. This can be found at a typical service station site without NAPL but high to moderate soil contamination. Catalytic oxidation of chlorinated hydrocarbons is best utilized at sites with moderate VOC impacts and moderate airflows. Off-site regenerable GAC is best utilized in instances where the mass loading is low. Steam regenerable GAC is best applied at high airflow situations (no need to heat the air) and moderate VOC concentrations (low regeneration frequency). Steam regenerable GAC is applicable for most VOCs but its isotherms should be checked for the individual compound in order to ensure that regeneration frequency is acceptable. Resin adsorption systems are best utilized for halogenated VOC air streams with moderate contamination. Bio-filters are best suited for the treatment of low to moderate concentrations of soluble and biodegradable VOCs such as BTEX and ketones at low to moderate airflows. Biofilters tend to be bulky and therefore large mass loadings require large units. Table 3 provides these general guidelines in a tabular format. Lifecycle concentrations may required a combination of these technologies for the most cost effective air emission control design.

Table 3   Air Treatment Technology Selection Guidelines

Technology	Most Applicable VOCs	Most Applicable Airflow	Most Applicable Mass Loadings
Off-Site Regenerable GAC	HC, Halogenated	All	Low
On-Site Regenerable GAC	HC, Halogenated	All	Moderate
Regenerable Resin	Halogenated	All	Moderate
Catalytic Oxidation	HC	Low-Moderate	Moderate
Thermal Oxidation	HC	Low-Moderate	High
Biofilters	Soluble HC	Low-Moderate	Low-Moderate

HC = Hydrocarbons
Low airflow considered in the 100 cfm range
Moderate airflow considered in the 500 cfm range
High airflow considered greater than 2000 cfm
Low mass loading considered in the 5 lb./day range
Moderate mass loading is considered in the 50 lb./day range
Moderate mass loading is considered greater than the 100 lb./day range

Reader is advised to use this table as a general guideline ONLY

Emission Control Case Study

Thus far, we have introduced the design considerations for chemical, physical, and biological processes typically used for emission controls at remediation sites. General guidelines of when and where to install these systems has also been provided. Following is an actual case study that will demonstrate the application of these concepts from a discussion of site characteristics, alternatives, and actual performance data. In particular, these performance data will highlight the lifecycle design concept using both predicted and actual system operation.

Site Description

The subject site for this case study is a former solvent and chemical waste recycling facility located in Michigan. The site occupies an area of approximately eight acres with operation of the facility discontinued in the early 1980s. During its operation, storage, handling, and processing of waste solvents and other chemicals resulted in releases of contaminants into the environment. Based on initial investigations conducted following discontinuation of facility operations, the site was added to the National Priorities List (NPL) in 1986.

The site was subdivided into three distinct units during the investigation. As shown on Figure 14, these areas are denoted as:

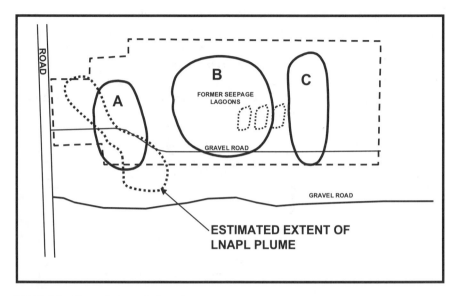

Figure 14   Case study – site layout.

- Zone A - solvent loading area
- Zone B - solvent processing area
- Zone C - soil berm area

Consistent between the areas, the primary constituents of concern are VOCs, in particular, tetrachloroethene (PCE), trichloroethene (TCE) and cis-1,2 dichloroethene (cis-1,2 DCE—a biological breakdown product of both PCE and TCE as described in Chapter 7). Trace concentrations of other VOCs and semivolatile organic compounds (SVOCs) were also found during the remedial investigation. These compounds, however, did not play a significant role in the selection of the vapor control system for the site.

In Zone A, a light LNAPL was also discovered. The LNAPL was found within an area of approximately 31,000 square feet (3/4 acre) with a maximum apparent thickness of nearly 5 feet. Through analysis, this LNAPL was found to be primarily mineral spirits (a hydrocarbon-based solvent) that also contained appreciable concentrations of dissolved chlorinated VOCs.

Soils at the site are typically fine grained sands extending from the ground surface to approximately 40 feet below grade. A confining bed of silty clay was found beneath this surficial sand that effectively limited the downward migration of contaminants released at the site. Within the surficial sand, a thin (5 to 20 foot thick) water table aquifer is present. Dissolved contaminants are found within that aquifer.

Remedial Approach

The remedial strategy adopted for this site combined several technologies to address the different phases of contamination. Specifically, a selective, product-only skimming system was installed within Zone A to remove the mobile fraction of the LNAPL plume. SVE was also chosen to remove the residual, immobile NAPL within that zone. SVE was selected to address the soil impacts found in each of the three zones at the site. An air sparging system installed into each of the zones (Chapter 5) will address the dissolved and residual contaminants below the water table.

The active phase of the remedial program at this facility will continue until contaminant mass recovery rates reach asymptotic levels. Provided conditions are protective of human health and the environment (consistent with the requirements of the ROD); natural attenuation processes will be relied upon to achieve final closure at the site. Until natural attenuation is applied, vapor withdrawal and treatment will be needed.

Emission Control Design Basis

Pilot tests were conducted at the site to define the radii of influence for both the SVE and AS systems. Given the size and complexity of the site, vapor flow modeling was used to optimize the number, placement, and operating conditions (applied vacuum and flow) for the SVE system. From the pilot testing/modeling, the radial influence of the SVE and AS wells were approximately 50 and 20 feet, respectively. A summary of the final design is as follows:

Zone A
    SVE Wells - 17 wells @ 35 scfm/well (595 scfm total extracted flow)
    AS Wells - 20 wells @ 8 scfm/well (160 scfm total injected flow)
Zone B
    SVE Wells - 20 wells @ 35 scfm/well (700 scfm total extracted flow)
    AS Wells - 27 wells @ 8 scfm/well (216 scfm total injected flow)
Zone C
    SVE Wells -10 wells @ 28 scfm/well (280 scfm total extracted flow)
    AS Wells - 10 wells @ 8 scfm/well (80 scfm total injected flow)

When operating together, the SVE system for all three zones will require a soil vapor extraction and treatment capacity of nearly 1,600 scfm. Soil vapor samples collected during the pilot tests were used to estimate the concentration of vapor-phase contaminants reaching the future emission control system. From those samples, a maximum VOC recovery rate of more than 40 lb/day was determined as the design criteria for the off gas control system.

Emission limits were determined based on the requirements of the local oversight agency (in this case the Michigan Department of Environmental Quality). Dispersion modeling was used to evaluate the impact of emissions from the SVE system on the surrounding area.

Alternatives Evaluation

Given the presence of LNAPL in Zone A and experience at many of the sites, the concentration of vapor phase contaminants from the SVE system were expected to remain elevated for an extended duration. While GAC alone was selected for use in Zones B and C, rapid and frequent exhaustion of GAC was expected in Zone A. Based on this expectation, several alternatives for emission controls were evaluated specifically for use in Zone A. After an initial screening, a detailed evaluation was conducted for the following emission control alternatives:

- Carbon (off-site regeneration)
- Carbon (on-site regeneration)
- Catalytic oxidation
- Catalytic oxidation with scrubber
- Thermal oxidation
- Thermal oxidation with scrubber

Each of the above systems would be able to achieve collection (carbon systems) or destruction (oxidation systems) of the recovered vapor phase contaminants in compliance with emission limits. The final system selection, therefore, was based on a cost-benefit analysis performed during the preliminary design phase of the project. This analysis was specifically based on the lifecycle design approach discussed in Chapter 2.

At this site, the use of off-site regenerable GAC will be ultimately used in the later stages of the project, when vapor concentrations decline to low levels. Based on experience and as summarized previously, oxidation technologies are inappropriate for high flow, low concentration vapor streams as will ultimately be the conditions at this site. The evaluation of alternative technologies is based on incremental costs with GAC as a baseline. On that basis, no capital costs are assigned to an off-site regenerable GAC system.

Table 4 provides a summary of the cost analysis conducted for the Zone A treatment system with graphical representation of the data provided on Figure 15. The cost analysis was conducted for a range of project durations, extending to two years (the maximum project duration anticipated for this area of the site). As shown on this table, thermal oxidation was projected to have the lowest overall cost (including both capital and operational expenses) at the end of a two year period. The margin between thermal and catalytic systems, however, was not significant ($125,000 for thermal and $132,000 for catalytic oxidation). Considering the lower operating temperature (less system fatigue) and fewer objectionable combustion byproducts, catalytic oxidation was selected for implementation.

As described previously in this chapter, both thermal and catalytic oxidation technologies, when applied with chlorinated VOCs, will result in the formation of hydrochloric acid in the exhaust stream. On that basis, oxidation systems combined with acid-gas scrubbers were included in the list of considered emission control systems. Based on the long-term projections for influent VOC concentrations (again the lifecycle approach) combined with dispersion modeling conducted for the permit application, the exhaust stream for this site did not require a scrubber system.

Table 4 Case Study—Off-Gas Cost Analysis Summary

Treatment Technology	Capital Costs[a]	O&M Estimated Costs per Week[b]	Cumulative Costs after 26 Weeks	Cumulative Costs after 52 Weeks	Cumulative Costs after 104 Weeks	Cumulative Costs after 156 Weeks
Carbon (off-site regeneration)[c]	$0	$2,521	$65,546	$131,092	$262,184	$393,276
Carbon (on-site regeneration)	$220,000	$360	$229,360	$238,720	$257,440	$276,160
PADRE (PURUS)	$209,000	$266	$215,916	$222,832	$236,664	$250,496
Catalytic Oxidation	$115,000	$164	$119,264	$123,528	$132,056	$140,584
Catalytic Oxidation w/Scrubber	$200,000	$216	$205,616	$211,232	$222,464	$233,696
Thermal Oxidation	$62,000	$605	$77,730	$93,460	$124,920	$156,380
Thermal Oxidation w/Scrubber	$137,000	$660	$154,160	$171,320	$205,640	$239,960

O&M Notes:
[a] Capital costs include equipment only.
[b] O&M estimated costs provided by vendor in response to *In Situ* Soil Vapor Extraction (ISVE) pilot test results.
[c] Assumes using proposed vapor phase carbon system for Zones A, B, and C of Phase III ISVE system.

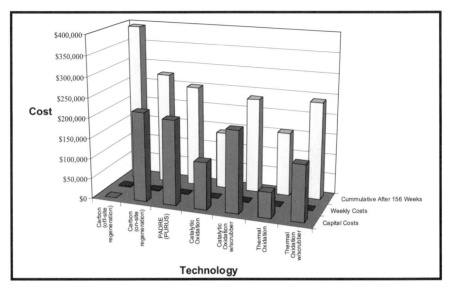

Figure 15   Case study – off gas cost analysis summary.

System Performance

Although the emission control alternatives evaluation and selection of catalytic oxidation was conducted specifically for Zone A at the site, the implemented design allowed the use of the oxidizer at any of the three zones. This added flexibility within the system allowed for the application of the most cost effective emission control scheme at any time during the project.

Since all three zones will have a lifecycle reduction in concentration over the life of the system, it was decided to optimize the use of the catalytic oxidizer. The carbon system was designed to be able to treat the entire 1800 scfm from the combined three zones. The catalytic oxidizer was designed to only treat the flow from one zone at a time. Initially, only one zone was turned on. Its high concentration stream was sent to the catalytic oxidizer. Once the concentration in that stream was below the cost effective use of the oxidizer, then the air stream was rerouted to the carbon system. A new zone was started up and that high concentration waste was now sent to the catalytic oxidizer. This lifecycle design maximized the use of the catalytic oxidizer while still using the more cost effective carbon for low concentration streams.

Table 5 provides a summary of daily costs for operation of the catalytic oxidizer at the site. In order to monitor the relative cost of competing emission control systems, Table 5 also includes calculated carbon consumption estimates using influent analytical data collected during system operation. Figure 16 provides a graphical presentation of these data covering the first year of operation.

As shown on Figure 16, the daily cost of operation for the catalytic oxidizer is generally consistent, within a narrow range from approximately $50 to $100 per day. As expected, operational costs for an oxidizer are not directly dependent on the

Table 5   Case Study—Summary of Daily Operating Costs

Date	Est. Catalytic Oxidizer Operating Costs ($/day)	Theoretical GAC Consumption Rate ($/day)
15-Aug-98	$61.03	$183
20-Aug-98	$64.65	$138
29-Aug-98	$51.60	$110
05-Sep-98	$52.33	$98
11-Sep-98	$61.75	$106
19-Sep-98	$59.58	$100
26-Sep-98	$60.30	$98
01-Oct-98	$78.42	$120
21-Oct-98	$105.09	$342
27-Oct-98	$90.60	$236
21-Nov-98	$72.91	$134
19-Dec-98	$85.52	$202
22-Jan-99	$55.81	$93
26-Feb-99	$107.27	$108
04-Apr-99	$86.68	$298
29-Apr-99	$83.50	$93
11-Jun-99	$111.62	$88
17-Jul-99	$101.18	$46
22-Aug-99	$102.92	$80

influent vapor composition within the range of concentrations processed by this system. Supplemental gas consumption, the primary system operation expense, is more a function of system flow rate and set point temperature than the VOC loading.

In contrast to the oxidizer daily costs, carbon costs are directly related to the VOC loading. For this reason, the estimated (calculated) costs for carbon treatment

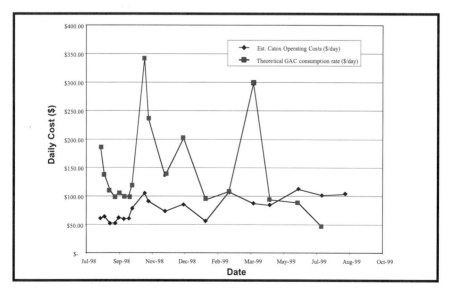

Figure 16   Case study – summary of daily operating cost.

have varied widely during the first year of operation. Excepting concentration spikes occurring following a period of extended down time and upon start up of the air sparging system, influent vapor concentrations have generally declined with time (Figure 17). As shown on Figure 16, the daily cost for carbon treatment reached levels that are consistently below the daily cost for operation of the oxidizer. The cross over point is displayed on Figure 16 at approximately May 1999. With this finding, operation of the oxidizer for Zone B was discontinued with that vapor flow stream sent to the on-site GAC system. Zone A was then started up with its vapor flow routed through the oxidizer to treat the initial, high concentration vapor stream. Daily costs will again be monitored for that area of the site to determine when carbon is more appropriate and cost effective for continued operation. Once Zone A has completed its catalytic operations, Zone C will be turned on and go through the same process.

By breaking the treatment system into three zones, this design was able to maximize the use of the most cost effective treatment technology. Normally, a designer has to decide between the technology that works best at high concentration and pay for increased operation costs as the concentration decreases, and the technology that works best at low concentrations and pay for the increased operation costs at the beginning of the treatment. This design avoided both of those problems. The only problem with this design solution was that the total time of operation for the entire system was extended due to phased approach. The economic analysis at this site showed that this was still the most cost effective approach. Each site will have to evaluate the various cost factors to determine if this phased approach will work for other sites.

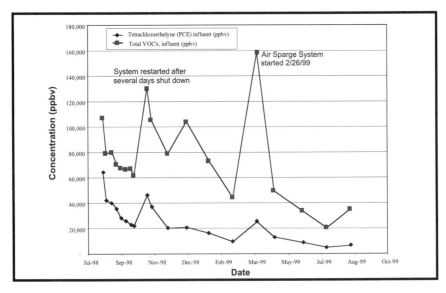

Figure 17   Case study – influent vapor composition.

REFERENCES

Bethea, R.M., *Air Pollution Control Technology,* Van Nostrand Reinhold Company, 1978.

Brusseau, M.L., Jessup, R.I., and Rao, P.S.C., "Transport of Organic Chemicals by Gas Advection in Structured or Hetrogeneous Porous Media; Development of a Model and Application to Column Experiments," Water Resources Research, 27,3189, 1991.

Faust, S. D., Aly, O.M., *Adsorption Processes for Water Treatment,* Buttersworth Publishing, 1987.

Gusler, G.M., Browne, T.E., Cohen, Y., "Sorption of Organics from Aqueous Solution Onto Polymeric Resins," Industrial Engineer Chemical Research 32(11), 2728, 1993.

Noll, K.E., Vassilios, G., Hou, W.S., *Adsorption Technology for Air and Water Pollution Control,* Lewis Publishers, 1992.

Rixey, W.G., and King, J.C., "Wetting and Adsorption Properties of Hydrophobic Macroreticular Polymeric Adsorbents," *J. Colloid and Interface Science,* 13(2), 320, 1989.

U.S. EPA, "Soil Sampling and Analysis for Volatile Organic Compounds," *Groundwater Issue,* Office of Solid Waste and Emergency Response. EPA/540/4-91/001 or NTIS DE91 016758, 1991.

Weber, W. J., *Physicochemical Processes for Water Quality Control,* Wiley Publisher, 1972.

Yao, C., and Tien, C., "Wetting of Polymeric Adsorbents and its Effect Affect on Adsorption Behavior of Hydrophobic Resin Particle," Research Polymers 13:121, 1990.

CHAPTER **7**

In Situ Bioremediation

Gary Boettcher and Evan K. Nyer

CONTENTS

1-56670-528-2/01/$0.00+$.50

INTRODUCTION

In situ bioremediation is a true *in situ* technology. In this process, biochemical reactions destroy or chemically modify compounds such that these compounds are no longer a threat to human health or the environment. The reactions occur below ground, making this one of the few in place or *in situ* technologies used today.

Chapters 3, 4, and 5 describe physical and chemical *in situ* technologies that are designed to actively remediate petroleum hydrocarbons, chlorinated hydrocarbons, pesticides, or inorganics (including heavy metals). As discussed in those chapters, the air used as a carrier during those processes can also stimulate biological reactions. Bioremediation is an important element of most of these remedial solutions where groundwater and soil are impacted with organic or inorganic compounds. However, not all remedial solutions require active soil or groundwater remediation. In recent years, protocols have been developed that are designed to evaluate the environment's ability to naturally attenuate impacts. This is a powerful remediation technology that will continue to be expanded, refined, and applied where soil and groundwater have been impacted with organic and inorganic compounds.

This chapter is divided into two main sections in order to understand bioremediation and implement natural attenuation. The first section will focus on biological and chemical processes that are important to understand when designing or operating all *in situ* biological remediation systems. The second section will focus on natural attenuation processes whereby the environmental conditions and these processes are able to achieve the remediation goals.

MICROBIOLOGY AND BIOCHEMISTRY BASICS

Microorganisms in soil and groundwater complete biochemical reactions. These reactions often directly or indirectly destroy or modify organic and inorganic chemicals. These microorganisms are living creatures, and as such, require favorable environmental conditions in order to complete their biochemical reactions and remediate organic and inorganic chemicals. Therefore, it is important to have a basic understanding of microorganisms, metabolism, growth, and microbial degradation processes in order to design or evaluate *in situ* bioremediation systems. This understanding will allow design engineers to exploit biochemical reactions in soil and groundwater environments and avoid potentially inhibitory conditions.

Microorganisms

Free-living microorganisms that exist on earth include bacteria, fungi, algae, protozoa, and metazoa. Viruses are also prevalent in the environment; however, these particles can only exist as parasites in living cells of other organisms and will not be discussed in this text. Microorganisms have a variety of characteristics that allow survival and distribution throughout the environment. They can be divided into two main groups. The eucaryotic cell is the unit of structure that exist in plants, metazoa animals, fungi, algae, and protozoa. The less complex procaryotic cell includes the bacteria and cyanobacteria.

Even though the protozoa and metazoa are important organisms that affect soil and water biology and chemistry, they do not perform important degradative roles. Therefore, this chapter will concentrate on bacteria and fungi.

Bacteria are by far the most prevalent and diverse organisms on earth. There are over 200 genera in the bacterial kingdom (Holt 1981). These organisms lack nuclear membranes and do not contain internal compartmentalization by unit membrane systems. Bacteria range in size from approximately 0.5 micron to seldom greater than 5 microns in diameter. The cellular shape can be spherical, rod-shaped, filamentous, spiral, or helical. Reproduction is by binary fission. However, genetic material can also be exchanged between bacteria.

The fungi, which include molds, mildew, rusts, smuts, yeasts, mushrooms, and puffballs, constitute a diverse group of organisms living sometimes in fresh water and marine water, but predominantly in soil or on dead plant material. Fungi are responsible for mineralizing organic carbon and decomposing woody material (cellulose and lignin). Reproduction occurs by sexual and asexual spores or by budding (yeasts).

Distribution and Occurrence of Microorganisms in the Environment

Due to their natural functions, microorganisms are found throughout the environment. Habitats that are suitable for higher plants and animals to survive will permit microorganisms to flourish. Even habitats that are adverse to higher life forms can support a diverse microorganism population. Soil, groundwater, surface water, and air can support or transport microorganisms. Since this text focuses on *in situ* treatment, the following briefly describes the distribution and occurrence of microorganisms in soil and groundwater only.

Microorganisms found in soil or groundwater represent the part of the entire population that has flourished under the environmental conditions that are present during the time of sampling. If the environmental conditions are changed by natural or man-made influences, then the microbial population will change in response to the new environment. Chapters 8 and 9 will show how to manipulate the environment in order to change the microbial population and promote new types of biochemical reactions. We will limit this chapter to mainly discussing what is naturally found in the soil and groundwater.

Soil

Bacteria outnumber the other organisms found in a typical soil. These organisms rapidly reproduce and constitute the majority of biomass in soil. It is estimated that surficial soil can contain some 10,000 different microbial species and can have as many as 109 cells/gm of soil. In addition, cellular biomass can comprise up to approximately 4 percent of the soil organic carbon (Adriano et al. 1999). Microorganisms generally adhere to soil surfaces by electrostatic interactions, London-van der Waals forces, and hydrophobic interactions (Adriano et al. 1999). Typically, microorganisms decrease with depth in the soil profile, as does organic matter. The population density does not continue to decrease to extinction with increasing depth, nor does it necessarily reach a constant declining density. Fluctuations in density commonly occur at lower horizons. In alluvial soils, populations fluctuate with textural changes; organisms are more numerous in silt or silty clay than in intervening sand or course sandy horizons. In soil profiles above a perched water table, organisms are more numerous in the zone immediately above the water table than in higher zones (Paul and Clark 1989). Most fungal species prefer the upper soil profile. The rhizosphere (root zone) contains the most variety and numbers of microorganisms.

Groundwater

Microbial life occurs in aquifers. Many of the microorganisms found in soil are also found in aquifers and are primarily adhered to soil surfaces. Bacteria exist in shallow to deep subsurface regions but the origins of these organisms are unknown. They could have been deposited with sediments millions of years ago, or they may have migrated recently into the formations from surface soil. Bacteria tend not to travel long distances in fine soils but can travel long distances in course or fractured formations. These formations are susceptible to contamination by surface water and may carry pathogenic organisms into aquifer systems from sewage discharge, landfill leachate, and polluted water (Bouwer 1978).

MICROORGANISM BIOCHEMICAL REACTIONS

Microorganisms responsible for degradation of organic environmental impacts obtain energy and building blocks necessary for growth and reproduction from degrading organic compounds. Energy is conserved in the C-C bonds, and during degradation, the organics are converted to simpler organic compounds while deriving energy. Ultimately, the organic compounds are degraded (mineralized) to carbon dioxide or methane, inorganic ions, and water. During the process, microbes use portions of these compounds as building blocks for new microbial cells.

As discussed above, microorganism populations can be numerous in soil and groundwater. These populations complete diverse biochemical reactions, and are able to thrive in wide ranges of environmental conditions. In addition, the presence of particular microorganisms and the biochemical reactions that they complete are influenced by the physical and chemical environment. These physical and chemical environments can be modified by organisms creating favorable conditions for a new consortium of organisms and biochemical reactions to occur. Often, different environmental conditions are created whereby new degradative pathways are induced resulting in the ability to biochemically degrade different organic pollutants, chemically modify inorganic compounds, or immobilize inorganic compounds such as heavy metals.

The following sections describe important biochemical reactions that site investigators and remediation design engineers should understand. Understanding these reactions will allow remediation teams to determine if biodegradation is likely occurring or if environmental conditions can be modified to create conditions favorable to degrade or modify environmental pollutants. Failure to understand these concepts can result in remediation systems that limit biological processes, and therefore minimize effectiveness.

Energy Production

Microorganisms derive energy by degrading a wide variety of organic compounds including man-made (xenobiotic or anthropogenic) compounds. Enzymes are induced, respiration occurs, organic compounds are cleaved releasing energy, intermediate compounds are produced, and growth and reproduction occurs. These processes allow microorganisms to thrive and contribute to the natural cycling of carbon throughout the environment (Figure 1). As seen in Figure 1, microorganisms perform a portion of the overall carbon cycling and it is this portion that bioremediation systems rely on to degrade or modify environmental pollutants.

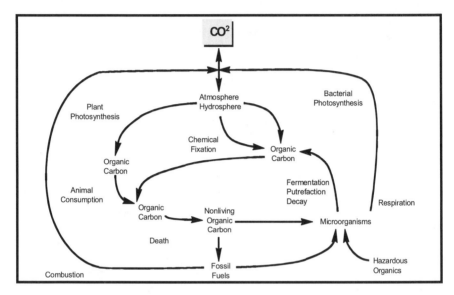

Figure 1   Carbon cycle.

Oxidation and Reduction

The utilization of chemical energy in microorganisms generally involves what are called oxidation-reduction reactions. For every chemical reaction, oxidation and reduction occurs. Oxidation of a compound corresponds to an oxygen increase, loss of hydrogen, or loss of electrons. Conversely, reduction corresponds to an oxygen decrease, an increase of hydrogen, or an increase in electrons. This process is coupled (half-reactions); if a target chemical is oxidized, another compound must be reduced. In this case, the reactant serves as the electron donor and becomes oxidized, while the other compound serves as the electron acceptor and becomes reduced. In terms of energy released, the electron donor is also an energy source (substrate), whereas the electron acceptor is not an energy source. Once the electron donor has been fully oxidized (lost all the electrons that it can loose) it is usually no longer an energy source but may now serve as an electron acceptor.

This is an important concept to understand because biochemical reactions and the ability to degrade or modify compounds are usually dependant on the oxidation state of the target compounds and the predominant biochemical processes that are occurring in soil and groundwater. For example, organic compounds that are in a reduced state, such as aliphatic hydrocarbons, are more likely to be oxidized in the environment. Chemicals that are in an oxidized state, such as highly chlorinated volatile organic compounds, are more likely to be reduced in the environment. In addition, because it is often difficult to directly confirm that degradation is occurring during remediation, it is often necessary to measure indicator parameters in order to determine the predominant biochemical processes that are occurring.

The types of electron acceptors used by microorganisms affect the quantity of energy that is available from organics. The energy available from the oxidation-reduction reaction is expressed as the standard electrode potential (oxidation-reduction potential [Eh]) (referenced to hydrogen at pH = 7). Common electron acceptors used to evaluate *in situ* bioremediation processes are shown in Figure 2. The electron accepting reactions are shown in order of decreasing energy availability. In addition, common organisms responsible for these reactions are also shown (Adriano et al. 1999 and Brock 1979).

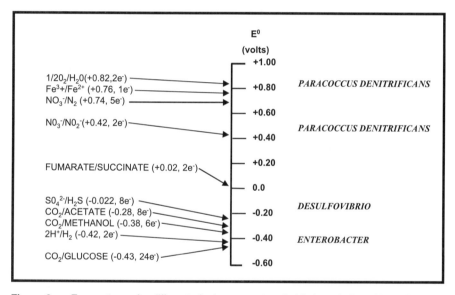

Figure 2   Energy tower for different electron acceptors in biodegradation pH = 7 (Adriano et al. 1999) (Adapted from Brock, 1979).

Aerobic Respiration

Aerobic microorganisms have enzyme systems that are capable of oxidizing organic compounds. The organic compound serves as the electron donor and the electrons are transferred to molecular oxygen (O_2). This is the most efficient (less energy required) biochemical reaction whereby the electron donor (organic substrate) is degraded producing biomass, carbon dioxide (CO_2), water, and potentially other organics as depicted by:

$$\text{electron donor (organic substrate)} + O_2 \text{ (electron acceptor)} \rightarrow$$
$$\text{biomass} + CO_2 + H_2O + \text{metabolites} + \text{energy.}$$

Facultative Respiration

In reduced or low molecular oxygen environments, facultative anaerobes are a class of microorganisms that are able to shift their metabolic pathways and use nitrate (NO_3^-) as a terminal electron acceptor. This process is called denitrification and is generally depicted as follows:

electron donor (organic substrate) + NO_3^- (electron acceptor) $\rightarrow$
biomass + CO_2 + H_2O + N_2 + metabolites + energy.

The reduction of NO_3^- to nitrogen gas (N_2) is completed through a series of electron transport reactions as follows:

NO_3^- (nitrate) $\rightarrow$ NO_2^- (nitrite) $\rightarrow$ NO (nitic oxide) $\rightarrow$
N_2O (nitrous oxide) $\rightarrow$ N_2 (nitrogen gas)

Most denitrifiers are heterotrophic and commonly occur in soil such as *Pseudomonas, Bacillus,* and *Alcaligenes* genera. A large number of species can reduce nitritate to nitrite in the absence of oxygen, with a smaller number of species that can complete the reaction by reducing nitrous oxide to nitrogen gas.

Anaerobic Respiration

Anaerobic respiration is completed by different classes of microorganisms in the absence of molecular oxygen. The anaerobic organisms important to environmental remediation include iron and manganese reducing bacteria, and sulfanogenic and methanogenic bacteria. Anaerobic growth in the environment is not as efficient as aerobic growth (less energy produced per reaction); however, these organisms complete important geochemical reactions including bacterial corrosion, sulfur cycling, organic decomposition, and methane production. These reactions are more complex than aerobic respiration and often rely on a consortium of bacteria to complete the reactions. In addition, these classes of bacteria are also capable of either degrading organic pollutants and/or alter environmental conditions whereby chemical reactions can occur. The following depicts the generalized (and simplified) reactions these classes of organisms complete:
Iron Reduction:

organic substrate (electron donor) + $Fe(OH)_3$ (electron acceptor) + H_2^+ $\rightarrow$
biomass + CO_2 + Fe^{2+} + H_2O + energy

Manganese Reduction:

organic substrate (electron donor) + MnO_2 (electron acceptor) + H_2^+ $\rightarrow$
biomass + CO_2 + Mn^{2+} + H_2O + energy

Sulfanogenesis:

$$\text{organic substrate (electron donor)} + SO_4^{2-} \text{ (electron acceptor)} + H^+ \rightarrow$$
$$\text{biomass} + CO_2 + H_2O + H_2S + \text{metabolites} + \text{energy}$$

Methanogenesis:

$$\text{organic substrate} + CO_2 \text{ (electron acceptor)} + H^+ \text{ (electron donor)} \rightarrow$$
$$\text{biomass} + CO_2 + H_2O + CH_4 + \text{metabolites} + \text{energy}$$

More detailed information regarding these biochemical reactions can be obtained by reviewing mircobiological texts such as Brock 1979, Stanier, Adelberg, and Ingrahm 1979, and Paul and Clark 1989.

Microbial Degradation and Genetic Adaptation

In the preceding sections, the reactions associated with degradation and growth were discussed. However, the susceptibility of an environmental pollutant to microbial degradation is determined by the ability of the microbial population to catalyze the reactions necessary to degrade the organics.

Readily degradable compounds have existed on earth for millions of years; therefore, there are organisms that can mineralize these compounds. Industrial chemicals (xenobiotic or anthropogenic) have been present on earth for a short time on the evolutionary time scale. Many of these compounds are degradable, and many are persistent in the environment. Some xenobiotic compounds are similar to natural compounds and bacteria will degrade them easily. Other xenobiotic compounds will require special biochemical pathways in order to undergo biochemical degradation.

Biodegradation of organic compounds (and maintenance of life sustaining processes) is reliant on enzymes. The best way to understand enzyme reactions is to think of them as a lock and key. Figure 3 shows how only an enzyme with the right shape (and chemistry) can function as a key for the organic reactions. The lock and key in the real world are three-dimensional. The fit between the two is precise.

Organic compounds in the environment that are degradable align favorably with the active site of specific enzymes. The microorganism will not affect compounds that do not align favorably or compounds that do not bind with the active site of their enzyme. Degradation of these compounds requires that the microorganism population adapt in response to the environment by synthesizing enzymes capable of catalyzing degradation of these compounds.

A few definitions would be helpful here in order to understand different levels of biological reactions. Biodegradation means the biological transformation of an organic chemical to another form with no extent implied (Grady 1985). Biodegradation does not have to lead to complete mineralization. Mineralization is the complete degradation of an organic compound to carbon dioxide or methane and inorganic ions. Recalcitrance is defined as inherent resistance of a chemical to any degree of biodegradation and persistence means that a chemical fails to undergo

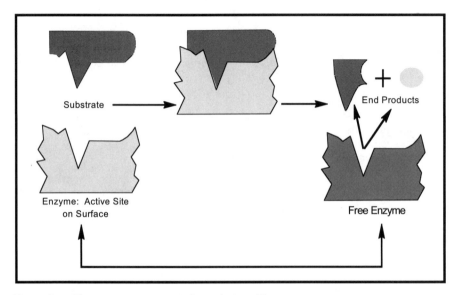

Figure 3   Enzymes are represented as a lock and key.

biodegradation under a defined set of environmental conditions (Bull 1980). This means that a chemical can be degradable but due to environmental conditions, the compounds may persist in the environment. With proper manipulation (or understanding) of the environmental conditions, biodegradation of these compounds can be demonstrated in the laboratory or field.

Gratuitous Biodegradation

Enzymes are typically described as proteins capable of catalyzing highly specific biochemical reactions. Enzymes are more specific to organic compound functional groups than to specific compounds. An enzyme will not differentiate between a C-C bond in a benzene molecule versus a C-C bond in a phenol molecule. The functional capability of enzymes depends on the specificity exhibited towards the organic compound. A major enzymatic mechanism used by bacteria to degrade xenobiotic compounds has been termed gratuitous biodegradation and includes existing enzymes capable of catalyzing a reaction towards a chemical substrate.

In order for gratuitous biodegradation to occur, the bacterial populations must be capable of inducing the requisite enzymes specific for the xenobiotic compound. Often times this occurs in response to similarities (structural or functional groups) with natural organic chemicals, for example, a bacterium producing the enzymes for benzene degradation. Chlorobenzene is introduced and is not recognized by the bacteria (its presence will not induce an enzyme to be produced). However, the enzymes already produced for benzene will also catalyze the degradation of chlorobenzene.

The capability of bacterial populations to induce these enzymes depends on structural similarities and the extent of substitutions on the parent compound. Generally, as the number of substitutions increases, biodegradability decreases unless a natural inducer is present to permit synthesis of required enzymes. To overcome potential enzymatic limitations, bacteria populations often induce a series of enzymes that coordinately modify xenobiotic compounds. Each enzyme will modify the existing compound such that a different enzyme may be specific for the new compound and capable of degrading it further. Eventually, the original xenobiotic compound will not be present and the compound will resemble a natural organic compound and enter into normal metabolic pathways. This concept of functional pathways is more likely to be completed through the combined efforts of mixed communities than by any one single species.

Cometabolism

Cometabolism has been defined as "the transformation of a nongrowth substrate in the obligate presence of a growth substrate" (Grady 1985). A nongrowth substrate cannot serve as a sole carbon source that provides energy to support metabolic processes. A second compound is required to support biological processes allowing transformation of the nongrowth substrate. This requirement is added to make a distinction between cometabolism and gratuitous biodegradation.

During cometabolism, the organism receives no known benefit from the degradation of the organic compound. In fact, the process may be harmful to the microorganism responsible for the production of the enzyme (McCarty and Semprini 1994). Cometabolism of chlorinated ethenes (with the exception of perchloroethene [PCE]) has been reported to occur in aerobic environments and it is believed that the rate of cometabolism increases as the degree of dechlorination decreases (Murray and Richardson 1993, Vogel 1994, and McCarty and Semprini 1994).

Microbial Communities

Complete mineralization of a xenobiotic compound may require more than one microorganism. No single bacterium within the mixed culture contains the complete genome (genetic makeup) of a mixed community. The microorganisms work together to complete the pathway from organic compound to carbon dioxide or methane. These associations have been called consortia, syntrophic association, and synergistic associations and communities (Grady 1985). We need to understand the importance of the community when we deal with remediation. Conversely, we need to understand the limitations of laboratory work with single organisms. This work does not represent the real world of degradation. Reviewing the strengths of the communities will also reveal the limitations of adding specialized bacteria that have been grown in the laboratory.

Community Interaction and Adaptation

Microbial communities are in a continuous state of flux and constantly adapting to their environment. Population dynamics, environmental conditions, and growth substrates continually change and impact complex interactions between microbial populations. Even though microorganisms can modify environmental disturbances, microbial ecosystems lack long-term stability and are continually adapting (Grady 1985). It is important to understand the complexities and interactions within an ecosystem to prevent failure when designing a biological remediation system.

Mixed communities have greater capacity to biodegrade xenobiotic compounds due to greater genetic diversity of the population. Complete mineralization of xenobiotic compounds may rely on enzyme systems produced by multiple species. Community resistance to toxic stresses may also be greater due to the likelihood that an organism can detoxify the ecosystem.

Community adaptation is dependent upon evolution of novel metabolic pathways. A bacterial cell considered in isolation has a relatively limited adaptive potential and adaptation of a pure culture must come from mutations (Grady 1985). Mutations are rare events. These mutations are generally responsible for enzymes that catalyze only slight modifications to the xenobiotic compound. An entire pathway can be formed through the cooperative effort of various populations. This is due to the greater probability that an enzyme system exists capable of gratuitous biodegradation within a larger gene pool. This genetic capability can then be transferred to organisms lacking the metabolic function that enhances the genetic diversity of the population. Through gene transfer, individual bacteria have access to a larger genetic pool allowing evolution of novel degradative pathways.

Genetic Transfer

Genes are transferred throughout bacterial communities by three mechanisms called conjugation, transformation, and transduction (Brock 1979, Stanier, Adelberg, and Ingrahm 1976, Moat 1979, Grady 1985, and Rittman, Smets, and Stahl 1990). Conjugation appears to be the most important mechanism of gene transfer in the natural environment. Conjugation involves the transfer of DNA from one bacterium to another while the bacteria are temporarily joined. The DNA strands that are transferred are separate from the bacterial chromosomal DNA and are called plasmids (Brock 1979, Stanier, Adelberg, and Ingrahm 1976, Moat 1979, and Rittman, Smets, and Stahl 1990). Plasmids exist in cells as circular, double-stranded DNA and are replicated during transfer from donor to recipient. Unlike chromosomal DNA that encodes for life sustaining processes, plasmid genes encode for processes that enhance growth or survival in a particular environment. Examples of functions that are encoded on plasmids include antibiotic resistance, heavy metal resistance, and certain xenobiotic degradation enzymes (such as toluene) (Rittman, Smets, and Stahl 1990).

There are many natural processes that the microorganisms employ to expand the type of compounds that they can use as an energy source. We can create environments and provide growth factors that facilitate these processes, or data can be collected that documents that the biochemical reactions are occurring without modifying the environment. The rest of this section will discuss these various processes.

GROWTH

Growth is defined as an increase in the quantity of cellular constituents, structures, or organisms (biomass). Growth is controlled by a complex interaction between food sources (usually organics), inorganic nutrients and cofactors, terminal electron acceptors, predators, physical conditions, and chemical conditions. It is the design engineer's objective to optimize these conditions in order to maximize the biological treatment system's effectiveness.

Growth Cycle

A microorganism growth cycle can be divided into several phases called the lag phase, exponential phase, stationary phase, and death phase (Figure 4). It is the remediation engineer's objective to design biological systems that maintain a high growth rate until the environmental pollutant has been degraded or modified. At this point, organic carbon (food) usually becomes limiting and the microorganism population proceeds into the death phase.

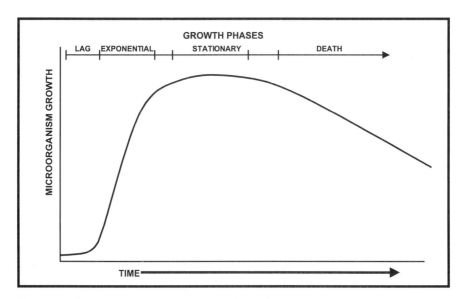

Figure 4   Typical growth curve for a bacterial population.

Important Environmental Factors that Affect Growth

As discussed earlier in this chapter, organic compound degradation and growth is completed in aerobic and/or anaerobic environments. These reactions occur only when the physical, chemical, and biological environment are conducive to supporting these reactions.

The following sections describe the most important factors that must be considered for every biological remediation design. Failure to include these factors in all biological designs can significantly limit remedial effectiveness, as these factors can control the type of bacteria that are prominent and the biodegradation rate. In addition to these factors, the physical environment associated with soil is important; however, these considerations are not included in this text. The authors highly suggest other sources such as Paul and Clark 1989 and Adriano et al. 1999 be reviewed to better understand biological associations in soil.

Water

Water is an important factor for biochemical reactions. In saturated and unsaturated conditions, the bacteria may have to expend energy in order to acquire the water that they require. In the aquifer, the availability of water to microorganisms can be expressed in terms of water activity, which is related to vapor pressure of water in the air over a solution (relative humidity). Water activity in freshwater and marine environments is relatively high and lowers with increasing concentrations of dissolved solute (Brock 1979). Bacteria can grow well in the saltwater of an ocean (or 3.5 percent dissolved solids). Therefore, groundwater, even from brine aquifers, will not pose any problems for bacterial growth.

In soil, water potential is used instead of water activity and is defined as the difference in free energy between the system under study and a pool of pure water at the same temperature and includes matrix and osmotic effects. The unit of measurement used is the *MPa*. As with water activity, this determines the amount of work that the cell must expend to obtain water. Generally, activity in soil is optimal at -0.01 MPa (or 30 to 90 percent of saturation) and decreases as the soil becomes either waterlogged near zero or desiccated at large, negative water potentials (Paul and Clark 1989).

pH

Microorganisms have ideal pH ranges that allow growth. Within these ranges, there is usually a defined pH optimum. Generally, the optimal pH for bacteria is between 6.5 and 7.5 standard units, which is close to the intracellular pH. A bacteria cell contains approximately 1000 enzymes and many are pH dependent (Paul and Clark 1989). Most natural environments have pH values between 5 and 9. Only a few species can grow at pH values of less than 2 or greater than 10 (Brock 1979). In environments with pH values above or below optimal, bacteria are capable of maintaining an internal neutral pH by preventing H^+ ions from leaving the cell or by actively expelling H^+ as they enter. The most important factor with pH is to not allow major shifts in pH during remediation.

Temperature

As the temperature rises, chemical and enzymatic reaction rates in the cell increase. For every organism there is a minimum temperature below which growth no longer occurs, an optimum temperature at which growth is most rapid, and a maximum temperature above which growth is not possible. The optimum temperature is always nearer the maximum temperature than the minimum. Temperature ranges for microorganisms are wide. Some microorganisms have optimum temperatures as low as 5° to 10°C and others as high as 75° to 80°C. The temperature range in which growth occurs ranges from below freezing to boiling.

No single microorganism will grow over this entire range. Bacteria are frequently divided into three broad groups: thermophiles, which grow at temperatures above 55°C; mesophiles, which grow in the midrange temperature of 20° to 45°C; and psychrophiles, which grow well at 0°C. In general, the growth range is approximately 30 to 40 degrees for each group. Microorganisms that grow in terrestrial and aquatic environments grow in a range from 20° to 45°C. Figure 5 demonstrates the relative rates of reactions at various temperatures. As can be seen in Figure 5, microorganisms can grow in a wide range of temperatures.

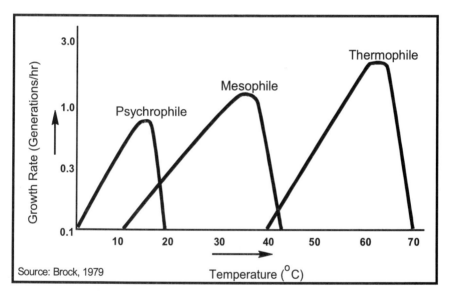

Figure 5   Relationships of temperature to growth rate of a psychrophile, a mesophile and a thermophile.

In general, biological reactions will occur year round in the aquifer due to the relatively constant temperature. Rates will be faster in warmer climates due to higher temperatures. Biological reactions in surface soils will be affected by temperature. Biological surface remediation will slow or not occur during the winter months in colder climates.

Hydrogen Ion Concentration

Hydrogen is a key component of anaerobic respiration and the terminal electron-accepting process. During the early stages of organic degradation, hydrogen is produced by a wide variety of microorganisms as part of normal metabolism. As the hydrogen is produced, anaerobic bacteria oxidize the hydrogen and reduce terminal electron acceptors. The rapid turnover of the hydrogen pool has been termed interspecies hydrogen transfer (Lovley and Goodwin 1988).

Nitrate- Fe(III)-, Mn(IV)-, sulfate-, and CO_2-reducing (methanogenic) microorganisms exhibit different efficiencies in using the H_2 that is continually produced. Nitrate reducers are highly efficient H_2 utilizers and maintain low steady-state H_2 concentrations. Fe(III) reducers are slightly less efficient and thus maintain somewhat higher H_2 concentrations. Sulfate reducers and methanogenic bacteria are progressively less efficient and maintain even higher H_2 concentrations. These terminal electron accepting processes generally result in characteristic H_2 concentrations in groundwater systems (USEPA 1998).

Oxygen

Oxygen is the most thermodynamically favored electron acceptor used by microorganisms to degrade organic compounds. Generally, an oxygen atmosphere in soil of less than 1 percent will change the predominant respiration reaction from aerobic to anaerobic (Paul and Clark 1989). In aqueous environments, oxygen concentration less than approximately 0.5 to 1.0 mg/l can switch metabolism from aerobic to anaerobic (Tabak 1981).

Nutrition

Up to this point we have discussed biochemical reactions responsible for deriving energy, namely respiration (organic degradation). These processes are called dissimilatory reactions where the chemical energy stored in C-C bonds is broken by enzymes producing energy and metabolites used to build biomass.

Microorganism growth also requires assimilatory reactions where the organism gathers carbon (C), nitrogen (N), phosphorous (P), sulfur (S), and micronutrients. The remainder of this section describes the role of nutrients in the degradation process.

Molecular composition of bacterial cells is fairly constant and indicates the requirements for growth. Water constitutes 80 to 90 percent of cellular weight and is always a major nutrient. The solid portion of the cell is made of carbon, oxygen, nitrogen, hydrogen, phosphorus, sulfur, and trace elements. The approximate elementary composition is shown in Table 1.

As can be seen from Table 1, the largest component of bacteria is carbon. The organic pollutants that we wish to destroy can provide this element. After carbon, oxygen is the highest percentage of the cell. When oxygen requirements of new cells are added to the required oxygen as an electron acceptor, large amounts of oxygen may be utilized in biological degradation.

The other major nutrients required by the microorganisms are nitrogen and phosphorous. The three main forms of nitrogen found in microorganisms are proteins, microbial cell wall components, and nucleic acids. The most common sources of inorganic nitrogen are ammonia and nitrate. Ammonia can be directly assimilated into amino acids. When nitrate is used, it is first reduced to ammonia and is then synthesized into organic nitrogen forms.

Table 1   Molecular Composition
of a Bacterial Cell

Element	Dry Weight (%)
Carbon	50
Oxygen	20
Nitrogen	14
Hydrogen	8
Phosphorus	3
Sulfur	1
Potassium	1
Sodium	1
Calcium	0.5
Magnesium	0.5
Chlorine	0.5
Iron	0.2
Other	~0.3

Phosphorus in the form of inorganic phosphates is used by microorganisms to synthesize phospholipids and nucleic acids. Phosphorous is also essential for the transfer of energy during organic compound degradation.

Numerous studies have been completed to determine the ideal C/N/P ratio of macronutrients to maintain or accelerate biodegradation. These studies evaluated microorganism composition, laboratory treatability studies, and field studies. In general, nutrients should be present in soil and groundwater and their approximate ratio should be 100/10/1. This ratio corresponds to the approximate ratio of these macronutrients in microorganisms. This ratio represents the macronutrient requirements for new microorganisms. When we do not need to grow new bacteria, then the requirements for macronutrients are much lower. Most natural attenuations do not require added nutrients. However, when large quantities of organics are present, then the addition of macronutrients will increase the rate of bacterial growth and the subsequent rate of organic destruction.

Micronutrients are also required for microbial growth. There are several micronutrients that are universally required such as sulfur, potassium, magnesium, calcium, and sodium. Sulfur is used to synthesize two amino acids, cysteine and methionine. Inorganic sulfate is also used to synthesize sulfur containing vitamins (thiamin, biotin, and lipoic acid) (Brock 1979). Several enzymes including those involved in protein synthesis are activated by potassium. Magnesium is required for activity of many enzymes, especially phosphate transfer and functions to stabilize ribosomes, cell membranes, and nucleic acids. Calcium acts to stabilize bacterial spores against heat and may also be involved in cell wall stability.

Additional micronutrients commonly required by microorganisms include iron, zinc, copper, cobalt, manganese, and molybdenum. These metals function in enzymes and coenzymes. These metals (with the exception of iron) are also considered heavy metals and can be toxic to microorganisms.

All of these factors are necessary to maintain a microorganism's metabolic processes. Often macronutrients (N and P) are limited in soil and groundwater, and it may be necessary to add these nutrients to enhance or accelerate biodegradation. Micronutrients, however, are usually present in soil and groundwater and amendment is usually not necessary. The design engineer should evaluate if nutrient amendments are required to complete soil or groundwater remediation. If amendments are required, the organic carbon should be the limiting factor in the biochemical reaction such that organic degradation occurs more completely.

Toxic Environments that Affect Growth

Many factors can render an environment toxic to microorganism. Physical agents such as high and low temperatures, high and low pH, sound and radiation, and chemical agents such as heavy metals, halogens, organic pollutants, and oxidants can inhibit microbial growth. In addition, oxygen, water, and nutrients can be toxic if added in too high of concentrations.

Chemical agents such as heavy metals and halogens can disrupt cellular activity by interfering with protein function. Mercury ions combine with SH groups in proteins, silver ions will precipitate protein molecules, and iodine will iodinate proteins containing tyrosine residues preventing normal cellular function. The effects of various metals in soil has been described (Dragun 1988) and is affected by the concentration and pH of the soil. Oxidizing agents such as chlorine, ozone, and hydrogen peroxide oxidize cellular components destroying cellular integrity.

It is also possible that the environmental pollutant will induce toxicity to microorganisms. These compounds can destroy cellular components such as cell walls, cause mutational changes and inhibit reproduction, or inhibit assimilatory or dissimilatory biochemical reactions. Often these toxic effects can be mitigated by reducing the concentrations of the toxicant such that the biochemical reactions will occur. It is important that the design engineer evaluate potentially toxic conditions, and if necessary, incorporate steps into the remediation process designed to reduce toxicity before relying on biochemical reactions to complete soil and groundwater remediation.

MICROBIAL DEGRADATION AND MODIFICATION

Much research has been completed to determine degradation pathways. Information sources such as Adriano et al. 1999 should be reviewed to obtain more information regarding specific studies and pathways. Many petroleum hydrocarbons, halogenated hydrocarbons, pesticides, and other anthropogenic organic compounds can be degraded biologically. In addition, microorganisms can also chemically modify inorganic compounds. As discussed in earlier sections, the ability of microorganisms to degrade or modify compounds depends on the ability to produce requisite enzymes and ideal environmental conditions for the reactions to occur. In addition, sufficient biomass and communication between the pollutant and the enzymes (intracellular or extracellular) is necessary.

As described in Adriano et al. 1999 (and others), degradation of organic compounds can be divided into three groups as follows (Figure 6).

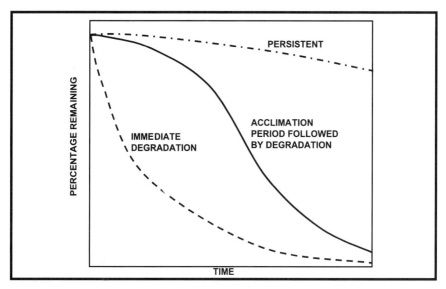

Figure 6   Degradation of organic compounds (Adriano et al. 1999).

1. Biodegradation starts immediately and the compounds are readily used as sources of energy and growth (immediate degradation).
2. Biodegradation starts slowly and requires a period of acclimation before more rapid degradation occurs.
3. The compounds are persistent and biodegradation is slow or does not occur.

There are general rules-of-thumb regarding degradation that remediation engineers should understand. Table 2 presents important physical, chemical, and structural elements that usually determine if an organic compound can be degraded.

Historically, *in situ* bioremediation has focused on treating organic pollutants. However, biological processes are also being used to modify inorganic compounds, particularly heavy metals, such that the compounds can be physically removed, made less toxic, or rendered immobile, and therefore, exposure is reduced.

The form of the metal (e.g., elemental, oxide, sulfide, ionic, inorganic complex, organic complex, coprecipitate), the availability of electron donors (C), nutrients (N and P), the presence of electron acceptors (O_2, NO_3^-, Fe^{3+}, Mn^{4+}, SO_4^{2-}, organic compounds), and environmental factors (pH, Eh, temperature, moisture) affect the type, rate, and extent of microbial activity, and hence, transformation of metals (Adriano 1999).

Oxidizing and reducing environments influence the mobilization and immobilization of metals. For example, in an anaerobic environment, certain metals are reduced enzymatically from a higher oxidation state to a lower one and this affects their solubility and bioavailability. The reduction of $Fe^{3+} \rightarrow Fe^{2+}$ increases its solubility, while reduction of $U^{6+} \rightarrow U^{4+}$ or $Cr^{6+} \rightarrow Cr^{3+}$ decreases their solubility (Adriano et al. 1999). These concepts along with mechanisms of microbial dissolution, stabilization, and recovery are explored in Adriano et al. 1999 with an emphasis on exploiting these processes for remediating impacted soil.

Table 2   Physical, Chemical, and Structural Properties That Influence Degradability of an Organic Compound

Property	Degradability	
	More Easily	**Less Easily**
Solubility in Water	Soluble in Water	Insoluble in Water
Size	Relatively Small	Relatively Large
Functional Group Substitutions	Fewer Functional Groups	Many Functional Groups
Compound More Oxidized	In Reduced Environment	In Oxidized Environment
Compound More Reduced	In Oxidized Environment	In Reduced Environment
Created…	Biologically	Chemically by Man
Aliphatics	Aliphatic up to 10 C-chains	High Molecular Weight Alkanes
	Straight Chains	Branched Chains
	Aromatic Compounds with One or Two Nuclei	Polyaromatic Hydrocarbons
Substitution on Aromatic Rings	-OH, -COOH, -CHO, -CO -OCH$_3$, -CH$_3$	-F, -Cl, -NO$_2$ -CF$_3$, -SO$_3$H, -NH$_2$
Substitutions on Organic Molecules	Alcohols, Aldehydes, Acids, Esters, Amides, Amino Acids	Alkanes, Olefins, Ethers, Ketones, Dicarboxylic Acids, Nitriles, Amines, Chloroalkanes
Substitution Position	*p*-position *o-* or *p*-disubstituted phenols	m- or o-position m-disubstituted phenols
Number of Hydroxy Group	Increasing	Decreasing
Biphenyl and Dioxins	One or Less Halogens	Two or More Halogens
Halogenated Alkanes	Few Halogenated Substitutions and Away from the C	Many Halogen Substitutions or Directly at C
Substitution of Halogen Derivatives	Asymmetrical Substitution	Symmetrical Substitution

DEGRADATION RATE

The ultimate goal of an effective *in situ* remediation design and operation is maximizing the rate at which organic compounds are degraded or inorganic compounds are modified. Maximizing the degradation rate is often balanced against economics and time requirements in order to implement the most cost effective solution.

The kinetics associated with degradation rates have been described in the literature (Suthersan 1997, and Suarez and Rifai 1999). In nature, degradation processes are complicated, variable, and depend on the physical, chemical, and biological properties of the environment. It is unlikely that degradation can be represented by one precise and consistent mathematical expression. Therefore, based on laboratory and field experience, it is commonly accepted that degradation rates are estimated using Monod kinetics, and zero- and first-order mathematical expressions. These expressions are used to simplify degradation rates and provide input for modeling the effectiveness of remediation systems.

Degradation rates are reported in various forms in the literature. In addition, site-specific degradation rates can be derived from laboratory, investigation, or remediation data. Half-life, rate constants, and percent disappearance are all used. These values are often used to determine if compounds are amenable to biodegradation under a defined set of conditions, and to predict the amount of time that may be required to complete the reactions. Therefore, it is important to understand and document that the environmental conditions from which these values were derived are representative of site conditions so degradation rates are not over or underestimated.

NATURAL ATTENUATION

Natural attenuation is, or should be, a component of all remedial solutions where groundwater is impacted. Few, if any, remediation technologies can achieve final site-specific remediation objectives like natural attenuation. Therefore, it is important to understand the basics of biochemical reactions, physical attenuation mechanisms, the regulatory basis for the technology, how natural attenuation should be applied, advantages and disadvantages, and the evaluation process.

The United States Environmental Protection Agency (USEPA) (USEPA 1998), United States Air Force, United States Navy, American Society for Testing Materials, and many state and local regulatory agencies have developed protocols for evaluating natural attenuation as a groundwater remedial solution. In addition, the Interstate Technology and Regulatory Cooperation Work Group, *In Situ* Bioremediation Work Team is a state led, national coalition of personnel from the regulatory and technology programs devoted to deploying and improving innovative environmental technologies. They have also produced technical requirements for evaluating natural attenuation (ITRC 1999). For the most part, all of these protocols are similar. In fact, many of the same experts in the field have contributed directly (or indirectly) to protocol development.

Natural attenuation includes several processes including: biodegradation, dispersion, sorption, and volatilization. When these processes are shown to be capable of attaining site-specific remediation objectives in a time period that is reasonable compared to other alternatives, they may be selected alone or in combination with other more active remediations as the preferred remedial alternative (USEPA 1998).

Monitored natural attenuation (MNA) is a term that refers specifically to the use of natural attenuation processes as part of overall site remediation. The USEPA (USEPA 1998) defines MNA as follows:

> The term "monitored natural attenuation," ..., refers to the reliance on natural attenuation processes (within the context of a carefully controlled and monitored cleanup approach) to achieve site-specific remedial objectives within a time frame that is reasonable compared to other methods. The "natural attenuation processes" that are at work in such a remediation approach include a variety of physical, chemical, or biological processes that, under favorable conditions, act without human intervention to reduce the mass, toxicity, mobility, volume, or concentration of contaminants in soil and ground water. These *in situ* processes include, biodegradation, dispersion,

dilution, sorption, volatilization, and chemical or biological stabilization, transformation, or destruction of contaminants.

Monitored natural attenuation is appropriate as a remedial approach only when it can be demonstrated capable of achieving a site's remedial objectives within a time frame that is reasonable compared to that offered by other methods and where it meets the applicable remedy selection program.... EPA, therefore, expects that monitored natural attenuation typically will be used in conjunction with active remediation measures (e.g., source control), or as a follow-up to active remediation measures that have already been implemented.

The intent of this section is to provide an overview of natural attenuation as it pertains to *in situ* bioremediation. Dispersion, sorption, and volatilization are important elements of natural attenuation; however, biodegradation (in most cases) is likely the most effective process and is the focus of this text.

This text is not intended to replace or reiterate all elements contained within the various protocols. Interested readers should at least obtain and review the USEPA protocol as a supplement to this text. A word of caution must be stated about the protocols, however. It is the authors' and others' (Norris 1999) opinions that the protocols should be used as a guideline, and that practitioners must have a complete understanding of the physical, chemical, and biological processes that control natural attenuation. If the processes are not understood, aspects of the protocols can be easily misapplied. For example, some protocols recommend that site conditions be scored. Inaccurate scores can lead to an erroneous decision regarding natural attenuation even when primary lines of evidence, such as the presence of degradation products, mass reduction, and plume stabilization data, indicate that natural attenuation is occurring. In addition, monitoring well placement, construction, and data collection procedures can introduce errors that if not considered, can lead to another erroneous score. Therefore, knowledge and common sense must be used to provide sound and defensible opinions regarding the technology.

Application of Natural Attenuation

Currently, natural attenuation is most routinely applied at sites where groundwater is impacted by petroleum hydrocarbon fuels and chlorinated hydrocarbons. These compounds are most frequently detected at impacted sites and their attenuation processes are best understood. Natural attenuation of other organic compounds and heavy metals can also occur, and the same protocols can be used to evaluate these processes. In addition, applying natural attenuation to soil is emerging and it is believed that MNA will become an important remedial solution for impacted soil.

Advantages and Disadvantages of Natural Attenuation

Natural attenuation has several advantages and disadvantages as compared to other remediation technologies. The advantages include the following:

- Less remediation waste is generated and cross-media transfer of contamination and human exposure is less as compared to typical *ex situ* technologies
- Less intrusion and few surface structures are required
- Can be applied to all or part of a site depending on-site conditions and cleanup objectives
- Can be used in conjunction with, or as a follow up to other remedial technologies
- Can lower overall remediation costs as compared to costs associated with active remediation

There are also potential disadvantages of natural attenuation as follows:

- Remediation time may be longer than time frames achieved by a more active remedial solution
- Characterization costs may be more complex and costly
- Degradation of parent compounds may result in the production of more toxic degradation products
- Long-term monitoring is often required
- Institutional controls may be required to ensure risk protection
- Contamination migration and/or cross-media transfer of contaminants can potentially occur if the site's hydrology and geochemistry changes
- There may be a negative public perception regarding the technology and public outreach and education may be required before the technology is accepted

While all of these advantages and disadvantages have to be considered, in the end, the designer may not have a choice but to use natural attenuation. As we discussed in Chapter 1, there are geological limitations to all active remediations. However, natural attenuation occurs throughout the aquifer. The bacteria are located everywhere (variable concentrations depending on the environment), and do not suffer from geological limitation. Once the active processes have accomplished all that they can, natural attenuation may be the only method to eliminate the last of the contaminants. During the diffusion controlled (Chapter 2) portion of the project, natural attenuation, and the enhancements discussed in Chapters 8 and 9, are the only remediation methods that can be successfully applied.

Lines of Evidence

Lines of evidence are used to evaluate natural attenuation. Lines of evidence are used because multiple processes can be effectively treating constituents, and it is difficult to prove that any one process is responsible for all treatment. Therefore, three primary lines of evidence are used to evaluate if natural attenuation is effectively treating groundwater impacts as follows (USEPA 1997 and 1998):

1. Historical groundwater and/or soil chemistry data that demonstrate a clear and meaningful trend of decreasing contaminant mass and/or concentration over time at appropriate monitoring or sampling points. In the case of a groundwater plume, decreasing concentrations should not be solely the result of plume migration. In the case of inorganic contaminants, the primary attenuating mechanism should also be understood.

This line of evidence is important and should be the first evaluation step. Groundwater concentrations must be stable or decreasing, and/or the dissolved contaminant plume no longer advancing. The processes controlling the plume may include volatilization, dilution, dispersion, advection, or biodegradation. Sufficient monitoring data over a period of time necessary to document anthropogenic or seasonal events must be collected. The intent of evaluating the first line of evidence is to demonstrate that the plume is stable, not to document the physical, chemical, or biological processes affecting plume stability.

In addition to understanding the fate and transport processes associated with the impacted plume, it is important to also use these data to evaluate potential human or ecological exposure pathways that may exist for current or future receptors. This is important because a natural attenuation remedial solution may not be the most expedient option, and if current or future human or ecological receptors are being exposed, a more active remedial solution may be required before natural attenuation is used to complete the process.

2. Hydrogeologic and geochemical data that can be used to demonstrate indirectly the type(s) of natural attenuation processes active at the site, and the rate at which such processes will reduce contaminant concentrations to required levels. For example, characterization data may be used to quantify the rates of contaminant sorption, dilution, or volatilization, or to demonstrate and quantify the rates of biological degradation processes occurring at the site.

The second line of evidence builds upon the first. Once it has been determined that the dissolved contaminant plume is stable, no longer migrating, concentrations are decreasing, or the contaminant mass is decreasing, then the mechanism by which the attenuation is occurring must be determined. Therefore, it is necessary to evaluate the likely mechanisms by which the contaminants are being destroyed. Biological degradation is likely the most predominant and important attenuation process; however, abiotic mechanisms must also be considered.

In order to evaluate if the impacts are being degraded, the second line of evidence is usually divided into two parts. The first includes completing mass balance calculations which include determining the likely environmental conditions and respiratory pathways occurring, and correlating concentrations of electron donors and acceptors to determine if it is likely that the processes will occur to completion. Computer modeling can be used as a tool, and many of these relatively new models are briefly described in the *Modeling Tools* section of this chapter. The second portion includes estimating the biodegradation rate constants that are important to predict when remediation will be complete. There are several methods to determine the biodegradation rate constant including comparing site conditions to published literature conditions and values, tracer studies, or using actual site data collected across a defined flow path.

3. Data from field or microcosm studies (conducted in or with actual contaminated site media) which directly demonstrate the occurrence of a particular natural attenuation process at the site and its ability to degrade the contaminants of concern (typically used to demonstrate biological degradation processes only).

When data from the second line of evidence is inadequate or inconclusive, laboratory or field treatability studies may be necessary. These studies are sometimes used to demonstrate that an attenuation process is effective (usually biological). It is usually necessary to complete field or laboratory treatability studies if biodegradability data for a particular compound is not available, potentially toxic and/or mobile degradation products may be produced, little monitoring data are available, or if the site-specific environmental conditions are such that degradability is questionable. In addition, there are situations where site boundaries, access, hydrologic barriers, monitoring network, or potential exposure to human or ecological receptors make it difficult to use historic and flow path data (see 2. above) to estimate biodegradation rate constants.

The USEPA provides guidance regarding interpreting lines of evidence as described below (USEPA 1997).

In general, more supporting information may be required to demonstrate the efficacy of monitored natural attenuation at those sites with contaminants which do not readily degrade through biological processes (e.g., most non-petroleum compounds, inorganics), at sites with contaminants that transform into more toxic and/or mobile forms than the parent contaminant, or at sites where monitoring has been performed for a relatively short period of time. The amount and type of information needed for such a demonstration will depend upon a number of site-specific factors, such as the size and nature of the contamination problem, the proximity of receptors and the potential risk to those receptors, and other physical characteristics of the environmental setting (e.g., hydrogeology, ground cover, or climatic conditions).

It is incumbent on the site owner or consultant to sufficiently demonstrate to federal or state regulators that natural attenuation is technically sound and will result in site remediation within a reasonable time frame.

Site Characterization

Before natural attenuation can be used as a component of the overall remedial solution, the site must be adequately characterized. Each site will have unique requirements for characterization and the following should be understood before natural attenuation can be selected as a remedial solution component:

- Lateral and vertical extent of soil and groundwater impacts
- Constituent's physical, chemical, and biological properties
- Fate and transport of constituents in soil, soil gas, air, surface water, and groundwater
- Potential current or future human or ecological receptors
- Groundwater geochemistry, environmental conditions—pH, temperature, total solids, etc., concentrations of electron donors and acceptors, nutrients, metabolic byproducts, and microorganism toxicity potential

These data are used to develop a conceptual site model whereby a three dimensional depiction of the site and processes controlling site impacts are understood. This conceptual model serves as the basis for determining if human or ecological receptors are or will be exposed, and ultimately serves as the basis to develop remedial action objectives, and select, design, construct, and operate remediation systems. If site data indicate that natural attenuation can achieve the site-specific remedial action objectives, is protective of human and ecological receptors, and can achieve cleanup in a reasonable time frame, then natural attenuation should be used as a component of the remedial solution.

Petroleum Hydrocarbons and Chlorinated Hydrocarbons

These sections focus on the biological attenuation aspects since degradation is likely the most predominant process affecting groundwater impacts. However, remedial design engineers and site investigators must also consider physical and chemical (abiotic) processes as they may influence contaminant attenuation where biochemical reactions are ineffective (toxic conditions), hydrogeologic conditions are unique, or the constituents are not readily biodegradable.

As described earlier, there are several protocols used to evaluate natural attenuation efficiency. These protocols should be used as guidelines. Remedial design engineers and site investigators should rely on practical experience and knowledge to formulate natural attenuation solutions.

Biodegradation of Petroleum Hydrocarbons

Petroleum hydrocarbons are biodegradable. These hydrocarbons include gasoline, diesel fuel, kerosene, oils, and many other petroleum-like solvents. Microorganisms are ubiquitous in soil and groundwater and are capable of degrading these compounds using aerobic, facultative, and/or anaerobic respiratory pathways. The generalized pathways are discussed in the energy production section of this chapter. The following depicts the stoichiometric ratio (mass balance) of a typical petroleum hydrocarbon (benzene) (electron donor) and terminal electron acceptors.

Aerobic Respiration (oxidation):
$7.5O_2 + C_6H_6 \rightarrow 6CO_2 + 3H_2O$
mass ratio of O_2 to + C_6H_6 = 3.1:1
0.32 mg/l benzene degraded per 1 mg/l of O_2 consumed.

Facultative Respiration (denitrification):
$6NO_3^- + 6H^+ \, C_6H_6 \rightarrow 6CO_2 + 6\,H_2O + N_2$
mass ratio of NO_3^- to C_6H_6 = 4.8:1
0.21 mg/l benzene degraded per 1 mg/l of NO_3^- consumed.

Anaerobic Respiration—Iron Reduction:
$60\,H^+ + 30Fe(OH)_3 + C_6H_6 \rightarrow 6CO_2 + 30\,Fe^{2+} + 78H_2O$
mass ratio of $Fe(OH)_3$ to C_6H_6 = 41:1
mass ratio of Fe^{2+} produced to C_6H_6 degraded = 15.7:1
0.045 mg/l of benzene degraded per 1 mg/l of Fe^{2+} produced.

Anaerobic Respiration—Sulfanogenesis:
$7.5H^+ + 3.75 SO_4^{2-} + C_6H_6 \rightarrow 6CO_2 + 3.75H_2S + 3H_2O$
mass ratio of SO_4^{2-} to C_6H_6 = 4.6:1
0.22 mg/l benzene degraded per 1 mg/l of sulfate consumed.

Anaerobic Respiration—Methanogenesis:
$4.5H_2O + C_6H_6 \rightarrow 2.25CO_2 + 3.75CH_4$
mass ratio of CH_4 produced to C_6H_6 = 0.8:1
1.3 mg/l benzene degraded per 1 mg/l of CH_4 produced.

Biodegradation of Chlorinated Hydrocarbons

A vast amount of work has been completed to determine the environmental fate of halogenated hydrocarbons. In particular, the biotic and abiotic fate of chlorinated hydrocarbons are becoming well understood and documented. As with petroleum hydrocarbons, biological degradation of chlorinated hydrocarbons also occurs using aerobic, facultative, and/or anaerobic respiratory pathways. However, several of these metabolic pathways are different and are more dependent on favorable subsurface environmental conditions including the presence of microorganisms capable of completing the reactions. In addition to requiring more favorable environmental conditions, the chemical structure, particularly the degree of chlorination, tends to affect how the compounds are degraded.

Several biological degradation mechanisms have been identified which are capable of transforming halogenated hydrocarbons. USEPA 1998 and Adriano et al. 1999 describe these processes in detail and practitioners are encouraged to review these bodies of work. Table 3 lists some of the more common chlorinated hydrocarbons found at impacted sites and notes the primary degradation mechanisms associated with each compound. These primary degradation processes are described below.

Biodegradation of the Chlorinated Hydrocarbon Used as an Electron Donor (Carbon and Energy Source)

During aerobic respiration (oxidation), the chlorinated hydrocarbons serve as the electron donor and provide energy and organic carbon to the organisms (primary substrate). Chlorinated hydrocarbons most susceptible to aerobic degradation include dichloromethane (methylene chloride), dichloroethenes (DCE), vinyl chloride (VC), 1,2-dichloroethane (1,2-DCA), and many chlorinated benzenes (Figure 7).

Anaerobic oxidation of chlorinated hydrocarbons is limited in the subsurface because the compounds are in an oxidized state; however, compounds such as methyl chloride, dichloromethane, and vinyl chloride are less oxidized and have been shown to oxidize in anaerobic conditions. In particular, vinyl chloride can be used as a sole carbon source (electron donor) and be degraded to carbon dioxide, chloride, and water via iron (III) reduction (Figure 7).

Table 3 Primary Degradation Mechanisms for Common Chlorinated Hydrocarbons

Degradation Process	Compound Name											
	PCE	TCE	1,1,1-TCA	DCE	DCA	VC	CA	CT	CF	DCM	CB	1,2-DBM
Aerobic Degradation												
As Primary Substrate (serves as electron donor and is oxidized)	N	N	N	Y	Y	Y	M	N	N	Y	Y	Y
Cometabolism[a]	N	Y	M	Y	N	Y	N	M	Y	Y	Y	N
Anaerobic Degradation												
As Primary Substrate (serves as electron donor and is oxidized)	N	N	N	N	U	Y	U	N	N	Y	U	U
Reductive Dechlorination[b]	Y	Y	Y	Y	Y	Y	Y	Y	Y	Y	Y	Y
Abiotic Transformation	N	N	Y	N	N	N	Y	M	N	N	U	Y

[a] Usually requires presence of growth or inducer compounds such as methane, alkanes, aromatic compounds (such as toluene), or ammonia.
[b] Compound serves as an electron acceptor and requires an alternative carbon source. Ideal conditions include denitrification, iron reduction, sulfate reduction, methanogenic.

PCE – perchloroethene or tetrachloroethene
TCE – trichloroethene
1,1,1-TCA – 1,1,1-trichloroethane
DCE – dichloroethenes
DCA – dichloroethanes
VC – vinyl chloride
CA – chloroethane
CF – chloroform

DCM – dichloromethane or methylene chloride
CB – chlorobenzenes
1,2-DBM – 1,2-dibromomethane
Y – Yes
N – No
M – May
U – Unknown

Adapted from ITRC 1999 and USEPA 1998.

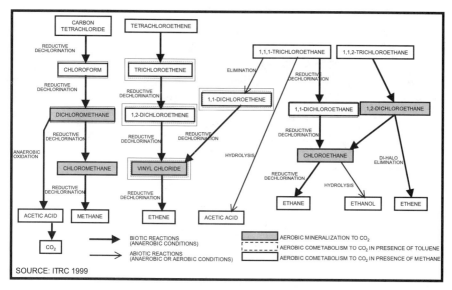

Figure 7   Common degradation pathways.

Biodegradation of the Chlorinated Hydrocarbon Used as an Electron Acceptor (Reductive Dechlorination)

Anaerobic reductive dechlorination is likely the most important mechanism whereby a large number of anaerobic bacteria, in the presence of electron donors and acceptors, cleave chlorine (or halogen) atoms which are replaced by hydrogen (Figure 8). In this case, the chlorinated hydrocarbons act as the electron acceptor and there must be an appropriate source of carbon (electron donor) to maintain microbial growth.

Halorespiration has recently been described whereby bacteria use the chlorinated hydrocarbon as an electron acceptor in reactions that provide growth and energy. Relatively little is known about halorespiration, and it has only been observed in anaerobic conditions and may lead to partial or fully hydrodehalogenated products.

Cometabolism

In aerobic cometabolism, the halogenated hydrocarbons are fortuitously transformed by bacteria in the presence of an alternative electron donor or growth substrate. The chlorinated hydrocarbon is neither an electron donor nor acceptor. Chlorinated ethenes, with the exception of perchloroethene (PCE) are susceptible to aerobic cometabolism and the oxidation rate increases as the degree of chlorination decreases. In addition, some chlorofluorocarbons may also be subject to aerobic cometabolism.

Anaerobic cometabolism may also occur. However, the process is much less understood and may not be distinguishable between other degradation mechanisms.

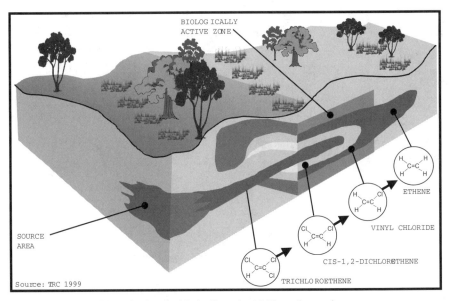

Figure 8   Anaerobic reductive dechlorination of a trichloroethene plume.

Abiotic Degradation

Abiotic degradation also occurs in the subsurface. These reactions are less understood but may be important attenuation mechanisms for compounds such as trichloroethane (TCA). Documents such as USEPA 1998 and Adriano et al. 1999 describe reactions such as hydrolysis, dehydrohalogenation, reactions with reduced sulfur compounds, reactions with natural organic matter and minerals, and reactions with transition metal coenzymes such as reduced vitamin B_{12}, coenzyme F_{430}, and hematin. Natural attenuation practitioners are encouraged to review these bodies of work and become familiar with the current knowledge of abiotic processes.

Natural Attenuation Data Collection and Evaluation

As described in the lines of evidence section of this chapter, there are three lines of evidence that are used to evaluate if natural attenuation is an appropriate remedial solution. The following section describes the generalized steps and data that should be collected. Keep in mind that the steps described below should be used as a guide since every site is different. In addition, depending on the regulatory requirements, there may be more or fewer requirements necessary to evaluate a site.

There are 10 general steps that should be completed in order to evaluate natural attenuation of petroleum and chlorinated hydrocarbons. In addition, these steps can be used to evaluate other organic and inorganic compounds not explicitly included in currently available protocols. Table 4 and USEPA 1998 describes data and data usage parameters typically used to evaluate natural attenuation. In addition, these data are grouped according to each of the three lines of evidence and can be used to plan, define, and execute natural attenuation evaluations.

Table 4 Data Collection Tiers for Evaluation and Implementation of Natural Attenuation

Parameter	Data Type	Ideal Use, Value, Status and Comments	Method	I	II	III	*
Geological							
Area Geology	Topography/Soil Type/Surface Water/Climate	Provides inferences about natural groundwater flow systems, identifies recharge/discharge areas, infiltration rates, evaluation of types of geological deposits in the area which may act as aquifers or aquitards.	Consult published geological/soil/topographic maps, air photo interpretation, field geological mapping.	✓	✓		
Hydrogeological							
Subsurface Geology	Lithology/Stratigraphy/Structure	Identify water bearing units, thickness, confined/unconfined aquifers, and effect on groundwater flow and direction (anisotropy).	Use published hydrogeologic Surveys/maps.	✓	✓	✓	
			Review soil boring/well installation logs.	✓	✓	✓	
			Conduct surface or sub-surface geophysics.			✓	*
Velocity	Hydraulic Conductivity (K)/Permeability (k)	Measure of the saturated hydraulic conductivity of the geological matrix. K times the gradient gives the specific discharge (v). If site is very layered or complex, measure the vertical/horizontal K.	Estimate range based on geology.	✓	✓	✓	
			Conduct: Pump, slug or tracer tests.		✓	✓	
			Estimate with grain size analysis.			✓	
			Permeability test.			✓	
			Downhole flowmeter/dilution test.			✓	*
	Gradient (h)	Measure of potential of the fluid to move (hydraulic gradient).	Water table and piezometric surface measurements.	✓	✓	✓	
	Porosity (n)	Measure of the soil pore space. Dividing the specific discharge by porosity gives the average linear groundwater velocity.	Estimate range based on geology.			✓	
			Measure bulk and particle mass density.			✓	*
Direction	Flow Field	Estimate of groundwater flow.	Water and piezometric contour maps.	✓	✓	✓	
			Downhole flowmeter.			✓	
Dispersion/ Sorption	Fraction of Organic Carbon (Foc)	Fraction of organic carbon: used to estimate the retardation of chemical migration relative to the average linear groundwater velocity.	Estimate or measure Foc in soil samples, estimate from published values, or compare migration of reactive and non-reactive (tracer) chemicals in the groundwater.	✓	✓	✓	*
	Dispersion	Longitudinal and horizontal dispersion (mixing) spreads out the chemical along the groundwater flow path.	Estimate based on distribution of chemicals or use tracer tests.	✓	✓	✓	

Note: *indicates parameter is optional depending on-site complexity.

Table 4   Data Collection Tiers for Evaluation and Implementation of Natural Attenuation (continued)

Parameter	Data Type	Ideal Use, Value, Status and Comments	Method	I	II	III	*
		Chemistry					
Organic Chemistry	Volatile Organic Constituent (VOC)	Identify parent solvents and degradation products; assess their distribution. Certain specific isomers/degradation products provide direct evidence of biodegradation (e.g., cis-1,2-dichloroethene [cis-1,2-DCE]), while others are formed due to abiotic degradation processes (e.g., formation of 1,1-dichloroethene [1,1-DCE] from 1,1,1-trichloroethane [1,1,1-TCA]). In addition, aromatic hydrocarbons (benzene, toluene, ethylbenzene, xylenes [BTEX]) and ketones can support biodegradation of chlorinated VOCs.	USEPA Methods.	✓	✓	✓	
	Semivolatile Organic Compounds (SVOCs)	Selected SVOC (e.g., phenol, cresols, alcohols) may support biodegradation of chlorinated VOCs.	USEPA Methods.			✓	
	Volatile Fatty Acids	Organic chemicals like acetic acid can provide insight into the types of microbial activity that is occurring and can also serve as electron donors.	Standard analytical methods or published modified methods using ion chromatography.			✓	*
	Methane, Ethene, Ethane, Propane, Propene	Provide evidence of complete dechlorination of chlorinated methanes, ethenes, and ethanes. Methane also indicates activity of methanogenic bacteria. Isotope analysis of methane can also be used to determine its origin.	Modified analytical methods, GC-FID.		✓	✓	
	Total Organic Carbon (TOC), Biochemical Oxygen Demand (BOD), Chemical Oxygen Demand (COD), Total Petroleum Hydrocarbons (TPH)	Potential availability of general growth substrates.	USEPA Methods.		✓	✓	
Inorganic/Physical Chemistry	Alkalinity	Increased levels indicative of carbon dioxide production (mineralization of organic compounds).	USEPA Methods.		✓	✓	
	Ammonia	Nutrient. Evidence of dissimilatory nitrate reduction, and serve as an aerobic co-metabolite.	USEPA Methods.		✓	✓	

Note: *indicates parameter is optional depending on site complexity.

Table 4   Data Collection Tiers for Evaluation and Implementation of Natural Attenuation (continued)

Parameter	Data Type	Ideal Use, Value, Status and Comments	Method	I	II	III	*
	Chloride	Provides evidence of dechlorination, possible use in mass balancing, may serve as conservative tracer. Road salts may interfere with chloride data interpretation	USEPA Methods.		✓	✓	
	Calcium/Potassium	Used with other inorganic parameters to assess the charge-balance error and accuracy of the chemical analysis.	USEPA Methods.			✓	*
	Conductivity	Used to help assess the representativeness of water samples, and assess well development after installation (sand pack development).	Electrode measurement in the field. Standard electrode.			✓	
	Dissolved Oxygen (DO)	Indicator of aerobic environments, electron acceptor.	Use flow through apparatus to collect representative DO measurements by electrode.	✓	✓	✓	
	Hydrogen	Concentrations in anaerobic environments can be correlated with types of anaerobic activities (i.e., methanogenesis, sulfate and iron reduction) and therefore this parameter is an excellent indicator of the redox environment. Hydrogen may be the limiting factor for complete dechlorination of chlorinated VOCs.	Field measurement. Flow through cell equipped with bubble chamber. As groundwater flows past chamber, hydrogen gas will partition into headspace. Headspace sampled with gas-tight syringe and analyzed in the field using GC. Equipment for analysis is not yet widely available. Relationship to dechlorination activity is still unclear and subject to further R&D.				*
	Iron	Nutrient. Ferrous (soluble reduced form) indicates activity of iron reducing bacteria. Ferric (oxidized) is used as an electron acceptor.	USEPA Methods and Field Test Kits.		✓	✓	
	Manganese	Nutrient. Indicator of iron and manganese reducing conditions.	USEPA Methods and Field Test Kits.		✓	✓	
	Nitrate	Used as an electron acceptor by denitrifying bacteria, or is converted to ammonia for assimilation	USEPA Methods.		✓	✓	
	Nitrite	Produced from nitrate under anaerobic conditions	USEPA Methods.		✓	✓	
	pH	Measurement of suitability of environment to support wide range of microbial species. Activity tends to be reduced outside of pH range of 5 to 9, and anaerobic microorganisms are typically more sensitive to pH extremes. pH is also used to help assess the representativeness of the water sample taken during purging of wells.	pH measurements can change rapidly in carbonate systems, and during degassing of groundwater. Therefore, pH measurements must be measured immediately after sample collection or continuously through a flow through cell.	✓	✓	✓	

Note: *indicates parameter is optional depending on site complexity.

Table 4   Data Collection Tiers for Evaluation and Implementation of Natural Attenuation (continued)

Parameter	Data Type	Ideal Use, Value, Status and Comments	Method	I	II	III	*
	Phosphorous	Limiting nutrient.	USEPA Methods	✓	✓	✓	*
	Oxidation-Reduction Potential (Eh)	Measure of oxidation-reduction potential of the environment. Ranges from +500 mV for aerobic environments to −500 V for anaerobic environments	Use flow through apparatus in the field to collect representative Eh measurements by electrode. Eh measurements can be affected by geochemical speciation of organic/inorganic chemical species. The measured Eh (using probes) can be confirmed by examining chemical speciation of Eh couples.				
	Sodium	Evaluate whether chloride may be associated with road salt.	USEPA Methods.			✓	*
	Sulfate	Used as electron acceptor. Changes in its concentration may provide evidence of activities of sulfate reducing bacteria.	USEPA Methods and Field Test Kits.		✓	✓	
	Sulfide	May provide evidence of sulfate reduction. May not be detected even if sulfate-reducing bacteria are active because it can react with various oxygenated chemical species and metals	USEPA Methods and Field Test Kits.		✓	✓	
	Temperature	Used to help assess the representativeness of water samples, and to correct temperature sensitive parameters/measuring devices. Microorganisms are active over a wide temperature range.	Field Measurement.	✓	✓	✓	
	Toxic Metals	The presence of metals (e.g., lead, copper, arsenic) can reduce microbial activity. Microorganisms are generally resistant	USEPA Methods.				*

Note: *indicates parameter is optional depending on site complexity.

Table 4   Data Collection Tiers for Evaluation and Implementation of Natural Attenuation (continued)

Parameter	Data Type	Ideal Use, Value, Status and Comments	Method	I	II	III	*
Microbiology							
Biomass	Microorganisms Per Unit Soil or Groundwater	Microbial population density between impacted and non-impacted/treated areas can be compared to assess whether microbial populations are responsible for observed degradation. The value of biomass measurements is still being explored for chlorinated VOC biodegradation.	There are three general techniques available: culturing (plate counts, BioLog, MPN enumerations); direct counts (microscopy); and indirect measurement of cellular components (ATP, phospholipid fatty acids).				*
	Biodegradation Rate and Extent	Demonstrate the indigenous microorganisms are capable of performing the predicted transformations. Determine nutrient requirements and limitations. Measure degradation rates and extent.	Varied. Shake flasks, batch, column, and bioreactor designs.				*
	Species/General/Functional Group	The presence of certain microbial species of functional groups (e.g., methanogenic bacteria) that have been correlated with chlorinated VOC biodegradation can be assessed. Research is being conducted to identify patterns of microbial composition that are predictive of successful chlorinated VOC biodegradation.	There are three general techniques available: culturing and direct counts; indirect measurement of cellular components; and molecular techniques (16s RNA, DNA probes, RFLP).				*

Note: *indicates parameter is optional depending on-site complexity.
Adapted from IDRC 1999.

Step 1. Delineate the Areal and Vertical Extent of Groundwater Impacts and Determine Hydrogeology

A sufficient number of monitoring wells should be constructed to delineate groundwater impacts. The monitoring array should be constructed, monitored, and sampled, such that background, impacted, side gradient, and down-gradient groundwater quality can be evaluated. In addition to constructing and sampling monitoring wells, geological and hydrogeological data should also be collected at this time in order to evaluate groundwater transport (Table 4).

Step 2. Determine the Potentiometric Water Surface of the Impacted Units, Evaluate Historic Water Level Trends, and Determine Hydraulic Gradient and Conductivity

Groundwater elevations should be measured and hydrogeological parameters measured or estimated. These data will be used estimate groundwater flow direction and velocity. It is necessary to confirm groundwater flow direction and velocity over time to determine if groundwater impacts are migrating or stable (Table 4). Effective natural attenuation solutions rely on demonstrating that groundwater impacts are stable (no longer migrating) and concentrations (and mass) are decreasing.

Step 3. Measure the Concentrations of Indicator Parameters

This portion of site investigation is likely the most important and unfamiliar step. The presence (or absence) of indicator parameters allows site investigators and remedial design engineers to determine the environmental conditions and likely degradation mechanisms occurring in groundwater. These indicator parameters can be divided into three main categories as follows:

- General groundwater quality
- Electron donors/acceptors
- Degradation or daughter products

These chemistry parameters are described in Table 4. These data are obtained by collecting and analyzing samples, analyzing groundwater in the field using direct-reading analytical instrumentation, or by using field test kits. Practitioners should consult appropriate natural attenuation protocol to refine field sampling and analytical protocols since there are numerous collection, analysis, and instrumentation options available.

Step 4. Determine the Distribution of Oxidation-Reduction Processes

The chemistry parameters described above and in Table 4 are evaluated relative to each other and relative to space (distribution). Isoconcentration maps or specialized visualization graphics (see modeling tools section, below) should be used to graphically display the presence and relationships between impacts, electron donors and acceptors, degradation products, groundwater flow, and proximity to potential human or ecological receptors.

It is important to show that the concentrations of organic constituents decrease and that terminal electron acceptors are being used. Table 5 and Figure 9 shows the relationships between oxidation-reduction potentials, respiration pathways, and the resulting byproducts of each respiration pathway. In addition, Table 6 shows the theoretical concentration of hydrogen as compared to predominant terminal electron accepting process. These concentrations are theoretically consistent site to site.

Table 5   Redox Versus Respiration Mechanism

Redox	Respiration	e⁻ Acceptor	Byproducts
+ 200 mv	Aerobic	O_2	CO_2
	Denitrification	NO_3^{2-}	NO_2^-, N_2, NH_3
	Manganese Reduction	Mn^{4+}	Mn^{2+}
	Iron Reduction	Fe^{3+}	Fe^{2+}
	Sulfanogenesis	SO_4^{2-}	H_2S
-400 mv	Methanogenesis	CO_2	CH_4

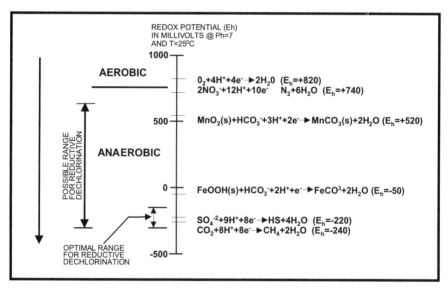

Figure 9   Oxidation-reduction potentials for various electron acceptors.

Table 6   Hydrogen Concentrations Characteristic of Different Terminal Electron Accepting Processes

Terminal Electron Accepting Process	Characteristic Hydrogen Concentration (nM)
Denitrification	<0.1
Fe (III) Reduction	0.2 – 0.8
Sulfate Reduction	1 – 4
Methanogenesis	>5

Currently, hydrogen data collection is challenging and before hydrogen data collection is attempted, the reader is directed to consult USEPA 1998 (and future publications) regarding hydrogen sampling and analysis.

Figure 10 shows the relationship between electron donors (organics) and electron acceptors relative to distance from the source and time. These processes and distribution patterns are consistent with most (if not all) petroleum hydrocarbon impacted sites.

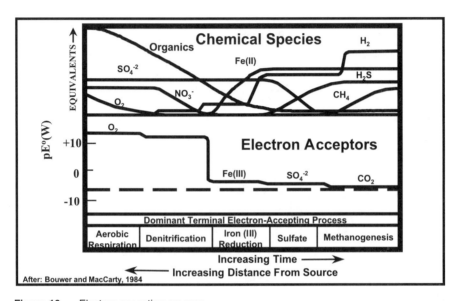

Figure 10   Electron-accepting process.

Data patterns associated with chlorinated hydrocarbon impacted sites are similar to petroleum hydrocarbon impacted sites with one primary difference. Petroleum hydrocarbons are always viewed as electron donors (organics, substrates, etc.). Chlorinated hydrocarbons (depending on the compound) can be either a donor or acceptor. If the chlorinated hydrocarbon is a donor, then the concentration decrease and associated electron acceptor patterns are generally consistent with Figure 10. However, if the chlorinated hydrocarbon is an electron acceptor, then a supplemental organic carbon source (electron donor) is usually required to effectively complete the biodegradation reactions. Figures 11a, 11b, 12a and 12b show common patterns associated with chlorinated hydrocarbon biodegradation in aerobic and/or anaerobic environments. Figures 11a and 12a show the pattern for VOCs and Figures 11b and 12b show the electron acceptor patterns respectfully. Comparing the concentrations of petroleum and chlorinated hydrocarbons to indicator parameters is used to document that the environmental conditions and pathways are favorable and consistent with known degradation pathways. Figure 13 shows a logic flow diagram that can be used to determine which degradation pathways are likely occuring.

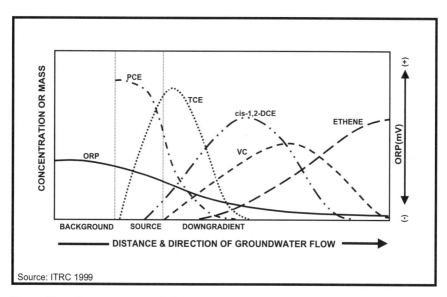

Figure 11a   Common patterns (VOC) of chlorinated solvent biodegradation in an anaerobic system.

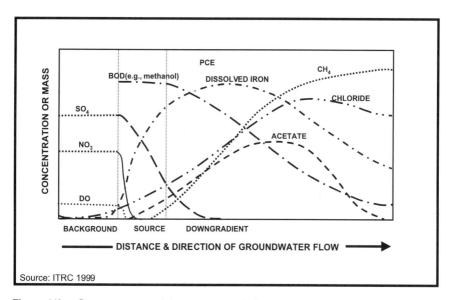

Figure 11b   Common patterns (electron acceptor) of chlorinated solvent biodegradation in an anaerobic system.

Step 5. Determine/Understand Likely Degradation or Transformation Processes and Estimate Biodegradation Rates

Degradation mechanisms for petroleum hydrocarbons, common chlorinated hydrocarbons, and other organic compounds, along with mechanisms to transform

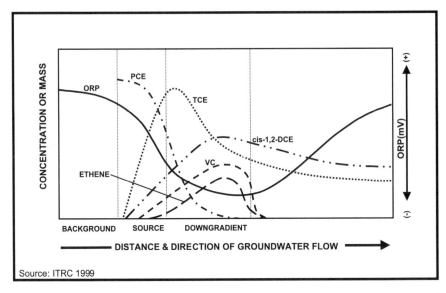

Figure 12a   Common patterns (VOC) of chlorinated solvent biodegradation in a sequential aerobic/anaerobic system.

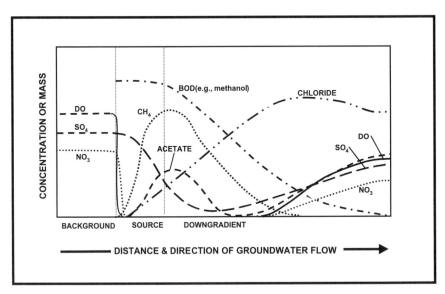

Figure 12b   Common patterns (electron acceptor) of chlorinated solvent biodegradation in a sequential aerobic/anaerobic system.

heavy metals can be found in numerous journals, tests, and protocols. It is important that site investigators and remedial design engineers have a basic understanding of these processes and the ideal environmental conditions required to complete the desired biochemical reactions.

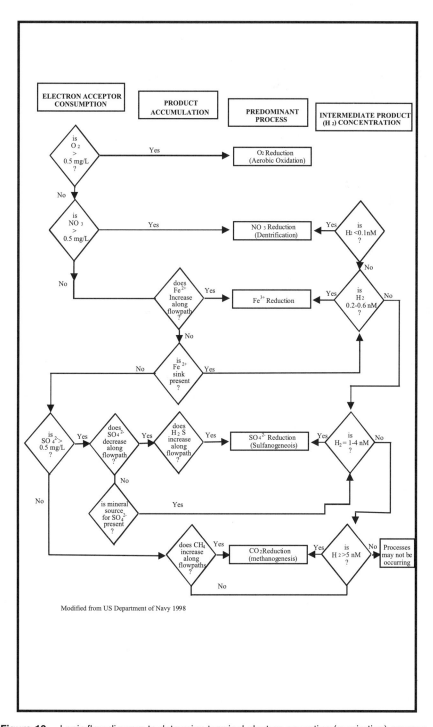

Figure 13 Logic flow diagram to determine terminal electron accepting (respiration) process.

Once a basic understanding of the degradation or transformation mechanism is understood, including knowledge of site-specific environmental conditions, biodegradation, or transformation rates should be estimated. Rate estimates can be made by evaluating site-specific field data, completing laboratory treatability studies, or by using published rates. Each approach has advantages, disadvantages, and limitations. USEPA 1998 and Wiedemeier 1995 describe several options for estimating biodegradation rates. Suarez and Rifai 1999 has compiled and statistically analyzed degradation rates for petroleum hydrocarbons and chlorinated hydrocarbons. In addition, several of the newer models developed for evaluating natural attenuation (see modeling tools, below) estimates biodegradation rates using site specific data. These sources should be consulted to determine an appropriate estimation method, or source of biodegradation rates.

Step 6. Evaluate Assimilative Capacity of Aquifer and/or Mass Flux of Contaminants

The assimilative capacity of an aquifer can be estimated for degradation of organic compounds that serve as electron donors including petroleum hydrocarbons, some chlorinated hydrocarbons, and other organic pollutants. During the degradation process, electrons are transferred from the donors to terminal electron acceptors and the organic compound is degraded. At the same time, the electron acceptors are reduced. The assimilative capacity of an aquifer is the theoretical ability of the aquifer to supply requisite terminal electron acceptors in order for the biochemical reactions to occur. This information is used to estimate the theoretical mass of organics capable of being degraded and to confirm that the concentrations of terminal electron acceptors are high enough to complete the biodegradation process and not become rate limiting. These calculations are simplified to assume that degradation only occurs and that biomass production and/or other geochemical processes are not occurring. Therefore, these calculations should be used with caution and site-specific monitoring data must be used to confirm that the contaminant mass is decreasing.

The assimilative capacity of an aquifer is calculated based on a stoichiometric mass balance between the electron donor(s) and electron acceptor(s) (Wiedemeier 1995) and can be simplified and described as follows:

$$AS = Organic_{ratio}(EA_b - EA_m) \tag{1}$$

where AS = the concentration of organic that the aquifer is capable of assimilating (assimilative capacity); $Organic_{ratio}$ = concentration of organic degradable per 1 mg/l electron acceptor consumed; EA_b = background electron acceptor concentration (mg/l); and EA_m = lowest measured electron acceptor concentration (mg/l).

For example, assuming the background dissolved oxygen concentration is 6 mg/l and the lowest dissolved oxygen concentration in the plume is 1 mg/l, then the assimilative capacity to degrade benzene is 1.6 mg/l, or

$$AS = 0.32(6 \text{ mg/l} - 1 \text{ mg/l})$$
$$AS = 1.6 \text{ mg/l}$$

The assimilative capacities for each respiration pathway being used are then typically added together to determine the overall assimilative capacity for a particular aquifer.

The assimilative capacity of an aquifer cannot easily be calculated for organic compounds that are servings as electron acceptors such as chlorinated hydrocarbons undergoing reductive dechlorination or cometabolism. This calculation is difficult because the pathways are more complex, less understood, and an alternative carbon source (electron donor) is often required as an energy source (described above), producing anaerobic conditions (no oxygen, negative oxidation-reduction potential, and elevated concentrations of hydrogen [H$^+$]). These reactions result in conditions favorable for the chlorinated hydrocarbon to be reduced (chlorine atom replaced by hydrogen). Therefore, mass flux calculations should be completed to confirm that site-specific mass losses are due to attenuation, not simply dilution.

It is important to confirm that the mass in groundwater (not just concentrations) is being decreased over a particular time period. This is applicable for petroleum hydrocarbons and chlorinated hydrocarbons. However, because an aquifer's assimilative capacity for degradation of chlorinated hydrocarbons cannot be readily calculated, an understanding of mass losses associated with chlorinated hydrocarbons (with time) is particularly important to document.

Mass losses with time can be calculated using hydrogeological and groundwater monitoring data based on the following equation

$$M = A \, b \, C \, V_{CF} \, M_{CF} \, \phi \tag{2}$$

where M = mass (lbs); A = lateral extent of impacted groundwater (ft^2); b = average thickness of the saturated zone (ft); C = arithmetic mean concentration of constituents in groundwater (μg/L); V_{CF} = 28.317 L/ft^3 (unit conversion); M_{CF} = 2.2046 x 10^{-9} lbs./μg (unit conversion); and ϕ = water-filled porosity.

The resulting estimate is compared to mass calculations completed using historical data to determine if the overall mass is decreasing.

Step 7. Compare Contaminant Transport Rate to Biodegradation Rate

If the dissolved contaminants in groundwater are not stable and are migrating, use the results of steps 1, 2, 5, and 6, and compare the contaminant transport rate to biodegradation rate. As described in USEPA 1998, there are several methods of estimating mass flux and transport. These calculations can be completed by hand or computer-based models can be used to estimate contaminant transport (see modeling tools section). Including the biodegradation rate in the evaluation, the results should determine the approximate distance that dissolved constituents may travel, the resulting concentrations, and the time frame.

Step 8. Evaluate Potential Receptor Impacts

A receptor survey should be completed to determine if potential exposure routes are complete. The exposure analysis should include potential human and ecological receptors. The analysis should evaluate current groundwater conditions, and using information derived from steps 1 through 8, determine if natural attenuation is likely to prevent contaminant transport to locations where exposure may occur.

Step 9. Determine Viability and Efficiency of Natural Attenuation

Natural attenuation can be an effective groundwater remedial solution. The technology can be used as the sole remedial solution or in conjunction with other more active remedial technologies. Natural attenuation is viable as follows

- As a sole remedial solution if biodegradation is readily occurring, remediation goals can be achieved in a timely manner, and current and future receptors will not be exposed (above acceptable concentrations) to contaminants in groundwater
- As a sole remedial solution if biodegradation is moderate, remediation goals can be achieved in a timely manner, and current and future receptors will not be exposed (above acceptable concentrations) to contaminants in groundwater
- In conjunction with more active remedial solutions if biodegradation is occurring (or expected to occur during or after active remediation), remediation goals can be achieved in a timely manner, and current and future receptors will not be exposed (above acceptable concentrations) to contaminants in groundwater

Step 10. Engineer a Site-Specific Natural Attenuation Remediation System

If the results of steps 1 through 9 indicate that natural attenuation is a viable remedial solution, then a monitoring and evaluation program should be designed. In addition, a contingency plan should also be designed in the event conditions change and natural attenuation is no longer effective.

The long-term engineered plan should be developed based on the locations of hydrogeological units, groundwater flow rates and trends, biodegradation rates, and the locations of potential human or ecological receptors. The ultimate objective is to determine if the plume conditions are changing. If change occurs, and there is potential for receptor exposure, then the contingency plan should be implemented.

If not already present at a site, groundwater monitoring wells should be located as follows

- Up-gradient of unimpacted groundwater
- Sidegradient of unimpacted groundwater
- Within the source area (preferably in the down-gradient portion of the source area)
- Down-gradient of the source area still within an area of high biological activity (approximately 1/3 of the way along the flow path)(anaerobic or anoxic zone)
- Down-gradient of the zone where a high rate of biological activity is occurring and ideally within a zone where oxygen is present and aerobic degradation is occurring (approximately 2/3 of the way along the flow path)(aerobic zone near the leading edge of the plume)
- Down-gradient of the plume where the concentrations are less than regulatory requirements and electron acceptor concentrations are depleted (as compared to up- and/or sidegradient groundwater quality) (sentinel wells)

Generalized monitoring well locations are shown on Figure 14. These locations should be used as a guide since every site is different. Contingency well locations should be identified; however, it is not necessary to construct these wells unless the contingency plan is implemented. In addition, non-functional or damaged monitoring wells should be properly destroyed at this time.

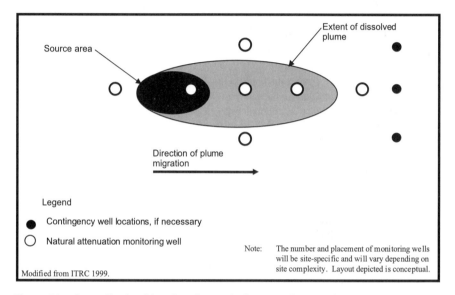

Figure 14   Generalized well locations for monitoring natural attenuation.

After the monitoring well locations have been determined, a monitoring plan should be developed. The monitoring plan should define (at least) the following:

- Monitoring time period (complete until the remedial action objectives have been achieved)
- Monitoring frequency (monthly, quarterly, semi-annual, or annual) (depends on hydrogeological conditions, receptor locations, and regulatory requirements). Frequency will change over time—after confirming that the original assumptions were correct, then the number of wells and the frequency of sampling can be reduced significantly
- Sampling procedures
- Field and laboratory methods of analyses (regulated constituents, general ground-water quality parameters, electron acceptors, and degradation products). Frequency is site specific and depends on plume stability and the quantity and quality of historical data
- Criteria and action levels for implementing contingency plan
- Contingency plan

The contingency plan should include proposed contingency well locations. The locations of contingency wells should be perpendicular to the groundwater flow path and up-gradient of the compliance boundary. The final number and locations will be based on site-specific features, access, numbers of source areas, and regulatory requirements.

The contingency plan should be implemented if monitoring results indicate that the groundwater conditions have changed and the dissolved constituents are advancing. The plan should include evaluation criteria to ensure that the conditions are statistically significant, to evaluate the nature of the changed condition, and a corrective action timeframe. In addition, the plan should identify corrective actions that may be completed to reestablish an effective remediation system. Possible corrective action steps could include the following:

- Continue to monitor using the contingency wells as sentinel wells
- Hydraulically contain impacted groundwater
- Review active remediation alternatives to reduce the mass of contaminants in the aquifer
- Enhance the rate of biodegradation by methods proposed in Chapters 8 and 9

MODELING TOOLS

Mathematical models are valuable tools for engineers and scientists to make remedial design decisions. A model integrates the available data enabling the quantification of patterns and trends that have been identified in site data. This integration of information can also simplify the interpretation of site conditions by demonstrating the interdependency of hydrogeologic conditions, the geochemical environment, constituent type and concentration, available electron acceptors on biodegradation, as well as to predict a reasonable estimate of future conditions.

There are always a few words of caution on the use of solute transport models about recognition of the uncertainties associated with predictions. Accurate modeling of even a conservative tracer using a solute transport model is a difficult problem. Even with high quality time varying contaminant data, a thorough understanding of the hydrogeolgic environment, with an emphasis on flow pathways, is required to understand the transport phenomena. The transport of reactive compounds is difficult because all of the processes are not well understood and often there are multiple chemical and physical attenuation mechanisms, which have similar effects on constituent concentrations.

It cannot be over stressed that bioattenuation modeling represents the leading edge of modeling technologies. It requires a complete understanding of the physical, chemical, and biological processes affecting the movement of contaminants. When properly constructed, these models can be used to visualize and understand complex processes occurring in the subsurface. During the last 15 years, numerous models have been developed that incorporate the biodegradation process. Some of these models only include a simple first-order degradation rate term while others (currently or will) include the interaction between parent compounds, degradation products, electron donors, and electron acceptors. The design tools for constructing these models have also advanced, thereby facilitating use and the design of complex applications.

Biological degradation modeling focuses on three fundamental concepts (Adriano et al. 1999), (1) microbial growth and survival; (2) efficiency of energy transfer from electron donating substrates to production of cells, reaction products, and heat; and (3) mass transfer of electron donors, electron acceptors, and nutrients from solution into dispersed or biofilm-based cells.

An exhaustive description of modeling and the availability of models to simulate *in situ* bioremediation is beyond the scope of this text. However, Adriano (1999) devotes an entire chapter to "Mathematical Modeling in Bioremediation of Hazardous Wastes" where theory, concepts, and actual models are described. Because natural attenuation has more recently been, and continues to be, better understood, some of the original and newer models used to simulate natural attenuation are described in Table 7. The models have been organized from simple to complex representing a wide range of potential applications. As the biological element of natural attenuation becomes better understood, it is anticipated that these models will be refined.

CASE HISTORIES

The following sections describe how natural attenuation is currently being applied at two sites. The first site is impacted with petroleum hydrocarbons, while the second site is impacted with chlorinated hydrocarbons. The examples are presented to provide a clearer understanding of how a site is characterized and natural attenuation is applied as a remedial solution.

Petroleum Hydrocarbon Site Application

A 25,000-gallon underground storage tank (UST) was removed from the infield area of an active municipal airport in 1988. Aviation fuel had been contained in the UST and its release was observed at the time of the removal. A total of 33 exploratory soil borings had been installed subsequently to characterize the impact, among which, 24 borings were converted into groundwater monitoring wells. Based on analytical data, it was estimated that approximately 15,000 to 60,000 gallons of hydrocarbons had been released from the UST. A VES was installed in 1992, and had been in continuous operation to address hydrocarbon impact in subsurface and NAPL observed in groundwater monitoring wells near the former UST via volatilization. The operation of VES was discontinued in 1996, based on low influent concentrations, and was restarted in 1998, to address the presence of NAPL in groundwater monitoring wells.

The site is situated in an alluvial sediment unit comprising of approximately 300 feet of basin fill. The unit is divided into three sub-units (sub-units A through C) based on the lithologic composition related to depositional environment of the sediments. Based on extensive hydrogeological investigations, the hydrocarbon impact is limited to within sub-unit A, extending from ground surface to approximately 120 feet below grade. Soil encountered within sub-unit A consists of alter-

Table 7   Description of Models Used to Simulate or Visualize Natural Attenuation

Model Name	Application	Level	Processes	Miscellaneous	Reference
BioTrends	Natural attenuation visualization.	Screening Level Tool / Data Interpretation	Uses BioRedox to simulate sequential decay reactions along a 1-D flow path line.	Completes natural attenuation trends, visualization, calculates degradation rates, and "scores" site.	Newell, Mcleod, and Gonzales 1996 *Proprietary Code*
SEQUENCE	Primary and secondary line of evidence visualization for biodegradation		Visual aids that may be used to simultaneously show spatial and temporal trends for multiple organic pollutants and geochemical indicators (including electron acceptors and metabolic by-products).	Visualization tool to better document lines of evidence that natural attenuation is occurring. Can eliminate need for multiple plume contours and data tables.	Carey, et. al. 1999a b *Proprietary Code*
BIOSCREEN	Three-dimensional contaminant transport for dissolved phase hydrocarbons in saturated zone under the influences of oxygen, nitrate, iron, sulfate, and methane limited biodegradation.	Screening Level Tools / Simple Analytical Tools / One-Dimensional Groundwater Flow	Advection, dispersion, adsorption, first order decay and instantaneous reactions under aerobic and anaerobic conditions.	An easy-to-use screening tool based on the Domenico analytical solute transport model. The program is written as a Microsoft Excel Spreadsheet	Newell, Mcleod, and Gonzales 1996 *Public Domain Code*
BIOCHLOR	Simulates remediation by natural attenuation of dissolved solvents at chlorinated solvent release sites. BIOCHLOR can be used to simulate solute transport either without decay or with biodegradation modeled as a sequential first-order process within 1 or 2 different reaction zones.		The program is based on the Domenico analytical solute transport model and includes 1-D advection, 3-D dispersion, linear adsorption and biotransformation via reductive dechlorination. Reductive dechlorination is assumed to occur under anaerobic conditions and solvent degradation is assumed to follow a sequential first-order decay process.	The program is written as a Microsoft Excel Spreadsheet. The program includes a Natural Attenuation Screening Protocol scoring system to help the user determine the potential for reductive dechlorination from site data.	Aziz, et. al. 1999 *Public Domain Code*
Bioplume I, II, III	Two-dimensional contaminant transport under the influence of oxygen, nitrate, iron, sulfate, and methanogenic biodegradation.	Intermediate Level Tools / Detailed Evaluation of Processes, Cannot Evaluate all Groundwater Flow Condition	Advection, dispersion, sorption, first order decay, biodegradation through instantaneous, first order, zero order, or Monod kinetics.	Model is based on 2-D solute transport code, USGS-MOC but has an integrated, Windows based, graphical user interface for seamless preprocessing and output review. Hydrocarbon source and each active electron acceptor are simulated as separate plumes.	Rifai et al. 1988 *Public Domain Code*

Table 7 Description of Models Used to Simulate or Visualize Natural Attenuation (continued)

Model Name	Application	Level	Processes	Miscellaneous	Reference
MT3D99	Three-dimensional contaminant transport in saturated zone under the influence of oxygen, nitrate, iron, sulfate, and methanogenic biodegradation.	MODFLOW Based Advanced Level Tools / Detailed Evaluation of Processes and Groundwater Flow Conditions	Codes capable of modeling advection in complex steady-state and transient flow fields, anisotropic dispersion, first-order decay and production reactions, and linear and nonlinear sorption. Can also handle monad reactions and daughter products. Also simulates multi-species reactions and simulate or assess natural attenuation within a contaminant plume.	Finite-difference solute transport codes linked with the USGS groundwater flow simulator, MODFLOW, and are designed specifically to handle advectively-dominatedtransport problems without the need to construct refined models specifically for solute transport.	S.S Papadopulous 1999 *Proprietary*
MT3DMS					Zheng and Wang 1998 *Public Domain Code*
RT3D					Clement 1997 *Public Domain Code*
SEAM3D					Waddill and Widdowson 1998 *Proprietary*
BIOMOD 3-D	Three-dimensional contaminant transport in saturated zone under the influences of oxygen, nitrate, iron, sulfate, and methane limited biodegradation.	MODFLOW Based Advanced Level Tools / Detailed Evaluation of Processes and Groundwater Flow Conditions	Advection, dispersion, diffusion, adsorption, desorption, and microbial processes based on oxygen limited, first order, Monod, anaerobic biodegradation kinetics, as well as anaerobic or first-order sequential degradation involving multiple daughter products are simulated	Three-dimensional finite-element fate and multicomponent transport model that is linked to the USGS finite-difference MODFLOW 3-D model	
BioRedox	BioRedox is used to model fate and transport of aqueous phase solids and interactions involving mineral-phase solutes.		Advection, hydrodynamic dispersion, equilibrium sorption, instantaneous or first-order reactions with optional representations of daughter product and halogen (e.g. chloride) transformations, coupled redox reactions and source attenuation.	Finite-difference solute transport codes linked with the USGS groundwater flow simulator, MODFLOW, and are designed specifically to handle advectively-dominatedtransport problems without the need to construct refined models specifically for solute transport	Carey, et. al. 1999a b *Proprietary*

nating layers and lenses of silts, sands, and gravel to approximately 90 feet below grade.

Depth to groundwater measurements collected between 1989 and 1994 ranged from approximately 47 feet to 60 feet below grade. The natural groundwater gradient is southwesterly with only slight variation over time. In 1993, the area experienced a period of unusually high rain fall and associated flooding in a nearby river that resulted in 6 to 10 feet rise in groundwater elevation at the site and in submerging some of the free product. In December 1994, groundwater was measured at 57 feet below grade and hydraulic gradient was approximately 0.003 feet per foot approximately 20 degrees south of normal ambient flow direction due to the stress in the aquifer from local pumping wells.

Based on soil samples analytical data, it was estimated that the hydrocarbon impact had extended to 80 feet below grade and spread over an area of approximately four acres. Within the extents of the impact, hydrocarbons were present in the forms of NAPL, adsorbed phase, and dissolved phase. Hydrocarbon NAPL was observed up to approximately 4 feet in thickness in the former UST area. Benzene and total hydrocarbons were detected at a maximum of 3,900 µg/L and 29,000 µg/L in the former UST area, respectively. As of May 1995, it was estimated that the VES had removed approximately 60,000 pounds of residual and separate phase hydrocarbons through volatilization. In addition, in excess of 20,000 pounds of hydrocarbons had been removed through degradation based on field measurements of oxygen depletion at the extraction wells. It was estimated that approximately 250,000 to 300,000 pounds of hydrocarbons and approximately 3000 pounds of benzene had remained in the subsurface after the initial VES operation in 1996. Average benzene concentrations in 1994 are depicted in Figure 15.

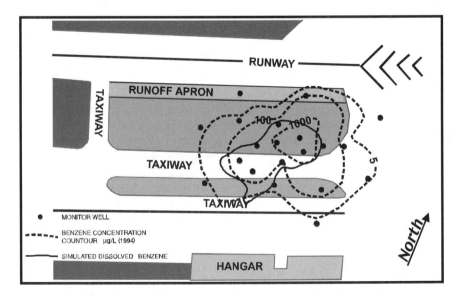

Figure 15   Average benzene concentrations (1994) and simulated dissolved benzene concentrations at water table two years after residual source of benzene eliminated.

In addressing the remediation of the hydrocarbon impact at the site, four corrective options were selected for evaluation: pump and treat for containment and removal, pump and treat dewatering with enhanced VES, air sparging, and *in situ* bioremediation. The options were evaluated for their technical, environmental, institutional, and cost aspects in order to select a remedial technology that would best address the site-specific conditions and impacts. Technical difficulties in achieving remedial goals and in the disposition of treated groundwater limited the remedial options involving pump and treat system. Inadequate demonstrability of implementing air sparging systems in large scale and the heterogeneity of impacted aquifer led to removal of that remedial option. Based on the evaluation, *in situ* bioremediation was selected as a remedial technology based on its demonstrated effectiveness, low cost of implementing, minimal impact on airport operations, and the ability to protect human health and environment.

The efficiency of *in situ* bioremediation had been demonstrated at the site by the evidence of naturally occurring biological reactions reducing hydrocarbon concentrations below regulatory cleanup standard within several hundred feet downgradient of the former UST area. Comprehensive site investigation indicated the occurrence of bioremediation at the site involving both aerobic and anaerobic respiration. The occurrence of aerobic respiration was determined by depleted dissolved oxygen levels within the impacted areas of the aquifer where oxygen is utilized by naturally-occurring microbes in oxidizing substrates such as benzene. The pattern of oxygen depletion is depicted in Figure 16.

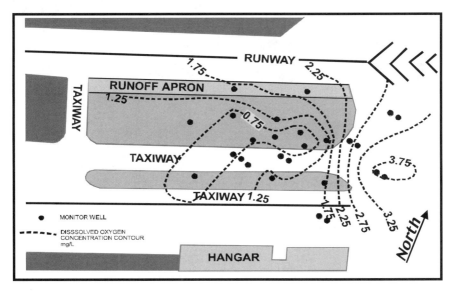

Figure 16   Average dissolved oxygen (DO) concentrations in 1994 (ppm).

The occurrence of anaerobic respiration was verified by the depletion of electron acceptors, including nitrate and sulfate within the impacted area. Figures 17 and 18 depict the occurrence of electron acceptors such as nitrate and sulfate within the

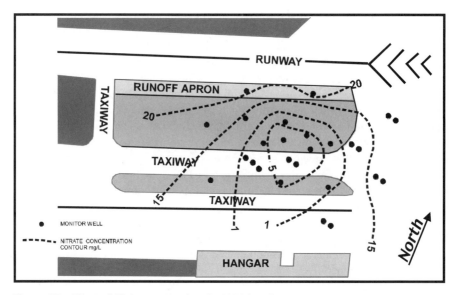

Figure 17   Nitrate (NO_3^-) concentrations in 1994 (ppm).

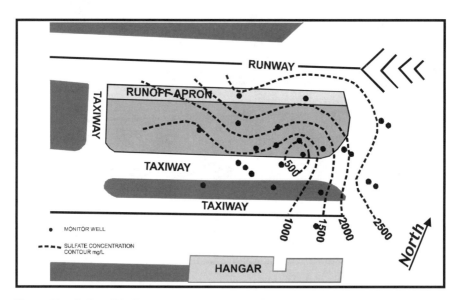

Figure 18   Sulfate (SO_4^{-2}) concentrations in 1994 (ppm).

impacted area relative to background levels. Elevated levels of ferrous iron near the former UST area also indicated that anaerobic respiration via iron reduction is occurring. Increased levels of ferrous iron were detected near the former UST area as depicted in Figure 19. Alkalinity was also measured in groundwater monitoring wells as an indirect indicator of biodegradation occurring at the site. Alkalinity was detected at an elevated level in a groundwater monitoring well located in former UST area when compared to the levels detected in up-gradient wells. Biodegradation

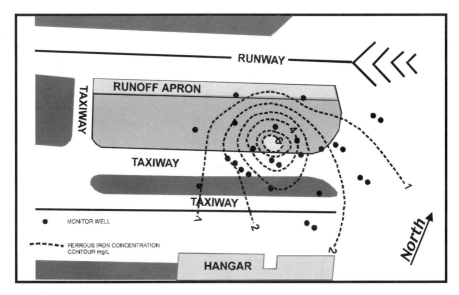

Figure 19   Ferrous iron (Fe^{+2}) concentrations in 1994 (ppm).

was evidenced by the elevated level of alkalinity since the CO_2, produced as a bi-product from biochemical reactions, forms carbonic acid that in turn dissolves the carbonate minerals in the aquifer matrix.

Sulfanogenesis, a biochemical reaction that oxidizes substrates while reducing sulfate, was determined to be the primary biodegradation mechanism occurring at the site because of its abundant supply ranging up to 2,500 mg/l in background wells. A specific occurrence of sulfanogenesis was evidenced in a groundwater monitoring well located in the former UST area where the lowest sulfate concentration was detected.

Based on the understanding of degradation mechanisms at the site, the assimilative capacity of the aquifer was estimated for the degradation of benzene, toluene, ethylbenzene, and xylenes (BTEX) as shown in Table 8. The assimilative capacity was calculated with the understanding of the stoichiometric ratio for biochemical reactions between BTEX compounds and electron acceptors. A ratio between the calculated assimilatory capacity and target BTEX compounds was calculated using the dissolved BTEX concentrations within the former UST area. The average BTEX concentration was calculated to be 3.77 mg/l, from the concentrations detected in groundwater monitoring wells in the former UST area and the total assimilative capacity relative to BTEX was determined to be 426 mg/l as in Table 8. The abundance in the assimilative capacity is largely due to the adequate supply of sulfate that will be used in degrading the BTEX compounds.

The project life for the remediation of hydrocarbon compounds was estimated in two steps: an estimate of the rate of transfer from the residual phase to the dissolved phase and an estimate of time for the degradation of hydrocarbon compounds where benzene was used as a surrogate for the calculation. In completion of the calculations, a period of 15 years was conservatively estimated for the remediation of hydrocarbon

Table 8   Calculation of Assimilative Capacity of Inorganics

Available Inorganic Constituent (mg/l)		Organic Constituent				Average (mg/l)
		Benzene	Toluene	Ethylbenzene	Xylene	
DO = 2.33						
Stoichiometric Capacity	(mg)	0.325	0.319	0.313	0.313	
Actual Capacity	(mg/l)	0.757	0.743	0.729	0.729	0.74
NO$_3$ = 18.8						
Stoichiometric Capacity	(mg)	0.21	0.206	0.203	0.203	
Actual Capacity	(mg/l)	3.95	3.87	3.82	3.82	3.86
Fe(II) = 4.2						
Stoichiometric Capacity	(mg)	0.046	0.046	0.045	0.045	
Actual Capacity	(mg/l)	0.193	0.193	0.189	0.189	0.191
SO$_4$ = 1982						
Stoichiometric Capacity	(mg)	0.22	0.21	0.21	0.21	
Actual Capacity	(mg/l)	436	416	416	416	421
CH$_4$ =< 0.2						
Stoichiometric Capacity	(mg)	1.3	1.28	1.26	1.26	
Actual Capacity	(mg/l)	<0.26	<0.26	<0.25	<0.25	<0.25*
				Total assimilative capacity relative to BTEX =		426

Notes: Stoichiometric Capacity = mg of organic constituent transformed by 1 mg of inorganic constituent.
      Actual Capacity = mg/l of organic constituent transformed by available mg/l of inorganic constituent.
      mg = milligram
      mg/l = milligram per liter
      DO = Dissolved Oxygen
      BTEX = benzene, toluene, ethylbenzene, and total xylenes
      * = value assumed to be zero for purposes of calculating the total assimilative capacity.

compounds. The project life includes 13 years for the transfer of benzene from the residual phase to the dissolved phase and two years for its degradation after the mass had been transferred to the dissolved phase. Figure 15 presents the simulated iso-concentration contour for dissolved benzene two years after complete dissolution of benzene from residual phase. As shown in the figure, significant reduction in the concentrations of benzene may be achieved via biodegradation in reasonably short period of time due to the large assimilative capacity of the aquifer.

A monitoring plan was developed to evaluate the intrinsic biodegradation process. The objectives of monitoring for intrinsic biodegradation were to monitor plume stability, document plume concentration trends, and confirm ongoing biodegradation processes. Specific sampling parameters and monitoring schedules were developed to accommodate the objectives of the monitoring and included tri-annual, annual, and semiannual monitoring of up-gradient, down-gradient, and source area wells for intrinsic biodegradation parameters as well as geochemical parameters.

Chlorinated Hydrocarbon Site Application

This site is a former aerospace machining facility located in Texas. The facility was originally constructed in the early 1960s for the manufacture of aircraft components, and was operated by various owners and operators for that purpose until 1994. The site is situated in an industrial area. Historic production processes at the site included the use of trichloroethene (TCE) until the mid 1980s, and the use of 1,1,1-trichloroethane (1,1,1-TCA) from the mid 1980s until 1994.

Beginning in 1988, numerous investigations were conducted to evaluate the site's hydrogeologic characteristics and to determine the extent to which the shallow groundwater beneath the site had been impacted by chlorinated hydrocarbons.

Geology and Hydrogeology

The topography at the site is relatively flat, with surface drainage flowing to a partially improved drainage ditch on the west side of the facility. The ditch flows south to a creek, which is a southeast-flowing perennial stream located immediately south of the site.

The site is situated on an outcrop of Quaternary-age alluvial deposits consisting primarily of clays and silty clays in the upper 15 to 20 feet and coarser grained silty sands and clayey gravels down to the underlying bedrock. The thickness of the alluvium at the site ranges from approximately 15 to 42 feet. The alluvial deposits were laid down by an ancient stream and are laterally discontinuous and lithologically heterogeneous. The sands and gravel at the base of the alluvial section are generally more porous and permeable than the overlying finer grained silts and clays, and provide a preferred pathway for groundwater flow in the area. The sand and gravel deposits are thin to absent in the area immediately south-southwest of facility's main building, and the bedrock surface in that area is relatively high. These variabilities cause preferential flow paths in certain areas of the site.

The alluvium is underlain by the Cretaceous-age Eagle Ford Shale, a weathered shale that is hard, dense, medium to dark gray, silty clay, with very thin sand and silt laminae. The Eagle Ford outcrops along the north side of the creek, southeast of facility. The shale units beneath the site generally measure 2 to 20 feet thick, and alternate with light gray, hard, well-cemented sandstone lenses approximately 2 to 5 feet thick. It is not known how thick the Eagle Ford is directly beneath the site, but it is expected to be greater than 100 feet thick.

Most of the groundwater monitoring wells are completed in the alluvium to the top of the bedrock shale formation. Water levels in these monitoring wells range from less than 3 feet below ground surface in the vicinity of the main building to greater than 24 feet below ground surface on the levee overlooking the creek. Groundwater flow is generally from the north-northeast to south-southwest across the site, with a southward deflection towards the creek. The hydraulic gradient at the site is relatively gentle on the north side of the site and becomes steeper towards the creek. The gradient is very steep in the southeast corner of the site, near the levee. The shallow groundwater flows primarily through the coarse grained sands and gravel located at or near the base of the alluvial section. Groundwater is also present in the shallow silty clays and clayey silts throughout much of the site; however, these sediments are relatively impermeable, and groundwater flows slowly through them. Due to the low production capacity of the alluvial groundwater, the groundwater is not used for drinking water or other water supply.

A bedrock water bearing zone is situated in a laterally continuous, 5 foot thick sandstone unit present across the site at a depth of approximately 50 to 60 feet. The sandstone bed is bounded above and below by the Eagle Ford Shale. Groundwater flow in the bedrock is to the east, and the gradient is significantly flatter than the gradient in the alluvial groundwater. The bedrock groundwater is used for irrigation in the vicinity of the site.

Groundwater Quality

The shallow alluvial groundwater beneath the site has been impacted by chlorinated hydrocarbons, primarily TCE, 1,1,1-TCA, and the daughter products of their degradation including 1,1-dichloroethene (1,1-DCE), cis-1,2-DCE, trans-1,2-DCE, 1,1-dichlorethane (1,1-DCA), and vinyl chloride (VC). Tetrachloroethene (PCE) is also present in the shallow groundwater, but in concentrations much lower than those reported for TCE and 1,1,1-TCA.

The center of mass of the chlorinated hydrocarbon plume is located near monitoring well MW-24 and the metal processing room, where a large vapor degreaser was operated. The exact release information for the source material (e.g., volume, date, duration, route, point of origin, etc.) is not known. Localized areas with high chlorinated hydrocarbon concentrations in soil have been identified; however, DNAPL has not been detected in these areas. In mid 1993, concentrations of TCE, 1,1-DCE, and 1,1,1-TCA in MW-24 were 35,900 µg/l, 24,300 µg/l, and 113,000 µg/l, respectively, and the concentration of total chlorinated hydrocarbons was approximately 173,200 µg/l. The plume extends west and south from MW-24, and low concentrations of chlorinated hydrocarbons have been detected in monitoring wells south of the creek. Groundwater flow south of the creek turns back and prevented the plume from migrating farther south.

Initial Evaluation of Remedial Technologies

In 1993, remedial technologies were evaluated and a risk assessment was conducted to determine what remedial technology or technologies would be appropriate
for the chlorinated hydrocarbons dissolved in the groundwater underlying the site.
Since the alluvial groundwater is not usable, potential exposure risks are limited to
contact by excavation workers and sampling personnel and to users of the creek.
Contact by excavation and sampling personnel will be minimized through institutional controls. Based on a conservative groundwater flow and transport model, it
was determined that concentrations of chlorinated hydrocarbons in the creek will
never exceed the applicable State of Texas surface water quality standards.

As previously discussed, the shallow alluvial groundwater flows primarily
through the sand and gravel unit at the base of the alluvium; however, the clayey
sediments directly above the sand and gravel unit are also saturated throughout most
of the facility. The hydraulic conductivity of the upper clay layer is orders of
magnitude less than that of the sand and gravel unit. In addition, there are low-
permeability clayey and silty lenses within the sand and gravel unit. The upper clay
layer, as well as silty or clayey lenses in the sand and gravel unit, contain chlorinated
hydrocarbons as a result of fluctuations in water levels and molecular diffusion from
the adjacent sand and gravel unit. The chlorinated hydrocarbons are adsorbed on
the clay and silt particles and are dissolved in the pore water of these sediments.
Due to the low permeability of the clays and silts, the hydrocarbons cannot be
effectively mobilized by traditional recovery methods (e.g., groundwater pump and
treat) that are dominated by advective transport.

Under active remediation scenarios, the chlorinated hydrocarbons entrained in
the clayey material would slowly diffuse to the sand and gravel layer in response to
a concentration gradient that would develop as groundwater is recovered from the
sand and gravel. As the chlorinated hydrocarbons concentrations decrease in the
sand and gravel unit in response to groundwater withdrawal and subsequent recharge
with fresher water, the hydrocarbons in the upper clayey unit would continue to
diffuse to the sand and gravel unit and effectively preclude remediation to drinking
water levels within a reasonable period of time.

Because the removal of chlorinated hydrocarbons at the site is rate limited by
diffusion, the cost effective and time efficient use of any remediation technology
based on water extraction is not practicable. Enhanced *in situ* remediation technologies are not feasible or effective because of the low permeability of the clayey
sediments and heterogeneity of the alluvial flow system. Therefore, in 1994, it was
proposed that natural attenuation was the most effective remedial technology for the
alluvial groundwater at the site.

Evaluation of Natural Attenuation

In 1996, the Texas Natural Resource Conservation Commission (TNRCC)
requested that the owner submit additional site characterization data in support of
natural attenuation as a long-term alternative for the remediation of the impacted
groundwater. Specifically, the TNRCC requested that the owner demonstrate that

degradation of the chlorinated hydrocarbons in the groundwater is occurring by using the first and second or first and third lines of evidence specified in "Technical Protocol for Evaluating Natural Attenuation of Chlorinated Solvents in Groundwater" (Wiedemeier et al. 1996) as stated below.

1. Document that reductions in contaminant concentrations along the flow path downgradient from the source of contamination are occurring
2. Document loss of contaminant mass at the field scale using
   a. Chemical and geochemical analytical data including decreasing parent compound concentrations, increasing daughter compound concentrations, depletion of electron acceptors and donors, and increasing metabolic byproduct concentrations
   b. A conservative tracer and a rigorous estimate of residence time along the flow path to document contaminant mass reduction and to calculate biological decay rates at the field scale
3. Microbiological laboratory or field data that support the occurrence of biodegradation and give rates of biodegradation

Site-specific conditions at the facility were documented and are consistent with Items 1 and 2a of the criteria specified in the technical protocol. The following site-specific data were used to demonstrate that natural attenuation of chlorinated hydrocarbons is occurring in the alluvial groundwater and that natural attenuation is an appropriate remedial alternative was documented:

- Reductive potential of the aquifer based on geochemical parameters
- Reduction of dissolved contaminant concentrations in monitoring wells along the groundwater flow path
- Plume retardation
- Reduction in total mass of contaminants

Geochemical Study

In general, geochemical data can be used to document that adequate electron acceptors (oxygen, nitrate, ferric iron, sulfate, and carbon dioxide) are available to facilitate the metabolic processes necessary to biodegrade organic contaminants and that geochemical parameters such as pH, oxidation-reduction potential, and temperature are conducive to microbial activity. In geochemical evaluations, data were collected from background locations, from within the contaminant plume, and down-gradient of the contaminant plume.

Consistent with the technical protocol, the reductive potential of the shallow water bearing zone was evaluated by collecting geochemical data from selected wells completed in the alluvial groundwater. Groundwater samples were collected from monitoring wells completed in background locations (MW-7, MW-9, and MW-14), in the area of highest contaminant concentrations (MW-24), in locations directly down-gradient of the area of highest contaminant concentrations (MW-6, MW-33, and MW-32), along the western edge of the contaminant plume (MW-22), along the southern edge of the contaminant plume (MW-27, MW-28, and MW-29), and in unimpacted areas beyond the western and eastern edges of the contaminant plume (MW-16, MW-17, and MW-18). The well locations are depicted in Figures 20 or 21.

Groundwater samples were collected with a micropurging technique using an adjustable rate, low flow pump and a flow through cell in an effort to minimize both the disturbance of stagnant water in the well casings and the potential for mobilization of particulate or colloidal matter. While purging each monitoring well, field parameters including dissolved oxygen, oxidation-reduction (redox) potential, pH, temperature, and specific conductivity were monitored. In addition, groundwater samples were collected for analyses of dissolved gases and chemical parameters indicative of the capacity of the groundwater to support natural attenuation of the constituents of concern.

Results

The technical protocol (Wiedemeir 1996) provides a weighting system that may be used as a preliminary screening tool to evaluate the potential for natural attenuation at a site. This weighting system and the scores obtained from evaluating the geochemical parameters measured in monitoring wells MW-24, MW-6, MW-33, and MW-27 are summarized in Table 9. Based on this screening methodology, the data measured at the site indicate that the evidence of biodegradation at this site is limited.

If this evaluation had been relied upon as the main indicator for biodegradation, natural attenuation may have failed as a remediation alternative. However, in considering the scores obtained using the screening tool, it should be remembered that the geochemical data collected at the facility are indicative of the groundwater characteristics within the larger soil pore spaces of the sand and gravel located at the base of the alluvium, where groundwater velocity is highest and would preferentially recharge groundwater being purged from a monitoring well. Within a soil microcosm, there are soil micropores where groundwater velocity is low and where groundwater is stagnant (Chapter 1). This phenomena is particularly prevalent in clayey soils such as those encountered in the upper two-thirds to three-quarters of the saturated soils. Therefore, the scoring system associated with the protocol had limited application to this site, and other evaluation methods had to be used.

Reduction of Contaminant Concentrations

Historical analytical data reported for the site indicate that contaminant concentrations along the flow path down-gradient from the center of mass of the plume have been reduced since monitoring groundwater began in 1993. Concentrations of TCE and 1,1,1-TCA have decreased in monitoring wells MW-8, MW-23, and MW-24 (located near the center of mass of the plume), and in most wells along the flow path down-gradient of these wells. As the concentrations of these parent products have decreased over time, the concentrations of their breakdown products, 1,1-DCE, *cis*-1,2-DCE, and 1,1-DCA, have increased moderately or decreased at a slower rate. As discussed previously, the detection of these breakdown products of TCE and 1,1,1-TCA indicates that the contaminants are degrading naturally.

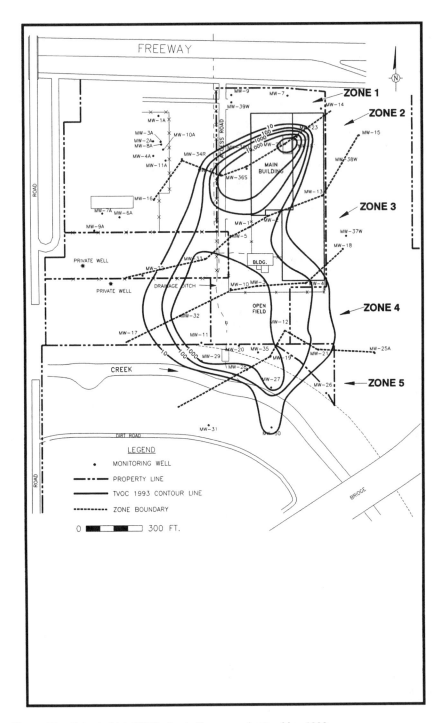

Figure 20   Extent of total VOCs in shallow groundwater, May 1993.

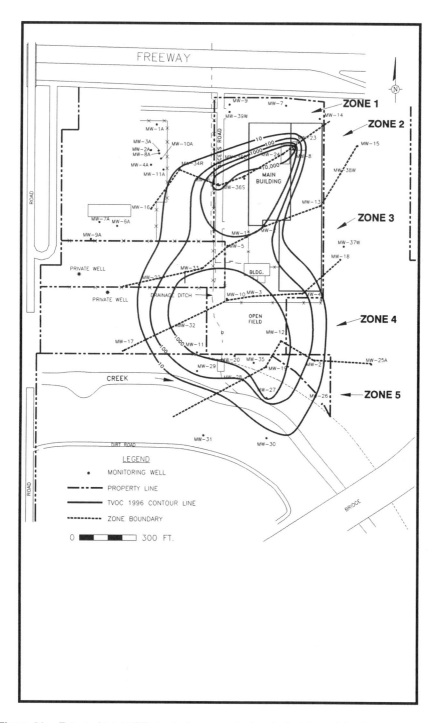

Figure 21 Extent of total VOCs in shallow groundwater, April, July, and October 1996.

Table 9   Chlorinated Hydrocarbon Natural Attenuation Case History; Screening of Biogeochemical Data

Analysis	Concentration in Most Contaminated Zone	Interpretation	Scoring Value	MW-24	MW-6	MW-33	MW-27
Oxygen	<0.5 mg/l	Tolerated, suppresses the reductive pathway at higher concentrations.	3	0	0	3	
Nitrate	>1 mg/l	VC may be oxidized aerobically	-3				-3
	<1 mg/l	At higher concentrations may compete with reductive pathway	2	0	2	2	2
Iron II	>1 mg/l	Reductive pathway possible	3	0	0	0	0
Sulfate	<20 mg/l	At higher concentrations may compete with reductive pathway	2	0	0	0	0
Sulfide	>1 mg/l	Reductive pathway possible	3	0	0	0	0
Methane	<0.5 mg/l	VC oxidizes	0	0	0	0	0
	>0.5 mg/l	Ultimate reductive daughter product, VC	3	0	0	0	0
Redox Potential	<50 mV	Reductive pathway possible	1	0	0	0	0
	<-100 mV	Reductive pathway likely	2	0	0	0	2
pH	5<pH<9	Optimal range for reductive pathway	0	0	0	0	0
	5>pH>9	Outside optimal range for reductive pathway	-2	0	0	0	0
TOC	>20 mg/l	Carbon and energy source; drives dechlorination; can be natural or anthropogenic	2	0	0	0	0
Temperature	>20C	Biochemical process is accelerated	1	1	1	1	1
Carbon Dioxide	>2x background	Ultimate oxidative daughter product	1	0	0	0	0
Alkalinity	>2x background	Results from interaction of carbon dioxide with aquifer minerals	1	0	0	0	0
Chloride	>2x background	Daughter product of organic chlorine	2	2	2	2	2
PCE		Material released	0	0	0	0	0
TCE		Material released	0	0	0	0	0
		Daughter product of PCE	2				
DCE		Material released	0				
		Daughter product of TCE (If cis is greater than 80% total DCE it is likely a daughter product of	2	2	2	2	2

Table 9　Chlorinated Hydrocarbon Natural Attenuation Case History; Screening of Biogeochemical Data (continued)

Analysis	Concentration in Most Contaminated Zone	Interpretation	Scoring Value	MW-24	MW-6	MW-33	MW-27
VC		Material released	0	0	0	2	2
		Daughter product of DCE	2				
Ethene/Ethane	>0.01 mg/l	Daughter product of VC/ethene	2	NA	NA	NA	NA
			3				
Chloroethane		Daughter product of VC under reducing	2	0	0	0	0
1,1,1-TCA		Material released	0	0	0	0	0
1,1-DCE		Daughter product of TCE or chemical reaction of 1,1,1-TCA	2	2	2	2	2
Final Scores				7	9	14	11

Final Score	Interpretation
0 to 5	Inadequate evidence for biodegradation of chlorinated organics
6 to 14	Limited evidence for biodegradation of chlorinated organics
15 to 20	Adequate evidence for biodegradation of chlorinated organics
>20	Strong evidence for biodegradation of chlorinated organics

NA: Not Analyzed

Source: Wiedemeier, Todd H., Swanson, Matthew A., Moutoux, David E., and Gordon, E. Kinzie, "Technical Protocol for Evaluating Natural Attenuation of Chlorinated Solvents in Groundwater," Draft – Revised Air Force Center for Environmental Excellence, Brooks Air Force Base, San Antonio, Texas, November 1996.

Although the concentrations of breakdown products have increased in most of the aforementioned monitoring wells, the total concentrations of chlorinated hydrocarbons in these wells have decreased over time.

Plume Retardation

By comparing the plume delineation maps using 1993 (Figure 20) and 1996 (Figure 21) analytical data, the chlorinated hydrocarbon plume underlying the facility has neither moved nor spread significantly since 1993. Isopleth maps were constructed depicting the extent of total chlorinated hydrocarbons, TCE, DCE, and VC using 1993 and 1996 analytical data. As shown, the areal distribution of total chlorinated hydrocarbons, observed in 1996, is comparable to the areal distribution of the contaminants observed in 1993.

The lack of plume movement and spreading observed is consistent with findings in research recently published by the Bureau of Economic Geology (BEG) in conjunction with the USEPA and the TNRCC (Mace et al. 1997). The BEG found that the lengths of benzene plumes from leaking petroleum storage tank (PST) sites stabilize once contaminant masses stabilize. Further, the BEG found that hydrocarbons adsorbed to aquifer soils slowly desorb into groundwater as the mass of dissolved hydrocarbons in the aquifer are removed. Thus, plume length will remain relatively constant for an extended period of time while total contaminant mass is decreasing.

Similar plume behavior at this site was observed although the constituents of potential concern are different than those addressed in the BEG document. The plume length has not increased; constituent concentrations have decreased rather than increased; and (as discussed below) contaminant mass has decreased. Therefore, it appears that the chlorinated hydrocarbon plume at the site has stabilized and will not move further south.

Reduction of Contaminant Mass

A reduction in contaminant mass was documented based on analytical monitoring data. The detection of breakdown products resulting from the degradation of TCE and 1,1,1-TCA indicates that contaminant mass is being degraded and not just diluted. Further, the total estimated chlorinated hydrocarbon mass in the shallow groundwater in 1996 was compared to the total estimated chlorinated hydrocarbon mass in the shallow groundwater in 1993 and found that a significant reduction in total chlorinated hydrocarbon mass occurred between 1993 and 1996.

The mass of VOCs present was determined by dividing the plume into five zones. Areas of the site with approximately the same saturated surficial aquifer thickness were grouped within the same zone. Generally, the saturated thickness of the surficial aquifer decreases towards the creek. The five zones are oriented approximately parallel to the creek and perpendicular to the general direction of groundwater flow (Figures 20 and 21). Zones based on saturated thickness minimize errors directly attributable to utilization of average aquifer thicknesses in mass calculations. In order to perform a direct comparison of calculated VOC masses, the same zones were used for each time period.

The initial step in calculating the mass of VOCs dissolved in the groundwater for each time period was to determine the area between adjacent concentration contours for each zone. The total number of individual areas to be determined for each zone was one less than one-half the number of contour lines that cross the zone boundaries.

The second step was to multiply the area between adjacent concentration contours by the arithmetic average concentration within the area, the arithmetic average saturated thickness of the surficial aquifer in the zone, the approximate soil porosity, and unit conversion factors to determine the pounds of VOCs dissolved in the groundwater

$$M = A \ b \ C \ V_{CF} \ M_{CF} \tag{3}$$

where M = mass (lbs.); A = area (ft^2) of specific average concentration; b = average thickness of saturated interval in zone (ft); C = arithmetic average concentration of area between concentration lines (g/L); V_{CF} = 28.317 L/ft^3; M_{CF} = 2.2046 x 10^{-9} lbs./g; and ϕ = porosity (0.17).

Since concentration data typically has a lognormal distribution, a simple arithmetic averaging scheme (rather than using the median value of a log normal distribution) results in an overestimation of calculated masses and conservative estimates of change in mass percentages between the two time periods.

The calculated masses between adjacent concentration contours were summed to provide the total VOC mass in the groundwater in each zone and all five zones were summed to determine the total VOC mass for each time period.

The average saturated thicknesses in Zones 1 through 5 used for 1993 and 1996 are shown in Table 10. Increased average saturated thicknesses were observed in 1996 and indicate this time period was wetter than 1993.

Table 10   **Average Saturated Surficial Aquifer Thicknesses for 1993 and 1996**

	1993	1996
Zone 1	15.08 ft.	15.45 ft.
Zone 2	13.86 ft.	14.17 ft.
Zone 3	10 ft.	10.29 ft.
Zone 4	7.28 ft.	7.71 ft.
Zone 5	5.9 ft.	6.72 ft.

The calculated groundwater VOC mass for Zones 1, 2, 3, 4, and 5 in 1993 were 234, 646, 47, 94, and 11 lbs, respectively, and totaled 1,032 lbs. For 1996, the calculated VOC mass in the groundwater for Zones 1, 2, 3, 4, and 5 were 111, 143, 38, 48, and 6 lbs., respectively, and totaled 346 lbs. The calculated mass of VOCs dissolved in the groundwater indicates a 66 percent reduction in mass or a reduction of approximately 230 lbs per year occurred between 1993 and 1996, Table 11. Since average saturated thicknesses in 1996 exceeded 1993 values, the calculated VOC mass reduction is not merely an artifact of different saturated thicknesses for the two time periods and would have rendered a more conservative calculated mass reduction.

Table 11   VOC Mass Present in 1993 and 1996 and Percent Change in Mass

Zone	1993	1996	% Reduction in VOC Mass since 1993
1	234	111	53%
2	646	143	78%
3	47	38	19%
4	94	48	49%
5	11	6	45%
Total	1032	346	66%

Conclusions

The historical analytical data collected at site demonstrate that:

- Constituent concentrations in groundwater in the area of highest known contaminant concentrations and along the flow path down-gradient of that area decreased over the 3.5 year period
- The groundwater plume has stabilized and will not move further south
- Chlorinated hydrocarbon parent products (TCE and 1,1,1-TCA) are degrading to their daughter products (1,1-DCA, 1,1-DCE, and cis-1,2-DCE)
- The estimated total mass of chlorinated hydrocarbons dissolved in groundwater was reduced from 1032 pounds to 346 pounds between 1993 and the end of 1996 (approximately 66 percent)

This project is a good example of the limitations of the scoring system with the protocol.

This and similar projects have resulted in the new protocols abandoning the scoring systems.

SUMMARY

Biochemical reactions are an important part of almost all remediations. The investigator must understand the potential bacterial effects at the site in order to correctly plan and interpret the remedial investigation. The remedial designer must include biochemical reactions in order to develop the most cost effective design. When microorganisms are used as the basis of a remediation, the designer must include all of the intrinsic reactions that are ongoing at the site.

REFERENCES

Adriano, D.C., Bollag, J.M., Frankenberger, Jr., W.T., and Sims, R.C., Coeditors, *Bioremediation of Contaminated Soils*, Agronomy No. 37, American Society of Agronomy, 1999.

Aziz, C.E., Newell, C.J., Gonzales, J.R., Haas, P.E., Clement, T.P., and Sun Y., "BIOCHLOR Natural Attenuation Model for Chlorinated Solvent Sites," in *Natural Attenuation of Chlorinated Solvents, Petroleum Hydrocarbons, and Other Organic Compounds,* 5(1): (Alleman, B.C., Leeson, A., Eds.), Battelle Press, Columbus, OH, 1999.

Bouwer, H., *Groundwater Hydrology*, McGraw-Hill, New York, 1978.

Brock, T.D., *Biology of Microorganisms*, Prentice-Hall, Englewood Cliffs, NJ, 1979.

Bull, A.T., *Contemporary Microbial Ecology*, D.C. Ellwood, J.N. Hedger, M.J. Lathane, J.M. Lynch, and J.H. Slater, Eds., Academic Press, London, 1980.

Carey, G.R., Van Geel, P.J., McBean, E.A., Murphy, J.R., and Rovers, F.A., "Modeling Natural Attenuation at the Plattsburgh Air Force Base," in *Natural Attenuation of Chlorinated Solvents, Petroleum Hydrocarbons, and Other Organic Compounds,* 5(1), (Alleman, B.C., Leeson, A., Eds.), Battelle Press, Columbus, OH, 1999a.

Carey, G.R., Van Geel, P.J., McBean, E.A., and Murphy, J.R., "Visualizing Natural Attenuation Trends at the Hill Air Force Base," in *Natural Attenuation of Chlorinated Solvents, Petroleum Hydrocarbons, and Other Organic Compounds,* 5(1), (B.C. Alleman, A. Leeson, Eds.), Battelle Press, Columbus, OH, 1999b.

Dragun, J., *The Chemistry of Hazardous Materials*, The Hazardous Materials Control Research Institute, Silver Spring, MD, 1988.

Grady, C.P., "Biodegradation: Its Measurement and Microbial Basis," Biotechnology and Bioengineering 27:660-674, 1985.

Holt, J.G., *The Shorter Bergy's Manual of Determinative Bacteriology, 8th ed,* Williams & Wilkins, Baltimore, MD, 1981.

Interstate Technology and Regulatory Cooperation Work Group (ITRC), "Natural Attenuation of Chlorinated Solvents in Groundwater: Principles and Practices," http://www.itrcweb.org, May 1999.

Lovley, D.R., and Goodwin, S., "Hydrogen Concentrations as an Indicator of the Predominant Terminal Electron-Accepting Reaction in Aquatic Sediments," Geochimica et Cosmochimica Acta, v. 52, p. 2993-3003, 1988.

Mace, Robert E., Fisher, R. Stephen, Welch, David M., and Parra, Sandra P., "Extent, Mass, and Duration of Hydrocarbon Plumes from Leaking Petroleum Storage Tank Sites in Texas," Geological Circular 97-1, Bureau of Economic Geology, The University of Texas at Austin, 1997.

McCarty, P.L., and Semprini, L., "Ground-Water Treatment for Chlorinated Solvents," in *Handbook of Bioremediation* (R.D. Norris, R.E. Hinchee, R. Brown, P.L McCarty, L. Semprini, J.T. Wilson, D.H. Kampbell, M. Reinhard, E.J. Bouwer, R.C. Borden, T.M. Vogel, J.M. Thomas, and C.H. Ward, Eds.), Lewis Publishers, Boca Raton, FL, 1994.

Moat, A.G., *Microbial Physiology*, Wiley-Interscience, New York, 1979.

Murray, W.D. and Richardson, M., Progress toward the biological treatment of C_1 and C_2 halogenated hydrocarbons, *Crit. Rev. Environ. Sci. Technol.,* 23(3): 195-217, 1993.

Newell, C.J., Mcleod, R.K., and Gonzales, J.R., *BIOSCREEN: Natural Attenuation Decision Support System,* User's Manual, Version 1.3, Air Force Ctr. For Environ. Excellence, Brooks Air Force Base, San Antonio, 1996.

Norris, R.D., "Benefits and Concerns with Application of the USEPA Protocol for Monitored Natural Attenuation," in *Natural Attenuation of Chlorinated Solvents, Petroleum Hydrocarbons, and Other Organic Compounds,* 5(1), (B.C. Alleman, A. Leeson, Eds.), Battelle Press, Columbus, OH, 1999.

Paul, E.A. and Clark, F.E., *Soil Microbiology and Biochemistry,* Academic Press, San Diego, 1989.

Rafai, H.S., Bedient, P.B., Wilson, J.T., Miller, K.M., and Armstrong, J.M., "Biodegradation Modeling at Aviation Fuel Spill Site," *J. Environ. Eng. Div.* 114(5): 1007-1029, 1988.

Rittman, B.E., Smets, B.F., and Stahl, D.A., "The Role of Genes in Biological Processes Part V," ES&T 24(1): 23-29, 1990.

Stanier, R.Y., Adelberg, E.A., and Ingrahm, J.L., *The Microbial World*, Prentice-Hall, Englewood Cliffs, NJ, 1976.

Suarez, M.P. and Rifai, H.S., "Biodegradation Rates for Fuel Hydrocarbons and Chlorinated Solvents in Groundwater," *Bioremediation Journal,* Vol. 3, Issue 4, 1999.

Suthersan, S.S., *Remediation Engineering: Design Concepts*, CRC Press/Lewis Publishers, Boca Raton, FL, 1997.

Tabak, H.H., Quave, S.A., Mashni, C.I., and Barth, E.F., Biodegradability studies with organic priority pollutant compounds, *J. Water Pollut. Contr. Fed.,* 53: 1503-1518, 1981.

United States Department of Navy, Technical Guidelines for Evaluating Monitored Natural Attenuation of Petroleum Hydrocarbons and Chlorinated Solvents in Ground Water at Naval and Marine Corps Facilities, September 1998.

U.S. Environmental Protection Agency, "Compilation of Groundwater Models," Office of Research & Development, Washington, D.C., EPA/600/R-93/11B, May 1993.

U.S. Environment Protection Agency, "Use of Monitored Natural Attenuation at Superfund, RCRA Corrective Action, and Underground Storage Tank Sites," Office of Solid Waste and Emergency Response Directive 9200.4-17, 1997.

U.S. Environmental Protection Agency, "Technical Protocol for Evaluating Natural Attenuation of Chlorinated Solvents in Ground Water," EPA/600/R-98/128, September 1998.

Vogel, T.M., "Natural Bioremediation of Chlorinated Solvents," in *Handbook of Bioremediation,* (R.D. Norris, R.E. Hinchee, R. Brown, P.L McCarty, L. Semprini, J.T. Wilson, D.H. Kampbell, M. Reinhard, E.J. Bouwer, R.C. Borden, T.M. Vogel, J.M. Thomas, and C.H. Ward, Eds.), Lewis Publishers, Boca Raton, FL, 1994.

Wiedemeir, T.H., Wilson, J.T., Kampbell, D.H., Kerr, R.S., Miller, R.N., Hansen, J.E. "Technical Protocol for Implementing the Intrinsic Remediation with Long-Term Monitoring Option for Natural Attenuation of Dissolved-Phase Fuel Contamination in Ground Water," Air Force Center for Environmental Excellence, Brooks Air Force Base, San Antonio, Texas, 1995.

Wiedemeier, Todd H., Swanson, Matthew A., Moutoux, David E., and Gordon, E. Kinzie, "Technical Protocol for Evaluating Natural Attenuation of Chlorinated Solvents in Groundwater," Draft -- Revision 1. Air Force Center for Environmental Excellence, Brooks Air Force Base, San Antonio, Texas, November 1996.

CHAPTER 8

Reactive Zone Remediation

Frank Lenzo

CONTENTS

1-56670-528-2/01/$0.00+$.50
© 2001 by CRC Press LLC

INTRODUCTION

One of the most important advances in the remediation of aquifers during the last few years has been reactive zone technologies. Before the development of these techniques we were limited to treatment methods that relied on advective movement of air and water. Once these processes removed the mass of contaminants that the carrier came into contact with, we had to rely on natural attenuation to remove the remaining contaminants. (This is the diffusion controlled portion of the project that we discussed in Chapter 2, Lifecycle Design.) While we could design mass removal techniques that would perform their function in several years, the natural attenuation would then take 25 to 100 years to complete the remediation. Even though we did not have any remediation equipment on the site, the project was not over, and we still had to monitor and report to the state. Reactive zone technologies increase the rate of remediation during the diffusion-controlled portion of the project. These techniques will finally allow us to remediate sites in a reasonable time frame.

The creation of subsurface (*in situ*) reactive zones was considered an innovative approach to remediation as late as 1997 (Suthersan 1997). As of the writing of this text, a handful of sites have been closed, dozens of sites have full-scale systems in place, and dozens more are in the midst of pilot demonstrations. The elegance of the approach and the focus on manipulating chemistry and microbiology *in situ* to achieve remedial goals make the technology appealing on many levels:

- Aesthetics
- Ease of implementation—less invasive and more natural
- Environmental compatibility—complements the natural environment
- Enhances and takes advantage of nature's capacity to remediate
- Regulatory acceptance
- Based on sound scientific principles
- Cost—capital and operating
- Holistic, environmentally, and economically sound solution

Reactive zones are simply treatment zones, developed *in situ*, using selected reagents that enhance, or modify, subsurface conditions in order to fix or degrade target contaminants (Figure 1). These zones are typically created to intercept and treat mobile groundwater impacts, but are now being applied to less mobile soil impacts as well. Ideally, reactive zones enhance natural conditions in order to speed up naturally occurring remedial processes (for example, enhancing an already reducing *in situ* environment can accelerate the natural attenuation of chlorinated compounds).

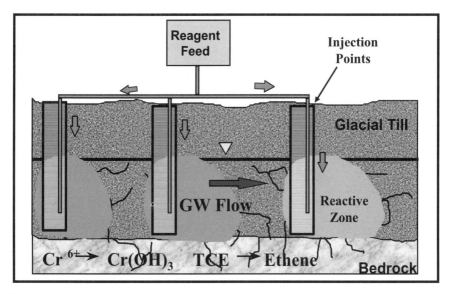

Figure 1 Reactive zones. Reprinted with permission from Power magazine, copyright McGraw-Hill, Inc., 1966.

In situ reactive zones are applicable to a wide range of target contaminants. They have been applied or are being tested on heavy metals (chromium, zinc, mercury, copper, arsenic, lead, and cadmium), chlorinated aliphatic hydrocarbons (CAHs) (trichloroethene, tetrachloroethene, 1,1,1-trichloroethane, carbon tetrachloride, and daughter products of these compounds), pentachlorophenol, and halogenated organic pesticides (1,2-dichloropropane [DCP] and 1,2-dibromo-3-chloropropane [DBCP]). Table 1 is a list of contaminants that have been treated using reactive zone technology, as well as several presently being tested.

Table 1   Compound and Representative Site List for *In Situ* Reductive Reactive Zone Technology

Location	Site Name/Description	Regulatory Authority	Target COCs	Status
Williamsport, PA	Textron/Manufacturing	CERCLA	Cd, Cr^{+6}, TCE, DCE, VC	Pilot 1995; Full-Scale 1997; closure proposed 1999
Saegertown, PA	Lord/Chemical	CERCLA	Chlorinated VOCs	Pilot 1998; FFS/Full-Scale Design
Reading, PA	Manufacturer/Textile Equipment	PA Act 2	TCE, Cr^{+6}, Pb, Cd	Pilot/Full-Scale 1997-1998 100% reduction in Cr^{+6}, 95% VOCs
Ambler, PA	Chemical Manufacturer	PA Act 2	TCE, TCA, Ni	Pilot ongoing
Central New Jersey	Pharmaceutical Manufacturer	NJDEP ISRA	PCE	Pilot 1998, Full-Scale 2000
Raritan, NJ	Pharmaceutical Manufacturer	RCRA	TCE	Site Screening, Pilot 1999
Utica, NY	Defense Contractor	CERCLA	TCE, DCE	Pilot 1999, ongoing
Jamestown, NY	Manufacturer	NYDEC	TCE, DCE	Pilot 1998, ongoing
Long Island, NY	Manufacturer	CERCLA	CR^{+6}, TCE	Pilot 1998-1999
Binghamton, NY	Landfill	CERCLA	TCA, TCE, BTEX, chlorinated propanes	Pilot 1998-1999, Full-Scale Design, ongoing
Wooster, Ohio	Sand and Gravel Distributor	Ohio EPA	TCE, DCE, VC	Pilot ongoing 1999
Greenville, SC	Manufacturer	SC DHEC	CT, CF, TCE	Pilot complete 1998, Full-Scale implementation ongoing
Oxford, NC	Manufacturer	CERCLA	TCE	Pilot 1999 ongoing
Myrtle Beach, SC	Chemical Manufacturer	SC DEQ	TCE	Site Screening, Pilot 1999
Blairsville, GA	Former Laboratory	GA DEP	PCE, TCE	Pilot 1998; Full-Scale Design ongoing
Crawfordsville, IN	Manufacturer	IN DEP	TCE, DCE	Pilot 1998, ongoing
Dallas, TX	Chemical Manufacturer	TNRCC	PCP, Cr^{+6}	Site Screening, Pilot 1999
Texas	Pharmaceutical Manufacturer	TNRCC	PCE, TCE, ketones, BTEX	Site Screening
Houston, TX	Drycleaner	TNRCC	PCE, TCE, DCE	Full-Scale 1997, 60% reduction in VOC mass
Houston, TX	Drycleaner	TNRCC	PCE, TCE, DCE	Full-Scale 1997, ongoing
Houston, TX	Shopping Mall	TNRCC	PCE, TCE	Pilot 1996; Full-Scale 1997 ongoing
Jasper, TN	Former Manfacturing Facility	TN DSWM/SRP	PCE	Site Screening, Pilot 1999
Oak Ridge, TN	Electronic Manufacturer	TN DSF/VOAP	TCE, radionuclides	Site Screening, Pilot 2000
Lansing, MI	Landfill	MI DEQ	PCE, TCE, TCA	Pilot 1998, ongoing
Grand Rapids, MI	Railcar Spill Site	MI DEQ	1,1 DCE, TCE, TCA	Site Screening, Pilot 1999

Table 1 Compound and Representative Site List for *In Situ* Reductive Reactive Zone Technology (continued)

Location	Site Name/Description	Regulatory Authority	Target COCs	Status
Emoryville, CA	Metal Plating Manufacturer	CA Central	TCE, DCE, Cr^{+6}	Pilot 1996, Full-Scale 1997, closure 1999
Fresno, CA	Pesticide Manufacturer	CA Central	DCP, DBCP, Cr^{+6}	Pilot 1999-2000
Monterey, CA	Manufacturer	CA Coastal	TCE, Cr^{+6}	Pilot 1996, Full-Scale 1997, ongoing
Santa Barbara, CA	Shopping Mall Development	CA Central	PCE	Pilot 1997, Full-Scale 1998, closure imminent 1999
Dominguez, CA	Chemical Manufacturer	CA LA	TCE, DCE	Screening; proposed parallel to P&T
Pinebend, MN	Landfill	MN DEP	PCE, TCE	Pilot 1996; Full-Scale 1997, ongoing
Washington, WI	Shopping Mall	WI DEP	PCE, TCE	Pilot 1998, Full-Scale ongoing
London, England	Automobile Manufacturer	Env. Agency U.K.	PCE, TCE	Full-Scale 1998, ongoing
Portsmouth, VA	Chemical Manufacturer	VA DEP	dissolved zinc	Pilot Study 1997-1998
Baraboo, WI	Badger AAP/ESTCP/AFCEE	USEPA	CVOCs, metals, and energetics	Site Assessment and Pilot Study Design underway
Bedford, MA	Hanscom AFB/ESTCP/AFCEE	USEPA	CVOCs	Site Assessment and Pilot Study Design underway
Lompoc, CA	Vandenberg AFB/ESTCP AFCEE	USEPA	CVOCs	Site Assessment and Pilot Study Design underway
San Francisco, CA	Former Naval Station Treasure Island	USEPA	CVOCs	Site Assessment and Pilot Study Design underway
Dallas, TX	NAS Dallas/NFESC	USEPA	CVOCs	Large-Scale Pilot underway
Moundsville, WV	Chemical Manufacturer	WV	dis. Hg	Pilot 1997-1998
Hampton, Iowa	Metal Plating Facility	USEPA	Cr^{+6}	18-Month Pilot/IRM reduced Cr concentration by 80% (11 mg/l to 2 mg/l)
Palatine, Illinois	Metal Plating Facility	Illinois EPA	CVOCs and Cr^{+6}	Pilot 1998-1999, Full-Scale 1999 ongoing

Reactive zones can take the form of a reducing zone or an oxidizing zone. The choice is driven by two factors: the natural environment and the nature of the contaminant being targeted. In addition, reactive zones can be created by taking advantage of the activity of indigenous microbial populations (through the injection of degradable organic substrates, other electron donors, or electron acceptors) or through the addition of chemical reagents (sodium sulfide, sodium bicarbonate, or sodium dithionite, for example).

This chapter will review the state-of-the-art of *in situ* reactive zones, by first covering the theory and basis of the processes applied for both metals and organics. The chapter will then present the application of the technology:

- Design considerations including hydrogeology, groundwater chemistry, subsurface microbiology, and reactive zone reagents
- Design criteria for wells, reagent feeds, system configurations
- Pilot testing including selection, set up and monitoring

Full-scale application of the technology will be covered using case studies from three sites in various stages of remediation. Finally, a section summarizing the limitations of the technology will close the chapter.

THEORY

Reactive zones can take a variety of forms, based on the target contaminants and the environment in which they are found. There are two basic types commonly applied: oxidizing and reducing. Both of these types can be applied by either changing the environment to produce the desired chemical and biochemical reactions or through a direct chemical reaction. Reactive zones can be created using two basic pathways: chemically induced and microbially mediated. These types of reactive zones, and the methods employed to create them, can be combined to suit the contaminant suite and the variability of site conditions as required. The next sections describe each individually and later sections describe how they can be combined to achieve the necessary remedial goals.

Before reading about the details of these treatment zones it is important to understand some basic elements of the chemical and microbial reactions that are taking place in the subsurface. It would be beneficial to the reader to review Chapter 7 for discussions related to chemical oxidation, microbial degradation processes and pathways, natural attenuation, and biogeochemical monitoring and sampling protocols. The following sections will review the more critical elements of each of these subjects, as they relate to reactive zones; however, a thorough understanding of the specific topics will be invaluable to the reader.

Oxidizing Reactive Zones

Oxidizing reactive zones can be simply defined as artificially enhanced subsurface treatment zones, in which the environment is maintained as strongly oxidizing (i.e., the redox conditions are maintained well above 0.0 mV and dissolved oxygen is maintained above 2.0 mg/l). This environment is created using the addition of air or oxygen—as in an air sparge system—or through the injection of chemical oxidants. The most common chemical oxidants include hydrogen peroxide, potassium permanganate, and ozone.

In an oxidizing reactive zone contaminants of concern (COCs) are specifically targeted to be chemically, or microbially, oxidized. The oxidant can be mild—for example, ORC$_{TM}$, air, or oxygen—and can be added to enhance the naturally aerobic environment in order to promote the biological oxidation of readily degradable compounds such as petroleum hydrocarbons, benzene, toluene, ethylbenzene, xylene, and vinyl chloride. This type of reactive zone application creates an aerobic environment that enhances bacterial growth that was limited in the original environment. These bacteria will degrade the COCs faster, and/or degrade certain organic compounds that only degrade under aerobic conditions. The oxidant can also take the form of a strong oxidant (peroxide, permanganate, or ozone, for example) that is injected specifically to chemically oxidize the COCs. The use of strong oxidants in reactive zones has the additional benefit of creating a down-gradient aerobic environment that can enhance aerobic microbial degradation of certain organic compounds. Both processes are discussed in more detail below.

Oxidizing reactive zones can be applied to the treatment of both organic compounds as well as metals. The suite of metals that can be treated using oxidizing reactive zones is small—primarily iron, manganese, and arsenic. It is important to keep the potential presence of metals in mind when evaluating the use of oxidizing reactive zones, particularly when chemical oxidants are applied. The concern revolves around the potential of releasing reduced forms of heavy metals such as chromium. One example involves the common historical use of TCE and chromium in plating operations, the release of which has led to the coincidental presence of both COCs in groundwater. As will be described in later sections of this chapter, chromium is naturally reduced from hexavalent to trivalent chrome. In the presence of a strong oxidant targeted at TCE, the trivalent chrome may be oxidized to reform the more toxic and mobile hexavalent form. Some forms of geological material (pyrite, for example) will release large amounts of iron and acid when exposed to oxidants. These potential negative results must be anticipated before application of aerobic reactive zones.

Reducing Reactive Zones

Reducing reactive zones can be simply defined as artificially enhanced subsurface treatment zones, in which the environment is maintained as strongly reducing (i.e., the redox conditions are maintained well below 0.0 mV and dissolved oxygen below 1.0 mg/l). This environment is maintained using the addition of chemical reductants or naturally degradable organic mass. In the case of the former, reductants such as sodium sulfide, sodium dithionite, L-ascorbic acid, and hydroxylamine are used to chemically reduce target metals in the subsurface (Suthersan 1997, and Khan and Puls 1999). As with the aerobic zones, the chemical addition can be used to change the environment to enhance new types of biochemical reactions, or the reductant can be used for direct chemical reduction of the metal. Both processes are discussed in more detail below.

In the latter case, degradable organic carbon—in the form of labile organic substrates such as sugars, lactate, or toluene—is added to the subsurface. The indigenous heterotrophic microorganisms present in the aquifer readily degrade the organic carbon resulting in the utilization of available electron acceptors present in the groundwater. Starting with dissolved oxygen, the microbial population then uses nitrate, manganese, ferric iron, sulfate, and finally carbon dioxide as electron acceptors. Depletion of these electron acceptors leads to successively more reducing conditions as the reduction-oxidation (redox) potential is lowered. Figure 2 summarizes the microbial respiration processes listed above.

Redox *	Respiration	e⁻ Acceptor	By-Products
+ 200 mv	Aerobic	O_2	CO_2
	Denitrification	NO_3^{2-}	NO_2^-, N_2, NH_3
	Manganese Reduction	Mn^{4+}	Mn^{2+}
	Iron Reduction	Fe^{3+}	Fe^{2+}
	Sulfanogenesis	SO_4^{2-}	H_2S
- 400 mv	Methanogenesis	CO_2	CH_4

* The redox values are for guideline purposes only. These values can vary by 2 to 3 times based on other factors.

Figure 2   Respiration processes/redox regimes.

Microbially mediated reducing reactive zones are being applied to treat both metals and organics in the subsurface. A variety of CAHs and heavy metals have been treated using reactive zones; recently, halogenated organic pesticides (HOPs) (ARCADIS Geraghty & Miller 1999) and pentachlorophenol (Jacobs et al. 2000) have been targeted for treatment using reducing reactive zones.

Chemically Created Reactive Zones

Reactive zones can be created using chemical reagents that impact the redox conditions in the subsurface or directly react with COCs present in the groundwater or soil matrix. Examples of oxidizing reactive zones that use chemical reagents to directly oxidize organic compounds were described earlier in this chapter.

Oxidizing chemical reagents can also be used to modify subsurface conditions to create conditions favorable to the aerobic degradation of organic compounds. As described earlier in this chapter, as well as in Chapter 5, biosparging is a form of reactive zone that uses the injection of air or oxygen to enhance aerobic degradation of organic compounds. ORC_{TM} is magnesium peroxide, a solid that hydrolyses to release oxygen. The resultant increase in dissolved oxygen in the groundwater provides a long-term source of oxygen to serve as an electron acceptor for indigenous aerobic and facultative aerobic bacteria present in the subsurface. These bacteria can then more rapidly metabolize aerobically degradable COCs.

Examples of reducing reactive zones that use chemical reagents are numerous. Several examples are listed below:

Hexavalent chromium reduction using sodium dithionite (Fruchter et al. 1999)
$S_2O_4^{2-} + Fe^{3+} \rightarrow Fe^{2+} + SO_4^{2-}$ (SO_2^- radicals reduce iron)
$Fe^{2+} + Cr^{6+} \rightarrow Fe^{3+} + Cr^{3+}$ (chromium reduced by iron)

Fruchter, et al. have demonstrated that in the presence of heat (>25 C) sodium dithionite can be used to reductively dechlorinate TCE and field demonstrations for TCE and TNT are planned for 2000 (Fruchter et al. 1999).

Cadmium precipitation using sodium sulfide (Suthersan 1997):
$Na_2S + Cd^{2+} \rightarrow CdS$

Hexavalent chromium reduction using ferrous sulfate (Walker and Pucik-Erickson 1999):
$3Fe^{2+} + Cr^{6+} + 3(OH)^- \rightarrow 3Fe^{3+} + Cr(OH)_3$ (neutral pH)

Zinc precipitation using sodium bicarbonate (Suthersan 1997):
$Zn^{2+} + NaHCO_3 \rightarrow ZnCO_3$

In each of these reactions the target COC metal is dissolved in groundwater. In the reaction that takes place with the reagent, the metal is reduced and precipitates out as a solid that is subsequently immobilized in the soil matrix. The solubility constant for the precipitated form is orders of magnitude lower than that of the dissolved form leading to much lower concentrations of the metal in groundwater.

Microbially Mediated Reactive Zones

As discussed in previous sections microbial populations can be used to create reactive zones *in situ*. The favored approach is to use indigenous microbial populations. The bacterial population may be stressed due to the COC impacts, or the ability of the microbial population to degrade the COC mass may be limited by a lack of electron acceptors (dissolved oxygen, nitrates, manganese, iron, sulfates, or carbon dioxide), or a lack of degradable organic carbon (electron donors). In order to take full advantage of the microbial population's ability to degrade organic mass, or to create the necessary conditions for the precipitation of metals, electron acceptors and electron donors can be added to the subsurface. In so doing, the microbial population is allowed to complete the remediation process *in situ*.

Both aerobic and anaerobic conditions can be enhanced and engineered to bring about the remedial goals for a site. Aerobic processes typically involve the addition of oxygen in the form of ambient air, pure oxygen gas, or chemical reagents with oxygen releasing compounds such as hydrogen peroxide. Nitrate addition can also be used to provide an alternate electron acceptor for the bacterial population to use in order to degrade organic compounds. Anaerobic conditions are favored when the target COC is a metal that must be reduced, in order to precipitate, or an organic compound that is in an oxidized state and can therefore be readily reduced under the right biogeochemical conditions.

As will be discussed later the microbial processes that enhance the degradation of organic compounds involve four potential pathways:

1. The COC may be used as a primary source of carbon for the bacterial population
2. The target COC can be used as an electron acceptor
3. The target COC may be fortuitously degraded in the presence of other readily degraded organic carbon sources
4. The COC may be degraded as a result of a strongly reducing condition caused by the anaerobic environment created by the bacterial population

Examples of this will be discussed in the following sections of this chapter and have been covered to some extent in Chapter 7.

Process Considerations

There are a number of specific process issues that must be considered when evaluating the use of, or applying, a reactive zone. These include the biogeochemical environment, COC chemistry, and microbiological degradation processes.

Biogeochemistry

The biogeochemical environment that has naturally evolved at a site impacted by COCs should always be understood before attempts are made to manipulate it using reactive zones. Since this environment is impacted by the COCs released as well as the geochemistry of the soil and groundwater in and around the area of concern, understanding the biogeochemical state of the environment is always the first step before implementing a reactive zone.

The biogeochemical environment was discussed in detail in Chapter 7, *In Situ* Bioremediation, in the discussions on natural bioremediation. Reactive zones, by definition, are applied *in situ*. Therefore, by their very nature they require a manipulation of the natural environment. In order to better understand that environment it is necessary to collect relevant analytical data related to the biogeochemical reactions taking place. These data will also be used as part of the reactive zone design. For example, the amount of electron donor needed to enhance and maintain a strongly anaerobic environment would be based upon the level of natural organic carbon and the extent to which the environment has already been reduced. The groundwater flow rate is also needed in order to calculate the mass of electron donor required. Specifically, it is necessary to collect the data summarized in Table 2 from three zones:

- Zone up-gradient of the impacted area—this provides a definition of the natural (i.e., unimpacted) environment that exists (or existed) when there are no COCs present in the subsurface
- Zone within the impacted area—this provides a definition of the state of the groundwater environment where COCs have become part of that environment. By comparing this zone to the up-gradient area, a measure of how much the COCs have influenced natural conditions can be made
- Zone down-gradient of the impacted area—this provides a definition of the residual impacts of COCs, COC degradation products and the byproducts of the biotic and abiotic reactions that have taken place within the impacted zone

The biogeochemical analyte list can be broken down into five major process related subsets:

1. Electron Acceptors: This includes dissolved oxygen, nitrate, manganese, ferric iron, sulfate and carbon dioxide. The microbial metabolic process is an oxidation-reduction process in which electrons are exchanged—electron acceptors being reduced and electron donors being oxidized. The electron acceptors listed above represent the most common electron acceptors utilized by the heterotrophic bacterial population in this metabolic process. Bacteria that use oxygen as an electron acceptor are considered aerobic, those utilizing the remaining electron acceptors are anaerobic. It should be noted that some COCs might also serve as electron acceptors for specific microbial populations (chlorinated aliphatic hydrocarbons for dehalorespirators have been reported) (Smatlak et al. 1996 and Yager et al. 1997).
2. Electron Donors: This includes total and dissolved organic carbon, some COCs, and methane. These are the microbial populations' source of nutrition, their food. This group is oxidized in the microbial metabolic process. A lack of electron donors will typically slow down the natural microbial populations' utilization of electron acceptors sometimes resulting in less reducing conditions. Also, electron donors may be plentiful in the source area but limited in the zone down-gradient of the source area. This will create different environments for the two areas. Dechlorination stalling at 1,2 cis DCE is one example of the type of impact that this situation can create.
3. Metabolic Byproducts: This subset includes carbon dioxide, nitrite, nitrogen, ammonia, dissolved manganese, dissolved iron, sulfide, and methane. These are all by-products of the natural respiration processes listed in Figure 2. The presence, or lack thereof, of each helps to define the metabolic processes that are taking place in the subsurface environment. For example, if the levels of nitrates and sulfates are depressed and the level of sulfide is elevated in the heart of a xylene plume when compared to the up-gradient zone, there is a strong indication that there is sulfanogenic bacterial activity and strongly reduced conditions within the plume. These conditions are likely the result of the degradation of xylene by the native bacterial consortia. As discussed in Chapter 7, these values can also be used to calculate the mass rate of destruction.

Table 2   Analyses, Methods, and Data Types for Monitoring Associated with Enhanced Bioremediation

Parameter	Method/ Reference	Technical Protocol	Method Detection Limit	Analytical Level	Holding Time	Data Use
Alkalinity	USEPA 310.1	HACH Test Kit Model AL APMG-L	50 mg/l	III	14 days	General water quality parameter used (1) to measure the buffering capacity of groundwater, and (2) as a marker to verify that all site samples are obtained from the same groundwater system.
Nitrate (NO_3)	USEPA 353.1	Method E300	0.2 mg/l	III	28 days	Substrate for microbial respiration if oxygen is depleted.
Nitrite (NO_2)	USEPA 354.1	None	0.1 mg/l	III	48 hours	May indicate anaerobic degradation process of nitrate reduction.
Sulfate (SO_4)	USEPA 375.4	Method E300	5 mg/l	III	28 days	Substrate for anaerobic microbial respiration.
Chloride (Cl^-)	USEPA 325.3	Method A4500	2 mg/l	III	28 days	General water quality parameter used as a marker to verify that site samples are obtained from the same groundwater system. Final product of chlorinated solvent reduction.
Methane (CH_4)	AM-15.01	SW3810	5 to15 ng/l*	III	14 days	The presence of methane suggests hydrocarbon degradation via methanogenesis.
Ethane & Ethene	AM-18	SW3810	5 ng/l*	III	14 days	Ethane and ethene are potential byproducts of chlorinated solvents suspected of undergoing biological transformation.
Nitrogen	AM-15.01	SW3810	0.4 mg/l*	III	14 days	May indicate anaerobic process of nitrate reduction.
Carbon Dioxide (CO_2)	AM-15.01	SW3810	0.4 mg/l*	III	14 days	Indicator of anaerobic degradation processes and aerobic respiration of organics.
COD	USEPA 410.4	None	10 mg/l	III	28 days	General indicator of organic substrates (electron donors).
Ammonia (NH_4)	USEPA 350.3	None	0.1 mg/l	III	28 days	General indicator of landfill leachate. May indicate reducing conditions.
Sulfide	USEPA 376.1	HACH Test Kit Model HS-WR	1 mg/l	I	7 days	May indicate anaerobic degradation process of sulfate reduction.
BOD	USEPA 405.1	None	2 mg/l	III	48 hrs	General indicator of organic substrates (electron donors).
TOC (dissolved)	USEPA 415.1	SW9060	1 mg/l	III	28 days	Used to evaluate the role of TOC and determine if cometabolism is possible in the absence of anthropogenic carbon.
Iron (total and dissolved)	USEPA 6010A	HACH Method DR/2000 10-phenanthroline	0.05 mg/l	I	6 months	May indicate an anaerobic degradation process due to depletion of oxygen, nitrate, and manganese.
Manganese (total and dissolved)	USEPA 6010A	None	0.01 mg/l	III	6 months	May indicate an anaerobic degradation process due to depletion of oxygen and nitrate.
Conductivity **			Manufacturers Range: 0 to 20 m5/cm	I		General water quality parameter used as a marker to verify that all site samples are obtained from the same groundwater system.

Table 2   Analyses, Methods, and Data Types for Monitoring Associated with Enhanced Bioremediation (continued)

Parameter	Method/ Reference	Technical Protocol	Method Detection Limit	Analytical Level	Holding Time	Data Use
Temperature **			Range: -5 to +45°C	I		Well development
Dissolved Oxygen (DO) **			Manf's Range: 0 to 20 mg/l	I		Concentration less than 1 mg/l generally indicate an anaerobic pathway.
Oxidation-Reduction Potential (ORP) **			Range: -999 to +999 mV	I		The ORP of groundwater influences and is influenced by the nature of the biologically mediated degradation of contaminants; the ORP of groundwater may range from more than 800mV to less than -400mV.
pH **			Range: 0 to 14 units	I		Aerobic and anaerobic processes are pH sensitive.

Notes:

Level III   Analyses are performed in an off-site analytical laboratory. Level III analyses do not use CLP procedures, but does utilize the validation or documentation procedures required of CLP analyses.

Level I   Field screening and analyses using portable instruments. Results are available in real time and the method is cost effective.

Technical protocols and data use from Wiedemeier et al (14).

* Method quantitation limit reported by Microseeps, Inc., Pittsburgh, Pennsylvania.

** Field parameters will be measured with a Yellow Springs (YSI) Model 600xl probe Method detection limits reports by YSI.

Source: (courtesy ARCADIS Geraghty & Miller, Inc.)

4. Indicator Parameters: This subset includes temperature, pH, conductivity, alkalinity, and oxidation-reduction potential (ORP)—also commonly referred to as redox potential. Natural changes to these analytes are an indication of microbial activity. Elevated temperature and conductivity, and depressed pH and ORP are typically seen in the presence of anaerobic bacterial activity. Alkalinity provides an indication of the buffering capacity of the ground water. Properly collected ORP (using the low-flow purge sampling technique [McCarty and Semprini 1994]) is a powerful tool for defining the anaerobic zones critical to the reductive dechlorination of CAHs. However, it should be noted that due to the complex geochemistry that results in the ORP value, it is difficult to compare ORP readings between sites.

5. Degradation Products: The last subset includes COC daughter products (such as TCE, cis-1,2-dichloroethene, and vinyl chloride for PCE; and chloroform, methylene chloride and chloroethane for carbon tetrachloride) and COC degradation end products (such as ethene, ethane, methane, and carbon dioxide). These are critical to understanding how far along the degradation pathway that the natural environment has carried the target COCs. These are also critical to the demonstration that degradation is in fact a component of the attenuation process and not that simply dilution, dispersion, and volatilization are taking place.

With a baseline snapshot of the biogeochemical environment available, it is possible to begin the reactive zone selection process and answer, to some extent, critical design questions:

- Is natural degradation taking place? If so, in what type of environment, aerobic or anaerobic?
- Is the natural degradation process stalled due to a lack of electron acceptors? Electron donors?
- Is there evidence that the necessary microbial population is present to degrade the target COCs?
- Are biotic or abiotic processes taking place? Or both?
- Should an oxidative or reducing zone be used?

Before more than a basic answer can be given for each of these questions it is necessary to consider two other elements in the reactive zone equation: COC chemistry and microbiological degradation processes.

COC Chemistry: Halogenated Aliphatic Hydrocarbons

One of the most common applications of the reactive zone technology is for the enhanced degradation of CAHs and other halogenated aliphatic hydrocarbons (HAHs). This is in part due to the widespread occurrence of CAHs in groundwater, but also as a result of the efficacy of the technology in handling CAHs. It was recognized in the middle to late 1980s that both aerobic and anaerobic bacterial populations naturally degraded petroleum hydrocarbons and BTEX, with aerobic populations providing the most rapid degradation rates. It wasn't until the early 1990s that the anaerobic degradation of CAHs was considered a viable mechanism in the remedial tool kit, this primarily due to the work of McCarty (1994), Wilson (1995), and Weidemeier (1996).

Enhancing an anaerobic environment using reactive zones in order to accelerate the degradation of CAHs has only become an accepted practice in the last two to three years (refer to Nyer et al. 1998, Suthersan 1997, Burdick and Jacobs 1998, Cirpka et al. 1999, and Lenzo 1999). In Chapter 7 the biological pathways available for several common CAHs are summarized (refer to Figure 7 in Chapter 7 for PCE, TCE, 1,1,1 TCA, and carbon tetrachloride). There are also potential abiotic pathways that can occur for these compounds; however, except for 1,1,1 TCA, the abiotic pathway is not practical unless chemical reagents are applied (refer to Chapter 7). The degradation pathways that are the focus of this discussion of reactive zones are biotic.

The primary pathways described in Chapter 7 for the CAHs are anaerobic pathways. For many of the more common COCs (such as TCE, 1,1,1TCA, and PCE) the anaerobic pathways are particularly efficient due to the fact that these compounds are relatively oxidized and are thus susceptible to reductive dechlorination in a reducing environment. As the chlorine atoms are stripped from the parent CAH, and there are fewer chlorine atoms attached to the base alkene or alkane molecule, the resultant chlorinated aliphatic is more reduced—less oxidized—and thus less susceptible to reductive dechlorination. Conversely, the more reduced forms (less chlorinated) are more easily oxidized than reduced and thus can be degraded under more highly oxidized conditions. An excellent example of this is the degradation sequence for PCE (Figure 3). PCE and TCE are readily reduced under anaerobic and reducing conditions; however, it takes more and more aggressively reducing conditions to achieve the degradation of TCE to DCE, DCE to VC, and finally vinyl chloride to ethene. The necessary reducing conditions for VC degradation may not be achieved naturally in the environment and thus a buildup of VC may be expected in a reducing environment. In an aerobic environment, however, VC, which is highly reduced,

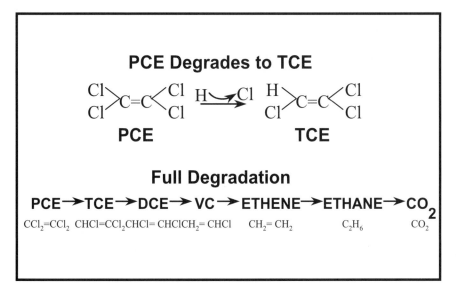

Figure 3 Reductive dechlorination of tetrachloroethylene.

can be readily degraded. In fact, VC can serve as a primary carbon source to aerobic bacteria (Guest, Benson, and Rainsberger 1995).

As the state-of-the-art of reactive zones for the CAH treatment has advanced, new compounds that can be remediated with this technology have been added to the list. Recently, work has begun in California to address HAH pesticides, specifically DBCP and DCP, using anaerobic reactive zones, and pentachlorophenol (PCP) has been targeted for treatment using *in situ* anaerobic reactive zones (Mueller et al. 2000, Jacobs et al. 2000). The most common degradation pathway for DCP is presented in Figure 4.

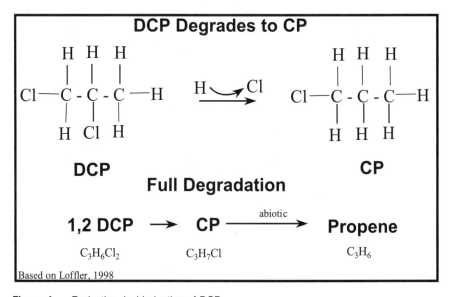

Figure 4   Reductive dechlorination of DCP.

Microbiology of Reactive Zones

The last consideration in the understanding of the reactive zone process is the role that microbiology plays. There are a number of microbial mechanisms that are naturally occurring in the environment that lead to the conditions necessary for the degradation of organic compounds or the precipitation of metals. Natural metabolic respiration processes are responsible for creating the reducing conditions necessary to reduce the valence state of chromium from +6 to +3 and in the process convert a soluble, toxic, and mobile form of chromium to a non-toxic form that can readily precipitate under typical groundwater conditions. These respiration processes have been covered in previous sections.

The key to the success of a reactive zone is the ability to manipulate the existing natural conditions in order to bring about the remedial ends sought. Thus if it is necessary to maintain a strongly reducing environment in order to reduce cadmium and form sulfide for the precipitation of cadmium sulfide, the natural conditions must be manipulated to allow the sulfanogenic bacterial population to flourish. Since sulfanogenesis takes place under reducing conditions near a neutral pH (Norris et al. 1994), the necessary conditions are created when applying the reactive zone. This assumes that there is an adequate supply of sulfate to produce sulfide and that the alkalinity is adequate to balance the pH of the ground water system.

The application of reactive zones for the treatment of HAHs involves several different microbial mechanisms. Each of the three major mechanisms is depicted in Figures 5, 6, and 7.

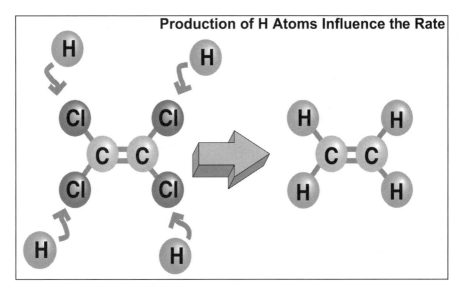

Figure 5   Hydrogenolysis.

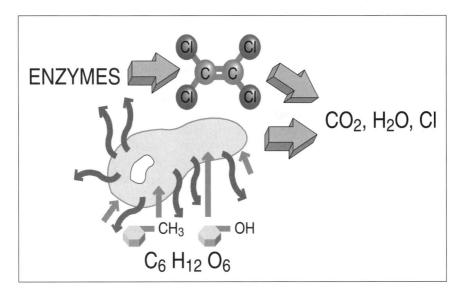

Figure 6   Cometabolic degradation.

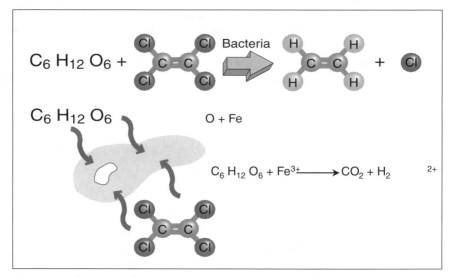

Figure 7   Dehalorespiration.

As with most microbially mediated processes in the subsurface, the microbial processes that take place in a reactive zone are a collection of processes that result from the natural selection of microbial populations that can survive and thrive in the conditions that exist at the site. As a result there is not any one mechanism, nor is there one bacteria that is responsible for the degradation process. It is in fact a consortia of microorganisms and a variety of mechanisms that bring about the results.

Figure 5 is a representation of the reductive dehalogenation of a HAH under reducing conditions. In this environment, an overabundance of hydrogen ions provides the opportunity for the dehalogenation process to take place. In the process, chlorine atoms are sequentially replaced by hydrogen ions, until finally the double-bond of the ethene molecule is broken and ethane is formed. The strongly reducing conditions created by the microbial population provides adequate hydrogen ions to bring about a rapid sequential dehalogenation of the HAH. Even the more reduced forms (fewer halogens on the base aliphatic molecule) are reductively dehalogenated. In this case the microbial population needs a readily degradable source of organic carbon in order to survive. This organic carbon can take the form of natural organic matter, anthroprogenic forms (like BTEX or phenol), or injected labile sources such as sugars or other easily degradable compounds. The bacterial population is living in an anaerobic environment.

Figure 6 is a representation of a cometabolic degradation process. In this mechanism the target COC does not serve as a primary source of organic carbon. Instead, the microbial population utilizes another source of organic carbon in order to survive. This carbon source could be natural, man-made, or a supplemental source. In the process of metabolizing the organic carbon, cofactors and enzymes are produced that are used by the bacterial population to degrade the organic carbon that is serving as the food for the bacteria. These enzymes fortuitously degrade the target COC, although the microbial population receives no direct benefit from the degradation of the COC. This mechanism can take place under both anaerobic and aerobic conditions. An example of an aerobic process is the cometabolic degradation of TCE in the presence of methanotrophic bacteria. The methanotrophs are provided with methane as a source of organic carbon. As the bacteria degrade the methane they cometabolically degrade the TCE. A similar process can take place in an anaerobic environment in which DCE is cometabolically degraded in the presence of xylene.

Figure 7 represents the degradation of a HAH by a dehalorespirator. The concept of a dehalorespiration process was introduced by McCarty (1997). In the process, certain microbial populations utilize the target HAHs as electron acceptors and in the process degrade the target compounds. The bacteria still require a source of electron donors in order to complete the metabolic redox reaction.

COC Chemistry: Metals

Reactive zones have been applied to the *in situ* treatment of metals in the groundwater for decades. One of the oldest applications of this process is related to the treatment of iron and manganese for drinking water supplies using the Finnish treatment process Vyredox$_{TM}$ (Zienkiewicz 1984). The Vyredox$_{TM}$ process utilizes an *in situ* oxidizing zone surrounding a groundwater production well to treat iron and manganese. The iron and manganese is oxidized *in situ* by creating an oxidizing environment through the introduction of aerated water in a series of injection wells surrounding the production well.

The *in situ* coprecipitation of arsenic and iron has been reported by Suthersan (1997), Whang (1997) and others, as a means of removing arsenic from groundwater via *in situ* oxidizing reactive zones. In the process an oxidant such as Fenton's reagent, peroxide, permanganate, or oxygen is used to oxidize arsenite (valence state 3+) to arsenate (valence state 5+) and to precipitate iron. In the process, iron and arsenic co-precipitate as an iron hydroxide arsenate complex and are removed from the groundwater (Figure 8). The efficacy of this process has yet to be tested in the field, however it holds a great deal of promise.

Some of the newest work in metals treatment using reactive zones is taking place in the arena of environmental cleanup and the application of reducing reactive zones for the treatment of heavy metals. The application of reducing reactive zones for metals will be examined through reference to a case study for a Superfund site in Pennsylvania (Lenzo 1999).

A Superfund site located in central Pennsylvania is impacted with chromium, cadmium, and CAHs. In the summer of 1996, an *in situ* reactive zone process was tested to demonstrate the ability to treat the heavy metals *in situ*. Molasses was used as an electron donor to create the necessary reducing conditions to bring about the reduction and precipitation of chromium and cadmium.

The goal of the reactive zone was to create the necessary reducing conditions to achieve precipitation of the target heavy metals as hydroxides, sulfides, or carbonates:

Hydroxide: $Me^{2+} + 2OH^- \rightarrow Me(OH)_2(s)$
Sulfide: $Me^{2+} + S^{2-} \rightarrow MeS(s)$

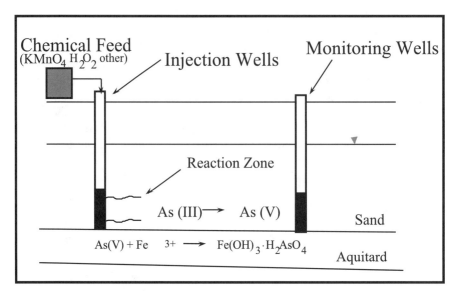

Figure 8   Arsenic oxidative IRZ.

Carbonate: $Me^{2+} + CO_3^{2-} \rightarrow MeCO_3\,(s)$
(Other metals that can be targeted using this approach include Fe, Hg, Mn, Zn, Ni,
    Cu, Ag, and Pb.)

The insoluble compounds that precipitate are immobilized in the soil matrix.
This precipitation process is essentially irreversible.

The creation of the desired reducing conditions in the groundwater were promoted by injecting a source of easily biodegradable carbohydrates, in the form of
dilute molasses solution, into the impacted saturated zone through a network of
injection wells. The carbohydrates (primarily sugars) present in the molasses were
degraded by the indigenous heterotrophic microorganisms present in the aquifer.
The biological degradation of the injected carbohydrates resulted in the utilization
of available electron acceptors present in the groundwater. Starting with dissolved
oxygen, the microbial population then used nitrate, manganese, ferric iron, sulfate
(sulfanogenesis), and finally carbon dioxide (methanogenesis) as electron acceptors.
Depletion of these electron acceptors leads to successively more reducing conditions
as the reduction-oxidation (redox) potential was lowered.

Following creation of the necessary reducing conditions in the groundwater,
dissolved cadmium in the groundwater reacted with available sulfide, and to a lesser
extent carbonate, in the aquifer to form the very stable cadmium sulfide and cadmium
carbonate precipitates. Sulfide was present in the groundwater as a result of the
microbial reduction of sulfate; sulfate was present naturally in the groundwater and
was also a component of the injected molasses solution. The precipitation of cadmium sulfide was also influenced by the presence of sulfate-reducing bacteria in the
subsurface which provided a biologically mediated pathway for the reduction of
sulfate to sulfide. Carbonate was naturally present in the groundwater.

    Creation of the reducing conditions also led to the reduction of hexavalent chromium to trivalent chromium. This reduction process yielded significant remedial benefits because trivalent chromium is less toxic, less mobile, and precipitates more readily than hexavalent chromium. The trivalent chromium created by hexavalent chromium reduction reacted with naturally occurring hydroxides to form chromium hydroxide precipitates.

    Both the cadmium and chromium precipitates that formed in the reactive zone have extremely low aqueous solubilities. As a result, the concentrations of cadmium and chromium dissolved in groundwater exiting the reactive zone were less than, or equal to, the target standards of 0.003 and 0.032 milligrams per liter (mg/l), respectively.

    Figure 9 is a summary of the chromium pilot test data gathered during the 6 month demonstration test. The test was conducted in two areas of the site. Concentrations of hexavalent chromium were reduced from 7 mg/l to nondetectable during the first 2 months of testing. Based on this pilot test data a full-scale system was designed and installed. Figure 10 is a simple process diagram for the treatment system, and Figure 11 is a photo of the inside of the 10 foot square treatment building.

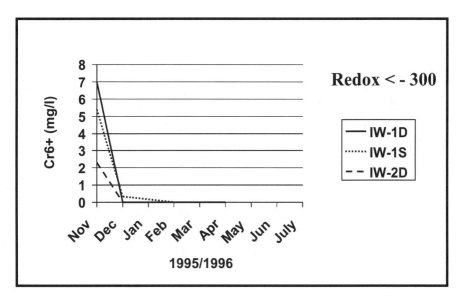

Figure 9   Chromium pilot data superfund site.

    The full-scale system consisted of 20 injection wells and 16 monitoring wells. A solution of molasses and water, varying in strength from 1:20 to 1:200, was injected twice a day into each injection well. The system went on line in January 1997. Data was collected on a quarterly basis for select biogeochemical parameters, chromium, cadmium, and CAHs.

    The baseline biogeochemical, chromium and cadmium conditions are shown in Figures 12, 13, and 14, respectively. The conditions after implementation and oper-

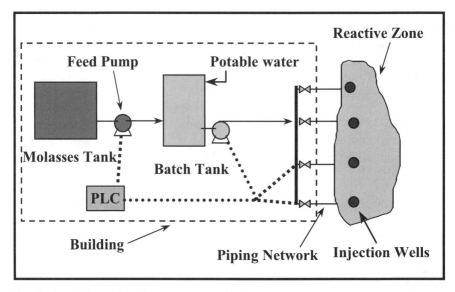

Figure 10   Molasses injection system: superfund site.

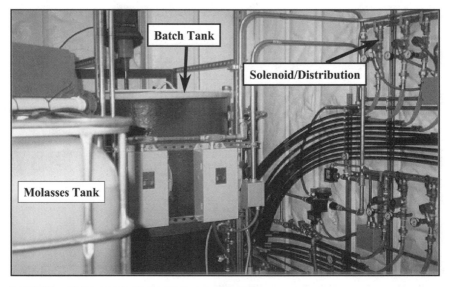

Figure 11   Molasses injection system: superfund site.

ation of the reactive zone system for 18 months are shown in Figures 15, 16, and 17 for the biogeochemical parameters, chromium and cadmium, respectively. The data was collected from monitoring wells located within and down-gradient of the injection well system.

As the data indicates (compare Figures 12 and 15), the establishment of a strongly reducing reactive zone led to the formation of sulfanogenic conditions and the

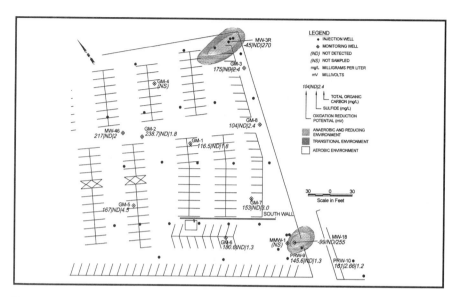

Figure 12 Distribution of groundwater indicator parameters, baseline conditions, January 1997.

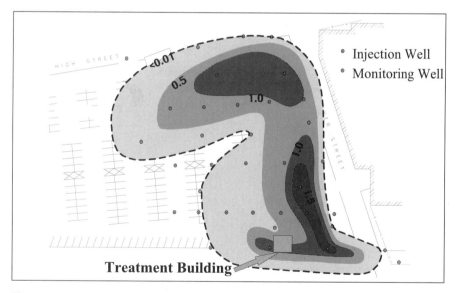

Figure 13 Distribution of hexavalent chromium baseline conditions.

reduction of sulfate to sulfide. The chromium and cadmium plume maps (Figures 13, 14, 16, and 17) highlight the effectiveness of the reactive zone technology in treating metals in groundwater.

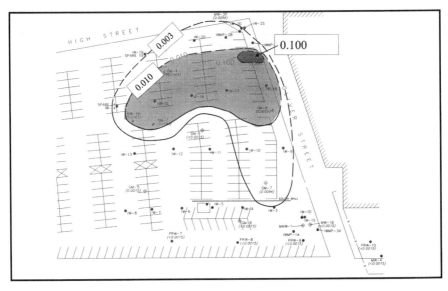

Figure 14   Distribution of cadmium: baseline conditions.

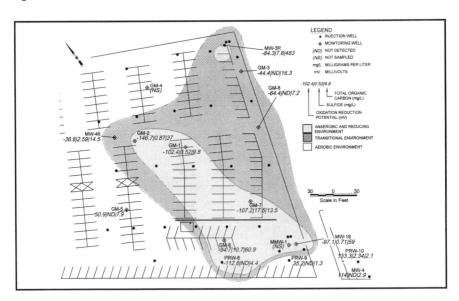

Figure 15   Distribution of groundwater indicator parameters, July 1998.

APPLICATION

The application of *in situ* reactive zones to the remediation of groundwater and soils must take into account all of the issues discussed above, as well as many issues not specifically discussed. Many of the latter items are common issues encountered

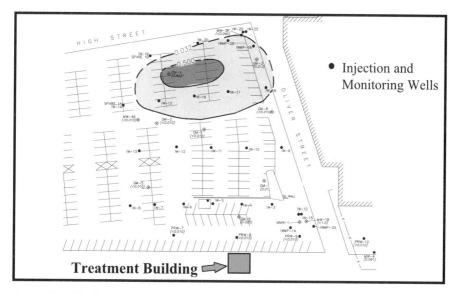

Figure 16 Distribution of hexavalent chromium, July 1998.

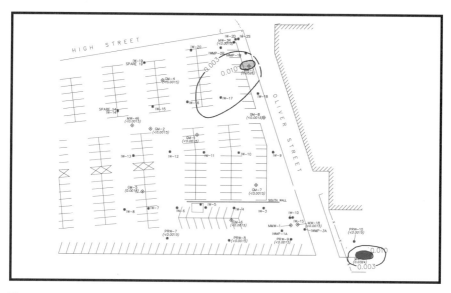

Figure 17 Distribution of cadmium, July 1998.

when applying any technology *in situ*. In this section particular attention will be paid to the following:

- Hydrogeologic Considerations: lithology, permeability, GW flow velocity, gradients, soil matrix conditions, saturated thickness, depth to groundwater, macro heterogeneitys

- Groundwater Chemistry: biogeochemical parameters, TOC, pH, temperature, HAH daughters, toxicity
- COC Characteristics: degradability, DNAPL

In addition to these design considerations are the issues associated with the system implementation, including the well layout, type of injection points, reagents to be used, frequency of injections, solution strength, and the maintenance and monitoring of the process. The need for, duration, and type of pilot testing required is also critical in understanding the application of the technology.

This section will include a review of these issues as they relate to the application of *in situ* reactive zones and will serve as an introduction to the case studies that follow.

Design Considerations

Any *in situ* technology is applied, de facto, out of sight. The understanding used to select, design, and apply the technology is projected based on limited data. The accuracy of the data we have describing the environment is a function of when, how, where, and often why it was collected. As a result, it is important to use layers of useful information to define the subsurface and the conditions in which we must apply an *in situ* technology. Equally as important is to collect relevant corroborating data after a specific technology is selected in order to ensure the proper application of that technology.

Among the critical design considerations for *in situ* reactive zones are hydrogeology, groundwater chemistry, microbiology, IRZ layout options, baseline definition, and reagents.

Hydrogeology

It is important to obtain specific hydrogeologic data in order to properly select and apply *in situ* reactive zones. The type of geology and the lithology of the subsurface are critical to the selection and placement of the injection points. While a complicated lithology can place constraints to the use of reactive zones, in most cases it will not completely eliminate reactive zones as a remedial option. By properly placing injection well screen zones to target specific impacted groundwater bearing layers, the technology can be effectively applied in most environments. Understanding the complexity and defining the lithologic variability as it relates to the groundwater impacts is an important first step.

A formation's permeability is important in that it impacts the ability of a single well point to serve as an effective reagent delivery mechanism. The higher the permeability the easier it is to deliver the reagent into the subsurface and the more effective a single delivery point can be. As the permeability decreases—all other factors remaining equal—the lateral distribution of reagent from a single injection point decreases.

Groundwater flow characteristics are another important consideration in the design of reactive zones. The groundwater velocity, flow direction, and the horizontal and vertical gradients impact the effectiveness of reagent injections and the speed with which the reagent will spread. Low velocity systems typically require lower reagent mass feed rates since the groundwater flux is reduced—all other conditions being equal. This is also an important criterion for the concentration and amount of additive that is needed to create the reactive zone.

Groundwater flow direction and gradients are also important to consider. They impact the groundwater velocity and also the direction in which groundwater flows. It is critical to understand the dynamics of groundwater flow to assure that the reagent injections will in fact form a reactive zone in the target area. Horizontal and vertical gradients are used to define the lateral (well point) and vertical (screen zone) location of the injection and monitoring wells. It is also important to understand the heterogeneity of the aquifer. The groundwater is the carrier that will take the additives throughout the aquifer. If there are areas where the groundwater does not flow, then the additives will not be able to reach those areas and the environment will not be changed. We will not be able to create the reactive zone in the no flow areas except through the slow process of diffusion.

The saturated thickness and depth to groundwater are also important characteristics to determine prior to applying reactive zones. The depth to groundwater will define well design and contribute significantly to determination of the cost of the full-scale system. The saturated thickness can also have a profound influence on cost, since there are practical limits on the maximum screened interval that can effectively be used in an injection well. Based on experience, a 25 foot screened interval represents a practical limit for an injection point. Of course, this limit will be impacted by the subsurface lithology, permeability, and groundwater flow characteristics. For example, if the lithology and resultant groundwater flow characteristics are such that there are variations in the flow characteristics that change within the target saturated interval, the use of multiple screened zones or multiple well points should be considered—even if the interval is less than 25 feet.

Finally the soil characteristics are important. The fraction of organic carbon (f_{oc}) will impact the amount of available organic carbon dissolved in groundwater, as well as the adsorptive capacity of the soil matrix. High f_{oc} soils are more likely to result in high DOC groundwater and high adsorptive capacity. The pH of the soil can affect the pH of the groundwater, and while microbial populations can endure a wide range of pH, ideally a pH close to neutral is the most conducive to healthy, diverse microbial populations.

Groundwater Chemistry

Groundwater chemistry includes an understanding of the target COCs, their daughter products, and the biogeochemical parameters discussed in the section entitled Process Considerations.

Understanding the conditions present in the groundwater will make the selection and application of reactive zones more likely to succeed. As discussed previously, the enhancement of natural conditions allows the designer to take advantage of natural processes that are already contributing to the degradation of the target compounds. Lacking that understanding one may end up trying to undo nature and find that it is necessary to spend twice the effort to bring about the desired result.

Of particular importance is the presence of degradation products, the presence and nature of electron acceptors, a definition of the redox conditions, and ORP in Figure 2, and the presence of electron donors. The presence of degradation products that indicate that a particular environment has established itself is typically easy to verify. For many sites historical COC data is available that may date back years—sometimes decades—and can be used to establish the presence of degradation products, as well as to evaluate trends in source and daughter products over time. This data can also provide information regarding historical impacts of variable organic species that may have served as electron donors. For example, at a site in central Pennsylvania, historical impacts of benzene provided a source of electron donors for indigenous microbial populations. This resulted in the degradation of TCE to ethene via the anaerobic reducing pathway described earlier (Figure 3). As the benzene source burned out the reductive dechlorination process stalled at DCE (Figure 18).

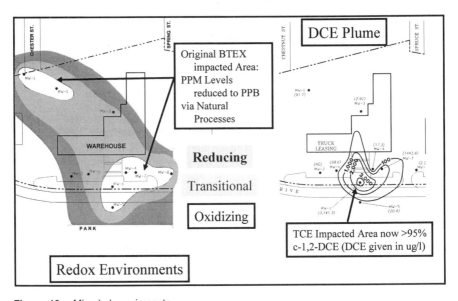

Figure 18 Mixed plume impacts.

Reducing reactive zones rely on the presence of an adequate source of electron donors (in the form of organic carbon) to establish and maintain a bacterial population that can maintain an anaerobic environment. The organic carbon may take the form of natural organic matter or anthroprogenic carbon sources—other organic COCs, such as BTEX, PHCs, ketones, or alcohols. Many times, as in the example in Figure 18, the source of organic carbon is weak or absent, and as a result supplemental sources must be considered to enhance the naturally reducing environment.

Microbiology

The presence of an indigenous microbial population is vital to the success of a microbially-mediated reactive zone. As discussed in Chapter 7, *In Situ* Bioremediation, this microbial population is a consortia of microbes, naturally adapted to the environment and thus ideally suited to surviving, and even thriving, in the impacted environment.

There are seldom situations that require the addition of specialty microbes to the subsurface to enhance degradation. Most "designer bugs" are developed in a controlled environment, in which they are allowed to evolve to treat specific target compounds. Often the selection process leads to select species of bacteria. The difficulty in using these bacteria in the natural environment relates to two major issues: survivability and distribution.

A single species targeted at a specific COC, or class of compounds, provides a uniquely focused means of attacking the target compound. With respect to survivability, however, it also represents an entire population that is vulnerable to the same stresses, diseases, and natural enemies. Once invaded, the entire population is at risk, and in a single event can be eliminated. In addition, the "designer bug" was developed under controlled conditions, not those found in the target groundwater environment to which they will be applied. Naturally evolved populations, on the other hand, are composed of multiple species. These natural populations have survived the precise environment they need to in order to bring about the remedial goals envisioned.

The distribution of the bacterial population is also critical. Microbial species spread slowly in the subsurface, therefore to effectively introduce a new population to a targeted subsurface zone, numerous injection points are required, making the practicality of the process suspect. Once again, the water would have to act as a carrier to move the bacteria throughout the affected area. This would not be an efficient method of distribution for the new bacteria. The only viable subsurface bacterial population is the natural population. However, as has been shown in this chapter, there are methods to enhance certain portions of the natural population so that they dominate.

Reactive Zone Layout

Reactive zones are applied in a number of different configurations and using a variety of approaches. The variety and combinations used are limited only by the variety of potential scenarios that may be encountered in the field and the ingenuity of the practitioner. For the purposes of this text, three basic layouts will be discussed: cutoff/barrier, plume-wide, and hot spots. Cutoff/barriers consist of a series of reagent injection points established perpendicular to the groundwater flow direction along a line that represents a critical boundary for remediation. This layout is commonly employed along a property line, or other artificial boundary established for the purpose of remediation or regulatory closure. In most cases the cutoff layout is less expensive to deploy since the density of injection points is not effected, but the number of points is typically significantly lower than the other layout options available.

Plume-wide reactive zones target a large portion of the impacted groundwater. Typically the injection points will be evenly spaced throughout the target impacted groundwater (Figure 19). By applying the reactive zone across the entire selected target isopleth, the speed and completeness of the remediation is enhanced. Obviously, there are cost implications with such an application; higher capital costs are traded for shorter remedial timeframes and the potential commensurate reduction in total O&M costs.

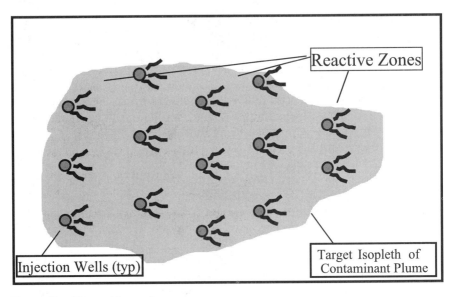

Figure 19   Plume-wide reactive zone.

Hot spot reactive zones target the source area. This layout is often employed in situations where the natural remediation process is successfully controlling the movement of the contaminant plume, but there is a need—regulatory or other—to speed up the overall remediation. In this case the source area is targeted with an area-wide reactive zone, in order to reduce contaminant mass quickly. Once the reactive zone has brought concentrations in the source area down to concentrations comparable with the remainder of the plume, the remediation can be allowed to progress at its natural pace.

These three basic layouts can be combined to suit the needs of the remediation. For example, if there is the potential for off-site movement of a plume in the short-term and a source area remains, it may prove most effective to establish a barrier reactive zone near the property line. At the same time a hot spot reactive zone can be established to reduce source mass and thus speed up the overall remediation.

Baseline Definition

It is critical that baseline conditions be established prior to initiating an active treatment program. Baseline conditions help establish a scale against which success can be measured. At a minimum, a complete suite of biogeochemical parameters and target COCs should be measured from wells in the up-gradient, impacted, and down-gradient areas of the site. During application of the reactive zone, field parameters such as ORP, DO, temperature, and pH should be monitored along with two or three electron acceptor species, TOC, DOC, two or three metabolic byproducts, and target COC end-products. Table 3 is a summary of the recommended analytical parameters for monitoring a reactive zone system.

Table 3   Typical Reactive Zone Monitoring Program

Event	Analytes	Schedule
Baseline	Biogeochemical, COCs (injection and monitoring wells)	Week 0
Monitoring Event 1	Abbreviated Biogeochemical (injection and monitoring wells)	Week 3
Monitoring Event 2	Abbreviated Biogeochemical (injection and select monitoring wells)	Week 8
Monitoring Event 3	Abbreviated Biogeochemical, COCs (monitoring wells)	Week 13
Monitoring Event 4	Abbreviated Biogeochemical (injection and select monitoring wells)	Week 19
Monitoring Event 5	Abbreviated Biogeochemical, COCs (monitoring wells)	Week 26
Monitoring Event 6	Abbreviated Biogeochemical (injection and select monitoring wells)	Week 33
Monitoring Event 7	Abbreviated Biogeochemical, COCs (monitoring wells)	Week 39
Monitoring Event 8	Abbreviated Biogeochemical, COCs (monitoring wells)	Week 52
Future Monitoring Events	Abbreviated Biogeochemical, COCs (monitoring wells)	Every 3 to 6 months

Notes: Biogeochemical Analytes includes complete suite listed in Table 2.
Abbreviated biogeochemical list includes DO, ORP, pH, T, TOC, light hydrocarbons, ferrous iron, sulfide.
COCs include the target organic compounds and daughter products.
All monitoring events should include monitoring of injection wells for TOC, pH, and ORP.

Reagents

There are a variety of reagents that are being used in reactive zone applications. Table 4 summarizes several of the available reagents reported in the literature, along with a brief description of the methods used to place the reagent and the means by which the reagent creates the reactive zone.

The most common reagents applied in the field today are molasses, Fenton's reagent/hydrogen peroxide, potassium permanganate, ORC_{TM}, and HRC_{TM}. Molasses and polylactate ester are used to create and maintain an anaerobic reducing zone for the reductive dechlorination of CAHs and for the precipitation of heavy metals. Magnesium peroxide is a slow release oxygen enhancement commonly applied to promote aerobic degradation of BTEX, petroleum hydrocarbons, and other aerobi-

Table 4 Selective Reactive Zone Reagents

Reagent	Type of Reactive Zone	Method of Delivery	Common Form of Reagent[a]	Mechanism	Comments[b]
Molasses	Anaerobic & reducing	Injection wells	Material is dissolved in water and then delivered via injection wells	Indigenous bacteria use as electron donor to create strongly reducing conditions	Numerous applications reported by ARCADIS Geraghty & Miller. (1, 15, 16, 26, 27) Applied under Patent #5,554,290 Cost: $0.30/lb molasses
Lactose	Anaerobic & reducing	Injection wells	Dissolved in water	Indigenous bacteria use as electron donor to create strongly reducing conditions	Reported applications by DiStefano (27) PCE reduced (27) Cost: $0.18/lb
Phenol	Anaerobic & reducing	Injection wells	Dissolved in water	Indigenous bacteria use as electron donor to create strongly reducing conditions	Reported applications in Europe Regulatory reluctance due to nature of reagent
Toluene	Anaerobic & reducing	Injection wells	Dissolved in water	Indigenous bacteria use as electron donor to create strongly reducing conditions	Reported applications by McCarty (28) Regulatory reluctance due to nature of reagent
Polylactate Ester	Anaerobic & reducing	Injection wells, direct-push points	Semi-solid, gel-like material placed in canisters in wells or directly delivered.	Lactic acid released upon hydration of material is metabolized by bacteria to release hydrogen.(29)	Slow-release material $12/lb Supplied by Regenesis under trademark HRC™
Fenton's reagent	Oxidizing and/or aerobic	Injection wells, direct-push points	Ferrous iron solution and variable strength hydrogen peroxide (10% to 50% soln)	Ferrous iron and peroxide injected as a means of oxidizing target compounds	Injected in a number of proprietary and non-proprietary formulas (30, 31, 32). Requires pH to be in the 4 to 5.5 range

Table 4   Selective Reactive Zone Reagents (continued)

Reagent	Type of Reactive Zone	Method of Delivery	Common Form of Reagent[a]	Mechanism	Comments[b]
Permanganate	Oxidizing	Injection wells, direct-push points, recirculation systems	Potassium permanganate dissolved in water (0.1% to 3% bw)	Reagent injected to oxidize target compounds	Wide range of practitioners report success (33, 34, 35). Raw material cost: $2/lb
Hydrogen Peroxide	Oxidizing and/or aerobic	Injection wells, direct-push points	Hydrogen peroxide injected at 1% to 35% solution strength	Reagent directly oxidizes target compounds or is a means to deliver oxygen	Wide range of practitioners report varying success (1, 13, 36). Most common use today is for direct oxidation 35% solution cost: $0.42/lb
Dithionite	Reducing	Injection wells, direct-push points	Sodium diothionite directly injected into groundwater.	Reagent reacts with iron and hexavalent chrome to form tri-valent chrome	Field demonstrations for chromium precipitation have been reported (6, 4) In the presence of heat, TCE reduction has been reported (6)
Sodium bicarbonate	pH adjustment	Wells or injection points	Mixed with water and injected.	Used to adjust pH as supplement to other reagents or stand-alone.	Used to maintain pH above 4.5 for chromium precipitation. Used for zinc precipitation
Magnesium Peroxide	Aerobic	Wells or injection points	Bulk powder, slurry or "socks"	Magnesium Peroxide is hydrolyzed, releasing oxygen into groundwater	Slow-release $10/lb Supplied by Regenesis under trademark: ORC$_{TM}$

a This represents the most common form used for injection.
b Costs are presented where available.

cally degradable COCs (e.g. ketones). The remaining reagents have been applied for the oxidation of a variety of COCs including BTEX and CAHs.

As previously discussed the goal of applying the reagent is to create a reactive zone in the subsurface that is sustained, easily maintained, inexpensive, and appropriate to the target compounds. As such, molasses and HRC have been successfully applied for the treatment of CAHs in groundwater. Molasses has also been applied for the treatment of heavy metals such as chromium, cadmium, mercury, and zinc. Polylactate ester is a slow-release material and, due to its viscous gel-like consistency, is most appropriate in low velocity groundwater systems, where the slow-release characteristic is particularly beneficial and where the consistency of the material will not lead to excessive implementation costs. Molasses is injected in a water solution and moves readily with groundwater. As a result the reagent can be effectively delivered across a wide area from fewer injection points. Molasses is readily degraded, thus leading to the rapid formation of anaerobic conditions. Over-delivery of the molasses, however, can lead to fermentation and pH-depression. Both materials are food-grade and acceptable to most regulatory agencies. Molasses and polylactate ester have been used in a number of different states in the United States and in countries around the world (Brazil, Canada, United Kingdom, Netherlands, and others).

As the efficacy of *in situ* reducing reactive zones has been demonstrated by many in the field, the variety of substrates applied has grown. Table 4 lists only a few of the many reagents tested in the laboratory or in the field. Other substrates include methanol, ethanol, sucrose, cellulose, edible oils, and proprietary blends of these and other sources of soluble organic carbon. The selection of the substrate is driven by the unit cost of the available soluble organic matter and the ease with which the material can be delivered to the subsurface. Many reagents are being selected from the wide variety of food-grade organic carbon sources (molasses, sucrose, and oils for example). The reagent selection process is often driven by how comfortable the regulatory agency is with the material and the variety and depth of practical applications available for a particular substrate.

Reagents for oxidative reactive zones have gained increasing attention recently, as the field data that has been developed has shown the efficacy of the application of oxidative reactive zones for the treatment of a wide range of organic compounds. Recent applications have included the treatment of adsorbed phase CAHs, BTEX, ketones and other common organic compounds (Geo-Cleanse Int. 1998, Tarr et al. 2000, Mott-Smith et al. 2000, Nelson et al. 2000, and Schnarr et al. 1998).

Hydrogen peroxide has been used as an oxygen supplement to enhance aerobic conditions since the mid 1980s. However, the recent applications of hydrogen peroxide (and Fenton's reagent—ferrous iron and hydrogen peroxide) in reactive zones has focused on the formation of hydroxyl radicals. Hydroxyl radicals are a very strong oxidizing agent and represent the strongest oxidant available other than fluorine. The resultant reactive zone is an environment in which target organic compounds are chemically oxidized to carbon dioxide, water, and chloride ions.

Potassium permanganate has been injected as a solution typically varying in strength from 0.1 percent to 3 percent. This reagent is not as aggressive as hydrogen peroxide; however, what it lacks in kinetics it makes up for in longevity. As a result, potassium permanganate can be as effective as hydrogen peroxide in an *in situ* reactive zone.

Regulatory Issues

Many states regulate the injection of materials into the subsurface. While this text does not attempt to cover the range of regulatory environments and issues that exist, it is important to consider the regulatory arena in which the remedial work is being conducted during the reagent selection process. Thus characteristics such as food-grade, no chemical residuals, and historic applications approved by other state or federal agencies all contribute to the rapid acceptance of an alternative substrate.

Design Criteria

This section covers the design criteria that define the full-scale system, including:

- well design—type, number, depth, screen zone(s), layout
- reagent feed—feed rate, solution strength, frequency of injection
- monitoring—analytes, frequency

Well Design

There are three basic types of wells (in reality delivery points) used in reactive zones: permanent wells, temporary well points, and direct-push delivery type well points. Permanent wells are constructed with materials appropriate for the geological formation, groundwater quality, and selected reagent. A typical injection well is shown in Figure 20. The wells in an unconsolidated formation are typically 2 to 4 inch diameter Schedule 40 or 80 PVC construction with slotted screens sized for the formation. In bedrock, an open bore hole in the target saturated zone is acceptable with a PVC or steel casing in the overburden. Wells constructed in this fashion provide a permanent means of delivering reagent to the subsurface. They are commonly applied in situations where readily soluble and degradable organic substrates—such as molasses—are applied. Permanent wells allow for multiple injections to establish and maintain the reactive zone. Permanent wells are also necessary in situations that involve impacts in bedrock geology, or where depths or soil strata make direct-push techniques impractical.

Direct-push techniques are also used for reactive zones in select applications. This type of well is limited to unconsolidated formations with impacts that are relatively shallow—depths can not exceed 50 to 75 feet. This technique is also constrained by the soil characteristics, particularly grain size. In some cases, where direct-push wells are used, a permanent well point is placed using a direct-push drilling rig such as a CPT (cone penetrometer type). This type of well is a small diameter point and is commonly applied where the number of injections will be limited and the need for well maintenance is minimal. This is a good design when the groundwater flow is shallow and slow. Under these conditions, a large number of injection points may be needed, and due to the slow groundwater flow few repeated

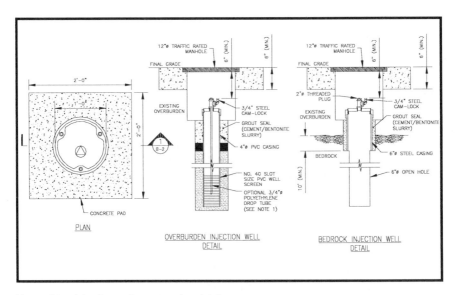

Figure 20   Injection well construction detail.

injections will be required. Another practical constraint to the application of a CPT well is availability of drilling equipment; while these are common in Texas, California, and Oklahoma (oil country), they are somewhat rare in the northeastern United States.

Another type of direct-push wellpoint is a temporary geoprobe or hydro-punch well. In this case, the well drilling process is an injection process. Thus as the well point is placed, the reagent is injected. When the injection process is complete, no well points remain. This approach is applicable for slow-release materials such as polylactate ester, which will dissolve slowly and require well points on 2 to 5 foot centers to treat the target zone. As with CPT, this technique is limited by depth and geology.

All reactive zone wells must be designed to target the impacted groundwater. Thus the depth and screened interval will be determined by the vertical delineation of the groundwater impacts—with all the limitations this implies. In addition, the lithology can have a profound impact on the well design and screened interval since the injected reagent will flow with groundwater, following the path of least resistance, which MAY NOT be where the target impacts are located. As a result, it is important to understand both the geology and the contaminant distribution when designing the wells. As a practical limit, screened zones in excess of 50 feet are not recommended due to the difficulty in evenly distributing the reagent across more than a 50 foot interval. Where the saturated thickness exceeds this limit, multiple well points are recommended.

The number of injection points and the spatial distribution of these points are a function of the contaminant distribution, the hydrogeology of the impacted zone, the type of injection point selected, and the type of reagent being used. The injection wells need to cover the entire area targeted by the reactive zone. The geology and groundwater velocity will control how wide an area a single point can impact. For example, in a tight geologic unit, groundwater is likely to move relatively slowly, and the ability to inject is limited by the permeability of the formation. As a result, the reactive zone developed from a single point will have a limited impact laterally from the injection point and in the direction of groundwater flow. Therefore more points will be needed on a tighter center to center spacing.

The type of reagent used can also effect the spatial distribution of the injection points. If a water soluble reagent, such as molasses, sucrose, or potassium permanganate is used, the reagent injectant will have flow characteristics, nearly identical to that of water and thus will move readily with the groundwater. As the reagent becomes more viscous, the ability to inject and achieve good lateral distribution will decline. As a result, in the latter case, more closely spaced well points will be required.

Reagent Feed

The feed rate, solution strength, and frequency of injection all relate to the target COC concentration. In Chapter 7, oxidant feed requirements were discussed, therefore in this chapter, comments will be limited to reducing reactive zones. For reducing reactive zones the reagent feed characteristics of rate, strength, and frequency tie back to the need to deliver adequate organic carbon, in the form of the selected reagent, in order to maintain reducing conditions in the subsurface. There are two criteria that must be met to create the reactive zone. First, enough organic must be added to ensure that the electron acceptors in the groundwater are utilized. This organic carbon feed rate can be derived based upon the total concentration of the various electron acceptors (primarily DO, nitrate, and sulfate) and the groundwater flow rate. The product of the electron acceptor concentration and the groundwater flow rate is the electron acceptor flux. The reactive zone design must supply enough substrate to overcome the electron acceptor flux.

Second, there must be enough substrate to drive the entire zone into highly reducing conditions. Typically the goal is to maintain between 50 and 100 times as much dissolved organic carbon in the reactive zone as there is COC in the target area (i.e., 100 ppm of DOC for every 1 ppm of COC). Based on experience, this translates to one or two orders of magnitude higher target DOC concentration in the injection wells. The reason higher concentrations must be fed in the injection wells relates to the fact that the organic carbon will be metabolized as it flows with groundwater, therefore it is necessary to establish a DOC gradient between the injection points and the rest of the reactive zone. It is important to keep in mind that the COC flux through a target zone is related to the groundwater flow velocity, the saturated thickness, and the COC concentration.

Both of these criteria are simply guidelines for a preliminary calculation of the donor feed rate. Experience has proven them to be adequate means to define a reasonable organic carbon feed rate to begin the reactive zone. Field data collected AFTER the reactive zone has been started is the true measure of the adequacy of the reagent feed. Field analytical data (in particular redox, pH, and TOC) from the injection wells and monitoring wells within the reactive zone should be used to confirm that the reactive zone has been established and is expanding with groundwater flow. Pilot and full-scale data has shown that during the initial stages of some reactive zones the organic carbon load to the injection well needs to be maintained between 100 and 9000 mg/l of organic carbon. In the long-term, once the reactive zone is established, these concentrations can be reduced somewhat, with a sustainable target of 50 to 100 mg/l of organic carbon within the reactive zone.

The solution strength, while related to the target organic carbon feed rate, is also impacted by other factors such as the groundwater flux, the ease with which a solution can be injected, and the cost to perform the injection. In the example above involving a tight geology, the amount of reagent is reduced since the groundwater flow and the contaminant flux is low, thus less reagent is required to deliver the required DOC load. In this situation a slow dissolving substrate may be more applicable. Conversely, if the groundwater velocity is high, the COC flux will be proportionally higher and the need for organic carbon higher. In this situation a readily soluble, easily injected reagent is more applicable. It is worth noting that the lateral, advective dispersion in a high groundwater velocity environment is relatively low, thus the reactive zone formed will be narrow, but long. Therefore more injection points will be required across the plume (perpendicular to groundwater flow), but the lines of injection wells can be spaced further apart in the direction of groundwater flow.

Monitoring

After the baseline biogeochemical environment has been defined and a reactive zone is being implemented, it is important to monitor the progress and health of the reactive zone. Early in the establishment of the reactive zone, possibly during a pilot, or demonstration phase, nearly the full suite of parameters (Table 2) should be monitored. After the zone has been established only the most critical analytes need to be monitored on a regular basis. In a reducing reactive zone the most important analytes include the redox (ORP), dissolved oxygen, pH, DOC, degradation end-products (ethene and ethane for CAHs), and COCs and their daughter products. In the case of reactive zones targeting metals, sulfides can also be important.

The frequency of monitoring is site specific, but generally the frequency should be adjusted over the lifecycle of the remediation. Initially, the sampling frequency will run between weekly and monthly—the former if no pilot test data is available to define feed rates. After the first two months of operation monthly or quarterly sampling will provide adequate data to monitor the reactive zone. Within the first 6 to 12 months of operation, quarterly monitoring should be sufficient. Long-term monitoring can be extended to semi-annual, however due to the short expected life of most reactive zones—less than 5 years—anything less frequent than semi-annual is unlikely to provide adequate data. Monitoring frequency should always remain flexible during the lifecycle of the reactive zone. This concept is critical to the health of the reactive zone and the success of the process. Disclosure of the need to be flexible is critical to the regulatory agency perceptions of the monitoring effort and the client's understanding of the budget and schedule.

Pilot Testing

The design criteria are most often defined in a pilot or demonstration test. Such a test is used to gather critical design data—well spacing and reagent feed rate, strength, and frequency—as well as to demonstrate the efficacy of the technology and satisfy regulatory agency concerns regarding the technique. The pilot test should always focus on an area of concern within the plume. Thus a successful pilot test can also be used as an interim remedial measure.

This section covers a typical field test for a reducing reactive zone. Keep in mind that before testing it is important to answer questions regarding the biogeochemical environment, hydrogeology, regulatory arena, etc. as discussed earlier in this chapter. This basis is critical to the pilot test program set up, focus, and endpoint.

Test Wells

To properly evaluate the *in situ* technology in the field, injection wells and a network of groundwater observation wells are required. The injection wells need to be located in an area of the site where sufficient impacts are present, and should be installed in a manner similar to wells that would be employed in a full-scale system. The groundwater observation wells for the field test should also be located within the impacted area. The observation wells should be located in a manner to evaluate both the performance of the degradation—or precipitation—process and the extent, both parallel and perpendicular to the direction of groundwater flow, of the *in situ* reactive zone.

The field test should employ at least one injection well, preferably two, that can be used to deliver reagent to the entire target zone. The screened interval should intercept the impacted zone, with consideration given to the lithology and groundwater flow conditions. The injection wells should be constructed as described above using appropriate drilling techniques. If there is little or poor quality geologic data available, consideration should be given to the need to gather supplemental geological data using split-spoon sampling techniques or other appropriate means. Following installation, the well should be developed to remove fine material and ensure hydraulic communication with the surrounding aquifer.

To provide the necessary level of performance monitoring for the field test, observation wells in the area down-gradient of the injection wells are required. A minimum of two and whenever possible as many as five observation wells can be included in the pilot program. The observation wells should be located at variable distances from the injection wells both parallel to the direction of groundwater flow and perpendicular to the direction of groundwater flow. This will allow the reactive zone to be defined in both directions relative to groundwater flow. Once again, consideration should be given to the variability of the site geology in locating screened intervals. It may be necessary to monitor multiple intervals if the geology dictates.

When possible, existing monitor or production wells may be used as injection or observation wells in order to speed up the pilot test program and control costs.

Reagent Injection

The composition of the reagent feed solution that will be used during the field testing, the solution injection rate, and the injection procedures are discussed in this section.

The composition of the reagent feed solution will need to be varied during the field test, based on field measurements made in the observation wells and the analytical results gathered during the groundwater monitoring program. The amount of reagent injected in the injection wells during the field testing will also be varied by increasing or decreasing the amount of solution injected, or by changing the frequency of injections.

An appropriate solution feed rate must be established and maintained in order to ensure that adequate electron donor solution is added and that the available electron acceptors are fully metabolized in the reactive zone in order to maintain a strongly reducing environment. At the same time, the feed rate needs to be controlled so as to minimize the amount of material that has to be injected into the subsurface. The proposed solution feed rate can be calculated based on achieving a sufficient DOC concentration in the groundwater that passes through the injection well area. For the field test, the volume of solution and solution strength should be calculated to achieve the DOC load discussed earlier using a reasonable number of injection events—in most cases pilot tests are conducted using a batch type injection program.

The frequency of injections will vary with the geologic, biogeochemical, and hydrogeologic conditions of each site. As a result the frequency of injections can vary from once a day to once every 6 months.

Duration of Field Study

Typical field tests last between 3 and 6 months. The rate of groundwater flow and the proposed observation well locations will determine the site-specific duration of the test—the closer the observation wells are to the injection wells and the faster groundwater moves, the sooner results can be expected and the shorter the pilot test needs to be. The testing is complete when the following criteria have been achieved:

- redox conditions down-gradient of the injection well are reduced (ORP less than −100 mV)
- the ratio of target COC to daughter products has declined—i.e., the source material is degrading
- the amount of end-product (ethene) has increased

These results may actually be achieved within 1 or 2 months of implementation of the pilot test program. Additional information may be necessary to satisfy regulators or clients and in order to optimize the full-scale design. For this reason the pilot is often extended beyond a simple demonstration. Additional criteria that are commonly applied include:

- reduction of target COC concentrations below a historical low baseline concentration

- a COC:daughter ratio less than 1
- definition of the lateral extent of the reactive zone based on a specified time frame
- rebound testing within the target zone (this applies to metals only) to demonstrate that precipitates do not remobilize after the reactive zone is terminated. This portion of the test will typically double the duration of testing

Many times the testing ends, but the injections continue. Once the reactive zone is established within the plume, maintenance dosages of the reagent will allow the zone to continue to serve as a remedial measure, until such time as the system is expanded or the cleanup goals are achieved.

Field Test Performance Monitoring

The most critical portion of the *in situ* field test is the performance monitoring. In this portion of the test, field monitoring of selected indicator parameters and groundwater sampling for field and laboratory analyses are conducted. The data collected from these performance monitoring activities are used to adjust reagent feed rates and, if necessary, the frequency of sample collection.

Performance monitoring takes the form of a baseline sampling event and periodic monitoring events during the period of field testing. The baseline monitoring event is used to establish the biogeochemical conditions and COC concentrations of pre-test groundwater and serves as the basis of comparison for all future data collected during the test.

To establish baseline conditions (i.e., groundwater conditions prior to the start of the field test), an initial round of groundwater samples should be collected from the injection and observation wells. These baseline samples should be analyzed for the following parameters:

- Metals: target metals (filtered and unfiltered), iron (total), iron (ferrous), manganese (total and dissolved)
- Biogeochemical Parameters: sulfate, sulfide, nitrate, total suspended solids (TSS), total and dissolved organic carbon (TOC/DOC)
- Dissolved gases and light hydrocarbons
- VOCs
- In some cases: bacteriological (total heterotrophic plate count, iron reducing bacteria, and/or sulfate reducing bacteria)

In addition, the levels of several indicator parameters should be collected in each of the wells. These indicator parameters include pH, redox, DO, and temperature.

In order to evaluate the extent of the *in situ* reactive zone, and the effectiveness of the *in situ* process, groundwater samples will also need to be collected as the field test progresses and at the end of the field test. The frequency will vary, but in most cases data should be collected within the first month of testing and then bi-weekly or monthly through the first 3 months of testing and bi-monthly to quarterly thereafter. The frequency is initially driven by the need to ensure that neither too much nor too little reagent is being added. After the first month or two of testing, sampling frequency will be driven by factors such as the groundwater flow velocity, the natural baseline biogeochemical environment, and the extent of groundwater impacts. The data should be collected from the two injection wells and observation wells. Data from the injection wells should be limited to fewer analytes than from the observation wells. All samples should be collected using low-flow purge techniques to ensure the integrity of the analytical data generated.

Field analysis for ORP, DO, pH, temperature, and conductance provides the most rapid and inexpensive water quality data and reactive zone definition. These parameters should be monitored with the greatest frequency. Once the field parameters change, the performance monitoring can be extended to the more costly parameters analyzed in the laboratory.

Test Results

The results from a pilot test can be used to:

- demonstrate the efficacy of the technology at a particular site by providing empirical, site-specific field data for the technology
- determine the reagent feed rate, the frequency of injections, and the solution strength required
- define the well spacing perpendicular and along the groundwater flow direction based on the extent to which the reactive zone was established during the test

The third case study reviews in detail a year-long pilot test program.

Full-Scale Application

Once the pilot test program has defined the critical design criteria, the full-scale design can be completed and the full-scale system authorizations obtained. Regulatory approval will likely be required along with well construction permits, injection permits, and possibly construction permits.

In order to demonstrate the application of the technique, several case studies are presented in this section.

Case Study 1: Federal Superfund Site, Pennsylvania

In the section COC Chemistry: Metals an example of an *in situ* reactive zone was introduced to describe how chromium and cadmium can be reduced and precipitated as hydroxides, carbonates, or sulfides in the subsurface. In Case Study 1 additional details of this site are presented.

Background

The site includes a 28-acre, operating aircraft manufacturing facility and the surrounding residential neighborhood, as well as down-gradient impacted groundwater (Figure 21). In the mid 1980s, the local water authority detected the presence of VOCs, specifically trichloroethene (TCE) and 1,2-dichloroethene (DCE), in the

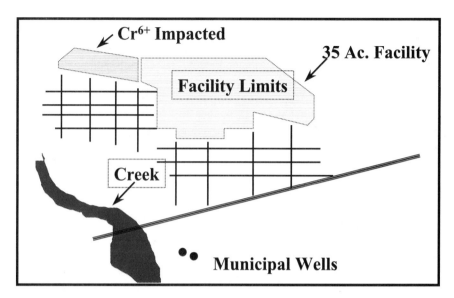

Figure 21 Case Study 1: Superfund site.

groundwater in the municipality's backup water supply well field, which is within 3,000 feet of the manufacturing facility. As a result, the facility owner entered into a Consent Order and Agreement (COA) with the state environmental agency to implement a remedial investigation and cleanup plan. The site was placed on the National Priorities List (NPL) in 1988; an Administrative Order of Consent (AOC) was signed for the purpose of conducting a Remedial Investigation (RI), Endangerment Assessment (EA), and Feasibility Study (FS) for the site. In 1991, a Record of Decision (ROD) was issued for the area of contamination within the plant boundaries. The ROD remedy consisted of:

1. A recovery and treatment system to contain and control off-site migration of groundwater
2. Institutional controls to limit future property use to those activities compatible with site conditions

In 1992 a Remedial Design Workplan (RDWP) was approved by the USEPA. The RDWP called for a line of pumping wells at the property boundary to contain and control off-site migration of the plume. The pumped water was to be treated for heavy metals (cadmium and chromium) and organic compounds (TCE, DCE, and vinyl chloride [VC]) using *ex situ* treatment; the treated water was to be discharged to a National Pollution Discharge Elimination System (NPDES) discharge point. In 1995, shortly after the NPDES permit was approved, the treatment system design and construction was bid. At that time an alternative treatment approach, which employed an *in situ* reactive zone, was proposed to the USEPA.

The primary target for the reactive zone at the time that the system was designed was a plume of hexavalent chromium and cadmium centered north and west of the main plant area. Overlapping the heavy metals plume was the northwestern leg of the CAH plume. The CAHs consisted of TCE and its degradation daughter products, DCE and VC. A pilot test program was initiated in late fall 1995 to evaluate the efficacy of the reactive zone technology for chromium and cadmium reduction. Two pilot test areas (one in the northeast corner and one in the southeast corner of the target area) were selected. A single injection well (existing 4 inch monitoring wells) was used in each area. Within the first 2 months of testing, the hexavalent chromium concentrations were reduced from as high as 7 mg/l to less than 0.05 mg/l in the test zone (Figure 9). Following the 6 month pilot program, a series of geoprobe samples were collected radially away from the injection wells in order to define the lateral influence of the injection well. This data indicated that the injection well achieved at least a 10 foot radius of influence.

Coincidental with the pilot test program, the ongoing quarterly CAH sampling program continued. Quarterly data collected from wells located near the pilot test wells during this test program indicated that the reactive zone was not only creating the conditions necessary for precipitating cadmium and chromium, but also the conditions needed to reductively dechlorinate the CAHs.

Full-Scale System

The full-scale remediation system was designed in the summer of 1996 and the ROD was rewritten to include the reactive zone for metals precipitation. The reactive zone system was constructed during the last four months of 1996 and went on line in January 1997. Based on the pilot study results that indicated enhanced reductive dechlorination of the CAH plume by the reactive zone system, the full-scale system design documents incorporated the use of the reactive zone as the technology to treat and contain the CAH contamination in the northwest corner of the property.

The geology at the site consists of a sandy-silt overburden overlying a weathered bedrock and a fractured limestone. The target area for the reactive zone is the shallow overburden to approximately 25 feet below land surface in an area covering approximately two acres. The maximum concentrations of hexavalent chromium, cadmium, and TCE were 3 mg/l, 0.8 mg/l and 0.7 mg/l, respectively, in the area being treated at the time that the treatment system was installed and went on line.

The use of the reactive zone to reduce chromium and cadmium is described in a previous section of this chapter. The reducing conditions created within the reactive zone are maintained to be strong enough to ensure that the enhanced reductive dechlorination of TCE and its daughter products results in the destruction of the CAHs of concern at the site to harmless end-products (carbon dioxide and water).

The full-scale system is summarized in schematic form in Figure 10. The system consists of twenty 2 inch diameter injection wells, completed in the unconsolidated sandy-silt overburden. A 2 inch diameter well was used to control capital costs and still insure ease of accessibility for well maintenance. The well spacing was based on the test data. Each well is connected to the reagent storage building via 1 inch diameter, buried, high density polyethylene piping. The reducing reagent (in this case molasses is applied under Patent #5,554,290) is added once or twice a day (comparable to the pilot test feed rates) at variable rates and concentrations based on the results of the system monitoring program. A total of 16 additional wells are used for monitoring of the treatment system. (Most of these wells were installed in order to satisfy regulatory and client requirements, since at the time this was the first full-scale treatment system of its kind.) A programmable logic controller monitors and controls the feed rate and frequency of the molasses feed and solution feed pumps, as well as the timing of the solenoid valve network that controls the metered flow to the injection wells. The molasses storage tank, mixing tank, feed pumps, control panel, and solenoid valve nest are housed in a 10 foot square pre-engineered building (Figure 11).

System Performance

The monitoring program was initiated when the treatment system went on line in January 1997. From that time, monthly to quarterly sampling has been conducted for redox, pH, chromium, cadmium, other metals, and CAHs. Quarterly sampling was dictated by the regulatory agency as part of the site-wide monitoring program; the initial monthly sampling was instituted to ensure that the process was properly controlled. The test results have closely matched or exceeded those collected during the pilot program:

- the concentrations of chromium have dropped in the hot spots from 2 to 3 mg/l to less than 0.05 mg/l (Figures 13 and 16)
- the overall chromium plume shrunk to approximately one-fourth its original area extent within 18 months of start up
- the peak chromium concentrations are isolated to one area at slightly above 0.5 mg/l
- the peak cadmium concentrations are isolated to two small areas at slightly more than 50 ug/l (Figures 14 and 17)
- TCE, DCE, and VC concentrations have steadily dropped in the majority of monitoring wells where CAH data is collected (Figure 22)

After 22 months of operation, the concentrations of TCE, DCE, and VC were near or below the regulatory limits of 5, 70, and 2 ug/l, respectively, indicating that the reactive zone technology can achieve concentrations in groundwater below mandated drinking water standards.

The operational efficiency of the reactive zone technology is also evidenced when the degradation rate is evaluated. Figure 23 summarizes the degradation rate for TCE and DCE before the reactive zone was implemented. Figure 24 presents the data at the site after the reactive zone was started. As the Figures illustrate, the rate of decline of TCE and DCE concentrations increased significantly once the reactive zone was implemented at the site.

The cost to implement the reactive zone and operate it for a little less than three years was approximately $400,000 including capital, operations, maintenance, and monitoring. This system replaced a pump and treat system that had an estimated present worth cost of more than $4,000,000. The cost for the pump and treat system included capital and operation and maintenance for a period of 20 years.

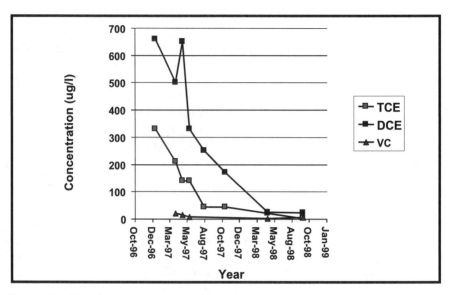

Figure 22   Case Study 1: chlorinated VOCs.

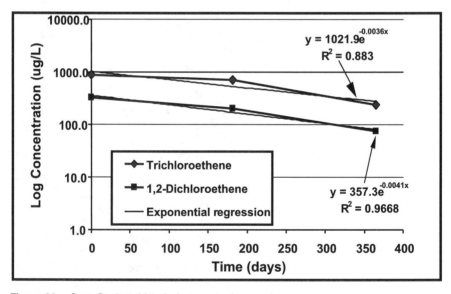

Figure 23   Case Study 1: historical concentration trends.

Case Study 2: State Voluntary Cleanup

The second case study involves a manufacturing facility located on the west coast. Facility operations involved metal plating and degreasing that led to the release of hexavalent chromium and TCE into the site groundwater.

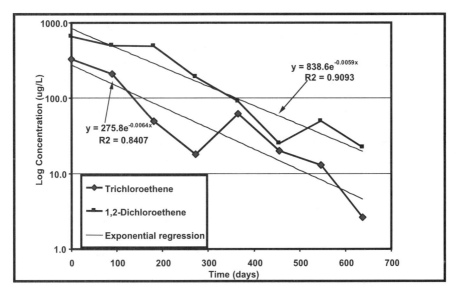

Figure 24 Case Study 1: concentration trends following implementation of full-scale *in situ* reactive zone.

Background

The site is slightly more than two acres in size (Figure 25). The geology is primarily silts and clays to 24 feet below land surface (bls), overlying an impermeable clay layer that serves as an aquitard. Groundwater flows from the area of MW-13 toward MW-10 at a relatively slow rate (less than 100 feet per year). The former chromium waste storage area is located near MW-13 and the former TCE degreasing area is located near the center of the plant.

Historic releases led to groundwater impacts in the forms of chromium and CAHs. As Figure 25 illustrates, the hexavalent chromium plume extends over most of the site. Initial peak hexavalent chromium concentrations in groundwater were more than 900 mg/l. The primary CAH was TCE, although degradation daughter products were also in evidence at the site, including DCE and VC. The initial maximum TCE concentrations in groundwater were approximately 24,000 ug/l, while DCE and VC concentrations were less than 500 and 100 ug/l, respectively.

A pilot study was completed in a small up-gradient portion of the site in late 1995 and early 1996 in order to demonstrate the efficacy of molasses injection for the creation of a suitable reactive zone. Due to the high concentrations of hexavalent chromium, there was concern that the natural microbial population might be too stressed to adequately address the needs of creating the reactive zone necessary for groundwater treatment. In order to supplement the natural population at this site, sludge from an anaerobic sludge digester was added to the molasses/water reagent mixture used. The pilot test program successfully demonstrated that the reactive zone technology would be successful. Based on the data collected, the full-scale system was implemented.

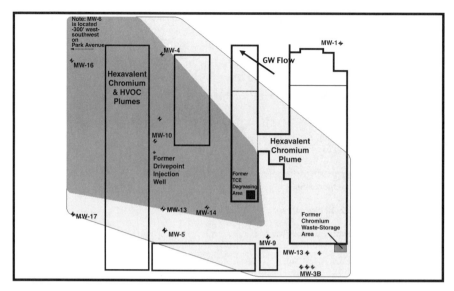

Figure 25   Case Study 2: site map, monitoring wells and plumes.

Full-Scale Implementation and Results

Ninety-one injection points were installed (Figure 26) to 24 feet bls at the site. Each point was 1 inch diameter with a 10 foot long, 0.010-inch slot screen. Each injection point was installed rapidly and at a relatively low cost using a direct-push approach with a Geoprobe™ drilling technique. Existing monitoring wells were used to track the progress of the treatment process.

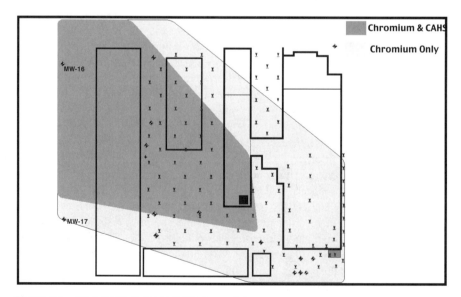

Figure 26   Case Study 2: reagent injection points.

Each injection point received an injection of a mixture of molasses, water, and anaerobic sludge. Unlike the daily dilute molasses/water reagent injections used in Case Study 1, the injections for the full-scale reactive zone at the second site were more concentrated and less frequent. The injections were initiated in the spring of 1997. A single injection took place in May 1997; the next injection occurred approximately one year later.

The theoretical groundwater flow for the site is suspected to be artificially high. Due to the site's proximity to the Bay margin and the very dense lithology, it was suspected that, in spite of the calculated low groundwater flow rate of 100 feet per year, the actual groundwater flow rate may actually be an order of magnitude less. Transport of the reagent throughout the subsurface in an acceptable time frame using dosed injections into fewer points/wells would probably not result in an acceptable rate of remediation. In fact, monitoring wells approximately 100 feet down-gradient from the on-site remediation area do not yet indicate that remediated groundwater has reached the off-site wells (three years after remediation commenced). In addition, it was determined that the periodic, site-wide injection events were more cost effective for this site—a key consideration for the client.

The theoretical basis for selecting both the concentration and the volume of injection was based on the above referenced pilot study, in which dosage concentrations and volumes were changed over time until an effective remediation environment was established at monitoring wells near the pilot study injection areas. The spacing of the injection points was likewise established from the data acquired during the pilot study, which indicated, in this lithology, a maximum 17 foot radius of influence.

Figure 27 summarizes the average chromium concentrations collected from eight wells across the chromium plume. Hexavalent chromium concentrations have been reduced at the site from initial concentrations in the range of 66 to 140 mg/l, to concentrations of 0.14 mg/l, to nondetect. Hexavalent chromium concentrations declined at a steady rate during the period between the first and second injections and are now nondetect (<0.05 mg/l) across the site. As the data in Figure 27 depicts, the majority of the chromium found in groundwater was hexavalent. In the reducing conditions created by the injection of the molasses/water/sludge mixture, the hexavalent chromium was reduced to tri-valent chromium and precipitated out of solution, most likely as chromium hydroxide. The precipitate was removed by the aquifer soils.

In the source area TCE concentrations were reduced from approximately 18 mg/l to 2 mg/l, while in the mid plume area, TCE concentrations were reduced from approximately 30 mg/l to nondetect. Figure 28 provides a summary of the data collected from MW-4 before the pilot test was initiated through March 2000. Prior to initiation of the reactive zone pilot the ratio of source CAH (i.e., TCE) to degradation product (i.e., DCE and VC) was approximately 9:1. After the first injection the concentrations of TCE initially dropped, while the concentrations of DCE and VC remained relatively unchanged. As a result, the ratio of source to daughter product declined to approximately 3:1.

Prior to the first full-scale injection the TCE concentrations increased again. This most likely occurred as a result of the release of microbial surfactants within the

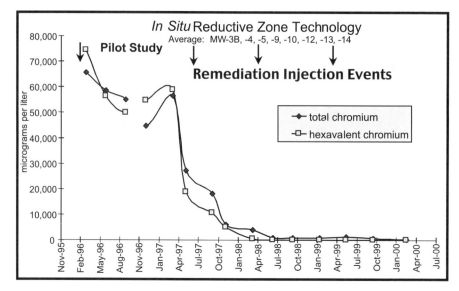

Figure 27   Case Study 2: chromium data.

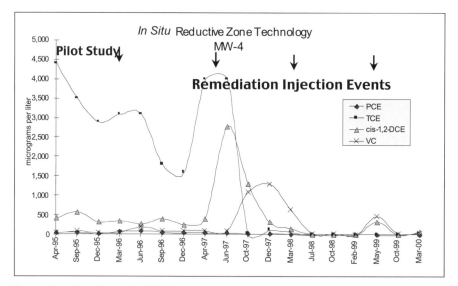

Figure 28   Case Study 2: CAH data.

reactive zone and/or transient increases in the groundwater elevation. Bacteria will generate surfactants in order to increase the amount of available organic carbon dissolved in the groundwater. The organic carbon must be in solution for the bacteria to metabolize the carbon. The surfactants are nondiscriminating, and thus aid in the desorption and dissolution of sorbed TCE. The resultant spike in TCE can be expected in some reactive zone sites.

Parallel to the TCE spike is an increase in daughter products, namely DCE and VC. As a result the ratio of source to daughter products continues to decline (less than 2:1 shortly after the first full-scale injection).

Another important observation that can be made using the data in Figure 28 is the lag between the peaks of the source and degradation products. The peak sequence follows the sequential dechlorination of the CAHs: the TCE peak is *followed* by the DCE peak; and the DCE peak is *followed* by the VC peak. It can also be concluded from the data that the reactive zone is reducing the concentrations of daughter products as well as the source product. Finally, the presence of daughter products of TCE also provides indication that the observed results are not a dilution phenomenon.

This site is presently being considered for the potential use of reactive zones to address vadose zone contamination at the site.

Case Study 3: PCE-Impacted Bedrock Pilot Test

This case study presents the results for a pilot test demonstrating the enhanced reductive dechlorination of tetrachloroethene (PCE) in bedrock groundwater. The pilot test was implemented in the mid-portion of an approximately 3000 foot long plume.

Background

A tetrachloroethene (PCE) spill was discovered in 1985 at a northeastern United States manufacturing plant. A subsequent groundwater investigation resulted in the installation of approximately 40 wells to delineate the PCE plume. PCE has been delineated horizontally and vertically at the site, and a dissolved PCE plume has been defined. Low concentrations of PCE (1 µg/l) extend approximately 3,000 feet (914 meters) down-gradient of the spill area. Historical groundwater monitoring indicates that the plume is currently at equilibrium. Hydraulic control has been established at the site through pumping. Low concentrations of PCE breakdown products (such as TCE and 1,2-DCE) were detected in the former source area during the ongoing groundwater monitoring at the site, indicating that natural reductive dechlorination of PCE was taking place prior to initiating the bioremediation pilot test. The observed reductive dechlorination in the source area was a result of favorable geochemical conditions: anaerobic and reducing conditions and the presence of organic carbon (electron donor) in groundwater. Less favorable conditions that exist outside the former source area (more aerobic and oxygenated groundwater and a lack of electron donor) do not promote continued degradation of the PCE plume.

Soil vapor extraction (SVE) had been employed in the former source area at the site, but the mass recovery was low and had reached asymptotic levels. Pumping of groundwater in the former source area and up-gradient of a stream has been ongoing at the site since 1987. The pumping remedy has been effective in protecting the stream and containing any further migration of PCE from the site. However, due to the inefficiency of pump and treat, and the elevated long-term costs associated with this technique, an *in situ* reactive zone was tested.

A pilot test was initiated in October 1998 to evaluate if the rate of reductive dechlorination down-gradient of the source area could be enhanced via an anaerobic *in situ* reactive zone. Field and laboratory data were collected prior to initiating the pilot test to establish baseline conditions, and were collected periodically during the pilot test to adjust the rates of reagent feed and document the progress of the reactive zone.

Geology/Hydrogeology

The overburden at the site consists of a thin layer of glacially deposited sands, silts, and clays and is approximately 10 to 20 feet (3 to 6 meters) thick. The overburden contains little groundwater and therefore the majority of the site wells were installed in bedrock, which consists of Triassic-aged, low permeability silt-stones and shales. Results of several pumping tests completed at the site indicate that the bedrock at the site has a hydraulic conductivity (K) in the range of 0.14 to 1.78 feet per day (0.04 to 0.54 meter per day). The groundwater velocity for the site is within a range of approximately 10 to 70 feet per year (3.1 to 21.3 meters per year).

The bedrock is primarily fractured along horizontal bedding planes that strike to the northwest with a slight dip (9°) to the northeast. Minor vertical fractures are also present in the bedrock. Groundwater flow is believed to be primarily along the horizontal bedding plane partings that are coincident with the strike of the rock. Groundwater flow is generally south-southeast with the orientation of the plume being coincidental with strike. There is historical evidence supporting the interconnection of the bedding plane fractures through previous pumping and injection tests on the aquifer.

Baseline Biogeochemical Assessment

A baseline biogeochemical sampling event was performed prior to initiating the pilot test to determine background conditions at the site, and to evaluate the biogeochemical environments present in various portions of the plume, including the pilot test area.

Groundwater samples were collected utilizing low flow sampling procedures from a background well located up-gradient of the assumed source area and the pilot test area. Field parameters (DO, ORP, pH, temperature, and conductivity) were collected at the well head using a flowthrough cell and a multimeter, and samples were collected for a full suite of biogeochemical parameters. The laboratory parameters included VOCs, alkalinity, ammonia, biochemical oxygen demand, chemical oxygen demand, chloride, total and dissolved iron and manganese, nitrate/nitrite, sulfate/sulfide, DOC, and TOC. Samples were also analyzed in the field utilizing a spectrophotometer for ferrous iron and sulfide, and submitted to a specialty lab for analysis of permanent gases (carbon dioxide, oxygen, nitrogen, methane, and carbon dioxide) and light hydrocarbons (ethene and ethane). Select baseline sampling results from the background and pilot test areas are provided in Table 5.

The background biogeochemical environment flowing onto the site was fairly aerobic (DO = 7.09 mg/l) and oxidizing (ORP +42.7millivolts [mV]). VOCs were

Table 5   Case Study 3 Performance Monitoring Data for Pilot Test Wells, MW-1 and MW-3

MW-1	DO mg/l	ORP mV	TOC mg/l	PCE µg/L	Ethene ng/l	MW-3	DO mg/l	ORP mV	TOC mg/l	PCE µg/L	Ethene ng/l
Apr-98	0.28	19.5	12	110	12	Apr-98	-	-	-	-	-
Dec-98	1.83	-73.3	27	-	-	Dec-98	-	-	-	-	-
Jan-99	3.33	-153.1	<1	-	-	Jan-99	-	-	-	-	-
Feb-99	0.35	-106.2	3.8	400	-	Feb-99	-	-	-	-	-
Mar-99	0.22	-90.4	4.4	-	-	Mar-99	-	-	-	-	-
Apr-99	0.19	-124.9	469	101	408	Apr-99	0.66	-41.2	1.9	17	92
May-99	0.08	-158.8	1,400	128	1,069	May-99	0.4	37.1	1.6	176	94
Aug-99	0.12	-167.4	1,110	250	2,115	Aug-99	0.52	-60.4	1.7	162	122
Sep-99	0.05	-309.2	814	134	3,651	Sep-99	1.11	-58.3	1.8	74.6	200
Oct-99	0.07	-194.8	1,180	156	2,937	Oct-99	0.29	129.3	1.6	39.6	122
Nov-99	0.25	-253.0	1,600	104	2,274	Nov-99	0.39	-14.9	<5	16.8	84
Dec-99	0.8	-198.6	1,430	56.6	2,332	Dec-99	0.27	134.5	2.6	104	189
Feb-00	0.29	-261.8		39.1	2,292	Feb-00	0.6	-76.1	1.6	20.9	157

not detected at this up-gradient location. Background groundwater contained low levels of some electron acceptors: non detectable levels of nitrate; < 1 mg/l of iron and manganese; and carbon dioxide concentrations of 9.7 mg/l. Ethene and ethane, the final products of reductive dechlorination of PCE were nondetect (< 5 nanograms per liter [ng/l]) in up-gradient groundwater.

The baseline biogeochemical environment in the pilot test area was transitional: DO levels indicated anaerobic conditions in two of the three pilot test wells, while ORP levels were in the +19 to +160 mV range. Total VOCs in this portion of the plume ranged from 1 to 813 µg/l. The only PCE degradation product detected was TCE in two of the three pilot test wells. Levels of ethene and ethane were low and not significantly above background, indicating that little natural reductive dechlorination was ongoing in this portion of the plume.

The pilot test wells showed a significant increase in the amount of nitrate in the groundwater during the baseline sampling event, which could also be a result of lawn watering (and fertilizer). Detectable concentrations of ammonia and the most elevated level of dissolved nitrogen detected during the initial baseline assessment at the site were also identified in this area, suggesting that some of the nitrate is being reduced in this area. ORP measurements suggest that the environment may be favorable for denitrification.

Pilot Study

Wells associated with the pilot test are shown on Figure 29. Reagent injections were initiated in October 1998. Approximately 200 gallons of reagent was initially injected under pressure on a weekly basis and consisted of a 10:1 ratio of molasses to water. Molasses was used as the electron donor due to its relatively low cost ($0.30/lb), high organic carbon content (approximately 60 percent by weight), and its ability to create a strong reducing environment in a short time period (as opposed to some other electron donors delivery techniques that rely on dissolution and diffusion). The frequency of injection was modified to biweekly in May 1999, and the ratio of molasses to water varied from a 10:1 to 20:1 ratio based on the performance monitoring performed during the pilot test. The reagent was injected under pressure (up to 30 psi) for a more thorough distribution into the bedrock aquifer system. Parameters associated with the performance monitoring focused on PCE and associated degradation products, DO, ORP, TOC, and ethene.

Performance monitoring was initiated in December 1998, approximately two months after commencement of the pilot test to monitor the development of the IRZ. Initial performance monitoring events focused on field parameters (primarily DO and ORP) and measuring TOC concentrations in groundwater. VOC monitoring was initiated after a TOC gradient had been established within groundwater in the pilot test area.

During the December 1998 performance monitoring event, anaerobic and reducing conditions had been established in the injection well and first down-gradient monitoring well (MW-1). TOC concentrations had significantly increased in the injection well, but little change in TOC concentrations were observed in the two

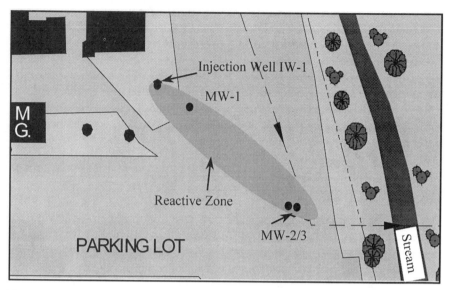

Figure 29   Case Study 3: pilot test area.

down-gradient wells. Little change was observed in the farthest down-gradient monitoring well (MW-2).

In January 1999, the injection well was deepened by 10 feet (3 meters) to encounter a more highly fractured bedrock zone, and increase the ability to delivery the reagent to the more impacted portion of the aquifer.

VOC concentrations in MW-1 significantly increased (PCE = 400 µg/l, TCE = 40 µg/l, and cis-1,2-DCE = 22 µg/l). This increase was due to a biological surfactant effect resulting from the increased microbial activity of the expanding microbial community. Reductive dechlorination was reducing an increased amount of TCE to cis-1,2-DCE, but increased rates of the reductive dechlorination of PCE to TCE were not occurring. The ratio of PCE to TCE during the background sampling event was 9:1; this ratio was 10:1 in February 1999, presumably due to natural surfactants. The ratio of PCE to TCE was 85:1 during the background sampling event, and this ratio was 18:1 during the initial surfactant effect. The increased degradation of TCE to DCE was due to the anaerobic and reducing conditions that had been established in the area of the well. However, significantly increased rates for the complete reductive dechlorination of PCE to ethene could not occur since the electron donor (carbon) injected in IW-1 had not reached this down-gradient location.

TOC concentrations continued to increase and the more strongly anaerobic and reducing conditions necessary for increased attenuation rates via reductive dechlorination were established in MW-1 through March 1999. Anaerobic and reducing conditions were present in IW-1, MW-1, and MW-3 during the April 1999 monitoring event, and aerobic and oxidizing conditions continued in MW-2. MW-3 is located adjacent to MW-2, and has a deeper open borehole interval. Monitoring was initiated in MW-3 since anaerobic and reducing conditions had not been observed in MW-2.

Elevated TOC concentrations continued to be present in the injection well (> 3,000 mg/l), and increased TOC concentrations (469 mg/l) were observed in down-gradient MW-1. PCE concentrations in MW-1 had been reduced from 400 µg/l to 101 µg/l. The reduced concentration was due to increased reductive dechlorination due to the availability of electron donor, and the ratio of PCE to TCE and DCE has also improved. The ratio of PCE to TCE was 4:1, and the ratio of DCE to PCE was 1.4:1. The most significant evidence for the increased rate of reductive dechlorination was the ethene data. Baseline concentrations in MW-1 for ethene were 12 nanograms per liter (ng/l). Ethene was detected at 408 ng/l during the April 1999 sampling event. Anaerobic and reducing conditions were present in MW-3 and ethene con-centrations were also higher than background conditions (92 ng/l).

Subsequent monitoring events performed between May 1999 and February 2000 have indicated that the anaerobic and reducing conditions present in IW-1, MW-1, and MW-3 have been maintained, and TOC concentrations in MW-1 have continued to increase. PCE and associated degradation products have continued to decline in MW-1. These declines are due to reductive dechlorination since significant increases in the concentration of ethene have been observed throughout this same time period.

Increased TOC concentrations have not been observed in MW-3. This appears due to the increased rates of microbial activity up-gradient of these wells. However, PCE concentrations in MW-3 continue to decline after a slight biological surfactant effect was observed in May 1999, and ethene concentrations continue to be more elevated than background conditions. PCE, cis-1,2-DCE, and ethene concentrations for MW-1 and MW-3 are shown on Figures 30 and 31, respectively.

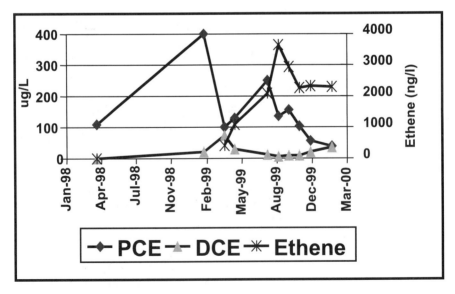

Figure 30   Case Study 3: MW-1 data.

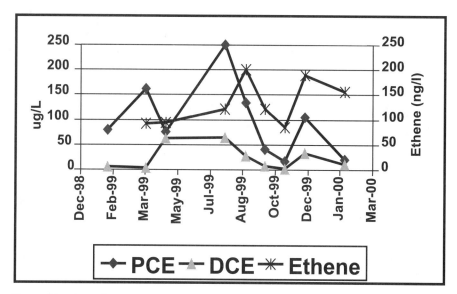

Figure 31 Case Study 3: MW-3 data.

Bulk Attenuation Rates

The rate by which a dissolved constituent attenuates at a particular site can be estimated through first order kinetics. It is important to note that the calculation of attenuation rates are only approximations of the complex processes that are occurring in nature. It should further be noted that the attenuation rates presented in this section consist of the effects of both destructive (biotic attenuation) and nondestructive (dilution) mechanisms. Based on the minimal volume of reagent delivered to the aquifer, and the increases in degradation daughter products that have been observed at the site, the lowering of PCE mass that has been observed is believed to be affected minimally by dilution.

PCE concentrations versus time from August 1999 to February 2000 are plotted on Figure 32. A trendline for the exponential regression of PCE is also presented. The equation describing the exponential regression is posted on the plot. This is the equation describing the exponential regression, where the first order attenuation constant (k) is –0.0099 and x represents time in days. The correlation coefficient (R^2) is also presented. Regressions with values of R^2 at and above 0.8 are generally considered to be useful.

Based on an attenuation constant (k) of –0.0099, the half-life for PCE is 70 days. This is significant due to the stable nature of the plume prior to the pilot test (little degradation) and in light of published half-lives for PCE in groundwater that range from 1 to 2 years (Howard et al. 1991).

The results of the pilot test demonstrate that the rate of reductive dechlorination in the bedrock aquifer was enhanced by the reagent injections associated with the reactive zone. The increased concentration of organic carbon quickly established the anaerobic and reducing conditions necessary for the complete degradation of PCE

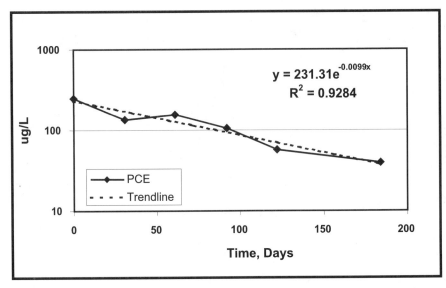

Figure 32   Case Study 3: reactive zone PCE degradation curve.

to ethene in the injection well. The anaerobic and reducing conditions migrated from the injection well to a well located approximately 240 feet down-gradient (MW-3). After initial desorption of PCE (presumably from a biological surfactant effect), PCE concentrations in the down-gradient wells have declined to concentrations significantly below the equilibrium conditions present prior to initiation of the pilot test. This is considered significant due to the elevated levels of ethene concurrently measured during the test.

The amount of time necessary to observe a significant reduction in overall source mass indicates some conditioning of the microbial population was necessary to enhance the natural degradation mechanisms. However, bioaugmentation was not necessary to provide the microbial populations to the bedrock groundwater environment, demonstrating that the microbial populations in bedrock at this site were ubiquitous.

A full-scale bioremediation treatment system was installed at the site during the first quarter of 2000. The reactive zone technology is also currently being evaluated in the source area (PCE concentration > 10,000 μg/l). The success of the bioremediation program has allowed for the discontinuation of groundwater extraction from some site wells, and will allow groundwater extraction to eventually be phased out for the site. This will result in a significantly more cost effective groundwater remedy for the treatment of chlorinated VOCs in bedrock groundwater at this site compared to traditional alternatives, limitations and lessons learned.

Limitations

As with any technology there are limitations to reactive zones. Although this approach represents one of the most cost effective ways of addressing impacts *in situ*, it still has limitations. The most important limitation is one suffered by all *in situ* technologies, the geologic conditions in which it must be applied. From the issue of variable lithologic units, to the difficulty in accurately defining the true variability in subsurface conditions, reactive zones are influenced by the nature of the geology in which it is implemented. For this reason, it is critical to define the geologic conditions as accurately as possible before applying a reactive zone. The most important issues were discussed in previous sections of this chapter, but three bear mentioning again.

1. Groundwater flow direction is so basic that sometimes its accurate definition can be overlooked, often to disastrous results. If the flow direction is not accurate, monitoring wells in a pilot study may be misplaced providing little or no data, or providing data that indicates that the process has failed, when in reality it has succeeded.

2. Groundwater flow velocity is critical to the selection and application of reagent feed rates, injection well locations, reagent selection, and monitoring frequency. Low groundwater velocity can lead to the development of uncontrolled fermentation, reduced pH, and the production of undesirable intermediates, such as alcohols, ketones, and mercaptans. High groundwater velocities can preclude the use of reducing reactive zones, by making the carbon demand so great to maintain the reactive zone that the injection rates and reagent requirements may no longer be cost effective.

3. The definition of the lithology and/or variability of permeability within the target zone impacts the well depths, screen locations, well design, and system capital costs. If the geology is not well defined, the reagent may be delivered to the wrong zone, or monitoring wells may be placed in zones where the reagent is not effectively being placed, leading to overfeeding of the reagent, based on poor data.

The variety of reagents available, the amount of testing conducted using any particular single reagent, and the regulatory hurdles associated with injecting foreign agents into groundwater are all issues that impact the applicability of this technique. These issues include public perception—a concern sometimes raised is the potential of creating a mutant super bug—which must be addressed anytime something is added to the environment.

Permitting requirements are another, often complicated, hurdle to clear. Most injection permits have been developed for waste disposal in deep injection wells, or for treated water being discharged to groundwater systems. These same permit programs are being applied to shallow injections of food-grade agents, many of which most people can find in their pantries at home. States have performed admirably in adjusting to the needs of reactive zones applications, by interpreting the existing permit programs to fit the needs of *in situ* reactive zones.

Coupling these issues with the typical issues of COC chemistry (degradability, solubility, volatility, DNAPL), groundwater chemistry (redox conditions, pH, buffering capacity) and microbiological factors (survivability, surfactant effects, nutrients, scavengers) makes reactive zones anything but a slam-dunk.

Listed below are several "lessons-learned rules" developed over the course of applying reducing reactive zones at more than three dozen sites. While they do not cover all the issues, they serve as a checklist of what to watch for:

Rule #1. Bacteria will utilize the easiest forms of carbon available to them; therefore, if they are successfully degrading BTEX and creating the necessary reducing conditions to co-metabolize CAHs, think twice before interrupting a good thing by giving the microbial population "candy" before they finish there "meal". By adding an easily degradable source of carbon (lactate, sucrose, etc.) the active degradation of COCs that are naturally serving as a primary substrate may be slowed. (Corollary to Rule #1: The best targets for reactive zones are plumes that have burned out their carbon source.)

Rule #2. Reactive zones should be used for VOCs when there are already indications of degradation. The remediation engineer should take cues from nature and enhance the natural processes whenever possible.

Rule #3. The microbial populations that are actively creating the reducing reactive zones are producing materials (surfactants) that can desorb mass from the soil matrix. This is necessary for the bacteria to metabolize the organics, but can result in some unwanted changes in plume character. In particular, COC concentration spikes should be anticipated and the disruption of plume equilibrium planned for. Surfactant effects can be detrimental without proper down-gradient control on the COC plume.

Rule #4. Plan for the byproducts of enhanced bacterial activity, namely the possibility of increases in methane, hydrogen sulfide, carbon dioxide, and COC vapors. Left unmonitored and uncontrolled these byproducts can lead to aesthetic nuisances (mercaptans, sulfide), explosive hazards (methane), and human health hazards (VC, hydrogen sulfide accumulation).

The treatment of DNAPL using reactive zones is a young and growing area of research and field application. The difficulty with using reducing reactive zones for DNAPL treatment is related to the need to make the target compound available to be impacted by the microbial population. In order to be degraded, the target COC must be dissolved in the groundwater. Thus it is necessary for the surfactants produced by the bacteria to dissolve the free-phase first. This is not a problem when the total mass of the DNAPL is small. This would be the case for DNAPL found in the form of residual DNAPL. However, when large pools of DNAPL are present, the process will be limited by the rate that the bacterial surfactants can interact with the DNAPL. Two problems can occur under these conditions. First, the CAH released from the DNAPL by the surfactants can cause a spike in the CAH concentration. Even though the process is removing mass from the site, the data shows an increase in concentration. This can be hard to explain to some regulators. Second, the process can take a long time. Due to the large mass of CAHs present in the form of the DNAPL, the reduction process will have to work a long time to eliminate all of the CAHs. Since we can only measure CAH concentration in the groundwater, progress will not be measurable. Once again, this can be hard to explain to a regulator, not to mention the client that provided the money for the IRZ effort.

CLOSING

The fact remains that despite the hurdles listed above, reactive zones are rapidly becoming commonplace and, in many instances, replacing conventional remedial approaches. These applications are having the result of shortening the total time that it takes to complete a remediation project. Reactive zones are being applied in more varied environments—bedrock geology, high and low permeability unconsolidated geologic conditions—to treat a wider spectrum of compounds—CAHs, HOPs, PAHs, PCBs, perchlorates—and in multiple medias—groundwater, soil and sediments. Additional enhancements are also being investigated, such as the use of biodegradable surfactants, operating reactive zones in series, and the addition of reducing reagents in combination with other reductive processes such as zero-valent iron.

As stated in the beginning of this chapter, the elegance of reactive zones is embodied in the application of the technique to enhance natural conditions in order to speed up natural processes, thus taking advantage of nature's natural cleansing processes. The more the remedial engineer can enhance ongoing processes, the less the effort and time that will be needed to achieve the remedial goals.

REFERENCES

ARCADIS Geraghty & Miller, "Enhanced Bioremediation Pilot Test Work Plan," Fresno, California, July 1999.

Barcelona, M.J., et al., "Reproducible Well-Purging Procedures and VOC Stabilization Criteria for Groundwater Sampling," Groundwater, 32: 1, pp. 12-22, 1994.

Burdick, J., Jacobs, D., Field Scale Applications to Demonstrate Enhanced Transformations of Chlorinated Aliphatic Hydrocarbons, Northeast Focus Ground Water Conference, October 1998.

Cirpka, O.A., Windfuhr, C., Bisch, G., et al., "Microbial Reductive Dechlorination in Large Scale Sandbox Model," *Journal of Environmental Engineering*, p. 861–870, September 1999.

Di Stefano, T.D., PCE Dechlorination with Complex Electron Donors, 2nd International Conference on Remediation of Chlorinated and Recalcitrant Compounds, May 2000.

Fruchter, J.S., Cole, C.R., Williams, M.D., Vermeul, J.E., Szecsody, J.E., Istok, J.D., Humphrey, M.D., *In Situ* Redox Manipulation Update, Abiotic *In Situ* Technologies for Groundwater Remediation, August–September, 1999.

Geo-Cleanse International, Inc., Product Literature, 1998.

Guest, P.R., Benson, L.A., Rainsberger, T.J. "Inferring Biodegradation Processes for Trichloroethene from Geochemical Data," *Intrinsic Bioremediation*, pp. 233-243, 1995.

Howard, P., Boethling, R., et al., *Handbook of Environmental Degradation Rates,* Lewis Publishers, 1991.

In Situ Oxidative Technologies, Inc., Product Literature, 1999.

Jacobs, D.L., et al., Field Demonstration – Enhanced Reductive Dechlorination of Pentachlorophenol in Ground Water, 2nd International Conference on Remediation of Chlorinated and Recalcitrant Compounds, May 2000.

Jerome, K.M., Looney, B.B., et al., Field Demonstration of *In Situ* Fenton's Destruction of DNAPLs, 1st International Conference on Remediation of Chlorinated and Recalcitrant Compounds, May 1998.

Khan, Faruque and Puls, Robert, *In Situ* Treatment of Chromium Source Area Using Redox Manipulation, Abiotic *In Situ* Technologies for Groundwater Remediation, August–September, 1999.

Lenzo, F.C., Remedial Technologys Evolution – Science Meets Strategy in the Third Millenium, Strategic Environmental Management, I: 3, 1999.

Lenzo, F.C., Metals Precipitation Using Microbial *In Situ* Reactive Zones, Abiotic *In Situ* Technologies for Groundwater Remediation, August–September, 1999.

Loffler, F.E., Sanford, R.A., and Tiedje, J.M., H_2 Threshold Concentration and f_e Values as Indicators of Halorespiration, Poster Board – EMSP Workshop (downloaded via Internet), 1998.

McCarty, P.L. and Semprini, L., "Groundwater Treatment for Chlorinated Solvents," *Handbook of Bioremediation*, Lewis Publishers, Boca Raton, FL, 1994.

McCarty, P.L., Breathing with Chlorinated Solvents, Science, 276:1521, 1997.

McCarty, P.L., et al., "Full-Scale Evaluation of *In Situ* Cometabolic Degradation of Trichloroethene in Groundwater through Toluene Injection," *Environmental Science and Technology*, 31:786-791, 1998.

Mott-Smith, E., Leonard, W.C., et al., *In Situ* Oxidation of DNAPL Using Permanganate: IDC Cape Canaveral Demonstration, 2nd International Conference on Remediation of Chlorinated and Recalcitrant Compounds, May 2000.

Mueller, J., et al., *In Situ* Source Management Strategies at Wood Treatment Sites, 2nd International Conference on Remediation of Chlorinated and Recalcitrant Compounds, May 2000.

Nelson, M.D., Parker, B.L., et al., Passive Destruction of PCE DNAPL by Potassium Permanganate in a Sandy Aquifer, 2nd International Conference on Remediation of Chlorinated and Recalcitrant Compounds, May 2000.

Norris, Hinchee, et al., *Handbook of Bioremediation*. CRC Press, Boca Raton, FL, 1994.

Nyer, E., Lenzo, F.L., Burdick, J., "*In Situ* Reactive Zones: Dehalogenation of Chlorinated Hydrocarbons," *Groundwater Monitoring Review*, Spring 1998.

Schnarr, M., Truax, C., et al., "Laboratory and Controlled Field Experiments Using Potassium Permanganate to Remediate Trichloroethylene and Perchloroethylene DNAPLs in Porous Media," *Journal of Contaminant Hydrology* 29, pp. 205-224, 1998.

Smatlak, C.R., et al., Comparative Kinetics of Hydrogen Utilization for Reductive Dechlorination of Tetrachloroethene and Methanogenesis in an Anaerobic Enrichment Culture, ES&T, 30:2850-2858, 1996.

Suthersan, Suthan. S., *Remediation Engineering Design Concepts*, CRC Press, Boca Raton, FL, 1997.

Tarr, M.A., Lindsey, M.E., et al., Fenton Oxidation – Bringing Pollutants and Hydroxyl Radicals Together, 2nd International Conference on Remediation of Chlorinated and Recalcitrant Compounds, May 2000.

Unger, M., Ph.D., Stroo, H., Gormley, J., Hill, S., Arguello, R., Assessment of *In Situ* Oxidation Technology, Partners in Environmental Technology, Technical Symposium & Workshop, Hyatt Regency Crystal City, Arlington, Virginia, November-December, 1999.

Using Organic Substrates to Promote Biological Reductive Dechlorination of CAHs, HRC Technical Bulletin #1.1.3, 1999.

Walker, William and Pucik-Erickson, Kara, *In Situ* Chromium Reduction: A Geochemical Evaluation. Abiotic *In Situ* Technologies for Groundwater Remediation, August 31–September 2, 1999.

Weidemier, T.H., Swanson, M.A., Moutoux, D.E., Wilson, J.T., Kampbell, D.H., Hansen, J.E., Haas, P., Overview of the Technical Protocol for Natural Attenuation of Chlorinated Aliphatic Hydrocarbons in Ground Water Under Development for the U.S. Air Force Center for Environmental Excellence, Symposium on Natural Attenuation of Chlorinated Organics in Ground Water, September 1996.

Whang, J.M., Adu-Wusu, K., Frampton, W.H., Staib, J.G., *In Situ* Precipitation and Sorption of Arsenic from Groundwater: Laboratory and Ex Situ Field Tests. 1997 International Containment Technical Conference, February 1997.

Wilson, J.T., Rifai, H.S., Borden, R.C., Ward, C.H., Intrinsic Bioattenuation for Subsurface Restoration, 1995.

Yager, R.M. et al., Metabolic and *In Situ* Attenuation of Chlorinated Ethenes by Naturally Occurring Microorganisms in a Fractured Dolomite near Niagara Falls, New York, ES&T 31:3138-3147, 1997.

Zienkiewicz, Andre W., "Removal of Iron and Manganese from Ground Water with the Vyredox$_{TM}$ Method," Second International Conference on Ground Water Quality Research, March 28, 1984.

CHAPTER 9

Phytoremediation

Eric P. Carman and Tom L. Crossman

CONTENTS

1-56670-528-2/01/$0.00+$.50

INTRODUCTION

Phytoremediation is a diverse and emerging technology that uses green plants to cleanup contaminated environmental media. As phytoremediation has been increasingly recognized, the technology has been applied both *in situ* and *ex situ* to contaminated soil, sediment, sludge, groundwater, surface water, and wastewater. In addition, the natural evapotranspiration process of vegetation has been recognized and harnessed as an alternative cover method to reduce landfill infiltration.

Although it is now being increasingly applied for environmental mediation, phytoremediation is not a new technology. The Roman civilization reportedly used eucalyptus trees to dewater saturated soils more than two thousand years ago. The excess water use by some plants, namely phreatophytic (waterloving) trees, has been long recognized as a nuisance in the agricultural industry, particularly in more arid regions. Water levels next to cottonwood and willow trees (two common phreato-phytes) in the southwestern United States are known to drop several feet during growing seasons. The principles of phytoremediation which are currently gaining acceptance for contaminant remediation have been reported in the scientific literature only since the late 1970s or early 1980s. The research, development, and application of this technology increased dramatically in the late 1980s and early 1990s because it is low cost and versatile, and in some cases has better public support as a method to cleanup contaminated media. Phytoremediation was first implemented and reported as an environmental cleanup technology for agricultural contaminants such as excess plant nutrients (nitrate, ammonia, and phosphate) and pesticides (Briggs, Bromilow, and Bromilow 1982), although the principles of phytoremediation have been applied in the wastewater industry for many years. USEPA has recently estimated that there are currently more than 100 sites around the world where phytoremediation is being implemented as a remedial technology.

Phytoremediation Applications

The USEPA has identified six broad applications of phytoremediation (USEPA 1998). These applications and their definitions include:

- Phytoextraction/Phytovolatilization—The uptake and translocation of organics and inorganics from the soil into the roots and above-ground portions of the plant. Organics that are extracted are subsequently either degraded within or volatilized from the plant tissue. Inorganics that are extracted accumulate and/or are methylated and volatilized from the plant tissue.
- Phytostabilization—The use of plants to immobilize organic and inorganic constituents in the soil and groundwater through adsorption and accumulation by roots, adsorption onto roots, or precipitation within the rhizosphere. Phytostabilization also includes site revegetation which reduces windblown dust and direct contact with contaminants.
- Enhanced Rhizosphere Degradation—The breakdown of organic constituents in the soil through microbial activity that is enhanced by processes within the rhizosphere.
- Rhizofiltration—The absorption, adsorption, or precipitation of contaminants that are in solution surrounding the roots.
- Hydraulic Containment—The use of plants, especially phreatophytes, to control the migration and flow of porewater, shallow groundwater, and contaminants dissolved in the groundwater.
- Alternative Covers (Phyto-Covers)—The use of vegetation as a long-term, self-containing cap growing in and/or over waste in a landfill.

Phytoremediation can be an effective technology to address both organic and inorganic constituents. Plants remediate organic compounds through several mechanisms. Organics can be taken up directly from the rhizosphere (defined as a zone of increased microbial activity at the root-soil interface that is under the influence of the plant root) and either metabolized by the plant, accumulated in the plant tissue, or transpired through the leaves (Schnoor et al. 1995 and Newman et al. 1999). These mechanisms are vital in the applications of phytoextraction, phytostabilization, and enhanced rhizosphere degradation. Figure 1 represents mass flow through a woody plant species.

Water and nutrients are taken up by the plant and carbon dioxide, oxygen, water, and photosynthates are released to the environment. In the case of phreatophytes, such as trees from the willow and poplar genus (*Salix* and *Populus*), the volume of water taken up by a single tree can be from several gallons to several thousand gallons of available water per day.

The processes occurring within the rhizosphere are integral to phytoremediation. Plants supply oxygen to the soil and release exudates, which include sugars, alcohols, amino acids, and enzymes. The exudates and enzymes enhance microbial growth and the growth of mycorrhizal fungi. The overall effect of the plant-microbe growth is an increase in microbial biomass by up to an order of magnitude or more, compared with microbial populations in the bulk soil. The microbes and mycorrhizal fungi

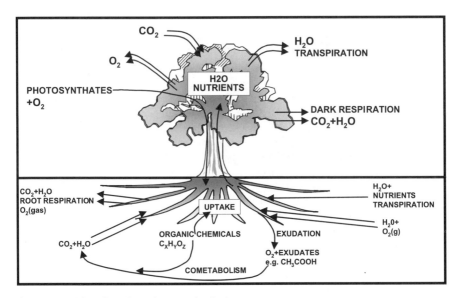

Figure 1   Mass flow through a woody plant.

subsequently promote degradation and co-metabolism of organics (Schnoor et al. 1995).

Organics are also taken up directly by plants and either accumulated, metabolized, or transpired through the leaf tissue. The fate of organics and inorganics in the rhizosphere, and the corresponding tendency of these constituents to be taken up by plants, can be predicted using the logarithm of the octanol-water partition coefficient (Kow) of the particular constituent. This relationship was reported by Briggs et al. (1982) and is commonly known as Brigg's Law. Table 1 illustrates fate of organics using Briggs Law.

Table 1   Organic Fate Predictions Using Briggs' Law

log K_{ow}	Mechanisms
<1.0	Possible Uptake & Transformation
1.0 - 3.5	Uptake, Transformation, Volatilization
>3.5	Rhizosphere bioremediation or Phytostabilization

* There are exceptions (1,4-dioxane) due to Brigg's emphasis on agricultural organics (pesticides, herbicides) not on soil, groundwater contaminants (BTEX, TCE, etc.) encountered in environmental remediation

Direct uptake of organics is an efficient process to remove moderately hydrophobic constituents, with a log Kow ranging from 0.5 to 3. Organics within this range include many of the volatile organic compounds (VOCs) including benzene, toluene, ethylbenzene, xylene, chlorinated solvents (such as trichloroethylene [TCE]), and aliphatics. Generally, constituents with log Kow less than 0.5 are too water soluble to be taken up into roots, and constituents with a log Kow greater than 3.0 are bound too tightly to the soil particles or roots to be taken up into the plant.

Examples of organic compounds with a log Kow less than 0.5 include methyl tertiary butyl ether (MTBE), and 1,4-dioxane. Constituents with a log Kow greater than 3.0 include most polycylclic aromatic hydrocarbons (PAHs).

It is also possible to predict the concentration of a contaminant that will absorb into the roots using the Root Concentration Factor (RCF). If the Kow is known, the RCF can be used to predict the ratio of the concentration in the roots, to the concentration in the external solution (Figure 2).

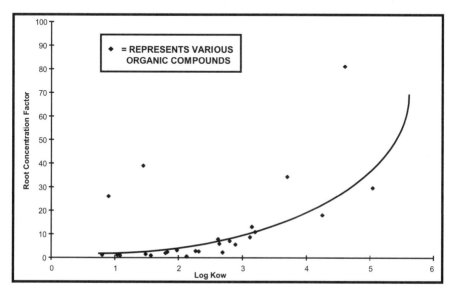

Figure 2   Root concentration factor (RCF) = (concentration in roots/concentration in external solution).

However, Briggs Law is only generalized, and as research in the field of phytoremediation increases, more constituents are likely to be found susceptible to treatment. Recent hydroponic studies at the University of Iowa suggest that 1,4-dioxane, a commonly detected solvent and suspected carcinogen with a log Kow of 0.27, is taken up and volatilized by the hybrid poplar *Populus deltoides x nigra*, DN34 (Kelley et al. 1999). In addition, recent laboratory tests and research hold promise that MTBE may also be susceptible to phytoremediation (Newman et al. 1998).

Metals have posed a considerable challenge to remediation by conventional technologies, which are generally expensive ex-situ processes that involve removal and transportation to cleanup soil. Exposure pathways from sites that are contaminated with metals include direct contact with the waste materials or soil/sediment contaminated by the metals, inhalation of windblown dust or particulate matter, and exposure to groundwater or surface water that has leached the metals. Remediation of metals-contaminated sites can include three possible changes in the chemical characteristics of the metal or the medium in which the metal is present. The concentration of the metal can be reduced by direct removal, the hazardous nature of the metal can be reduced without removing any of the metal (for example through an *in situ* method like solidification or vitirification), or the metal bioavailability can be reduced.

The applications of phytoremediation that directly address metals and other inorganics include phytoextraction, phytostabilization, and rhizofiltration. Phytosta-bilization involves covering a site with vegetation, thereby reducing erosion, enhancing soil nutrients, and eliminating direct contact and transport off-site of metals containing media such as wind and water. The effectiveness of phytostablization in limiting direct contact, and transport by wind and water of metals, was demonstrated at two Superfund sites in the Midwest (Pierzynski et al. 1994).

Phytostabilization can also refer to the use of rhizosphere processes to tightly bind metals to soil within the rhizosphere, or to the root tissue itself. Exudates released in the rhizosphere can increase the soil pH up to 1.5 pH units and increase soil oxygen content, having a significant effect on the redox conditions of the soil and promoting oxidation, reducing mobility and bioavailability of metals (Azadpour and Matthews 1996).

Phytoextraction is the use of vegetation to uptake and accumulate inorganics into plant tissue, both from the soil and from metals dissolved in pore water or shallow groundwater. Plants that accumulate high concentrations of metals are known as hyperaccumulators. Certain plant tissue and tree sap may contain up to 3 percent zinc and 25 percent nickel by dry weight, without apparent harm to the plant. Certain metals, including selenium and mercury can also be taken up, methylated, and volatilized (Meagher et al. 1998 and Banuelos et al. 1998)

Types of Vegetation Currently Used in Phytoremediation

As the technology of phytoremediation expands, the types of plants identified for applications of phytoremediation for organic and inorganic compounds has expanded. Early efforts were focused on utilization of hybrid poplar trees, fast growing phreatophytic trees which have a well-documented physiology and genetic characteristics from their use in the pulp and paper industry and fuel from biomass research. Currently numerous types of vegetation, including trees and grasses, have been applied in phytoremediation to address VOCs, PAHs, radionuclides, pesticides, and herbicides. In addition, geobotanical exploration has revealed many more metal hyperaccumulators than were previously identified. Approximately 400 plant taxa are now known for Cd, Co, Cu, Pb, Ni, Se, and Zn hyperaccumulation (Flathman and Lanza 1998).

Benefits and Limitations to Phytoremediation

Phytoremediation is becoming recognized as a cost effective remedial method to address contaminated sites and landfills. Advantages to phytoremediation are its low capital cost, generally about one third to one fifth of the cost of more conventional technologies. In addition, this technology tends to have low costs for ongoing operation and maintenance (O&M), although it should not be construed as maintenance free. The combination of effectiveness, low cost, and low O&M make phytoremediation attractive for non-point source contamination, such as nitrates and pesticides in agricultural settings and parking lot runoff in urban areas. Phytoremediation also minimizes wind and water erosion and minimizes the production of undesirable waste by-products. Some plant species can also reduce the net infiltration of surface water, which minimizes the potential for leaching of contaminants into groundwater. Phreatophytes can take up large volumes of available water, and can be used to capture shallow groundwater (less than about 20 feet below land surface), in a manner analogous to conventional pump and treat systems. In addition, the technology can greatly improve soil conditions by increasing soil organic carbon, enhancing microbial and fungal populations, and humifying metals and recalcitrant organics by complexing the metals with soil organics (Schwab and Banks 1994).

Phytoremediation has been accepted by the public, since it is environmentally compatible and can improve the long-term aesthetics of a site. Phytoremediation can be used as a single treatment technology, or it can be coupled with more aggressive conventional technologies. For example, contaminated soils from a site can be excavated and treated in engineered phytoremediation treatment units (EPTUs), rather than thermally treated or taken off-site and disposed of in a landfill. Contaminated groundwater can also be pumped from a site using conventional methods, and then used to irrigate trees or grasses, rather than treated using conventional technologies (e.g., air stripping or bioreactor). At a landfill in Oregon, the City of Beaverton uses effluent from the publicly owned treatment works (POTW) as irrigation water for hybrid poplars which have been planted as an alternative cover to their city landfill. Furthermore, these trees are periodically harvested and sold to a nearby paper mill for a net profit for the landfill (Madison, Licht, and Ricks 1991). Phytoremediation can also be integrated with landscape design practices, so that the remediation system is an attractive addition to the property.

Despite the benefits of phytoremediation, there are disadvantages to the technology that make it unsuitable or undesirable for some environmental applications. Phytoremediation is a long-term remedial technology at most sites, with treatment times on the order of several years. In addition, the technology can be directly implemented only where the contaminants are present at depths within about 20 feet of the land surface. If vegetation is used for the purpose of extracting groundwater, the contaminants must be located within a few feet of the water table surface. Plants have adapted to grow in some of the most inhospitable conditions known to exist. However, phytoremediation will not be successful if soil conditions or contaminant characteristics/concentrations prove to be phytotoxic. In addition, some types of vegetation, while suitable for phytoremediation, may not be desirable or acceptable in certain applications.

Phytoremediation of metals poses special considerations that can make its use impracticable at the current time. For example, the consequences of transferring contamination from soil or groundwater into plants that can enter the food chain must be considered, particularly for heavy metals such as lead and cadmium because of their known human health aspects (Mench et al. 1994). Research has focussed on improving the efficiency by which plants can uptake metals by introducing synthetic complexing agents. Although the addition of the agents can enhance root uptake, the complexing agents can also increase downward mobility of the metals away from the rhizosphere, such that contamination spreads and poses a new threat to groundwater. As the technology matures, some of these limitations may be overcome, and other limitations will undoubtedly be identified.

There is considerable overlap between several of the phytoremediation applications, and in many cases the distinction between the applications becomes blurred. Furthermore, most phytoremediation projects will generally harness more than one of these six broad applications to achieve site remediation. A more detailed description of each of the six USEPA-described applications of phytoremediation and several case histories are presented below.

PHYTOEXTRACTION/PHYTOVOLATILIZATION

Phytoextraction refers to the uptake and translocation of contaminants into the roots and above-ground portions of plants. Phytovolatilization refers to the gaseous discharge of methylated inorganic or organic compounds from the plant tissue. Phytoextraction/phytovolatilization in particular is an example of how the distinction between phytoremediation applications can overlap. For example, phytoextraction has been used to describe the uptake and translocation and accumulation of inorganic compounds by plants, specifically metals or radionuclides. However, some organic compounds (e.g., TCE) can also be extracted from the subsurface, and subsequently degraded within or volatilized from the plant tissues. For the purpose of this chapter, we have grouped phytoextraction/phytovolatilization of inorganics and organics together, and we will present two case histories that highlight applications for both types of compounds.

Although both inorganics and organics can be extracted by plants, the fate of the compounds once extracted by the plant are very different. Inorganics, such as metals, tend to accumulate in the roots and shoots and phytoextraction of metals capitalizes on the tendency of some metals to relocate from soil or water to plant tissue. When in plants, the metals can be more cost effectively disposed of than in soil, sediment, or groundwater. Inorganics can also be methylated and volatilized from leaf tissue. Meagher et al. (1998) have shown that engineered plant species containing bacterial genes allowed plants to convert root extracted ionic mercury and methyl mercury to metallic mercury. The metallic mercury is then volatilized from the plants at rates which are below those that would cause airborne mercury hazard.

The relative tendency of plants to uptake, immobilize, or exclude metals is highly contaminant specific and soil specific. Soil factors that influence the tendencies include:

- Soil pH—increases in soil pH generally reduce the solubility of metals and the uptake of plants
- Cation exchange capacity (CEC)—increases in CEC of soil reduces plant uptake
- Organic matter—inorganic forms of metals are generally taken up more readily than organic forms

- Natural and synthetic complexing agents—the presence of complexing agents such as ethylene-diaminetetra-acetate (EDTA) and diethylene-triaminepenta-acetic acid (DPTA) generally increases the solubility of metals, making them more available to roots and more likely to be taken up and accumulated in a plant

Plants that grow in environments with high concentrations of metals can either adapt to accumulate the metals, or exclude, or avoid the metals. Hyperaccumulators avoid the toxic effects of metals, such as clorosis, necrosis, disruption of chlorophyll synthesis, alteration in water balance, and stunted growth by binding the metals to cell walls, pumping metal ions into vacuoles, or complexing heavy metals by organic acids (Azadpour and Matthews 1996, and Pierzynski et al. 1994). Excluder plant species may absorb heavy metals, but restrict their transport to the shoots of the plants. This type of heavy metal tolerance does not prevent uptake of heavy metals, but restricts translocation, and detoxification of the metals takes place in the roots. Mechanisms for excluder detoxification include immobilization of heavy metals on cell walls, exudation of chelate ligands, or formation of a redox or pH barrier at the plasma membrane (Taylor 1987).

Organics, once extracted by a plant, tend to be broken down and metabolized, or volatilized from the leaf tissue. Whether or not organics are extracted by the plant is generally dictated by Brigg's Law, which was discussed previously in this chapter. Regardless, the reader needs to be aware that most phytoremediation sites incorporate more than one of the six applications. Even if the main design application for the plants is something other than extraction/volatilization, these processes may also be occurring during the remediation.

Phytoextraction Case Histories

Phytoextraction and Accumulation of Lead, Magic Marker Site, Trenton, New Jersey

This Brownfield site located in Trenton, New Jersey has been the focus of a Superfund innovative technology evaluation (SITE) demonstration project that addresses lead contaminated surface soils in a residential/commercial part of the city. Contamination of the Magic Marker site resulted from various manufacturing processes, including lead-acid battery production between 1947 and 1987.

The site soils consist of gravelly sand and miscellaneous debris, and site investigations identified lead in the upper 0.61 meters (2 feet) of soils that exceed the regulatory limit of 400 mg/kg. Lead contamination, ranging from 200 to 1,800 mg/kg, exhibited considerable variation across the site. The demonstration project evaluated a total of three crops grown in a 9.1 x 17.4 meter (30 x 57 foot) plot and compared the results to a 9.1 x 12.2 meter (30 x 40 foot) control plot. Two crops of *Brassica junacea* (Indian Mustard) plants were grown for a 6 week period and harvested over the spring and summer of 1997. One crop of sunflower plants was grown in the summer of 1998. Harvested plant tissue samples were collected to evaluate the amount of lead uptake in each crop, and soil samples were collected to evaluate the change in lead concentrations in the root zone. EDTA and other amendments were added to the soil between the crops to solubilize the metals and facilitate uptake and absorption by plants, resulting in increased efficiency of the phytoextraction and accumulation process. Plants that were grown on the site were dried and removed from the site.

The distribution of soil lead concentrations before and after phytoextraction was applied is presented in Figure 3. After the phytoextraction program, the treated area with soil concentrations of lead below the 400 mg/kg cleanup criteria increased to 57 percent of the plot area from 31 percent of the plot area.

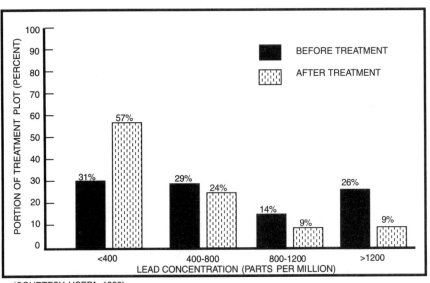

(COURTESY, USEPA, 1999)

Figure 3   Distribution of lead concentrations in the top 6 inches of soil, Magic Marker site.

The average lead concentrations accumulated in the above-ground plant tissue samples from the two *Brassica* crops were 830 mg/kg and 2,300 mg/kg, respectively. The increase in the concentrations of lead in the above-ground portion of the plants were attributed to the soil amendments (USEPA 1999).

USEPA estimates that phytoextraction for soil covering a 10 acre site typically requires six to eight crops over three growing seasons. Harvesting these crops is expected to produce an estimated 500 tons of biomass from the upper 0.3 meter (1 foot) of top soil. This represents a substantially lower mass (0.25 percent) of the 20,000 tons of contaminated soil that would otherwise require excavation and disposal (USEPA 1999).

Phytoextraction and Degradation of TCE, Controlled Field Study, Washington

A significant amount of research on phytoremediation of organics, specifically for TCE and other VOCs, has been completed at the University of Washington. Although TCE has been previously shown to be reduced using phytoremediation, the mechanisms and fate of the TCE are still not known with certainty. Between 1995 and 1997, a controlled field study using TCE was performed at a site outside of Fife, Washington. The objective of that study was to determine the fate of TCE after the interaction of the contaminant with a poplar clone, H11-11 (*Populous trichocarpa x P. deltoides*). H11-11 was chosen based on its demonstrated ability to take up and degrade TCE within the laboratory and greenhouse (Newman et al. 1999).

A series of artificial aquifers, or cells, was constructed at the Washington site using double-walled 60 mil polyethylene liners for the study. The approximate dimensions of each cell were 1.5 meters deep x 3.0 meters wide x 5.7 meters long (4.9 x 9.8 x 18.7 feet). The cells contained coarse sand, overlain with silty clay loam topsoil native to the site. Each cell had an influent well on one end with a distribution pipe to allow the addition of controlled amounts of either water or water containing TCE to the sand layer. The bottoms of each cell were sloped to the effluent ends where extraction wells were installed (Newman et al. 1999). Rooted cuttings of the poplar clones were planted during May 1995. Prior to planting, the roots and tops of the cuttings were pruned to a length of 45 centimeters (18 inches). The trees were planted in each cell with a spacing of 1 meter (3.3 ft).

Cells were dosed during the initial year with a TCE and water mixture with an average concentration of 0.038 millimolar (mM) (4,993 micrograms per liter [ug/l]), and during the second year the cells were dosed with an influent concentration of 0.11 mM TCE (14,453 ug/l). The goal of the water management program was to maintain water levels in the bottom of the cells within a range of 15 to 25 cm (5.9 to 9.8 in). Water was either introduced through surface irrigation or removed from the extraction well in each cell to maintain the target water level. However, after the first year of the field test, water had to be added to the cells through surface irrigation, since the transpiration rates of the poplars exceeded the dosing rates.

Transpiration of TCE from the leaves was determined by two methods. These included a leaf bag technique and open path Fourier transform infrared (OP-FTIR) spectroscopy. The OP-FTIR measures ambient TCE concentrations in the vicinity of the tree. Degradation of TCE in the rhizosphere soil was also determined in the second year (1996) and chloride testing of soils for mass balance data was performed near the end of the test in 1997. Plant and leave tissue samples were collected during each growing season and were analyzed for TCE and the products of the reductive dechlorination of TCE (TCE/R).

The trees in the artificial aquifer grew rapidly, attaining a mean height of 3 m (10 ft) by the end of the first season, 7 m (23 ft) by the end of the second season and 11 m (36 ft) by the end of the third season. Growth of the trees was not significantly affected by the TCE in the dosed concentrations.

TCE/R were detected in the effluent water of planted cells nine weeks following the initial dosing. Over the three year test, the mass of TCE/R recovered in the effluent from an unplanted control cell was 67 percent of the mass of TCE added. In contrast, the mass of the TCE/R from the planted cells was only 1 to 2 percent of the TCE that was dosed. The recovery of TCE/R in the effluent water from the planted cells was low when the transpiration rates of water were high (during the growing seasons), and higher when the transpiration was low (beginning and end of the growing seasons). The apparent relationship between transpiration and low recovery of TCE/R suggests that the majority of TCE loss from the artificial aquifer was associated with plant uptake of water (Newman et al. 1999).

TCE and TCE metabolites were also detected in the leaves, branches, and roots of the poplars. TCE was the major chlorinated compound that was detected in the branches or trunks of the trees. The higher proportion of TCE in the branches and trunks was attributed to these parts of the tree being less metabolically active than the leaves or roots, which had higher proportions of metabolites (Newman et al. 1999). The total mass of TCE that was transpired was estimated for the three years. The total TCE lost through transpiration was estimated to be only 9 percent of the TCE lost from the cells during 1996. No transpired TCE was detected in tests conducted in 1997. The results from the rhizosphere study conducted in 1996 did not indicate the presence of rhizosphere degradation of TCE in soil.

Chloride concentrations in soil from cells that were planted with the hybrid poplars and dosed with TCE contained higher concentrations of chloride than soil from cells that were not exposed to TCE. Newman et al. (1999) attributed the higher chloride results to TCE being taken up in plants, dechlorinated in the plant tissue, and subsequently excreted by the roots.

Overall, the trees were able to remove more than 99 percent of the TCE added through dosing in the three year study. Less than 9 percent of the TCE was transpired to the atmosphere during the second and third year and examination of the tissue showed low levels of TCE metabolites. Chloride accumulated in soil in amounts that generally correspond to TCE losses, demonstrating the TCE was broken down through metabolism in the plant tissue, rather than degraded in the rhizosphere soil or volatilized through the leaves.

The mass balance for chloride in soil in one of the TCE exposed cells from 1995 to 1997 is presented in Table 2. Included in the table is a summary of the mass of chloride in the TCE which was lost in the cell (i.e., influent TCE chlorine minus effluent TCE chlorine), the estimated TCE chloride lost to due to transpiration, the estimated chloride lost to oxidative metabolites in the plant, and the chloride present in the soil of the TCE exposed cell, compared to the control cell. The total recovered chloride was 70 percent of the lost TCE chlorine (Table 2).

PHYTOSTABILIZATION

The application of phytoremediation via phytostabilization refers to two different processes. One aspect of phytostabilization is the use of plants to immobilize contaminants in soil and water. The stabilization of contaminants can be achieved through a number of plant and soil processes including absorption and accumulation by roots, adsorption onto roots, and precipitation within the rhizosphere.

According to Cunningham et al. (1995), plant processes that aid in phytostablization include:

- Transport of ions across root-cell membranes
- Water flux to the plant driven by plant transpiration

Table 2   Mass Balance for Chlorine in Cells with Hybrid Poplars, Washington Field Study

	mol of chlorine or chloride ion			3-year total loss	3-year total recovered
	1995	**1996**	**1997[b]**		
TCE-chlorine lost from water in system[a]	2.75	13.4	11.32	27.47	
TCE-chlorine recovered from transpiration[d]	0.28[e]	0.87[e]	0[f]		1.15
TCE-chlorine recovered from oxidative metabolites in plant tissue[d]					
leaves	0.03×10^{-3}	0.006	0.002		0.008
branches	0.03×10^{-3}	0.005	0.002		0.007
trunk	0.05×10^{-3}	0.01	0.003		0.013
roots[g]	0.03×10^{-3}	0.006	0.002		0.008
excess chloride ion in soil[b]	ND	ND	18		18
chloride balance				27.5	19.2
recovery efficiency					70%

[a] Chlorine added in the form of TCE-chlorine was balanced against the amount of TCE-chlorine, metabolite-chlorine and free chloride ion recovered from the system. Masses given cover the three years that the experiment ran. ND, not determined.

[b] Prior to 15 August 1997.

[c] Corrected for presence of reductive dechlorination products.

[d] Leaf areas and mass of tree tissues per cell were adjusted by the ratio of tree heights at the end of the respective growing season to that measured at the end of 1996.

[e] Calculated using average of leaf bag assays in 1996, 1.5×10^{-11} mol h^{-1} cm^{-2} leaf.

[f] No TCE was recovered from the leaf bag assay in 1997.

[g] Mass of roots was calculated as 0.40 times the mass of the above-ground woody parts of the plants.

Source: Reprinted with permission from Newman et al., 1999.

- Absorption of organics into the roots
- Entrapment of organics in the plant lignin (lignification)

Soil processes that aid in phytostabilization include:

- Biochemical fixation (humification)
- Chemical fixation (precipitation)
- Physical fixation (solid-state diffusion into soil structures and formation of oxide coatings)

Phytostabilization of inorganics in soil can be achieved by the addition of soil amendments that reduce contaminant solubility. For example, lead solubility in soil was reduced by adding alkalizing agents, phosphates, mineral oxides, organic matter, and biosolids, making it unavailable to leaching, mammalian ingestion, and plant uptake (Cunningham et al. 1995). Data show that stabilization of the soil with amendments also reduces plant shoot uptake of lead by 90 percent and increases the general tillability of the soil (Cunningham et al. 1995).

A second aspect of phytostabilization refers to the physical process of establishing or re-establishing a vegetative cover on sites that have lacking a natural vegetation. It is a technique that borrows practices commonly used in the field of mining reclamation where the lack of vegetation can be due to high concentrations of contaminants or from physical disturbances that have taken place. Risk assessments have shown that windblown contaminants actually constitute the greatest threat to human health at many sites, including mine tailings sites (Pierzynski et al. 1994). Phytostabilization reduces the mobility of the contaminant, prevents migration into surrounding media (air, groundwater, surface water, and sediments), and reduces bioavailability. Phytostabilization through revegetation can also enhance the *in situ* humification of both organic and inorganic contaminants.

The type of vegetation selected for phytostabilization of a site will be dependent on the nature of the site. For example, metal-tolerant species can be used to restore vegetation on sites with high metals concentrations, such as abandoned smelters or mine sites. These plants will perform their function by decreasing the potential migration of contamination through wind and decreasing leaching into groundwater and surface water.

Phytostabilization Case History

Whitewood Creek Site, South Dakota

An 18 mile stretch of Whitewood Creek was contaminated with arsenic and cadmium from 130 years of gold mining near Whitewood, South Dakota. Tailings from the gold mine contained an average concentration of 1250 mg/kg total arsenic and 9.4 mg/kg of total cadmium. The pH of the tailings ranged from 3.9 to 5.4 (Pierzynski et al. 1994).

An experimental plot of 3,100 hybrid poplar trees was planted along Whitewood Creek in 1991. The goal of the experimental plot was to vegetate the mine tailings, thereby reducing wind-blown dust and decreasing the vertical and lateral migration potential of the arsenic and cadmium into nearby Whitewood Creek. Samples were collected from trees in the plot at the end of the first growing season to determine the concentrations of arsenic and cadmium in the leaf, stem, and roots of the trees. A commercial fertilizer was used at recommended rates to ensure vigorous early growth of cuttings.

Genetically identical cuttings were also established in a laboratory plant incubator. Cuttings were planted in a mixture of tailings (ratios of 100 percent, 50 percent and 0 percent tailings) and Hoagland R growth fertilizer. The mixture of 50 percent tailings was grown in tailings and vermiculite (50:50 by mass mixture) and the mixture of 0 percent tailings was composed only of peat. In general, the poplar cuttings grew in all three of the mixtures, although cuttings rooted in the 100 percent tailings grew more slowly than those of the 0 percent and 50 percent tailings mixtures.

The trees in the experimental field plot along Whitewood Creek grew to a height of 12 meters (39 feet) at the end of the first growing season (Pierzynski et al. 1994). Roots formed along the entire length of the cutting in the soil so that a dense root mass was established to intercept infiltration and flow of water toward Whitewood Creek. Samples of leaves, stems, and roots were collected from the field and the laboratory to compare translocation and uptake of arsenic and cadmium. The results from the analyses are presented in Figure 4. Samples of poplar leaves in the experimental plot did not accumulate significant amounts of arsenic or cadmium, and the reported rates of the field accumulation were generally lower than the rates determined in the laboratory (Figure 4).

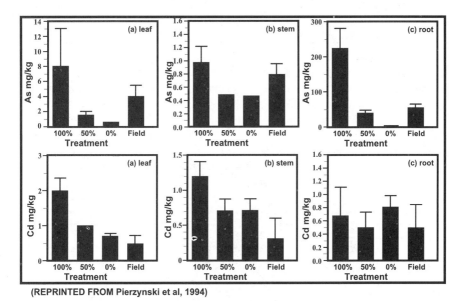

(REPRINTED FROM Pierzynski et al, 1994)

Figure 4   Total arsenic and cadmium in leaves, stems and roots in fertilized laboratory (0, 50, 100% tailings) and field, Whitewood Creek, South Dakota.

Concentrations of arsenic and cadmium were also measured in native vegetation at the site and were reported to be of the same order of magnitude as those in the poplar trees. However, the leaves of lambsquarter were high in arsenic (14 mg/kg) and the leaves of native cottonwoods were somewhat higher in arsenic (1.6 mg/kg) than the experimental plot of the poplars. The field and laboratory investigation showed that the mine tailings could be vegetated and stabilized with hybrid poplar trees without an unacceptable uptake of arsenic and cadmium in leaves (Pierzynski et al. 1994).

ENHANCED RHIZOSPHERE DEGRADATION

Enhanced rhizosphere degradation refers to the process of biologically breaking down constituents in the soil through microbial activity near the plant roots. This application of phytoremediation is becoming an increasingly accepted method to treat soil contaminated with petroleum hydrocarbons (Banks et al. 1999), PAHs (Aprill and Sims 1990, and Schwab and Banks 1994), insecticides and herbicides

(Ferro, Sims, and Bugbee 1994), and munitions, as well as macronutrients such as nitrogen and phosphorous (Schnoor et al. 1995).

The rhizosphere, first described by Lorenz Hiltner in 1904, is a complex zone of increased concentrations of nutrients and oxygen, microbial activity, and biomass at the root-soil interface (Hiltner 1904). Although the rhizosphere is commonly referred to as the root zone, it is not a uniform, well-defined volume of soil but a zone of soil that has a maximum microbiological gradient adjacent to the root, that declines with distance away from the root (Rovira and Davey 1974). Excellent overviews of the rhizosphere and rhizosphere processes are available (Rovira and Davey 1974, and Anderson, Guthrie, and Walton 1993) and only a brief description of rhizosphere characteristics is presented in this chapter.

The overall effect of the plant root-soil interaction is an increase in microbial biomass by an order of magnitude or more compared with microbial populations in bulk soil away from the roots. This rhizosphere effect is often expressed as the ratio of the number of microorganisms in the rhizosphere (R) soil to the number of microbes in the non-rhizosphere soil (S), or R/S ratio. Although the R/S commonly range from 5 to 20, they can run as high as greater than 100. An example of increased microbial populations in vegetated and nonvegetated soils contaminated with an aliphatic hydrocarbon is shown in Figure 5.

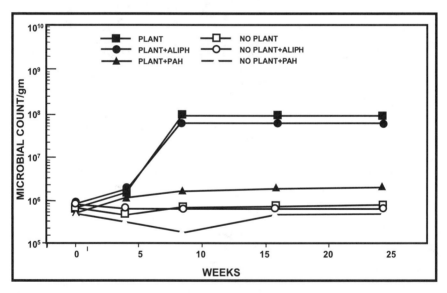

Figure 5   Increasing microbial concentrations.

The microbial composition in the rhizosphere is also complex, and is a function of the plant species, soil type, and growth period of the plant. The principal microbial assemblage generally includes bacteria, actinomycetes, and mycorrhizal fungi, in that order of predominance (Rovira and Davey 1974). Mycorrhizal fungi are particularly important in the rhizosphere processes because of their larger size and ability to grow at greater distances from the roots, although they are fewer in number and the populations grow slower than bacteria.

The increased microbial populations and diversity in the rhizosphere is a result of plants releasing exudates, which include oxygen and nutrients, sugars, alcohols, amino acids, and enzymes (e.g., dehalogenase, nitroreductase, peroxidase, laccase, and nitrilase). Roots also release organic material in the form of decaying roots and mucigel, a gelatinous substance that is a lubricant for root penetration through soil during growth. The volume of exudates and enzymes released is quite large, and has been estimated to be 10 percent to 20 percent of the total photosynthesate production of a hybrid poplar tree. It is the exudates that are the primary source of energy for microbes in the rhizosphere. In the process of metabolizing these substances, environmental contaminants can be either metabolized directly, or co-metabolized by the microbes. The degradation of isotopic-labeled benzo(a) pyrene in the rhizosphere of a plant, compared with unplanted soil, is presented in Figure 6.

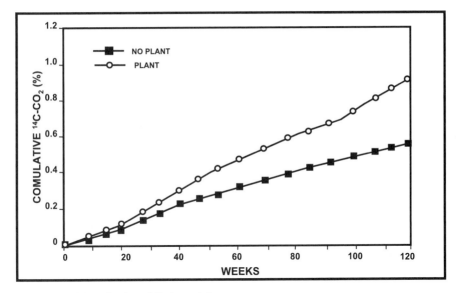

Figure 6   Mineralization of [14]c-benzo[a]pyrene.

Enzymes also have a causative effect on the degradation of environmental contaminants. Plants are known to release specific enzymes that can be especially effective for enhanced rhizosphere degradation. For example, parrot feather (*Myriophyllium spicatum*) contains the enzyme nitroreductase, which has been shown to be effective in reducing concentrations of the munition trinitrotoluene (TNT). Dehalogenase, an enzyme in hybrid poplars (*Populus* sp.) has been shown to reduce concentrations of chlorinated solvents, such as TCE (Schnoor et al. 1995).

Many plant enzymes have been identified in phytoremediation and are currently being researched in their roles in catalyzing beneficial degradation, metabolism, immobilization, and accumulation. Genetic engineering may, in the future, allow transfer of genes to supply various plant species with a capability to contain, exude, and express a variety of enzymes beneficial for soils and groundwater remediation.

In the absence of bacteria and fungi, plant exudate production can decrease, subsequently providing fewer organic substrates for microbial growth. Findings also suggest increased biomass may be the cause of decreased persistence of several compounds that are toxic to plants, including diazinon and 2,4-D (Sandmann and Loos 1994). This increase in biomass suggests that the consortia of rhizosphere microorganisms can actually adapt to protect plants from injury (Anderson, Guthrie, and Walton 1993).

Enhanced Rhizosphere Degradation Case Histories

Craney Island Fuel Terminal, Virginia

The Craney Island Fuel Terminal (CIFT) in Portsmouth, Virginia is the Navy's largest fuel storage facility in the United States and is the location of a Department of Defense environmental remediation technology demonstration project for phytoremediation. The phytoremediation project was conducted at a bioremediation treatment cell that is approximately 15 acres in area, and is underlain by layers of compacted clay, synthetic geogrid, sand, and polyethylene (Banks et al. 1999). Contaminated material for the phytoremediation study was excavated in 1995 from lagoons that had been used at the CIFT from 1940 to 1978 for gravity-oil separation of ballast and bilge waste from ships. The phytoremediation study took place in a 0.5 acre portion of the bioremediation treatment cell that was filled to a depth of 2 feet.

Prior to implementing phytoremediation, soil conditions within the bioremediation treatment cell were sampled for agronomic characteristics and total petroleum hydrocarbons (TPH). The sandy loam soil had relatively low moisture-holding capacity, but a relatively high organic matter content. Measurements of electrical conductivity indicated relatively high salts, but not high enough to pose phytotoxicity or growth limitations. Concentrations of TPH were relatively consistent across the cell, with an average concentration of 4551 mg/kg and a standard deviation of 1045 mg/kg (Banks et al. 1999).

A greenhouse study germination assessment was performed using soils from the site to determine the potential for phytotoxic effects. Plant species for the germination assessment were selected based on adaptation to climate, local soil conditions, tolerance to growth in contaminated soil, and shallow rooting that would be compatible with the design of the bioremediation unit. Species selected for the germination assessment included Bermuda grass (a warm season perennial grass), tall fescue (a rapidly growing cool season grass with an intensive root system), and white clover (a shallow rooted legume). The results from the greenhouse germination assessment indicated that all plants germinated and grew in the contaminated soils, with only white clover showing apparent stunting of growth from the contamination.

The study area was divided into distinct plots of approximately equivalent size, with six replicate plots per treatment. The grasses were planted in September 1995, with one of the following plant types: tall fescue, a mixture of Bermuda grass and rye grass (a cool season annual), or white clover. All three vegetation treatments grew well, but the pattern of growth varied for each treatment. Bermuda grass was established by sod and provided the most rapid cover and rooting for all species. Rye grass was seeded into the Bermuda grass to help provide winter growth and rooting while the Bermuda grass was dormant. By the seventh month sampling in May 1996, Bermuda/rye grass had the most root production of all treatments. Tall fescue grew well, but took longer to establish as it was planted from seed. At the 12 month sampling, tall fescue had the highest mass and density of roots. White clover grew well during the first year, but did poorly during the second year due to a lack of moisture.

Microbial numbers and diversity were initially higher in the vegetated plots, but were approximately the same as the unvegetated plots in the second year of the study. The clover plots, which had the highest number of microbial petroleum degraders and the lowest TPH concentrations at the last sampling event, also had the most root turnover.

There was a statistically significant reduction in TPH in all vegetated plots, as compared to unvegetated plots (Table 3). White clover showed the highest rate of TPH degradation, followed by tall fescue and Bermuda grass. There was no evidence of a plateau having been reached at the 24 month sampling date. In addition, petroleum hydrocarbons did not leach from the root zone into the underlying materials and polynuclear aromatic hydrocarbons (PAHs) did not accumulate in the plant shoots.

Table 3   TPH Dissipation for the Craney Island Phytoremediation Field Study

Treatment	13 Months	24 Months
Clover	29%	50%
Fescue	33%	45%
Bermuda	27%	40%
Bare	21%	31%

Reprinted with permission from Banks et al, 1999.

Trends observed in the TPH degradation were not directly transferable to rates of PAH degradation. For nearly all of the individual PAHs studied, the percent degradation was highest in the tall fescue and least in the unvegetated plots. Although TPH degradation was highest in the white clover, PAH degradation in white clover was often less than in the tall fescue and the same as the unvegetated control plot. In addition, the relationship between plant growth and hydrocarbon degradation in the study did not have a direct correlation. Among the plant treatments, white clover had the lowest root and above-ground biomass production, yet it had the highest rate of hydrocarbon degradation. The study concluded, among other things, that while it is necessary to establish a healthy vegetation at phytoremediation sites, other factors such as rate of root turnover, root exudation patterns, and the influence of vegetation on soil physical properties appear to be just as important as the quantity of vegetation produced (Banks et al. 1999). Costs for the phytoremediation program at Craney Island, as reported in Banks et al. (1999), is presented in Table 4.

Table 4   Costs for Implementing Phytoremediation, Craney Island
             Demonstration Project, Virginia

	Craney Island, Virginia
Implementation	$900,000
Operation & Maintenance (per year)	NS
Treated Volume (yd3)	45,000
Implementation Costs (Plus 2 years O&M)	$20/yd3

NS Not Specified
Implementation costs for Craney Island included 2 years O&M, based on full-
   scale design presented in Banks et al, 1999.

Active Industrial Facility, Wisconsin

During the late 1970s, a section of below-ground piping transferring No. 2 fuel oil from a larger above-ground storage tank (AST) failed at this active industrial facility in Wisconsin, resulting in a subsurface release. Remediation was immediately implemented. Approximately 15,000 gallons of fuel oil were subsequently recovered from shallow trenches installed at the site. Hydrocarbon constituents in groundwater were also reduced to below regulatory standards.

The site is underlain by 3 to 15 feet of heterogeneous fill material comprised of wood timbers, sawdust, construction debris, clay, sand, and gravel. The depth of the water table ranges from 3 to 9 feet, and the water table slopes west toward the adjacent river. Several investigations at the site have determined that highly contaminated soil (concentrations greater than 1,000 mg/kg diesel range organics [DRO]) remain in four generalized hot spots at the site (Figure 7). Three of the hot spots are below a hard-packed gravel equipment storage area, and a fourth is located below a vegetated area along the river.

The objective of this project was to remediate soil and fill materials contaminated with DRO within the four hot spots at the facility to below 1,000 mg/kg DRO. A series of feasibility tests were initially performed on soil samples beginning in 1994. These feasibility tests included collecting soil samples from the hotspots and performing microbiological enumerations, accelerated bioventing tests, respirometry testing, and agronomic testing on the samples. Overall, the results from the feasibility testing show that concentrations of the DRO exhibit a wide degree of variability both across the site and within each of the four hot spots, ranging from 40 mg/kg to 5,000 mg/kg. Although there did not appear to be a lateral trend in elevated concentrations of DRO, the highest concentrations were present within the capillary fringe. A viable population of microbes capable of degrading fuel oil were found to be indigenous to soil at the facility and reductions of between 40 percent and 90 percent in the concentrations of the DRO were observed over the course of a 24 week accelerated bioventing study (Carman, Crossman, and Gatliff 1998).

Soil samples were also collected from two hotspots during June 1995 to determine phyto-toxic effects on tree root development and concentrations of agronomic constituents of interest. A bench top root development study was performed using soil from hot spots, cuttings from willow trees indigenous to the site and hybrid poplar cuttings obtained from a nursery. After two months of

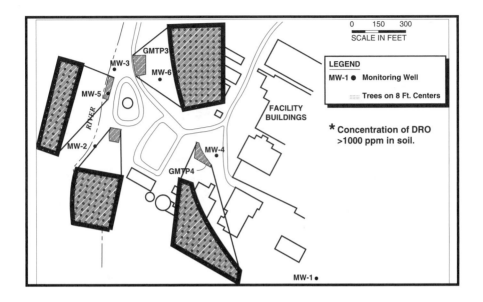

Figure 7   Location of soil hotspots and Prairie cascade willow trees, Wisconsin.

development, willow and hybrid poplar plants were transplanted into soil from the hot spots. Both hybrid poplars and willows exhibited good aerial growth during the root development portion of the agronomic assessment. The willows, however, demonstrated a more pronounced tendency to establish rooting within the DRO-contaminated soil. With the exception of the relatively high concentration of soluble salts in subsurface soil, the constituents of concern were within acceptable ranges for tree growth.

Hybrid willow trees (*Prairie cascade*) were planted in the four hot spots during May 1996. A total of 300 trees were planted at a spacing of 8 feet. Trees were planted using *TreeMediation*™, a patented process which focuses rooting activity and rhizosphere development in the zone of contamination. *Treemediation*™ was developed by Applied Natural Sciences of Fairfield, Ohio. Figure 8 presents a photograph of the hybrid willow trees in Hot Spot 2 following planting in May 1996 and near the end of the fourth growing season in 1999.

Operation and Maintenance

Site visits were made periodically during the first growing season during 1996 to monitor the growth of the trees, precipitation at the site, and general health of the trees. Precipitation was monitored using data from a local meteorological station and plant tissue samples were collected from trees in the four hot spots during October 1996 to determine fertilizer requirements. Yellowing of leaves was noted

A) After planting in May 1996   B) During July 1999

Figure 8   Photographs of Prairie cascade hybrid willow trees at an active industrial facility in Wisconsin.

in a portion of the trees and insect damage was apparent in several of the trees, especially along the adjacent river. Nitrogen, phosphorous, and potassium were detected at concentrations comparable to reported concentrations in leaves of orange, almond, and apple (trees with commonly reported leaf nutrient content). Samples collected from leaves that exhibited yellowing, however, contained lower concentrations of nitrogen than leaves that exhibited normal green color. Based on the results from tissue sampling, a fertilization program was initiated beginning in the fall of 1996 and has continued through 1999. The program consists of surface applications of a high nitrogen fertilizer in the rows between the trees. Subsequent tissue analysis indicates that levels of primary nutrients in trees with leaves that had formerly yellowed have increased, and the number of trees with yellowing leaves has been greatly reduced. Insecticide has been applied as needed (once or twice per year) and appears to have substantially reduced the insect damage. Trees planted as part of the phytoremediation program have grown an average of 5 to 6 feet in the initial four growing seasons of the program.

Confirmatory Soil Samples

Soil samples were collected during October 1999 (the end of the fourth growing season) to determine to effectiveness of the phytoremediation program in reducing the concentration of DROs in the hotspots. A total of 33 samples were collected from near the locations of samples that were collected at the onset of the phytoremediation program. Based on the results of the samples, concentrations of DRO decreased 66 percent to 68 percent in two of the four hotspots, although the trend in DRO within a third hotspot was not clear, and a fourth hotspot was not accessible at the time of sampling during 1999. Soil samples were also collected outside of the hotspots near previously sampled locations, as a control. Concentration of DRO outside of the hotspots have not decreased as significantly as DRO inside the hotspots, suggesting that activities associated with the phytoremediation program have been responsible for accelerating the decrease in DRO.

Total costs for the phytoremediation program at the site in Wisconsin are shown in Table 5. A benefit of the phytoremediation program is that the majority of the O&M activities, including fertilization, watering, and insecticide applications can be performed by local operations personnel who have been properly trained.

Table 5   Costs for Implementing Phytoremediation, Active Facility in Wisconsin

	Active Facility, Wisconsin
Implementation	$80,000
Operation & Maintenance (per year)	$7,000
Treated Volume (yd3)	4,500
Implementation Costs (Plus 2 years O&M)	$21/yd3

RHIZOFILTRATION

Rhizofiltration is the process of using roots from plants to absorb, concentrate, and precipitate metals, radionuclides and organics, usually from effluent waste waters. Rhizofiltration is an application of phytoremediation that is similar to phytoextraction in that it removes, rather than degrades, contaminants.In contrast to phytoextraction, which uses plants grown in soil or sediments, rhizofiltration is generally implemented in an aquatic environment or within an effluent stream for the purpose of contaminant removal. Generally, rhizofiltration can use either terrestrial plants or aquatic plants, although terrestrial plants that are used have fibrous root systems that are capable of removing more mass. Phreatophytic terrestrial plants offer the advantage of taking up large volumes of water and developing long root systems.

Disposal methods for plants that are saturated with contaminants are dependent on the type of contaminant, concentration within the plant tissues, and vegetative volume. Roots and/or plant tissue can be harvested and dried, with the residual material treated with acid to reclaim the metals, combusted to further reduce the volume or disposed of in a landfill. Plants that are used to treat organics, munitions, or excess nutrients may not necessitate disposal, since these contaminants may actually be metabolized within the plants.

Rhizofiltration has been shown to be an effective method on pilot scale systems for removing cadmium, lead, copper chromium, nickel, and zinc (Ensley et al. 1994). In a pilot scale test, sunflowers (*Helianthus annuus* L.) were cultivated hydroponically and in an aerated nutrient solution. Solutions of deionized water with heavy metals were introduced at concentrations ranging from over 600 mg/l of zinc and chromium to more than 1.5 mg/l of cadmium and lead. The results indicated that metals were removed to low ug/l concentrations within 24 hours from being introduced into the system. Removal of metals from the solution involved a rapid initial phase, probably due to surface absorption and a second, longer duration phase. The initial phase was attributed to physical and chemical processes including chelation, ion exchange, and chemical precipitation. The second phase, which was longer, was attributed to biological processes including intracellular uptake and translocation of the metals into the plant shoots (Ensley et al. 1994). Uranium concentrations in water from a Department of Energy (DOE) site in Ashtabula, Ohio were also reduced by 95 percent within 24 hours from initial concentrations ranging from 100 ug/l to 400 ug/l to below regulatory standards of 20 ug/l, using a sunflower cultivar (Ensley et al. 1994).

Laboratory experiments have also been conducted by Argonne National Laboratories to determine the potential for hybrid poplar trees to uptake and accumulate high concentrations of zinc (Negri et al. 1996). In greenhouse tests, it was shown that hybrid poplars grown in inert quartz sands were able to uptake and translocate concentrations of zinc in solutions of up to 800 mg/l in a single pass. Poplars that received numerous doses of the zinc solution accumulated zinc at concentrations up to 38,000 mg/kg in the root tissue (Negri et al. 1996).

Phytoremediation by rhizofiltration offers potential to remediate water effluent at many sites. Below is a case history that highlights rhizofiltration in a constructed wetland.

Rhizofiltration Case History

Rhizofiltration, Milan Army Ammunition Plant, Milan, Tennessee

A field demonstration was conducted at the Milan Army Ammunition Plant near Milan, Tennessee, to evaluate the feasibility of treating explosives contaminated groundwater with two types of constructed wetlands. Both a surface flow (lagoon) and subsurface flow (gravel wetland), were evaluated in this demonstration project.

The project was executed in three phases. Phase I included conducting plant screening studies, conducting treatability studies, and designing and constructing the demonstration facility. Phase II included operation of the demonstration, monitoring, and evaluating the performance of both the technical and economic aspects of the project. Phase III included collection of additional data to improve the design, operation, and economic success of a scaled up gravel based system.

During Phase I, standard methods were developed to evaluate the ability of aquatic macrophytes (large aquatic plants) to lower the contaminant levels of TNT, RDX, and related compounds in explosives contaminated waste. After a variety of submergent and emergent aquatic macrophytes were screened for their ability to remediate the contaminated water, treatability studies were undertaken to test performance, and the systems were designed and constructed.

Both the gravel and lagoon based systems were designed for a total hydraulic retention time of approximately 10 days at an influent flow rate of 5 gpm per system. The lagoon based system consisted of two lagoons connected in series. Each lagoon had dimensions of 24 x 9.4 x 0.6 m (79 x 31 x 2 ft). The gravel based system consisted of two gravel filled beds connected in series. The first bed was maintained in an anaerobic condition by adding milk replacement starter (MRS) to the water every two weeks. The second cell was maintained in an aerobic condition via a Tennessee Valley Authority patented process. The anaerobic cell had dimensions of 32 x 11 x 1.4 m (105 x 36 x 13 ft) and the aerobic cell had dimensions of 11 x 11 x 1.4 m (36 x 36 x 13 ft). Both wetlands contained plants specifically selected to ensure explosives degradation. The lagoon based system was planted with sago pond weed, water stargrass, elodea, and parrotfeather. The gravel based system was planted with canary grass, wool grass, sweetflag, and parrotfeather.

The primary objective of the Phase II demonstration was to evaluate the technical feasibility of using the wetlands for remediating explosives contaminated water. The goal was to reduce TNT concentrations to levels less than 2 ug/l and total nitrobody (including TNT, RDX, HMX, TNB, 2A-DNT, and 4A-DNT) concentrations to levels less than 50 ug/l. The system operations began during June 1996 with the introduction of explosives contaminated water. The systems were operated until September 1997, at which time the lagoon based system was retired and the gravel based system's operations were continued for Phase III.

Groundwater from two wells was used over the course of the Phase II demonstration. The first well had an average total nitrobody concentration of 3,250 ug/l. The second well had an average nitrobody concentration of 9,200 ug/l. Average influent concentrations of explosives in the water from each well are shown in Table 6.

Table 6   Average Influent Concentrations in Wells Used For Constructed
          Wetlands Demonstration Project, Milan Tennessee

Explosive	First Well (Before 11/21/96)	Second Well (After 11/21/96)
TNT	1,250 ug/l	4,440 ug/l
RDX	1,770 ug/l	4,240 ug/l
TNB	110 ug/l	330 ug/l
HMX	110 ug/l	91 ug/l

Reprinted with permission from U.S. Army Environmental Center, 1999.

While the Phase II demonstration results indicated that both the gravel and lagoon based systems could degrade explosives, the gravel based system was clearly superior. The lagoon based system met the goal of reducing TNT concentrations below 2 ug/l only during the first 50 days of the demonstration and was unable to satisfactorily degrade RDX or meet the total nitrobody removal goals during the demonstration. In addition, it was difficult to maintain an adequate plant population within the lagoon based system. Problems encountered included a tadpole infestation which defoliated the plants, difficulty in reestablishing plant growth, and a loss of vegetation in a hailstorm.

The gravel based system was able to degrade TNT and RDX, was able to meet the demonstration goals during all but the coldest months, and was able to establish a sustainable ecosystem. During winter operations, the gravel based system had difficulty meeting the total nitrobody reduction goal due to reduced microbial activity. However, the design and cost analysis indicate that a gravel based system could be economically resized to overcome the winter performance issues by increasing the water's retention time from a total of 10 days to a total of 14.5 days (12 days in the anaerobic cell and 2.5 days in the aerobic cell). (U. S. Army Environmental Center 1999).

Few explosives were observed to accumulate in the gravel, sediment, and plants of the gravel based system. The quantity of total nitrobodies (RDX, TNT, TNB, HMX, 2,4-DNT, and 2,6-DNT) and total explosives (nitrobodies plus measured byproducts) on the gravel and sediments were always less than 1 percent to 1.4 percent of the mass of nitrobodies entering the lagoon and gravel based wetlands, respectively. The low accumulation of explosives in the wetland cells and the observation of explosive byproducts indicated that explosives were being removed from the water via biological degradation.

The primary objective of the Phase III demonstration was to collect additional data to improve the design, operation, and economic success of scaled up gravel based systems. To conduct Phase III, the gravel based system's operating parameters were modified to:

- Use a less expensive carbon source (molasses syrup as opposed to MRS)
- Allow frequent addition of the carbon source (twice daily versus biweekly)
- Decrease the rate of carbon addition (carbon rate cut in half)
- Lower the influent flow rate from 5 to 3 gpm

The Phase III demonstration was conducted from September 1997 to July 1998. The gravel based system's performance during Phase III was about equal to its Phase II performance. As in Phase II, the gravel based system was generally able to meet the demonstration goals. However, during Phase III, the gravel based system was unable to meet the 50 ug/l total nitrobody limit from early December 1997 through mid June 1998 due to the combined effects of decreased microbial activity resulting from low water temperatures, an increase in influent nitrate concentrations which compete with the reactions leading to explosive degradation, and an increase in the anaerobic cell's redox potential. The system's higher redox potential was attributed to the absence of sufficient carbon to support the optimum level of microbial activity.

Operationally, the gravel based system performed better in Phase III than it did in Phase II. During this period, the system did not experience ponding. Two changes probably kept the gravel from becoming blocked with solids: MRS was no longer the carbon source, and the molasses was added regularly via an automated pumping system. A cost assessment was developed based on a conceptual design for a 10 acre, full-scale, gravel based system which had been designed to remediate 200 gpm of groundwater from another area of the Milan Army Ammunition Plant. The estimated costs for constructing the 10 acre, full-scale, gravel based wetland was $3,466,000 in 1998 dollars. The estimate includes all costs associated with constructing the wetland and should be considered "turnkey." However, these costs did not include items such as groundwater extraction wells, utilities other than electricity, postconstruction sanitation facilities, roads or parking lots, and operator training.

Assuming a 95 percent system availability and 30 year life, the total cost (operation and maintenance cost plus capital cost) for treating groundwater with this gravel based system was estimated at $1.78 per thousand gallons of groundwater (U.S. Army Environmental Center 1999).

HYDRAULIC CONTAINMENT

Hydraulic containment is an application of phytoremediation to capture groundwater or soil pore water. It harnesses the ability of plants, especially phreatophytes, to root deeply and take up large quantities of water. This application of phytoremediation has also been referred to as hydraulic pumping or solar pumping.

The water uptake of certain plants can be significant enough to suppress the water table, and in some instances, create zones to capture contaminated groundwater. Poplar trees, for example, are known to transpire between 50 and 300 gallons of water per day. It has been reported that a single cottonwood in a hot climate can transpire several thousand gallons of water per day, the equivalent of approximately one-half acre of alfalfa (Gatliff 1994 and Miami Conservancy District 1991). Heat pulse techniques have shown that two eucalyptus species (*Eucalyptus wandoo* and *Eucalyptus salmonophloia*) transpire up to 13 gallons of water per day (18,000 liters/year) in western Austrialia (Farrington et al. 1994).

The tendency of phreatophytes to root deeply and uptake large volumes of water has actually made them a nuisance in semiarid regions of the southwest United States, where aquifer levels can drop several feet during growing seasons. However, areas of Australia have been densely reforrested with eucalyptus trees to lower the aquifer levels and improve the quality of shallow groundwater. For example, at a site in western Australia, removal of native vegetation for farming resulted in a disturbance to the natural hydrologic balance and rising aquifer levels. Salts, principally sodium chloride, which had been previously immobilized in the vadose zone were dissolved by rising aquifer levels, resulting in increasing salinity of the groundwater. Densely planted *Eucalyptus* and *Pinus* trees lowered groundwater levels more than 20 feet and salinity of the groundwater decreased an average of 11 percent (Bari and Schofield 1992).

Hydraulic containment is an application suited to relatively shallow groundwater or pore water, and is restricted to depths generally less than 30 feet below land surface, or the deepest rooting depth of the vegetation. Rooting depth is a function of plant species, soil conditions, and water availability. Roots will generally extend only to a depth necessary to maintain plant viability and will not extended more than a foot or two into the water table. Therefore, special planting techniques need to be considered if depths greater than 10 feet are needed.

Hydraulic containment using vegetation can actually be superior to traditional pump and treat systems since traditional pump and treat requires that individual pores be interconnected to allow removal by pumping. If pores are not interconnected (e.g., capillary pores in the vadose zone), water and contaminants dissolved in the water will not be affected by pumping. In contrast, roots can penetrate the non-interconnected pores and remove the water within the pores. In addition, root penetration can also be achieved in very dense soils (nonpermeable) that are not conducive to pump and treat systems.

Sites where phytoremediation has been used for hydraulic containment include riparian corridors (applied along a stream or river bank), areas that have been deforested (such as the example discussed in western Australia), and other sites where shallow groundwater is present. Two case histories where hydraulic containment with hybrid poplar trees has been demonstrated are discussed below.

Hydraulic Containment Case Histories

Gasoline Station, Ohio

A gasoline release from an underground storage tank (UST) occurred at a small active gasoline station in the upper midwest. Investigations at the station began in 1996 and a total of 14 groundwater monitoring wells and piezometers were installed to determine the magnitude and extent of the release.

The geologic materials at the station include fine grained sands and silty clay materials. The depth to groundwater is shallow, ranging from 3 to 6 feet below land surface. The hydraulic conductivity of the saturated materials range from 2.7×10^{-5} cm/sec to 5.2×10^{-4} cm/sec. The results of the work indicated that gasoline constituents, including BTEX and petroleum hydrocarbons, were present in the groundwater.

As we discussed above, many phytoremediations include multiple uses of the properties of the trees. In addition to hydraulic control of the groundwater movement, this case history had the following additional goals:

- Harness the large evapo-transpiration capacity of hybrid poplar trees to create a hydraulic barrier to off-site migration of contaminated water (e.g., establish hydraulic control)
- Use the trees to uptake and degrade the organic constituents present in the groundwater (phytoextraction)
- Enhance microbial degradation of the gasoline constituents in the rhizosphere of the trees (enhanced rhizosphere degradation)

During March 1997 an agronomic assessment of the soil conditions was performed, which included a field analysis of pH, nitrogen, phosphorous, and potassium. A total of 162 hybrid poplar trees were purchased for the phytoremediation program at the station. Both whips and rooted stock of male *Populus deltoides nigra* (DN-34) were selected, based on their fast growing nature, hardiness, and ability to root deeply without occupying a large amount of space.

Trees were planted during early April 1997. A standard trencher was used to plant the poplar whips to a depth of approximately 3 feet below land surface in a plot that was 10 feet wide and 75 feet long. Between every three to five whips, a 2 year old poplar with established roots was planted. The whips were intended to provide a dense cover of trees and the more mature poplars were intended to provide a more immediate impact on shallow groundwater. Based on the results of the agronomic testing, soil amendments consisting of topsoil and two fertilizers were placed into the trenches. The two fertilizers used for the amendments contained ratios of nitrogen, phosphorous, and potassium (N:P:K) of 12:12:12 and 29:3:4. Additional trees to serve as reserve stock were purchased and planted alongside the station building.

The trees began to leaf during May 1997 and grew approximately 2 feet within 8 weeks of planting. Monthly inspections of the trees have been performed throughout the growing seasons, and the trees have grown to a height of approximately 30 feet through November 1999. Water levels in monitoring wells are also measured monthly, and groundwater samples for laboratory analysis of BTEX are collected on a quarterly basis. O&M activities for the trees, including pruning and fertilization, and application of insecticides has been performed on a monthly basis since 1997.

Water levels after planting in 1997 and during June 1999 are shown on Figure 9 and hydraulic gradients across the site are shown on Figure 10. The potentiometric surface map of May 5, 1997 (one month after the trees were planted), compared to the potentiometric map from June 30, 1999, shows a significant difference in the hydraulic gradient across the phytoremediation system. The increase in hydraulic gradient across the system is a result of the pumping action of the hybrid poplar trees. Photographs of the trees from August 1999 are presented in Figure 11.

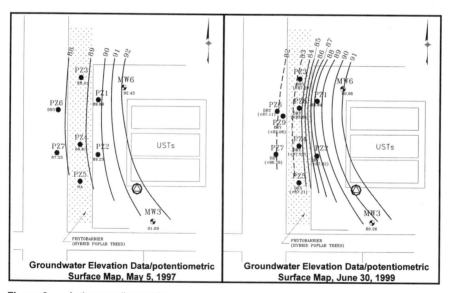

| Groundwater Elevation Data/potentiometric Surface Map, May 5, 1997 | Groundwater Elevation Data/potentiometric Surface Map, June 30, 1999 |

Figure 9   Active gasoline station.

The water elevation data provide strong indication that the system is effectively controlling groundwater migration along the down-gradient portion of the site. This is indicated by an approximate groundwater table drawdown of 3.5 to 4.5 feet beyond the effects of seasonal groundwater fluctuations and natural groundwater flow.

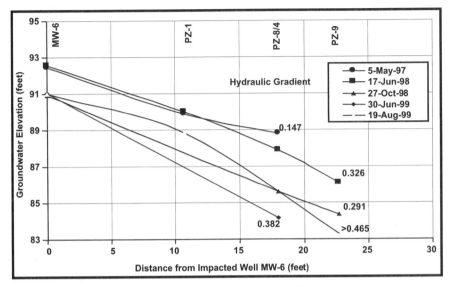

Figure 10   Groundwater elevations across the phytoremediation system.

Wood Preservative Site, Tennessee

This site consists of an approximately two acre area that was impacted by 90 years of creosote treatment of railroad ties. The site is the first State of Tennessee Superfund site with an approved phytoremediation system.

The site is underlain by unconsolidated residuum and/or fluvial deposits, which in turn overlie shallow shale bedrock. Shale bedrock is shallow and forms the stream bed of an adjacent creek which is a local hydrologic boundary for groundwater flow near the stream. The water table averages approximately 7 feet below land surface and shallow groundwater flow is southeast toward the adjacent creek. Laboratory results from soil and groundwater samples indicated that BTEX and PAHs have impacted these media at the site. Site details are shown in Figure 12.

The scope of activities associated with the project included source delineation and characterization, recommendation of remedial alternatives, and implementation of a remedy. Phytoremediation and operation of a shallow groundwater interceptor trench was recommended and approved for the site. The goal of the phytoremediation program was to aid in hydraulic containment of shallow affected groundwater (along with the interceptor trench) and to enhance the uptake and degradation of BTEX and PAHs.

Over 1000 hybrid poplars were planted in a 1.8 acre area in May 1997 and May 1998. Based on the results from an agronomic assessment, trees were planted in trenches across the site. To evaluate the effectiveness of the combined phytoremediation system and interceptor trench after 2 years operation, water levels and hydraulic gradients were measured and calculated and a water balance and groundwater flow model for the site was performed.

Figure 11   Photograph of trees taken during August 1999, active gasoline station, Ohio.

Hydraulic Gradients

Water level elevations in monitoring wells located away from phytoremediation and interceptor trench influence (MW-1 and MW-2) were compared to wells located near the influence of the phytoremediation and trench (P-2, P-15, and P-16). From June 3 to July 29, 1998, water levels decreased 1 foot in the wells closer to the trench (Loftis 1999). Figure 13 presents a cross section of water level elevations across the site from west to east. This figure presents data from August 1998 through June 1999 and illustrates higher water levels on the west end of the property and

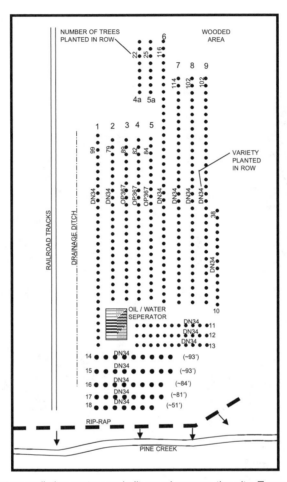

Figure 12   Phytoremediation system as-built, wood preservative site, Tennessee.

declining water levels on the east end where the phytoremediation system and trench are located.

This hydraulic information suggests that the phytoremediation system and trench are effective in stopping flow of affected groundwater into the stream which is located east of the trench.

Water Balance and Flow Model

Rainfall from the first 6 months of 1999 (32.85 inches) that fell on the 1.8 acres of phytoremediation system equates to 1.6 million (M) gallons. Assuming that 25 percent is runoff (0.4 M gallons), and the interceptor system recovered 0.65 M gallons, the phytoremediaton system (trees and grasses) has captured approximately 0.55 M gallons via evapotranspiration and uptake. Overall, the water balance also suggests that the phytoremediation system and trench are effectively intercepting surface water infiltration and groundwater.

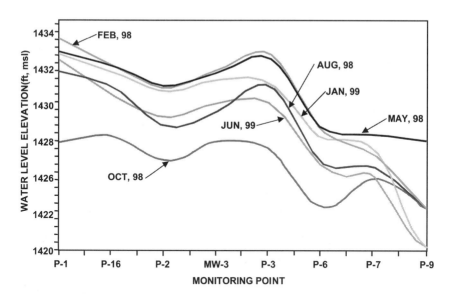

Figure 13   West to east gradients. Note increasing gradients near phytoremediation system and recovery trench.

Three steady state single layer finite difference models were constructed to represent site conditions. The first model was calibrated with actual field data and did not take into account evapotranspiration from the poplars or effects from the interception system. The second model simulated site conditions while the interceptor trench was operating and the third model simulated site conditions during the hybrid poplars growing season. The goal of this model was to determine the minimum rate of water each tree could uptake. Results from that model indicate a water uptake rate of 4.6 gallons of water per day per tree (Loftis 1999).

Operation and Maintenance

The trees and understory grasses are routinely inspected for insects, disease, leaf color, dead limbs, drought stress, and overall health. Trees and/or understory grasses showing worm and/or insect infestation were sprayed with appropriate insecticide (Sevin) during April 1999. In addition, the area was mowed and trimmed to keep weed growth and water competition to a minimum and fertilizer (N:P:K 10:10:10) was spread along the tree lines in April 1999. Future O&M activities will include watering (if necessary), continued inspection and fertilization and excavation of several trees to determine root growth and penetration.

Laboratory Results

Total BTEX concentrations were reported as estimated values just above laboratory detection limits in MW-5 (0.0047 mg/l) and MW-6 (0.0031 mg/l). Laboratory results from the trench sump appears to fluctuate an order of magnitude between spring and fall sampling events. Total PAH constituents were detected as estimated values in MW-1 (0.002 mg/l), MW-5 (0.031 mg/l), MW-6 (0.765 mg/l), and the sump (4.015 mg/l). Total PAH concentrations in MW-2, MW-3, MW-4, and MW-7 remained below detection or increased only slightly from the previous sampling event. However, concentration of total PAHs has decreased an order of magnitude since installation of the remediation system and the horizontal extent of PAHs in groundwater across the site has decreased and does not extend beyond the phytoremediation system and interceptor trench.

ALTERNATIVE COVERS (PHYTO-COVERS)

A phyto-cover is an engineered vegetative system that is constructed as an alternative to a traditional landfill cap. Typical cover systems minimize the infiltration of rainfall and snowmelt into waste and protect groundwater by reducing leachate production. Phyto-covers harness the natural evapotranspiration process of shallow and deep rooted plants to control the infiltration of water into waste materials and create a zone from which the plants can extract pore water within waste materials.

The purposes of a landfill cover are to protect groundwater quality by reducing water infiltration and leachate production, prevent surface exposure to waste, control surface water runoff, and meet established regulatory criteria. Typical traditional single barrier landfill covers will consist of clay, geocomposite, or geosynthetic membrane.

Properly designed phyto-covers can meet the objectives of a traditional landfill cover by creating a storage and pump water extraction system. The water holding capacity of the soil and the waste penetrated by the root system provides the storage capacity, storing water infiltration during periods of precipitation and during the dormancy period of the vegetation. The evapotranspiration capacity of the phyto-cover provides pumping from the storage, reducing soil moisture and water that may have migrated into the shallow waste materials during the dormancy period of the vegetation (Stack, Potter, and Sutherson 1999). An example of an established phyto-cover is presented as Figure 14.

Design criteria that must be considered when designing a phyto-cover system include:

- Preparing a site-specific water balance to determine the amount of additional material required to balance the inflow of precipitation with the outflow of the system, that is, the thickness of material required to create the storage
- Determining the types of vegetation and the soil and soil amendments for the phyto-cover. The vegetation must be appropriate to access the excess water stored in the soil and waste material during the dormancy period
- The characteristics of slopes and drainage required for the site and the system
- Gas generation and emissions from the landfill and whether supplemental gas emission controls are necessary

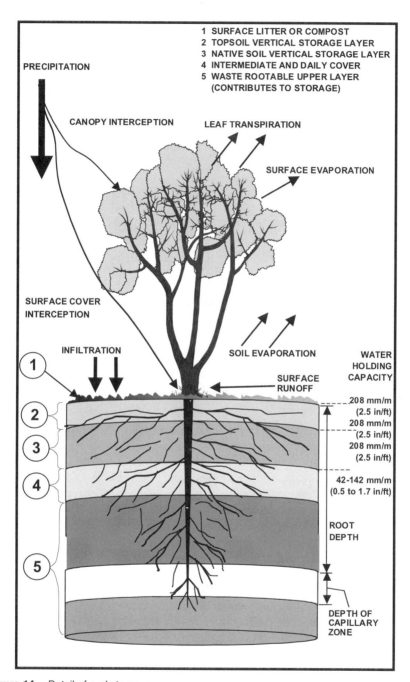

Figure 14   Detail of a phyto-cover.

Conducting a water mass balance is the fundamental way to calculate the water-storage capacity necessary to avoid leachate generation (Stack, Potter, and Sutherson 1999). Water balance parameters are presented in Table 7.

Table 7   Water Balance Analysis

Precipitation
Surface Runoff (R/O)
Potential Evapotranspiration (PET)
Infiltration (I)
Soil Moisture Storage (ST)
Actual Evapotranspiration (AET)
Potential Flux of Water Through the Cover System

The primary elements of a water balance include precipitation, surface runoff, potential evapotranspiration, and potential water flow through the cover system. The water balancing processes are typically evaluated using the Hydrologic Evaluation of Landfill Performance (HELP) model developed by the Waterways Experiment Station (WES) in Vicksburg, Mississippi. However, the HELP model was developed for use in low permeability environments, with shallow rooted landfill cap grasses. For the design of a phyto-cover, the HELP model is used in conjunction with other evapotranspiration models that can accommodate higher evapotranspiration values typical of deeply rooted plants and trees. PHYTOSOLV (Potter, Stack, and Suthersan 1999), a water balance model that augments the HELP model, has been designed to evaluate the functionality and economics of a phyto-cover versus a traditional landfill cap.

Benefits of a Phyto-Cover

There are a number of pollution control, environmental, and economic benefits to a properly designed phyto-cover, when compared to traditional covers. A phyto-cover has the potential to enhance the biodegradation of waste materials, through enhanced rhizosphere degradation and phytostablization processes. By contrast, a traditional cap provides no stimulation of natural biodegradation, and can actually limit biological growth by altering biogeochemical conditions in the waste and reducing natural attenuation processes. Phyto-covers are porous and permeable to gases, and can allow passive gas venting if appropriate, whereas traditional caps are essentially impermeable and may require elaborate gas venting systems.

Phyto-covers and traditional caps are both designed to be vegetated on the surface; however, traditional caps are designed to inhibit root growth into the waste, and require that small burrowing animals be kept away because of the breaches they can create in the barrier. Phytocaps are more of a natural system and are designed to allow the deep rooting vegetation to penetrate the waste. Native species including grasses, shrubs, and trees that have shallower root penetration are allowed to colonize within the phyto-cover vegetation, if deemed appropriate under site management guidelines. Burrowing animals are unlikely to penetrate into the rooting depths of the phyto-cover system and therefore it is not necessary to control their migration onto or within the landfill cover. A phyto-cover system can provide nest sites for birds and other arboreal species, and can serve as a noise and dust pollution control measure for nearby residents. The phyto-cover also resists wind and water erosion, in contrast to a traditional cap, which is prone to cracks, rips, and tears.

Traditional caps are also more material and maintenance intensive during construction than an equivalent phyto-cover systems. EPA studies have shown that implementing any type of remediation poses risk to site workers, neighbors, and the public; however, there is less of a threat for these types of injuries with a phyto-cover than a traditional cap.

The capital cost per acre for installing a phyto-cover system is less than a traditional cap, generally about 40 percent to 60 percent less (Stack, Potter, and Sutherson 1999). O&M costs for a phyto-cover are also lower than for a traditional cap, due to less inspection and repair costs. Although O&M costs for a phyto-cover system are lower, they require routine maintenance (watering, fertilizing, and insecticide applications), especially in the first four growing seasons.

There can be significant benefits of a phyto-cover, but like any technology, this application is not suitable for all situations. Its use should be evaluated on a site-by-site basis. There must be clearly defined benefits to installing a phyto-cover, including reduced construction and/or O&M costs, enhanced biodegradation, or stabilization of the wastes, and equivalent protection of human health and the environment.

Case Histories for Phyto-Covers

Municipal and Industrial Landfill, Tennessee

This two acre site was used for municipal and industrial waste disposal activities from about 1956 to 1963. Disposal activities at the site reportedly involved digging pits, filling them with waste, and covering the waste with soil.

Soils at this site are a mixture of clay and silt, with fragments of chert bedrock. The thickness of soil ranges from 0 to 25 feet across the site. A stream, which is fed by groundwater discharge and overland runoff, is present on three sides of the site. Depth to groundwater is variable across the site, ranging from less than 5 feet to nearly 25 feet with the most shallow groundwater unit discharging into the stream. Impacts to media at the site were minimal. Although VOCs and PAHs were detected in soil and groundwater at the site, the results from a risk assessment indicated that the only unacceptable risks were hypothetical future risks associated with exposure to thallium in soil. The occurrence of thallium in soil was attributed to the previous use of rodenticides at the site.

As part of a focused feasibility study (FFS), several remedial alternatives were evaluated. The objectives of the remedy were to protect future residents from the risks associated with ingestion and contact with impacted soils or waste, control infiltration of surface water, and reduce generation of leachate. A quantitative water balance, demonstrating the effects of the water holding capacity of the soil cover and increased evapotranspiration, was prepared prior to recommending a phyto-cover for the site.

The phyto-cover features a soil and compost layer and soil amendments with a stand of fast growing hybrid poplar trees and grasses that quickly established a dense canopy and deep rooting system. The soil cover, trees, and grasses serve to eliminate potential direct contact with waste and impacted soil. Infiltration of precipitation is limited via two processes. First, the canopy of leaves intercepts rainfall and reduces the amount reaching the ground surface. Second, the moisture that does enter the ground is extracted by the roots during plant respiration. Water that falls on the cover during the dormant season of the trees is stored in the compost layer for evapotranspiration during the following growing season.

Construction of Phyto-Cover

Site preparation activities began in February 1998, and included clearing of all existing vegetation from the site and construction of berms to prevent wind and rain erosion during site work. The site was subsequently graded, and a 2 foot layer of topsoil, woodchips, and fertilizer were spread over the surface using a bulldozer. The topsoil and amendments were placed in two separate lifts.

The tree planting was initially scheduled to begin during mid April 1998, but rains delayed the activities until mid May. Using a planter attachment to a tractor, 2235 1 year old hybrid poplars were planted to depths ranging from 12 inches to 14 inches below land surface. Trees were typically placed 3 feet apart in rows, with a distance of 10 feet between rows. A map showing the final layout of the site is presented in Figure 15.

A grass cover consisting of a mixture of perennial ryegrass, red clover, and oats was subsequently planted using a broadcast spreader, followed by disking. Overall construction costs for the phyto-cover system were approximately half that of a traditional single barrier cover and O&M are significantly lower than a traditional landfill cap (e.g., the cost of the phyto-cover system was $143,000, compared to an estimated $280,000 for a traditional single barrier cover).

O&M activities are ongoing and consist of inspection of the site every 2 weeks, mowing, watering and trimming the trees after heavy rains or storms. O&M activities also include application of fertilizer and insecticides on an as-needed basis. Natural predation of the trees by beaver was problematic during the first year after the trees were planted in 1998. However, the predation problem was alleviated through the construction of iron fencing, 6 feet in length and buried to a depth of 1 foot. As of November 1999, the trees had grown from an initial height of 4 feet to between 8 and 10 feet in height, and a dense canopy of leaves and grasses cover the former waste disposal site.

Lakeside Reclamation Landfill, Beaverton, Oregon

This existing landfill located within the Tualatin River riparian border in Oregon was used for the disposition of residential and commercial wastes, including demolition debris, tree stumps, excavated soil, and yard wastes. In 1990, 7455 hybrid poplar trees were planted by the Lakeside Reclamation Landfill in a 0.6 acre closed cell. The goal of this phyto-cover was to remove water at rates greater than precipitation while growing a marketable crop of pulp wood. A final cap was installed on the landfilled materials, in accordance with State of Oregon regulations. An addi-

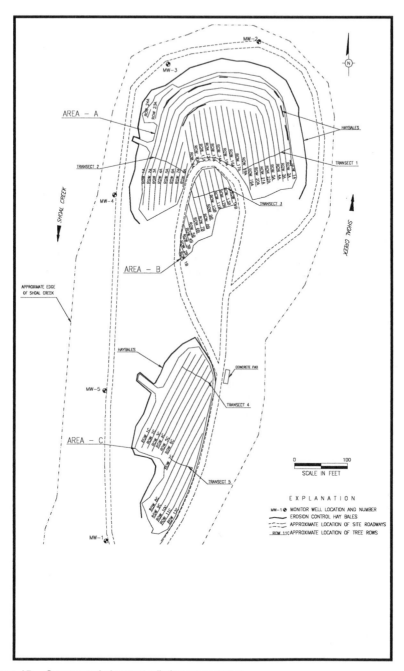

Figure 15   Constructed phytoremediation cover sytem.

tional 2 feet of soil was placed over the cover, and a planter was constructed by modifying an existing soil ripper that was pulled behind a crawler tractor. Poplar whips that were 2 feet in length and 5 feet in length were planted. The 5 foot trees were planted to a depth of 40 inches and the 2 foot trees were planted to a depth of 15 inches in more shallow ripper cuts (Madison, Licht, and Ricks 1991). Three different poplar species were planted, including Imperial Carolina, DO-1, and NE-19, to test the survivability and growth rates of the different species. Tensiometers were also installed on the landfill cap and in a background location to assist in determining the impact of the phyto-cover on soil moisture within the cap. These tensiometers were installed in nests at depths 1, 3, and 5 feet.

Survivability of the poplars during the first year of growth (1990), ranged from 54 percent (NE-19) to 93 percent (Imperial Carolina). The low survivability of the NE-19 variety was attributed to a heavy freeze. The trees grew to an average of 6.8 feet, and had an average trunk diameter of 0.7 inches. Although the trees appeared to be in good health, a portion of the trees were heavily foraged by grazing deer, and drift of a herbicide applied to a portion of the site resulted in the death of several nearby trees. After the trees had grown to a height above foraging levels of the deer and the foraging stopped, their growth rates improved.

The results from the tensiometer testing were difficult to interpret the first year. The tensiometers installed at a depth of 1 foot showed both wet and dry conditions during the growth season. The results from tensiometers installed at a depth of 3 feet showed dry conditions during the growth season and the results from the 5 feet interval showed a presence of moisture, but not saturation during the growth season. After the growth season ended, saturated conditions occurred at all three depths, indicating that the phyto-cover was effective the first year in reducing infiltration of water during the growth season. Generally 5 foot cuttings kept soils drier longer than the 2 foot cuttings (Madison, Licht, and Ricks 1991).

PHYTOREMEDIATION ENGINEERING CONSIDERATIONS

Phytoremediation is an active approach to achieving site remediation. Although it is an approach that utilizes natural processes, a successful phytoremediation system must have similar design considerations as more conventional approaches. A common pitfall is to oversimplify the technology or ignore basic engineering and management concepts when implementing a phytoremediation program. Phytoremediation engineering considerations include:

1. Fully characterized site conditions, including geology and agronomic conditions. Site characterization has to go above and beyond defining the lateral and vertical extent of contamination, since constituents other than regulatory driven contaminants can significantly impact a phytoremediation project. For example, the presence of high concentrations of salts, low or high soil pH, lack of nitrogen, etc., can be phytotoxic even though the regulated parameter, such as BTEX, may not be phytotoxic. Thorough agronomic testing can minimize the potential for phytotoxicity. Potting studies or greenhouse studies and/or pilot testing are recommended prior to the full-scale implementation of projects involving phytoextraction and rhizofiltration.

2. Clearly defined remedial objectives. As with any technology, the remedial objectives, including the cleanup goals and site end use, need to be clearly defined prior to implementing the technology. In the case of phytoremediation, goals can be quantitative (e.g., 1000 ppm DRO), or more qualitative (established vegetative cover to reduce windblown dust).

3. Timeframe for remediation. This is a new technology, and many of the specific cleanup timeframes have not been fully established. In addition, sites suited for this technology must have several years to meet remedial objectives. However, one approach to overcome this apparent time constraint is to implement a combination of phytoremediation and non-phytoremediation technologies, with the non-phytore-mediation technologies operating during the first several years that it takes the phytoremediation system to become fully operational.

4. Known fate of contaminants. The fate of contaminants needs to be understood and factored into the program for each site. The fate of some constituents such as nitrates, petroleum hydrocarbons, and selected VOCs is becoming more well established in plant systems, and can be anticipated by a review of the published literature. However, the fate of some contaminants in phytoremediation applications is not well established. In particular, the fate of heavy metals is highly specific to the type of plant species and the site conditions. A well-designed laboratory potting study, greenhouse experiment with soil from the site or on-site pilot study can document the fate of contaminants.

5. Species of vegetation and planting techniques. Selecting the appropriate type and species of plant will largely define whether a phytoremediation project is a success or a failure. As simple as this may seem, site constraints often result in less than optimal species being selected. Other factors that should be considered when selecting the species include water requirements and availability, tolerance to agro-nomic conditions, disease and insect resistance, acclimation, and use of native versus non-native species. Planting techniques can dictate where roots will penetrate, and can be the difference between success and failure. Planting can be as simple as trenching rows of poplar whips near the surface, to planting technologies such as Treemediation™, a method patented by Applied Natural Sciences of Hamilton, Ohio.

6. O&M. A common pitfall in the engineering community is the misconception that phytoremediation is free of maintenance. This is not true, and has lead to the failure of numerous projects. Although the costs of O&M for phytoremediation are less than costs for conventional remediation systems, O&M is essential or the project will fail. O&M requirements include watering, fertilization, and insecticide appli-cations. In addition, natural predation (by insects and animals) can lead to entire programs being wiped out in a short period of time. For example, a beaver colony nearly destroyed a program in Ohio, and a deer predation caused significant damage to trees at a site in Maryland. Where wildlife predation is possible, fencing or other access barriers to wildlife should be considered.

THE FUTURE OF PHYTOREMEDIATION

The future for phytoremediation holds promise in the overall remediation mar-ketplace as the trend toward more passive remedial technologies and recognition of the importance of natural processes continues. There will be growth in three areas,

(1) the types of vegetation used for phytoremediation; (2) expanded applications of phytoremediation and integration of phytoremediation with natural processes; and (3) engineered systems, architectural design, and site planning.

Currently the types of vegetation that have been shown to be effective in phytoremediation are growing rapidly, but are still relatively limited in the published literature. As phytoremediation matures, additional species of plants will be identified for contaminant specific applications and site-specific applications. Genetically engineered plants show promise for use on expanded types of contaminants and for obtaining higher valued lumber, and biochemicals/pharmaceutical products. Engineered phytoremediation treatment units (EPTUs), constructed wetlands, and alternative caps will be increasingly used on a wider variety of sites.

Phytoremediation will be used increasingly as a remedial alternative for brownfields redevelopment, based on its aesthetic appeal and overall low cost. Phytoremediation, natural processes, and conventional remedial technologies will be merged through innovative design for an effective, low-cost site remediation. For example, EPTUs, vegetation, and constructed wetlands could be integrated with conventional excavation and groundwater extraction for the remedial approach illustrated in Figure 16. Such an approach was proposed at an urban former MGP site with organic contamination of soil and shallow groundwater. This approach would have used excavation and soil redistribution to construct EPTUs, which could enhance biodegradation of organics in soil. Trees planted down-gradient on the property would be coupled with natural attenuation to remove and treat shallow contaminated groundwater on a seasonal basis. More conventional groundwater extraction would supplement the trees on a seasonal basis, and water pumped would be treated with a constructed wetlands and used for irrigation. Vegetation planted on the EPTU and

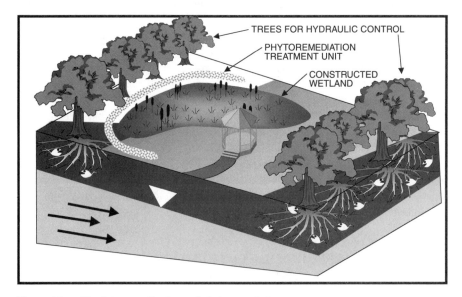

Figure 16   The future applications of phytoremediation.

in the constructed wetland would be specifically selected to address contaminants present at the site. The estimated cost for the integrated system was approximately $1,000,000 which is 20 percent of the cost of soil excavation and thermal treatment.

Further References

Phytoremediation is a rapidly evolving treatment technology, and readers should keep current with the changes. Sources of additional information include the journals such as the *International Journal of Phytoremediation* (published by CRC Press) and *Environmental Science and Technology* (published by the American Chemical Society). In addition, numerous internet sites (e.g., remediation technologies development forum, www.rtdf.org), internet newsgroups, and conferences are available for up-to-date research and developments on this promising technology.

REFERENCES

Anderson, T. A., Guthrie, E. A., and Walton, B. T., *Bioremediation: ES&T Critical Review*, Environ. Sci. & Technol., 27(13) 2630-2636, 1993.

Aprill, W., and Sims, R.C., "Evaluation of the Use of Prairie Grasses for Stimulating Polycyclic Aromatic Hydrocarbon Treatment in Soil," *Chemosphere* 20: 253-266, 1990.

Azadpour, A. and Matthews, J. E., *Remediation of Metal-Contaminated Sites Using Plants,* Remediation, Summer, pp. 1-18, 1996.

Banks, M.K., Govindaraju, R.S. Schwab, A.P., Kulakow, P. and Finn, J., *Phytoremediation of Hydrocarbon-Contaminated Soil,* Edited by Stephanie Fiorenza, Carroll L. Oubre and C. Herb Ward, Lewis Publishers, Boca Raton FL, 1999.

Bari, M.A. and Schofield, N.J., "Lowering of a Shallow, Saline Water Table by Extensive Eucalypt Reforestation, *Journal of Hydrology,* 133, pp. 273-291, 1992.

Briggs, G.G., Bromilow, R.H., and Bromilow, A.A., *Pesticide Science*, 13, 495-503, 1982.

Banuelos, G.S., Ajwa, H.A. Wu, L. and Zambrzuski, S., "Selenium Accumulation by Brassica Napus Grown in Se-Laden Soil from Different Depths of Kesterson Reservoir," *Journal of Soil Contamination*, 7(4):481-496, 1998.

Carman, E. P., Crossman, T. L., and Gatliff, E. G., "Phytoremediation of No. 2 Fuel Oil-Contaminated Soil," *Journal of Soil Contamination*, 7(4):455-466, 1998.

Cunningham, S., D., Berti, W. R., and Huang, J. W., "Remediation of Contaminated Soils and Sludges by Green Plants", Battelle Press, Third International *In Situ* and On-Site Bioreclamation Sympnosium, 1995.

Ensley, B. D., Dushenkov, V., Il. Raskin, and Salt, D.E., "Rhizofiltration: A New Technology to Remove Metals from Aqueous Streams," *New Remediation Technology in the Changing Environmental Arena*, B.J. Scheiner, T.D. Chatwin, H. El-Shall, S.K. Kawatra and A.E. Torma (Editors), Society for Mining, Metallurgy, and Exploration, Inc., Littleton, CO, 1994.

Farrington, P., Bartle, G.A, Watson, G.D., and Salama, R.B., "Long-Term Transpiration in Two Eucalypt Species in a Native Woodland Estimated by the Healt-Pulse Technique," *Australian Journal of Ecology,* 19: 17-25, 1994.

Ferro, A.M., Sims, R.C., and Bugbee, B., "Hycrest Wheatgrass Accelerates the Degradation of Pentachlorophenol in Soil, *J. Environ. Qual.,* 23:272-279, 1994.

Flathman, P. E. and Lanza, G. R., "Phytoremediation: Current Views on an Emerging Green Technology," *Journal of Soil Contamination* 7(4) 415-432, 1998.

Gatliff, G. "Vegetative Remediation Process Offers Advantages Over Traditional Pump and Treat Technologies," *Remediation,* pp 343-352, Summer 1994.

Hiltner, L., Ueber neuere Erfahrungen und Probleme auf dem Gebiet der Boden-bakteriologie und unter besonderer Berucksichtingung der Grundingung und Brache., Arb. Deut. Landw. Ges. 98:59-78, 1904.

Kelley, S. L., Aitchison, E. W., Schnoor, J. L., and Alvarez, P. J. J., "Bioaugmentation of Poplar Roots with Amycolata Sp. CB1190 to Enhance Phytoremediation of 1,4-Dioxane., Proceedings," The Fifth International *In Situ* and On-Site Bioremediation Symposium, Andrea Leeson and Bruce C. Alleman, Editors, 5(6) pp. 139-144.

Loftis, D. R. Hydrogeologic Analysis and Data Collection for the Oneida Tie Yard Site, Unpublished Master's Degree Thesis, Virginia Polytechnical Institute, May 1999.

Madison, M. F., Licht, L. A., and Ricks, F. B., "Landfill Cap Closure Utilizing a Tree Ecosystem," American Society of Agricultural Engineers, Paper 912137, Presented at International Summer Meeting, Albuquerque, New Mexico, 25 pages, 1991.

Miami Conservancy District, Aquifer Update No 1, Vol 1, 1991.

Negri, M. C., Hinchman, R. R. and Gatliff, E. G., Phytoremediation: Using Green Plants to Cleanup Contaminated Soil, Groundwater, and Wastewater. Argonne National Lab., IL. 6 pages, 1996.

Newman, L. A., Doty, S. L., Gery, K. L., Heilman, P. E., Muiznieks, I. A., Shang, T. Q., Siemieniec, S. T., Strand, S. E., Wang, X., Wilson, A. M., and Gordon, M. P., "Phytoremediation of Organic Contaminants: A Review of Phytoremediation Research at the University of Washington," *Journal of Soil Contamination,* 7(4):531-542, 1998.

Newman, L. A., Wang, X., Muiznieks, I. A., Ekuan, G., Ruszaj, M., Cortellucci, R., Domroes, D., Karscig, G., Newman, T., Crampton, R. S., Hashmonay, R. A., Yost, M. G., Heilman, P. E., Duffy, J., Gordon, M., and Strand, S. E., "Remediation of Trichloroethylene in an Artificial Aquifer with Trees: A Controlled Field Study," *Environ, Sci and Technol,* 33(13) 2257-2265, 1999.

Pierzynski, G.M., Schnoor, J.L., Banks M.K., Tracy, J.C., Licht, L.A., and Erickson, L.E., "Vegetative Remediation at Superfund Sites," from "Mining and its Enviromental Impact" Issues, *Environmental Science and Technology,* No. 1., R.E. Hester and R.M. Harrison Editors, pp. 49-69, 1994.

Potter, S., Stack, T. M., and Suthersan, S., "Landfills: Another Look at Cover Design," *Civil Engineering,* 1999.

Rovira, A.D. and Davey, C.B., *The Plant Root and Its Environment,* University Press, Charlottesville, VA, 1974.

Sandmann, E.R.I.C. and Loos, M.A., Chemosphere 13: 1073-1084, 1994.

Schnoor, J., L., Licht, L., A., McCutcheon, S. C., Wolf, N. L., and Carreira, L. H., "Phytoremediation of Organic and Nutrient Contaminants," *Environ. Sci and Technol.,* 29(7), 1995.

Schwab, A.P. and Banks, M. K. "Biologically Mediated Dissipation of Polyaromatic Hydrocarbons in the Root Zone." T.A. Anderson and J.R. Coats (Eds), *Bioremediation Through Rhizosphere Technology.* American Chemical Society, ACS Symposium Series 563, pp. 132-141, 1994.

Stack, T. M., Potter, S. T., and Suthersan, S. S., "Putting Down the Roots; Phytocovers Use Trees Shrubs and Crops to Lock Waste into Landfills," *Civil Engineering,* April, pp. 46-49, 1999.

Taylor, G.T., *Journal of Plant Nutrition,* 10, 1213-1222, 1987.

U.S. Army Environmental Center, "Demonstration Results of Phytoremediation of Explosives-Contaminated Groundwater Using Constructed Wetlands at the Milan Army Ammunition Plant, Milan, Tennessee," Report Number SFIM-AEC-ET-CR-95090, 1999.

United States Environmental Protection Agency, *A Citizen's Guide to Phytoremediation*, EPA 542-F-98-011, 1998.

United States Environmental Protection Agency, *Tech Trends*, USEPA Document EPA 542-N-99-005, August Issue No. 34, 1999.

CHAPTER **10**

Fracturing

Donald F. Kidd

CONTENTS

1-56670-528-2/01/$0.00+$.50
© 2001 by CRC Press LLC

INTRODUCTION

Regardless of the carrier fluid, low permeability, fine grained soils and rock represent a significant challenge to *in situ* contaminant remediation alternatives. We have already discussed the problems of moving air and water carriers through these types of geologic conditions. Without the movement of these carrier fluids, *in situ* remediation methods are severely limited in their effectiveness. Despite the low permeability of clays, silts, and competent rock, these geologic formations can still become impacted. Over time, organic contaminants can permeate throughout a wide area of the subsurface in vapor, nonaqueous (NAPL), and aqueous phases, migrating through natural fractures and by diffusion into the fine grained soils. Once in these zones, rapid or sufficient removal of the contaminants is difficult to achieve, if not impossible.

Within low permeability settings, excavation and above-ground treatment/disposal or encapsulation are commonly selected remedies. As with all above-ground remediation, the excavation process may actually enhance the potential exposure of the population to the subsurface contaminants during the process. This is always an objectionable consequence of the cleanup process. The excavation process is also disruptive to ongoing facility operations, and impacted soil transported to a landfill poses some long-term liability.

Another technology, fracturing for permeability enhancement, is rapidly being developed to address these low permeability zones. The limitations on achievable contaminant reduction *in situ* are mainly due to inadequate carrier fluid exchange frequency and/or nonuniform distribution of the carrier fluid. The fracturing process seeks to increase soil permeability within discrete zones through the production of high permeability fractures. Both hydraulic and pneumatic fracturing are designed with this purpose in mind.

APPLICABILITY

Almost any rock or soil formation can be fractured, given enough time, energy, and determination. The key aspects that have to be considered for remediation purposes are: Will the benefit derived from fracturing offset the cost of the process, and what are the risks and benefits of the process? Armed with the answers to these two questions, the decision to proceed with testing and, ultimately, full-scale application of the technique can be made on an informed basis.

Fracturing is most appropriately applied to soils where the natural permeability is insufficient to allow adequate carrier movement to achieve project objectives in the desired time frame. The following soil types and rock are generally treatable with the fracturing technologies (Schuring and Chan 1993):

- Silty clay/clayey silt
- Sandy silt/silty sand
- Clayey sand
- Sandstone

- Siltstone
- Limestone
- Shale

Fracturing a sand or gravel formation, while possible, is probably not justified because the increase in soil permeability would likely be incremental.

Fracturing technologies are equally applicable to both vadose zone (unsaturated) soils and saturated soils within an aquifer. The idea is to improve the flow of carrier fluids for contaminant removal, or delivery of nutrients or reactive agents.

By itself, fracturing is not a remediation process. There is no inherent advantage to having contaminants in contact with a high permeability formation and, in fact, there can be disadvantages to this situation. Fracturing has to be combined with some other technology to be of benefit for reducing contaminant concentration, mobility, or both. Essentially, fracturing serves solely to engineer changes in the subsurface so that carriers can more effectively reach the contaminants of concern. Contaminant removal and encapsulation processes by degradation, volatilization, dissolution (leaching), and stabilization are still controlled by the characteristics of the contaminants and the impacted media.

The important process of diffusion has been previously discussed in terms of how it impacts the spreading of contamination in the subsurface and the implications on the cleanup process. Controlling or limiting the role of diffusion on remediation is perhaps the most promising aspect of induced fracture formation. The importance of this understanding cannot be overemphasized to those responsible for developing remediation programs. Chapter 1 describes diffusion as a process by which dissolved chemicals move independently of the primary, advective flow path of impacted water or soil vapors. The chemicals can move into materials of low flow or even noflow (stagnant) conditions. Diffusion based flow occurs due to molecular movement and is enhanced by differences in dissolved concentrations between areas of high flow into the relatively clean, low flow materials. When trying to reverse the contamination process (cleaning the impacted aquifer or vadose zone), these areas of diffusion created pockets of contaminants can significantly extend the life of a project.

By fracturing, not only do we create higher permeability zones for enhancement of advective flow through the impacted material, we also shorten the pathway for diffusion controlled flow of the carrier fluid. The creation of advective flow channels and shortened pathways for the lower velocity diffusive flow result in enhancement of the carrier delivery (or recovery) process. The final cleanup level attainable by fracturing and the associated remediation process will still be governed by characteristics of the soil/rock and contaminant. Diffusion limited extraction will still influence the rate of contaminant recovery even after fracturing, and contaminant/media attractive forces will still influence the final concentration. This is important to understand.

As discussed in Chapter 3 (Vapor Extraction and Bioventing), the flow volume of the carrier fluid has a significant impact upon the rate of contaminant removal. The flow volume of the carrier is in turn a function of soil permeability and the *in situ* pressure gradient (pressure values between points of reference). The vapor velocity in the subsurface is then limited by the achievable volumetric flow rate at the extraction well. The following equation can be used to approximate this volumetric flow rate given knowledge or estimation of soil permeability and radial influence

$$Q/H = \pi * (k/\mu) * P_w * [1-(P_{atm}/P_w)^2]/\ln(R_w/R_I)$$
(Johnson, Kemblowski, and Colthart 1990)

where, Q = flow rate (ft³/min); H = screen length (ft); k = soil permeability (ft² or darcy); P_w = well pressure (atm); P_{atm} = atmospheric pressure (atm); R_w = well radius (ft); and R_I = radial influence (ft).

Figure 1 illustrates the relationship between pressure gradient (applied vacuum levels), sediment permeability, and vapor withdrawal rates (Johnson, Kemblowski, and Colthart 1990).

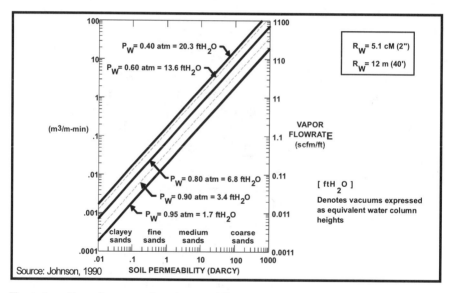

Figure 1   Vapor flow vs. sediment permeability.

The vapor flow velocity diminishes rapidly with distance from the point of extraction, again as discussed in Chapter 3. This occurrence results from expansion of the area through which the vapors pass. Mathematically, the vapor velocity through the soil or rock is calculated as flow (Q) divided by area (A). As an example of this relationship, Figure 2 illustrates a typical extraction well constructed with a 10 foot screened section within a low permeability clayey sand (k = 0.1 darcy). As illustrated on this figure, the area of flow is defined as:

$$A = 2 * \pi * r * L$$

where, A = area (ft²); r = radial distance (ft); and L = screen length (ft).

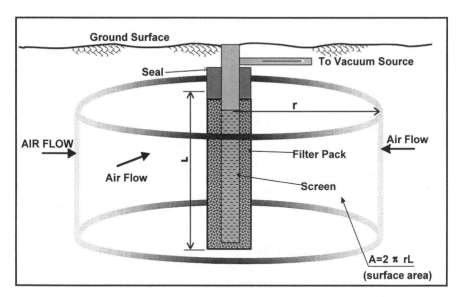

Figure 2   Illustration of radial flow and area of flow.

Note this estimation of flow area assumes that flow into the extraction well is primarily horizontal such that $k_v \ll k_h$. (K_v is vertical permeability and K_h is horizontal permeability.) This approximation is especially valid for materials with a significant fraction of fine grained particles.

For our hypothetical situation, an applied vacuum of 12 inches of mercury (in. Hg.) is predicted to result in a flow of 1.1 scfm based on the above equation. Figure 3 illustrates the predicted vapor velocity versus radial distance from the extraction well, again assuming that the vertical flow component is negligible. A horizontal line is also placed at approximately 0.01 ft/min which represents the minimum critical velocity described in Chapter 3. In summary, this critical velocity represents the minimum recommended velocity to optimize contaminant recovery rates. As shown on Figure 3, for a clayey sand, the critical velocity occurs at distances less than two feet from the extraction well. To maintain vapor velocity in excess of the critical value, well spacings would be tight. The benefit of fracturing in the situation described above is that more vapor can be withdrawn from the subsurface and the desired velocity profile can be extended further out into the contaminated formation.

Fracturing can also expand the applicability of other *in situ* remedial technologies beyond vapor and liquid extraction. As an example, hydraulic or pneumatic fracturing of a low permeability vadose zone overlying a more transmissive geologic unit can allow the use of air sparging (detailed in Chapter 5) in a geologic setting which is normally unsuitable. Generally, the injection of air beneath a low permeability formation can initially result in organic-laden vapor accumulation beneath this zone, and eventually the lateral migration of these vapors. With the uncontrolled migration of contaminant vapors beyond the influence of a collection system, errant emissions can result leading to unforeseen exposure routes and/or the contaminants

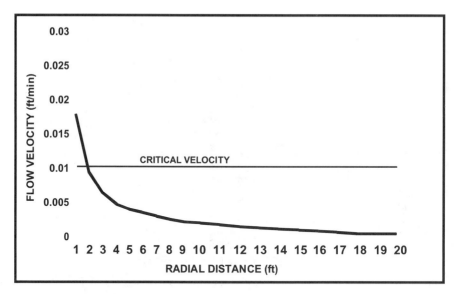

Figure 3   Flow velocity vs. distance.

may re-enter the groundwater through dissolution. The latter case would result in the expansion of the dissolved plume initially targeted by the remedial action. By installing a fracture network above the zone of aeration (sparging), the vapor collection system can recover the stripped contaminates thereby avoiding these two undesirable occurrences.

Figure 4 provides a conceptual illustration of this application of fracturing. For the case illustrated, the formation overlying the impacted water producing sand zone is fractured and connected to a vapor collection system. Through the fractures, the vapors containing elevated contaminant levels are provided a capture zone.

Geologic Conditions

As with all remedial techniques, fracturing is beneficial for environmental remediation only for a range of site conditions. In addition to the consideration of soil/rock types described in the previous section, the mode of deposition and changes occurring after deposition affect the effectiveness of fracturing. Most notably, the state of *in situ* stresses has long been characterized as the primary variable in the orientation of fracture formation (Hubbert 1957).

When fractures are formed by the injection of fluids, they are oriented perpendicular to the axis of least principal stress with propagation following the path of least resistance. For environmental remediation, horizontal fractures are of the greatest benefit. Vertically oriented fractures offer limited additional benefit to remediation as the fractures will tend to reach the ground surface at a relatively short distance from the injection point. Normally consolidated formations and/or fill materials have been found to produce vertically oriented fractures. For vapor extraction technologies, described in detail in Chapter 3, the short circuiting of

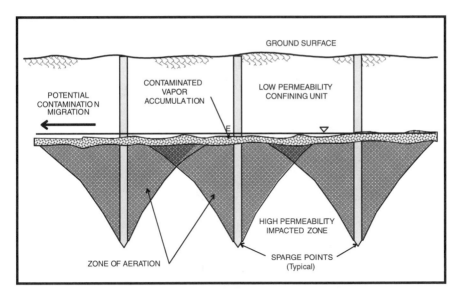

Figure 4   Sparging under low permeable soil.

vapor flow resulting from vertical fractures would actually be detrimental to the cleanup effort. Essentially, the soil vapor will follow the path of least resistance through the fractured media with little influence occurring beyond these engineered, preferential pathways.

In situ stress fields are subdivided into horizontal (x and y direction) and vertical (z direction) components. When initially deposited, sedimentary formations represent essentially hydrostatic conditions whereby the three principal stresses are in equilibrium and are equal to the weight of overburden. External forces (tectonics, burial/excavation, glaciation, and cycles of desiccation/wetting) after deposition then modify these stress fields.

Over-consolidation is defined as compaction of sedimentary materials exceeding that which was achieved by the existing overburden. Again, changes to the *in situ* stress fields after deposition have imparted a residual stress component to the formations. Over-consolidation of soils specifically results in stress fields favorable for fracturing. In this instance, the least principal stress in the vertical direction. The induced fractures would again be created perpendicular to this stress and be horizontally oriented. Figure 5 illustrates the concept of stress fields showing both equal and unequal stresses and the resulting orientation of fractures.

The formation and later retreat of glaciers is one condition which results in over-consolidation. The weight of the ice on the soil initially compacts the sedimentary grains. When the ice melts, the vertical stress is relaxed but the horizontal stress still maintains a residual component of the loaded conditions. This is not the only condition which results in over-consolidation, however. Erosion or overburden removal by excavation also presents conditions which relax the vertical stress field. Additionally, the cyclic swelling and desiccation of clay-rich formations can also create conditions of over-consolidation.

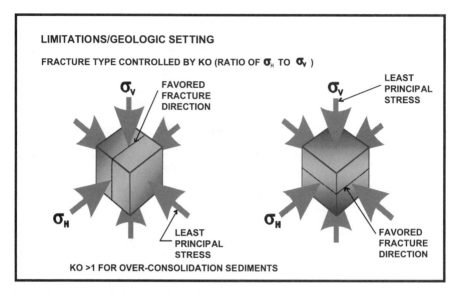

Figure 5   Pneumatic/hydraulic fracturing.

TECHNOLOGY DESCRIPTION

With the current state of the technology described in this chapter, there are two types of fracturing methodologies employed for environmental applications. Hydraulic (water based) and pneumatic (air based) fracturing variants of permeability enhancement are described in the following sections. The selection between these two types of fracturing are based on these considerations:

- Soil structure and stress fields
- Contractor availability
- Target depth
- Desired areal influence
- Acceptability of fluid injection by regulatory agencies.

Hydraulic Fracturing

Hydraulic fracturing was first developed as a means of enhancing oil and gas production. The first successful fractures completed for this purpose are credited to the Hugoton gas field in Grant County, Kansas in 1947 (Gidley et al. 1989). Early fracturing fluids were a gasoline-based, napalm gel and contributed significantly to the hazards of fracture installation. Since its beginning, more than 1 million fracture treatments have been completed. Currently, 35 to 40 percent of all production wells are fractured to enhance production rates. The process is reportedly responsible for making 25 to 30 percent of United States oil reserves economically viable. In other words, many oil and gas reserves would not be produced with only naturally occurring pressure distributions and gravity drainage controlling recovery rates. The parallels between economic recovery of petroleum hydrocarbons and viability of *in situ* treatment alternatives are evident.

As the names imply, the primary difference between hydraulic and pneumatic fracturing for permeability enhancement is the penetrating fluids used by each technology to create the subsurface fractures. Hydraulic fracturing fluids are characteristically viscous, produce minimal fluid losses to the formation, and have good post-treatment breakdown characteristics. High viscosity is desirable for the fluids to create a wide fracture and transport the proppants into the formation. Low fluid loss is important to minimize the volume of injected fluid while achieving the desired penetration. Finally, post-treatment breakdown is necessary such that the injected fluids do not clog the formation. Cross-linked guar is an example of a common fracture fluid used for both petroleum reservoir stimulation and for environmental applications. This fluid is a common thickener used in the food production industry which essentially breaks down to water with very little residual materials deposited into the formation. A food-grade carrier fluid minimizes the potential for regulatory objections to the process.

Because of the characteristic high viscosity, fracture fluids are capable of transporting particles (termed propping agents or proppants) through the fractures out into the formation. These proppants then support the fractures upon relaxation of the injection pressure and, to some degree, prevent closure of the fractures. Silica sand is most commonly used in both environmental and petroleum applications due to its relatively low expense, range of particle size, and general availability. The use and applicability of proppants other than sand are described in the section on proppants later in this chapter.

Hydraulic fracturing is a sequenced process in which multiple fractures can be generated within the impacted soil or rock formation. The separation between fractures is dependent upon an economical evaluation and the physical characteristics of the soil. The desired result of the fracturing process is a formation which allows for the effective delivery of carrier fluids and results in either a more rapid reduction of contaminant concentration, minimization of project costs, or ideally, both of these occurrences. The mechanics of the fracturing process are described in a later section.

Pneumatic Fracturing

Fracturing of soil or rock formations can also be accomplished using a compressed air or other gas source. As with the hydraulic variant, pneumatic fracturing proceeds by isolating discrete zones of the formation and applying energy (in this case compressed gas). Inflatable packers with delivery nozzles within the isolated intervals of the formation are typically used (Figure 6).

To create the fractures pneumatically, compressed air is supplied at a pressure and flow that exceed both the *in situ* stresses and the permeability of the material. This energy then fractures the material and creates conductive channels radiating from the point of injection (Schuring, Jurka, and Chan 1991/1992). Injection pressures on the order of 150 psig and flow rates as high as 800 scfm or higher are used to create the fractures.

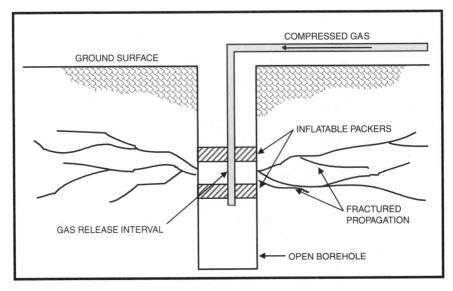

Figure 6   Pneumatic fracturing schematic.

The pneumatic fracturing procedure typically does not include the intentional deposition of foreign proppants to maintain fracture stability. The created fractures are thought to be self-propping. Essentially, disruption of the soil or rock structure during the injection of pressure results in localized realignment of grains within the fracture preventing closure after relaxation of the pressure. Testing to date has confirmed fracture viability in excess of two years, although the longevity is expected to be highly site-specific (Schuring and Chan 1993).

Without the carrier fluids used in hydraulic fracturing, there are no concerns with fluid breakdown characteristics for pneumatic fracturing. There is also the potential for higher permeabilities within the fractures created pneumatically (compared to hydraulic fractures) as these are essentially air space and devoid of proppants.

SCREENING TOOLS

Fracturing success is dependent on the application of both sound engineering and sound judgment. The data base of cleanup sites for which fracturing has been applied for testing and more so for full-scale remediation is limited. With continued testing and reporting of both successes and failures, our understanding of the technology will develop to the point where geologic conditions favoring the technology will become better understood.

Screening a site for possible application of fracturing first requires that the project team not only understand the mechanics and applicability of fracturing to enhance permeability, but also the implications for the ultimate cleanup technology. By the point in the project when remedial alternatives are being considered, the extent of the contaminant impact should be defined. From this information, a preliminary estimate of the number of fractures necessary to provide coverage can be assessed from previous case studies. As a general rule of thumb, fracture formation in the range from 20 to 35 feet or more is possible for near-surface soils and, with all other factors remaining the same, increases with depth of burial. The relationship between burial depth (loading) and fracture dimensions needs to be considered for a full-scale application. Specifically, more closely-spaced shallow fractures may need to be created to achieve the desired end result. Fracture propagation in rock formations has been found to be greater than in soil formations, primarily due to the competence and cohesion of these units.

Knowing the limitations of the technology may result in its early disqualification based on site conditions and proposed objectives. As with any technology, the timely elimination of an alternative on a sound basis may result in overall project cost reductions and heightened focus on the final remedy.

Geologic Characterization

A primary step in the evaluation of fracturing applicability is through an examination of detailed and accurate geologic cross-sections illustrating sediment layering and the relationship between contaminants (target zone) and the different grain sizes. Because contaminants often reside within low permeability, fine-grained soils, this relationship is important to understand.

At least one continuous core boring should be installed during the remedial investigation phase of the project to characterize minor changes in lithology. Cores collected during continuous and depth-specific sampling should also be examined for factors contributing to secondary permeability (i.e., coarse-grained sediment inclusions, bioturbations, and naturally-occurring fractures). These secondary permeability characteristics of the soil or rock formation may influence the creation of engineered fractures. Pneumatic fractures, in particular, may propagate along existing fracture patterns. Hydraulic fractures have been found to be less influenced by existing fractures (Murdoch 1993).

Geotechnical Evaluations

In addition to qualitative site evaluations described above, target zone soil samples can be submitted for geotechnical evaluations of grain size analysis, Atterberg limits (plasticity and liquid limit testing), moisture content, cohesion, and stress evaluations (unconfined compressive strength). Details and implications of these tests are as follows:

- Grain size analysis: Although fractures can be created in sediments and rock of nearly any grain size, the highest degree of permeability improvement can be expected from the finer grained soils.
- Permeability: As discussed previously, fracturing is generally applied at sites with characteristically low permeability. A baseline estimate of permeability (vapor and/or liquid) is often available from testing conducted at the site during site

investigation. This baseline estimate of permeability provides a basis of for evaluating the necessity, benefit, and effectiveness of the fracturing process. In general, greater improvement of carrier fluid flow or radial influence (in terms of percentage) is observed in formations with lower initial permeability.

- Atterberg limits: This parameter characterizes the plasticity of a soil. In general, fractures created in highly plastic clays will not propagate as well as in more brittle materials. Formations having $W_n < W_l$ are most suitable for artificial fracturing, where W_n is the natural moisture content and W_l is the liquid limit. Soils having $W_n > W_l$ (or liquidity index >0) may liquify under a sudden shock imparted during the fracturing process. The estimation of Wn and Wp (plastic limit) would also give an indication of the degree of consolidation of soil. If Wn is closer to Wp than to Wl, the soil may be over-consolidated. If Wn is closer to Wl (or larger), the soil may be normally consolidated.

- Moisture content: Overall soil permeability improvements are achievable with fracturing; however, vapor flow in particular is also controlled by soil moisture. Improvements in vapor flow through highly saturated soils (at or near field capacity) will not be achieved by the production of fracturing alone. Additional means of moisture removal may be required to obtain the desired effect from fracturing in these instances.

- Cohesion: Generally, the more cohesive the soil, the more amenable to fracturing and, upon relaxation of fracture stresses, longevity of the fractures. Fracturing in cohesive soils, such as silty clays, has been particularly successful.

- Unconfined Compressive Strength: The unconfined compressive strength can be used for predicting the orientation and direction of propagation of fractures. As evident from previous discussions, the state of *in situ* stresses plays a key role in the orientation and ultimate utility of engineered permeability enhancement. The artificially induced fractures are assumed to be vertical in normally consolidated soil and horizontal in over-consolidated (or preconsolidated) deposits.

Pilot Testing

Upon completion of the preliminary screening and geotechnical testing, pilot testing is typically conducted for further performance evaluation and to provide a design basis for a full-scale system. Pilot testing is by far the most powerful and useful means of screening a site for a full-scale remedial program incorporating fracturing. Experience has shown that preliminary screening of a site cannot always accurately predict the applicability or performance of either hydraulic or pneumatic fracturing

To conduct pilot testing, it is wise to select a contractor who not only understands the applicability and limitations of fracturing, but also has specific experience in the field application. As the technology advances from infancy to a more mature status, the field of contractors meeting this criteria will increase along with the state of the art.

The pilot test plan should incorporate the following:

- Area selection
- Baseline permeability/mass recovery estimation (if not already done)
- Fracture point installation
- Test method and monitoring

Area Selection

Area selection of the cleanup site is the first step in designing the pilot program. The decision must be made whether to test the technology within the impacted areas of the site or conduct testing outside the contaminant zone. It is generally preferred to test within the area of impact to reduce the impact of lateral inhomogeneities and to provide data on contaminant recovery rates before and after fracturing. After all, the second criteria is the primary benchmark of fracturing performance: the influence and removal of contaminants.

For pilot testing of a single fracture well installation, an area of approximately 4,000 to 8,000 square feet should be sufficient. This area would encompass the anticipated maximum limits of the fracture propagation. Figure 7 provides a typical pilot test configuration. Note that both existing monitoring wells, as available, and specially installed monitoring points are included within the test area. Testing procedures are briefly summarized below.

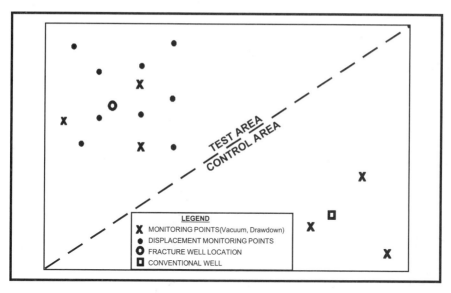

Figure 7   Pilot test configuration.

Baseline Permeability/Mass Recovery Estimation

To aid in the evaluation of fracturing benefits versus the costs and risks of the technology, a baseline estimate of soil permeability and contaminant mass recovery rates are typically conducted prior to attempting to fracture the formation. Testing for permeability and mass recovery potential are described in detail in Chapter 3 (Vacuum Extraction and Bioventing) and Chapter 4 (Vacuum-Enhanced Recovery). Because fracturing is generally considered for low permeability formations (i.e., $k_{air} < 1$ darcy, $k_h < 10^{-4}$ cm/sec), the procedures necessary to evaluate these important parameters would most likely follow those outlined for vacuum-enhanced remediation.

Fracture Point Installation

A specifically designed fracture point installation program is required for pilot testing of the fracturing technologies. The fracture intervals are selected to coincide with the known occurrence of the contaminants. Fracture locations are also targeted for the low permeability sediments or rock within a layered setting. The relationship between contaminant distribution and media permeability underscores the importance of a focused investigation program which provides this information during the assessment phase of the project.

With the variation between hydraulic and pneumatic fracturing, the field procedures for fracture point installation are briefly discussed separately.

Hydraulic Fracturing

Hydraulic fracturing for environmental applications initially involves drilling to near the depth of the fracture interval, advancing a lance to the depth of injection, and fracture initiation with a high pressure water nozzle (to create a notch with the desired horizontal orientation), followed by the controlled injection of the fracturing fluid and proppants. This multi-stage process is illustrated on Figure 8 (Murdoch 1991).

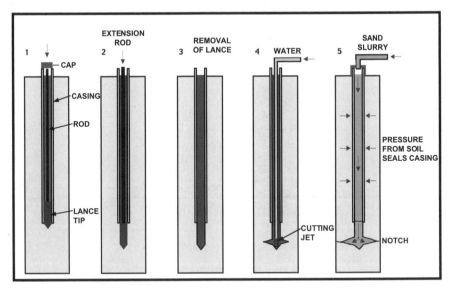

Figure 8   Multi-stage hydraulic fracturing process.

Because the fractures will naturally migrate toward the ground surface, the volume of each injection is restricted to minimize this occurrence. At sites where hydraulic fracturing is being tested for the first time, this may present a trial-and-error sequence where the optimum injection volume is only discovered after daylighting of a near-surface fracture has occurred. Daylighting is defined as the observation of fracture media at ground surface. Continue injection of fracture fluids beyond this point will likely not result in additional subsurface fracturing. Once established, the fracture which has reached ground surface provides a relatively easy escape route for the fluids.

Injection pressures for hydraulic fracturing are typically less than 100 psig with this pressure applied by pumps specifically designed for high viscosity, high solids fluid handling. The injection pressure is monitored throughout the process using a continuously recording transducer or similar measuring device. An example of a pressure curve generated during fracturing is illustrated on Figure 9. As shown on this figure, the pressure initially builds up as injection is initiated. The pressure reaches a peak value and drops off dramatically. This drop in injection pressure signals the initial propagation of the fracture. Once this fracture is completed, the lance is withdrawn from the soil, the borehole is advanced and the process is repeated. Successive fractures are created in this manner and illustrated on Figure 10. The optimum separation between fracture intervals is site-specific and best determined during field testing, but is typically on the order of 2 to 5 feet.

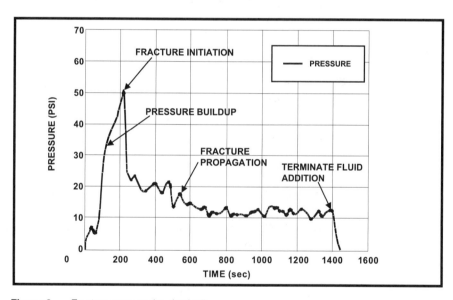

Figure 9   Fracture propagation (typical).

The high viscosity fracture fluid must break down in place so that the benefit of a relatively high permeability zone can be realized. A common fracture fluid, guar gum, is often mixed with an enzyme to hasten the breakdown process, although the guar will break down on its own, given sufficient time.

Upon reaching the desired maximum depth of fracture installation through the process described above, the bore hole is often completed using convention well installation techniques. The placement of a well central to the point of radiating fractures allows for the withdrawal of vapors and liquids through the relatively permeable zones containing secondary permeability and proppants. Testing of the

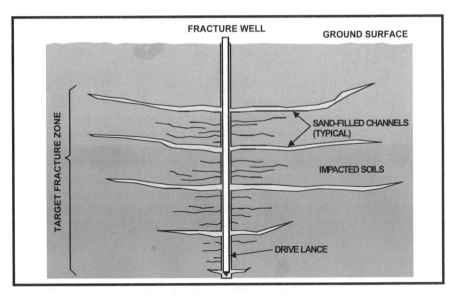

Figure 10   Successive fracturing of target zone.

fracture performance is then conducted, similar to design-testing of vapor extraction, vacuum-enhanced recovery, or conventional aquifer analysis.

Pneumatic Fracturing

For the pneumatic variant of the fracturing process, a bore hole is generally advanced to the desired depth of exploration and the augers are withdrawn. Variations on this technique may be required should noncohesive (e.g., flowing sands) be encountered, however. With the low permeability formation exposed within the borehole, a dual-interval packer is inserted and inflated sealing off a discrete zone (Figure 6). With this zone isolated, a high pressure, high volume compressed air source is introduced between the packers. After completion of this fracture, the packers are deflated and moved to the next interval.

Compression of the borehole wall may occur during inflation of the packers. A tight seal must be obtained or the injected air will simply bypass the packers and vent through the bore hole. With this compression, the bore hole may not be suitable for use in the completion of a well as can be done with hydraulic fracture points. Carrier fluid flows passing through the damaged bore hole may actually be lower after fracturing than a well completed in a more standard manner. The advantage gained during fracturing, therefore, would be eliminated.

Injection pressures as high as 150 psig and flows in excess of 500 scfm are used to initiate and propagate the fractures (Schuring, Jurka, and Chan 1991) for pneumatic fractures. In contrast to the relative quiescent injection process for hydraulic fracturing occurring over a period from 10 to 60 minutes per fracture, pneumatic fractures are created over a period of approximately 20 seconds.

The moderately high flows and high pressures for pneumatic fracturing can be supplied by a bank of compressed air cylinders which are recharged between fractures. This configuration can meet the specific requirements of the process and still be relatively transportable: a criteria which has to be met for the economic viability of the technology. Compressors capable of this air delivery rate and pressure, while available in an industrial setting, are not applicable for pneumatic fracturing field work.

Test Method and Monitoring

Pilot testing of the fracturing technologies is generally a two-step process. The first step is conducted during the actual installation of the fractures during which time the approximate dimension and orientation of the fracture pattern is determined. The second step in the testing process is to determine the influence to carrier fluid movement within and beyond the area of fracture propagation and the corresponding contaminant removal rate improvement achieved through the process.

Fracture Orientation

Ground surface displacement (heave) is generally recorded during fracturing by an array of survey points which are monitored in real time. The change in ground surface elevation during fracturing has been found to provide a reasonable approximation of the fracture locations in the subsurface.

For pneumatic fracturing, the surface heave during pressure application is substantially higher than after pressure relaxation. The residual heave is generally 10 to 20 percent of the maximum displacement (typically less than a few inches) and provides evidence of fractures creation. For hydraulic fracturing, the ground displacement is directly related to the volume of proppant injected into the soil with the displacement a close approximation of the fracture dimensions (thickness of fracture zones). For this method of fracturing, proppant layers up to an inch thick have been created at each depth of injection. The thickness of the fractures diminishes with distance from the point of injection.

For both pneumatic and hydraulic fracturing, the fracture zones are generally asymmetric about the point of injection, following the path of least resistance. Heterogeneity within the soil matrix, naturally occurring fracture patterns and, to a lesser degree, bedding planes appear to influence the orientation of the created fracture zones. Surface loading also influences the pathway of the fracture front. High surface loading created by man-made structures or changes in topography can also influence the fracture patterns. This tend can be beneficial in some instances, as temporary surface loading can be used to steer the fractures toward the desired location. Vehicles have been used successfully for this application.

Because of the displacement caused by the fracture formation, care must be exercised when working adjacent to buildings or other structures. While some structures can withstand these moderate displacements, the integrity of others may be compromised. A careful evaluation of the structure's strength and flexibility must be conducted prior to implementing a fracture program under circumstances such as these. This type of evaluation is beyond the scope of this text.

Carrier Fluid Influence

The dimension of fracture propagation is only one aspect of the pilot testing program, and probably not the most important consideration. With the underlying, driving force behind the technology being to improve the efficiency of carrier fluid transport and/or fluid delivery, the second stage of a typical testing program is to measure and quantify how the transport has been modified. The enhanced flow characteristics are then compared with the baseline estimate of carrier fluid flow characteristics, as discussed earlier, and a determination is made as to the relative benefit of the process. With this key information, the applicability of the process and, ultimately, the number, locations, and depth intervals of a full-scale fracture program can be designed.

As described in detail in previous chapters, the second phase of the pilot testing program may consist of a groundwater pumping test, vapor extraction test, or vacuum-enhanced pilot test.

PROPPANTS

In contrast to proppants utilized solely for maintenance of fracture aperture or viability, reactive or conductive agents have also been recently promoted due to their characteristic properties. For example, a time-release oxygen source was developed in cooperation with the EPA consisting primarily of sodium percarbonate (Vesper et al. 1993). This proppant, when injected into the impacted soils, reportedly will slowly release oxygen over a 4-month period. The advantage of this occurrence is that the aerobic conditions can be locally maintained in the soil without active vapor withdrawal or injection.

Similarly, testing is currently underway for the injection of iron filings for the creation of subsurface conditions favoring dehalogenation of chlorinated solvents. A flat-lying reactive wall is thus created which testing indicates can promote the accelerated attenuation of chlorinated solvents such as PCE and TCE without the high cost of extraction, above-ground treatment and disposal. The concepts behind reactive wall applications are presented in Chapter 11.

A graphite-based proppant is also being tested for the enhancement of electro-osmotic dewatering and *in situ* resistive heating due to its electroconductivity. This process, named the "Lasagna Process" by its developers, creates a sequence of flat-lying conductive beds within the impacted soil or rock. With the cyclic application of low voltage DC current to these beds, osmotic flow is induced within the impacted formation. An acid front is created within the treatment zone. The contaminants are desorbed and removed from the permeable fracture zones by vapor extraction or direct pumping. While this technology may be beneficial for relatively rapid con-taminant removal from low permeability formations, it has not been tested under field conditions as of this publication.

FULL-SCALE DESIGN

Upon completion of site screening, and pilot testing, a full-scale fracturing program may be implemented. A full-scale fracturing program (beyond the installation of a few fracture points) should not be attempted without field verification of its performance. The implementation of the full-scale program would be based on economic and feasibility evaluations. In essence, fracturing would be selected as a component of a final remedy if the cost of fracture creation is less than would be required for multiple well installations with lower well spacings or alternate strategies such as excavation and above-ground treatment/disposal.

Based on the field testing, a fracture well pattern would be selected to encompass the area of known contaminant impact. This pattern must take the asymmetric orientation of the fracture propagation. For instance, the fractures may be more closely spaced north to south than east to west due to asymmetry. To control the role of diffusion and the possible creation of low flow zones, an engineering safety factor should be applied such that the fractures overlap.

The depth intervals for the fractures should correspond with the known distribution of contaminants. This requirement again emphasizes the importance of site characterization.

Because of heterogeneity present at almost every site, the full-scale fracturing program should be designed with some flexibility in mind. In most instances, it would be wise to specify a range of possible fracture point placements with adjustments made during implementation of the program to optimize the performance of the fractures. For example, it may be necessary to tighten the fracture spacings and reduce fracture volume (for both pneumatic and hydraulic fracturing) if daylighting is found to occur in one or more areas of the site, even if pilot testing of the same parameters did not produce this result.

Depending on the size of the project and number of fracture points, it may also be advisable to implement the fracturing program in a phased approach. For example, fracture wells could be installed on a one-week cycle. During the first week, fracture points could be installed, followed by testing of these points for performance (carrier fluid extraction/movement rates). Adjustments can then be made for the next cycle of fracture installations.

The limiting role of diffusion cannot be over-emphasized. Even with fracturing, contaminant removal rates will be rate-limited by diffusional flow between the areas of high, advection-controlled flow. When compared to contaminant removal rates before fracturing, post-fractured rates will be higher, if the process is successful and applied under the right conditions. Eventually, however, diffusion-controlled flow will again predominate. The impacted soils may not have reached the target concentration by this time.

CASE HISTORIES

Fracturing tests, and to a lesser degree full-scale remediations, have been conducted at a number of facilities for both saturated and unsaturated zone soils. The following section provides a brief synopsis of some case studies conducted under the direction of the SITE program sponsored by the EPA. The case studies illustrate the performance of both pneumatic and hydraulic fracturing.

Pneumatic Fracturing Air Phase

A site in New Jersey was selected for a demonstration of the effectiveness of pneumatic fracturing. The impacted horizon at this site is characterized as siltstone and shale with naturally occurring fractures. Fractures were installed between 9 and 16 feet below grade. The primary contaminant was TCE.

Before fracturing, the vapor extraction rates from each of the tested wells was below the sensitivity of the measuring instrument (<0.6 scfm) at an applied vacuum of 136 inches of water. A single fracture well was installed central to the monitoring points, as illustrated on Figure 11. The distances between the fracture well and the monitoring points was 7.5 to 20 feet.

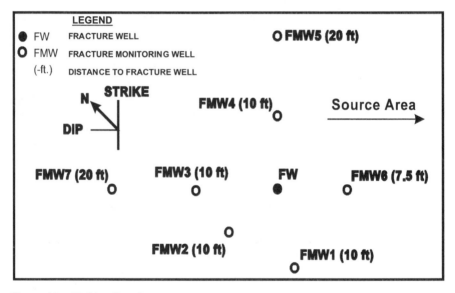

Figure 11   Well location plan.

Based on elevation measurements recorded by an electronic tiltmeter during fracturing, surface heave was observed up to 35 feet from the fracturing well. The flow rates from each of the test wells surrounding the fracture well increased substantially after fracturing. Specifically, the flow rate increased from 3 to more than 15 fold after fracturing as illustrated on Figure 12. Vacuum measurements within the monitoring points also increased post-fracture from 4 to almost 100 fold. Contaminant recovery rates increased by a factor of approximately 8 to 25 as a result of the permeability enhancement.

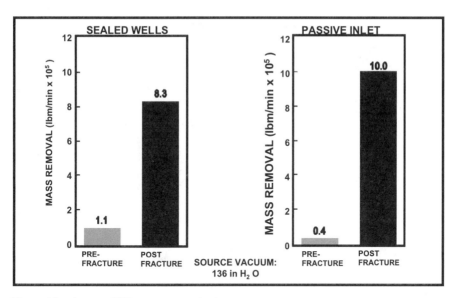

Figure 12   Average TCE mass removal rate.

A down hole video camera was used to observe the fracture patterns created in the formation. The inspection of the fracture well found that the primary mode of propagation was through existing fractures, although some new fractures were created.

EFFECTIVENESS—HYDRAULIC FRACTURING AIR PHASE

Hydraulic fracturing tests were conducted on vadose zone soils at a site in Illinois which were impacted with TCE, 1,1,1-TCA, 1,1-DCA, and PCE (USEPA 1993). The soils are characterized as a silty/clayey till to a depth of approximately 20 feet below grade. The conductivity of the soil was estimated at 10^{-7} to 10^{-8} cm/s. The pilot-scale demonstration created six fractures in two wells at depths of 6, 10 and 15 feet below grade over a one day period. A site plan showing well and monitoring point locations is provided as Figure 13.

At an applied vacuum in excess of 240 inches of water, the vacuum influence in unfractured soil was negligible, decreasing to a few tenths of an inch of water at a distance of 5 feet from RW-1 and RW-2, clearly demonstrating the limitations of a standard vapor extraction system under the site conditions.

The flow rates in unfractured soils were also low, at approximately 1 scfm. In the fractured soil, flow rates from approximately 14 to 23 scfm were achieved under similar vacuum levels (isolating one interval of RW-4 in which fractures vented to surface). Suction head measurements in the fractured soil also increased dramatically, up to 25 feet from the fracture wells. Contaminant recovery rates similarly increased in fractured soil from 7 to 14 times after fracturing.

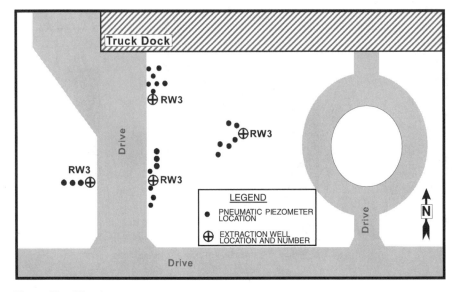

Figure 13   Site plan.

REFERENCES

Gidley, J. L., Holditch, S. A., Nierode, D.E., Veatch, R. W. Jr., *Recent Advances in Hydraulic Fracturing,* Society of Professional Engineers, 1989

Hubbert, M. King, "Mechanics of Hydraulic Fracturing," *Petroleum Transactions,* AIME, Vol 210, pg 153-166, 1957.

Johnson, P.C., Kemblowski, M.W., and Colthart, J.D., "Quantitative Analysis for the Cleanup of Hydrocarbon-Contaminated Soils by *In Situ* Soil Venting," *Groundwater,* 28(3), 1990.

Murdoch, L., "Feasibility of Hydraulic Fracturing of Soil to Improve Remedial Actions," EPA/600/S2-91/012, 1991.

Murdoch, L., "Hydraulic and Impulse Fracturing Techniques to Enhance the Remediation of Low Permeability Soils," Unpublished Paper, 1993.

Schuring, J. R., Jurka, V., Chan, P. C., "Pneumatic Fracturing to Remove VOC's," *Remediation,* Winter 1991/1992.

Schuring, J. R., Chan, P. C., "Pneumatic Fracturing of Low Permeability Formations - Technology Status Paper," Unpublished Paper, 1993.

USEPA Risk Reduction Environmental Laboratory, Superfund Innovative Technology Evaluation (SITE) Applications Analysis and Technology Evaluation Report, "Hydraulic Fracturing Technology," EPA/540/R-93/505, 1993.

Vesper, S.J., Murdoch, L.C., Hayes, S., Davis-Hooper, W.J., "Solid Oxygen Source for Bioremediation in Subsurface Soils," *Journal of Hazardous Materials,* 1993.

CHAPTER 11

Permeable Treatment Barriers

Peter L. Palmer

CONTENTS

1-56670-528-2/01/$0.00+$.50

INTRODUCTION

Permeable treatment barriers are an innovative technology that show a lot of promise for remediating shallow groundwater plumes. In principle, a permeable treatment barrier (also referred to as permeable reactive barriers) containing the appropriate treatment material is placed across the path of a contaminant plume. As contaminated groundwater moves through the barrier, the contaminants are removed or degraded, allowing uncontaminated water to continue its natural course through the flow system. Much of the early work on permeable treatment barriers has been performed by the Waterloo Center for Groundwater Research, University of Waterloo. Their work focused primarily on the use of a reactive material, zero-valent granular iron, to degrade halogenated organic compounds in groundwater. Although much of the focus on permeable treatment barriers today is on the application of zero-valent iron, some of the concepts developed have been applied to permeable treatment barriers that use media other than zero-valent iron to remediate impacted groundwater. A summary of early developments using zero-valent iron as permeable treatment barriers has been compiled by Gillham and Burris (1994). More recent developments, including application of zero-valent iron for remediating chlorinated solvents as well as other constituents including metals, inorganics, nutrients, and radionuclides have been compiled by the Environmental Protection Agency (EPA 1999).

Permeable treatment barriers are gaining a lot of attention not necessarily because they speed up the remediation process, but because they recognize the limitations of groundwater cleanup programs and factor these limitations into minimizing the lifecycle costs of remedial programs. Permeable treatment barriers generally rely on the natural movement of water to carry the contaminants through the treatment barrier where they are removed or degraded. By doing so, permeable treatment barriers eliminate or at least minimize mechanical systems, thus minimizing long-term operation and maintenance costs that so often drive up the lifecycle costs of remedial projects. Long-term operation and maintenance costs are reduced because the site generally does not need a continuous input in energy and manpower. Failures due to mechanical breakdowns are also reduced. In addition, technical and regulatory issues concerning discharge of treated groundwater are avoided or minimized.

This chapter will focus on concepts, applications, and methodologies for installing and using permeable treatment barriers in remedial programs.

DESIGN CONCEPTS

Permeable treatment barrier systems generally rely on groundwater to carry the contaminant to the barrier where it is removed or degraded. Since the vast majority of permeable treatment barriers rely on natural groundwater flow, the focus of this section on design concepts will be on these types of systems. Permeable treatment barriers can be constructed of a variety of materials or use natural geology. The system area can be used to create chemical or biochemical reactions or simply to facilitate a process to remove the contaminant. In its simplest form, the treatment barrier is similar to that shown in plan view in Figure 1 and in cross section in Figure 2. As the reader can see in these figures, a plume is migrating down-gradient from a source and an *in situ* permeable barrier is present to remediate the plume constituents *in situ*. For example, if the plume contained VOCs, the treatment barrier could be a series of air sparging points, which would introduce air into the plume and rely on the air to carry the contaminants vertically for release to the atmosphere or for capture by a vapor extraction system. This design could be used for degradable

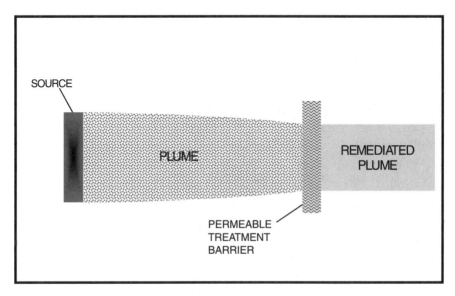

Figure 1   Plan view of permeable treatment barrier.

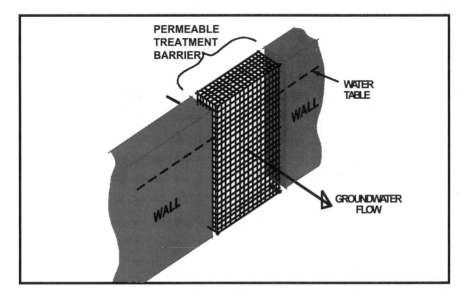

Figure 2   Cross section of permeable treatment barrier.

organics (oxygen transfer for the natural bacteria to degrade the compounds) or nondegradable volatile organics (air carrier to remove the compounds). This simple design would only be viable in geology that could facilitate air sparging.

Many times the geology prevents the application of a technology, or we need to add treatment material to the barrier. In these cases, we must remove the natural geology and create a permeable barrier out of the required material. To continue with the above example, if the native materials are low permeable sediments, then an additional purpose of the permeable barrier could be to change the geology by excavating the native material and back filling with more porous material which would be amenable to air sparging. Another method that is more widely applied for chlorinated hydrocarbons is the use of zero-valent iron to dehalogenate the compounds. In the above example, the natural geology would be removed so that zero-valent granular iron could be placed into the path of the groundwater. There are a number of ways that this could be achieved in order to minimize construction costs, and these are discussed later in this chapter.

To successfully remediate a plume, a permeable treatment barrier must be large enough to remediate the entire plume. For large or deep plumes this becomes impractical. To overcome this problem for large shallow plumes, a system can be installed consisting of low-permeability barriers, which funnel flow to a smaller permeable treatment barrier (referred to as a gate) to treat the plume (Figure 3). This concept was developed by the University of Waterloo and is referred to as the Funnel-and-Gate System™. There are a number of combinations/configurations that can be used to effectively control and remediate a plume (Starr and Cherry 1993). For instance, Figure 3 shows a single gate system, and Figure 4 shows a system consisting of three gates. When dealing with funnel and gate systems, in all cases, the sole purpose is to use the gate to pass contaminated groundwater through the treatment

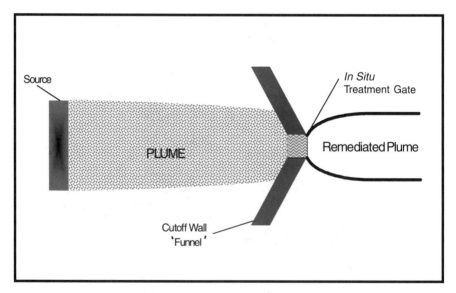

Figure 3   Funnel-and-Gate System™ using a single gate.

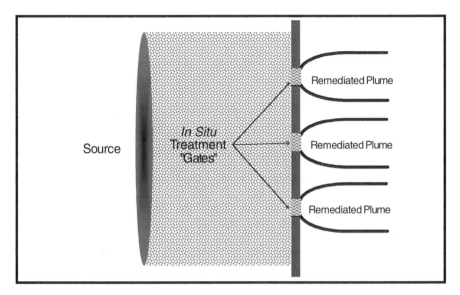

Figure 4   Funnel-and-Gate System™ using multiple gates.

barrier that remediates the groundwater. The funnel is integrated into the system to force water through the gates and is used for practical and economic reasons. Slurry walls, sheet piling, and other materials which could form the funnel are often easier and/or more economical to install than the treatment barrier. Consequently, the design is focused on balancing the ratio of funnel to gate areas to achieve remedial objectives at the least cost. It must be remembered that groundwater travels at a relatively low speed. Therefore, the residence time in the reactive portion of the treatment barrier (gate) can be significant even when the treatment portion of the barrier is not continuous.

Conceptually, plumes with a mixture of contaminants can be funneled through a gate with multiple treatment barriers in series. For instance, one treatment barrier could be used for degrading hydrocarbons and a second treatment barrier in series could be used for precipitating metals. This concept is illustrated in Figure 5. In addition, if the gate needs to be removed at some point during or after remediation such as the case with sorption processes (activated carbon, ion exchange, etc.), then considerations should be given to installing a retrievable treatment barrier. These could take the form of different shapes, but they would each have sufficient permeability to allow migration of groundwater through the container holding the reactive material, and the container would have sufficient strength to maintain its shape and structural integrity during placement and removal.

There are numerous nuances when installing a treatment barrier. For instance, Figure 6 shows a treatment barrier designed to remediate a shallow plume that is located in the uppermost portion of the aquifer. Since the plume remains shallow, the treatment barrier does not need to penetrate the entire thickness of the aquifer and is referred to as a hanging barrier (gate). In situations where a hanging barrier is under consideration, it is important to understand contaminant transport to ensure

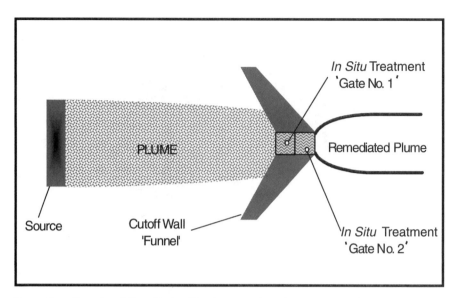

Figure 5   Funnel-and-Gate System™ using gates in series.

that the contaminants remain in the upper portion of the aquifer even after the subsurface is disturbed as a result of installation of the treatment barrier. If the barrier has a higher resistance to flow than the original geology, then the groundwater may flow around it and no treatment will occur.

In other cases where the saturated thickness is relatively thin compared to the overall aquifer thickness, a buried treatment barrier may be used. In these applica-

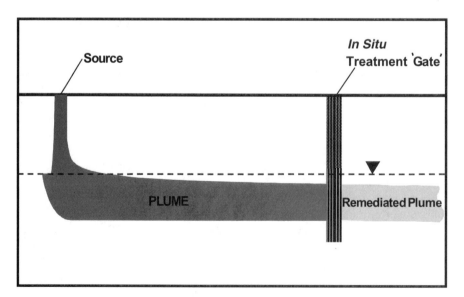

Figure 6   A hanging gate system for remediating shallow plumes.

tions, the treatment barrier is limited in height to that portion of the zone in which groundwater is moving. A low permeable barrier, usually a synthetic liner, can be placed above the treatment barrier as a precaution to ensure that groundwater flow over the permeable treatment barrier does not occur.

Geosiphons are now being used to entice groundwater to move through a permeable treatment barrier. This is a relatively new application and is somewhat limited in that it requires that the hydraulic head between the contaminant plume and a discharge point be significantly different. The discharge point could be a surface water body at lower elevation, or it could be an underlying confined aquifer with a lower potentiometric surface. Geosiphon™ technology was developed at the DOE Savannah River Technology Center. This technology consists of a large diameter vertical well that is designed either with, (1) a gravel pack containing material that will degrade or absorb contaminants (for applications where treated groundwater is discharged above-ground), or (2) wells with removable, flow-through, permeable treatment barrier canisters (for discharges to underlying aquifers) that are positioned between the upper and lower well screens (Phifer et al. 1999).

Installation Methodologies

There are a number of methodologies for installing permeable treatment barriers including continuous trenching, excavation and backfilling, overlapping caissons, soil mixing, and hydrofracturing. Each of these is discussed in more detail below.

Continuous Trenching

Continuous trenching is generally used when placing a permeable treatment barrier across the entire length of a plume. Specialized equipment is used to remove the soil from the trench and replace it with the treatment material in one pass. This has some obvious advantages in that the equipment itself keeps the side walls of the trench open until the treatment material is in place. Continuous trenching is normally applied to depths of less than 20 feet.

Excavation and Backfilling

Excavation and backfilling is normally used for installing funnel and gate configurations. The funnel is installed via one of two methods. The first method is to construct a slurry wall as the funnel. A slurry wall is constructed by using a backhoe to dig a trench down to the desired depth and to stabilize the side walls of the trench with a bentonite slurry mixture. As the trench is dug, the bentonite slurry is pumped in to maintain a positive head on the side walls and to act as a sealant to prevent groundwater from moving into the trench and undermining its stability. As trenching progresses, the soil removed from the trenching is mixed in with bentonite (± 5 percent) at land surface and placed back into the trench. Care must be taken to ensure adequate mixing of the bentonite with the soils removed so that the desired hydraulic conductivity is achieved throughout the length of the slurry wall.

The other method used to construct the funnel is to use a traditional system of sheet piles driven along the length of the funnel. The sheet piles are installed using a vibratory or pneumatic hammer to the desired depth. For this type of application, special sheet piling with interlocking and sealable joints are available. This helps ensure that the funnel is leak proof and will effectively direct the flow of groundwater for treatment in the gate.

To construct the gate, several methods are also available. The most widely used method involves digging an open trench with a backhoe and stabilizing the side walls until after the trench has been backfilled with permeable and/or treatment material. Similar to the funnel installation methods, traditional methodologies such as steel sheet piling can be used. In this application, two rows of sheet piling are driven to the desired depth with a vibratory or pneumatic hammer, and the soil between them excavated. After placement of the treatment material, the sheet piling is removed. Biodegradable polymer slurry such as guar gum, which has been used in drilling water wells for decades, can also be used to stabilize the excavation. Similar to a slurry wall, during barrier construction the biodegraded slurry would provide physical support to stabilize the trench walls until the treatment material is in place. Where it differs from a slurry wall is that after barrier completion, the polymer is flushed out. Any material remaining should biodegrade over a short period of time. This method is generally applied to depths up to about 50 feet. It should be determined beforehand that the treatment material and polymer are compatible; otherwise, the polymer could degrade too quickly, which could cause the side walls to collapse prematurely.

Steel Caissons

Overlapping steel caissons are also used in some applications. Under the right conditions, steel caissons could be installed to depths up to 100 feet. Similar to steel sheet pilings, the caissons are driven or vibrated down to the desired depth, and the soil within them is removed with an auger and replaced with the treatment material. The caissons are then withdrawn. Steel caissons do not require the bracings that are needed between steel sheet pilings, and the selection of the preferred methods would probably be based upon site constraints (available space, proximity to structures whose foundations could be sensitive to displacement, etc.) and costs.

Soil Mixing

Soil mixing has been used in the environmental field primarily for *in situ* mixing of solidification/stabilization agents to physically or chemically bind the contaminants of concern (primarily metals) to minimize their mobility. For permeable treatment barriers, a similar process would be used whereby large diameter mixing augers are drilled into the subsurface and during the process, treatment barrier additives are injected through the hollow stem. This method is also good to depths of about 100 feet under the right conditions, and is desirable because soil management costs are minimized. The biggest concerns involve the ability to get complete mixing between the native soil and the treatment material. For this methodology to be effective, the native material must have sufficient permeability and must be compatible with the treatment material. In addition, the treatment material generally must maintain its effectiveness throughout the duration of the remedial program.

Hydraulic Fracturing and Permeation Infilling

More recently, hydraulic fracturing and permeation infilling has been used. Both of these technologies are similar in that they rely on water or biodegradable polymer under pressure to place the treatment material across the path of groundwater flow. In the case of hydraulic fracturing, the technology is used to create a vertical barrier in unconsolidated sediments consisting of the treatment material, such as zero-valent granular iron. This is achieved by simultaneously creating a vertical fracture and placing the treatment material within that fracture. Orientation and depth of the fracturing is critical to ensure that a continuous treatment barrier is created. The hydraulic fracture is constructed using a series of PVC casings installed along the wall alignment and which are grouted in place and cut along the fracture orientation using a special down hole tool. A packer is then set and injection well heads attached to the packer assembly and injection hoses connected to the pumping unit. Hydraulic fracturing is then initiated using methods similar to that described in Chapter 10 (Hocking, Wells, and Ospina 1998).

Permeation infilling is used in fractured bedrock where most of the groundwater flow occurs within the fractures. The equipment that is used is similar to that used in hydraulic fracturing. In this situation the fracture already exists, so the objective is to use water and biodegradable polymer as a means to slurry the treatment material into the fractures.

PERMEABLE TREATMENT BARRIER PROCESSES

There are a variety of processes that could be integrated into permeable treatment barriers including:

- Transformation processes
- Physical removal
- pH or Eh modification
- Metals precipitation
- Sorption processes
- Nutrient stimulation

Presented below is a brief description of each of these processes.

Transformation Processes

Waterloo Center for Groundwater Research has conducted extensive research on the use of zero-valent metals to promote transformation of various chlorinated organic compounds. They have concentrated on using zero-valent iron to dehalogenate chlorinated aliphatic organic compounds both in the laboratory and in field tests (Gillham, O'Hannesin, and Orth 1993). In the laboratory they studied various metals including stainless steel, copper, brass, aluminum, iron, and zinc using 1,1,1 TCA and found iron and zinc to be most effective. Because of lower costs and availability, they focused their follow-up tests on a range of chlorinated organic compounds using only iron.

The use of zero-valent iron to degrade chlorinated aliphatic organic compounds is an abiotic process. Studies performed by Matheson and Tratnyek (1994) suggested that direct electron transfer on the iron surface was the most probable degradation mechanism. This has since been confirmed and the reaction rates are generally considered to be directly proportional to the surface area of the iron. Studies performed by Matheson and Tratyek (1994) and Gillham and O'Hannesin (1994) suggest that degradation is the result of two reactions. The first reaction is the corrosion of iron by water that produces ferrous iron (Fe^{2+}), hydrogen gas (H_2), and hydroxide (OH^-). The second reaction is between the organic compound and the iron and again produces Fe^{2+}, along with halogen ions (such as chloride) and the nonchlorinated hydrocarbon. Two consequences of these reactions are, (1) dissolved iron is produced which can react with other compounds and precipitate on the zero-valent iron; and (2) hydroxide is formed which can cause a significant increase in pH. The effect of these by-products on the life of the zero-valent iron will be discussed later in this section.

Gillham (1996), Johnson, Sunita, and Tratyek (1996), and Tratyek et al. (1997) have published summary tables of reported degradation rates for common groundwater contaminants. The half-lives vary based on the compound, the source of iron, and the geochemistry of the groundwater (Gillham 1999). For chlorinated ethenes, Tratnyek et al. (1997) suggest the following half-lives: tetrachloroethylene (PCE)—20 minutes; trichloroethylene (TCE)—110 minutes; 1,1-dichloroethylene (1,1-DCE)—650 minutes; trans 1,2-dichloroethylene (t1,2-DCE)—350 minutes; cis 1,2-dichloroethylene (c1,2-DCE)—1000 minutes; and vinyl chloride (VC)—830 minutes. These rates were normalized to 1 m^2 of iron surface area per milliliter of solution. Gillham (1999) suggests that since surface area of commercially available iron materials varies, for design purposes for a specific site the half-lives of the contaminants present should be determined in the laboratory using the iron that will be used and the groundwater from the site.

Because the degradation process is first-order, Gillham (1999) provides a useful rule of thumb for designing zero-valent iron application: a decrease in concentration of three orders of magnitude requires 10 half-lives. For example, using the half-lives above, to reduce PCE from 1,000 ppb to 1 ppb would require a residence time of about 200 minutes. Thus if the groundwater velocity is 1 foot per day, then a wall thickness of about 2 inches would be required. Of course, factors of safety need to be considered in the design to account for the formation of degradation products, groundwater temperatures (compared to laboratory), and precipitation on the iron surface.

In 1992, the University of Waterloo commercialized using zero-valent iron in permeable treatment barrier applications by forming a company called EnvironMetal Technologies Inc., (ETI). ETI reports that more than 75 sites are in the design or implementation stage using this technology. Since this is a new technology, long-term performance of this application is not available. Factors that would affect the long-term performance of this technology include, (1) consumption and replacement of the iron; (2) loss of activity of the iron surface; and (3) potential clogging of the pores. Clogging of the pores would increase the velocity through the wall, or even reduce the permeability such that contaminated groundwater can no longer pass through the permeable barrier. Gillham (1999) looked at each of these factors and concluded the following:

- Based on the expected rate of iron consumption, permeable barriers of only a few tens of centimeters in thickness should persist for many decades.
- Siderite ($FeCO_3$), iron hydroxide ($FeOH_2$), and/or magnetite (Fe_3O_4) will likely precipitate on the iron. In addition, the increase in pH will likely cause calcium and bicarbonate dissolved in the groundwater to precipitate as calcium carbonate. For water high in calcium and with a high alkalinity, and considering the iron precipitation, annual porosity losses of as high as 2 to 5 percent could be expected. The loss in porosity would occur predominantly in the up-gradient end of the barrier.

Other constituents in the groundwater could interfere with the degradation kinetics. For instance, nitrate in solution can lower the degradation rate for TCE, most likely due to competition of reaction sites on the iron surface. Reduced degradation rates have also been observed for groundwater high in dissolved organic carbon.

The first field demonstration of this technology was at a test site at Canadian Forces Base Borden in Ontario that was initiated in 1991 and was terminated in 1996 (O'Hannesin and Gillham 1998). The treatment barrier was constructed into medium- to fine-grained sands by driving sheet piling to form a cell about 5 feet wide by 17 feet long. The barrier was formed using 22 percent iron grindings and 78 percent coarse sand by weight with a permeability greater than the native sands. The sheet piling was then removed and a plume of dissolved PCE and TCE was allowed to move through the barrier.

Over the five years that the study was conducted there was no evidence of a decline in the performance of the barrier in degrading PCE and TCE. Approximately 90 percent of the TCE and 86 percent of the PCE were removed from groundwater passing through the barrier. Figure 7 shows the maximum concentrations of TCE, PCE, and chlorides 299 days into the test. The barrier resulted in significant mass transformation of TCE and PCE, particularly within about the first 1.5 feet (5 days residence time) of the barrier. Although the barrier dehalogenated approximately 90 percent of the mass, ± 10 percent of the mass was not affected. It was concluded that had more reactive iron been used or had a greater percentage of iron been used in the iron/sand mixture, complete removal of TCE and PCE would have been achieved. The increase in pH (from 7.9 to 8.7) was less than what has been experienced at other sites (typically 9 to 10) which was believed to be due to the buffering effect of the sand. Core samples from the barrier were collected periodically during the study. These were taken at an angle such that a cross section of the barrier could be obtained. It was concluded that although precipitates were forming, only minor

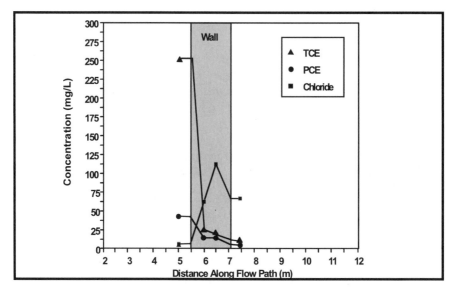

Figure 7   Concentrations of chlorinated compounds across the *in situ* treatment wall.

changes (believed to be iron oxide) in the up-gradient few millimeters of the barrier occurred.

EPA has compiled information on a number of sites using this technology (EPA 1999). Review of this document indicates that most if not all of the applications for the chlorinated aliphatic hydrocarbons are performing well with good contaminant degradation. Some problems have been encountered. At some sites the groundwater flow direction has varied or changed direction from that designed for. In other cases, an increase in groundwater levels up-gradient of the barrier has resulted in migration of contaminated groundwater around or beneath the barrier. As with any remedial technology, it is important to have a good understanding of the hydrogeologic and hydrogeochemical conditions in the contaminant zone so that these factors are well understood during the design phase. Some flexibility is also required in the design, in case the groundwater flow conditions change over the operational life of the treatment barrier.

At the Caldwell Trucking site in New Jersey, a permeable treatment barrier was installed in unconsolidated sands and a fractured basalt. The hydrofrac and infilling methodology was selected to form the treatment barrier, in part due to the depth of the plume (50 feet). The treatment barrier contains zero-valent iron and began at a depth of 15 feet. It consists of two 3 inch thick permeable barriers, 150 and 90 feet in length. The 150 foot barrier was constructed into the unconsolidated zone by using 15 hydrofrac/infilling wells at 15 foot intervals to emplace zero-valent iron using vertical hydrofracture technology. The barrier in the lower fractured bedrock was constructed by pumping guar gum containing zero-valent iron through an open borehole (referred to as infilling). Although there were some difficulties initially in the enzymatic degradation of the guar gum, the guar gum has not interfered with the barrier's permeability nor impacted the iron's reactivity (EPA 1999).

Physical Removal

The best example of physical removal was discussed earlier, whereby air is used to physically remove VOCs from groundwater. This is an adaptation of air sparging; however, it differs in that in a treatment wall, the geology is changed to increase soil permeability, in order to make sparging applicable. Using the funnel and gate concept, a funnel would be constructed to direct groundwater flow to a gate and the gate would consist of materials more permeable than the native material with air sparging points placed within to remove the VOCs from the groundwater.

A pilot scale test using this type of application was performed at the East Garrington gas plant in Alberta, Canada. The target compounds were benzene, toluene, ethylbenzene, and xylene (BTEX). Air sparging was the preferred technology for remediation; however, the BTEX was located in a low hydraulic conductivity glacial till at shallow depth. To overcome the geology limitation, two 145 foot long trenches were dug and backfilled with pea gravel. Air sparging was then performed within these trenches with good results (EPA 1999).

Modify pH or Eh Conditions

Modification of pH conditions occurs when the pH in the contaminant plume is raised or lowered. The main use of this application to date has been to precipitate dissolved metals in a plume. However, this technology could also be used as part of a multiple barrier application. The first barrier could be used to create the necessary environment for the reaction in the second barrier. A biological reaction zone may require a prebarrier to adjust the pH into a range in which the bacteria may grow. While organic destruction would be the main purpose of the design, the pH would have to be modified before the bacterial reaction could occur.

The main use of this application is for metal removal. For instance, at an abandoned mine site in the west, a plume of groundwater with a pH less than 3.0 was discharging elevated concentrations of copper and zinc to surface waters. To solve the problem, a permeable treatment barrier was installed adjacent to the seepage face of the plume. The barrier was designed to increase the pH of acidic groundwater by conveying the groundwater through a limestone bed. As a result of the increase in the pH, metal hydroxides precipitated, and the metals were removed from the groundwater. This was designed to have a contact time of one hour and has successfully operated for several years.

Similarly, a permeable treatment barrier was installed at the Tonolli Superfund Site in Pennsylvania. The barrier was designed to remediate lead, cadmium, arsenic, zinc, and copper. The barrier was constructed by using the continuous trench method. It was approximately 3 feet wide, 20 feet deep, and 1,100 feet long, and was backfilled with limestone (EPA 1999).

Modification of Eh involves the exchange of electrons between chemical species that effects a change in the valence state of the species involved. To be successful in permeable barriers, the redox (oxidation/reduction) reaction must occur readily under the natural temperature and chemical setting of the groundwater.

Blowes and Ptacek (1992) proposed to incorporate solid-phase additives as the reactive material in treatment walls to control redox conditions to promote removal of electro-active metals. Their work focused on using reduced iron (fine grained zero-valent) for removal of dissolved hexavalent chromium. Column studies in the lab demonstrated that highly reducing conditions generated by the reduced iron caused the chromium to change its state to the less soluble trivalent chromium that then precipitated (chromium hydroxide). Additional column studies were performed by Petrie and others (1993) on samples from the U.S. Coast Guard Station near Elizabeth City. Their work confirmed the findings of Blowes and Ptacek (1992). A pilot study was initiated at the facility in 1995 by constructing a treatment barrier 2 feet thick. The results indicate that dissolved chromium was reduced to acceptable levels within the first 6 inches of the barrier.

Precipitation of Metals

Precipitation of metals could be accomplished by altering pH or Eh conditions as discussed above or by chemical additives to a treatment barrier. For instance, in a plume containing dissolved lead, hydroapatite added to a treatment barrier could serve as a source of phosphate to precipitate lead phosphate. Studies performed by Schwartz and Xu (1992) demonstrate its potential applicability to treatment barriers.

Other chemical reactions are possible to precipitate metals depending on the type of dissolved metal that is involved. The key to the design for precipitation of metals is to have the necessary chemical in a solid phase form so that it can be part of barrier construction. The chemical must still be able to interact with the dissolved metal in the plume. Another example would be to use sulfide to precipitate metals. Metal sulfides are generally less soluble then metal hydroxides. Sulfides will also transfer from one precipitate to another. The barrier could be constructed of calcium or iron sulfide. Metals that have lower solubility as sulfides (silver, mercury, etc.) would take the sulfide from the calcium or iron and precipitate. The calcium or iron would be released to the water.

In Fry Canyon, Utah, a field demonstration was initiated to determine the technological and economic feasibility of using a permeable treatment barrier to remediate uranium. The barrier was constructed in 1997 and comprised three different treatment materials including bone char phosphate (PO_4), zero-valent iron, and amorphous ferric oxide (AFO). After one year of operation all three were removing uranium. Zero-valent iron has demonstrated the highest removal efficiencies at more than 99 percent. Both the AFO and the PO_2 barriers removed about 90 percent of the dissolved uranium (EPA 1999).

Contaminant Removal via Sorption or Ion Exchange

Contaminant removal via sorption also shows promise, but as with any developing technology a number of issues need to be resolved. Sorption could be accomplished using a number of commonly used above-ground sorptive material such as activated carbon, zeolites, or ion exchange resins.

Activated carbon, for instance, is used to physically absorb a variety of organic compounds in groundwater. The activated carbon physically absorbs the organics due to the attraction caused by surface tension of the activated carbon, and it has gained widespread use in above-ground applications for groundwater treatment because of the large internal surface area that activated carbon possesses. Since physical adsorption is a reversible process, its application in reactive walls requires that at some point, either after the carbon is spent or the plume is remediated, the carbon must be removed from the subsurface and managed accordingly. If the carbon is left below ground, it will release the absorbed organics to the clean water carrier. This requires that the activated carbon be installed in retrievable containers, discussed earlier in this chapter, or is installed so that a vacuum truck, or other extraction device, can easily remove it. Activated carbon treatment barriers require consideration of other factors that could interfere with their adsorptive abilities. For instance, dissolved iron could be present which could compete with organics for absorptive surfaces within the activated carbon. In addition, biological growth could also "blind" the carbon, significantly reducing its effectiveness to remove organics.

A permeable treatment barrier was installed at the Marzone Superfund site in Georgia using activated carbon to remediate groundwater contaminated with pesticides including BHC, beta-BHC, DDD, DDT, lindane, and methyl parathion. A modified funnel and gate system was constructed by installing a 400-foot barrier using vibrating beam technology. Up-gradient of and parallel to the barrier, a collection trench was installed using coarse-grained material. Collected groundwater was routed to treatment vaults containing 1,800 pounds of activated carbon. Treated groundwater was conveyed beyond the barrier wall by a pipe that terminated into a distribution trench constructed perpendicular to the wall. The results of preliminary testing indicate that contaminant concentrations are below detection levels (EPA 1999).

If the contaminants are inorganics, ion exchange may be applicable. Ion exchange is the exchange of an ion with high ion exchange selectivity for an ion with a lower selectivity. Solid phase materials that can exchange an ion dissolved in groundwater for one of its ions may be suitable for incorporation into a treatment barrier. Any divalent ion will usually have higher ion exchange selectivity than will a monovalent ion. Most metals present in groundwater are in the divalent or trivalent state and are amenable to ion exchange. However, ion exchange treatment can be expensive. As in the case of activated carbon it is a physical process which is reversible, so the spent material must eventually be removed and managed properly. Also, naturally occurring constituents such as dissolved iron and bacteriological growth can "blind" ion exchange material reducing its effectiveness.

A pilot-scale demonstration project using a surfactant-modified zeolite (SMZ) was performed by the Large Experimental Aquifer Program near Portland. The project was designed to demonstrate this technology for remediating hexavalent chromium. The results indicated that the barrier performed as designed with retardation factors on the order of 50 (EPA 1999).

Biological Degradation

Biological degradation applications to permeable treatment barriers are generally aerobic and can be accomplished in many different ways. The two main criteria for an aerobic biological treatment barrier are surface area and oxygen. The bacteria need the proper amount of surface area in order to attach and interact with the plume. Too little surface area (i.e., large gravel) and an insufficient amount of bacteria will be in the treatment barrier. Too much surface area (i.e., silt and clay) and the water will not travel past and interact with all of the bacteria.

The oxygen can be supplied in many ways. An air sparging curtain can be installed in natural or geologically altered (to increase permeability) material to supply oxygen to enhance degradation of aerobically biodegradable compounds such as BETX. Early research was performed by Bianchi-Mosquera (1993) on using solid phase oxygen release compounds (ORC) to enhance degradation of BETX. ORCs are commonly in the form of concrete briquettes that slowly release oxygen to groundwater, although slurries are now being used that can be injected into an aquifer. Numerous full-scale applications of ORC have demonstrated their effectiveness in degrading BETX.

At a site in Alabama, a pilot study was conducted to remediate methyl isobutyl ketone (MIBK) using an *in situ* biological reactor. The reactor was designed with two sequential chambers. In the first chamber oxygen was added via a sparging system to raise the dissolved oxygen level and precipitate dissolved iron in groundwater passing through the chambers. The second chamber was filled with oyster shells, which have a high surface area, to promote biological growth of MIBK degrading bacteria. The results showed that after the initial start-up period, the *in situ* biological reactor achieved 95 percent reduction in MIBK levels.

DESIGN CONSIDERATIONS

There are a number of considerations that must be carefully thought out in the design of permeable treatment barriers, particularly ones incorporated into funnel and gate configurations. Of utmost importance is that the system manages the entire plume. Since groundwater flow directions can vary, it is extremely important that these fluctuations are accounted for. Secondly, the residence time in the treatment barrier must be sufficient to achieve the desired reduction in contaminant concentration. Lastly, the ratio of cut off walls to treatment barrier in a funnel and gate system should be optimized to minimize costs. For funnel and gate systems, flow and transport modeling is almost always needed to evaluate these considerations and optimize the design.

One of the most critical factors in permeable treatment barrier design is the relationship between residence time of contaminated groundwater in the treatment zone (reactor) and the rate of contaminant degradation that occurs within the gate. On the one hand, the designer is trying to maximize discharge through the gate to maximize the width of the capture zone. On the other hand, the designer wants to maximize residence time in the gate to ensure adequate treatment. Since these are inversely related, they must be properly balanced. Again, flow and transport modeling can be very useful in evaluating these relationships.

As one might expect, the discharge through the gate increases with increased hydraulic conductivity of the gate; however, relatively few increases occur with hydraulic conductivities greater than ten times the aquifer. This is important because high hydraulic conductivity porous media usually have large grain sizes and low surface area to mass ratios. Since reaction rates in most *in situ* media are proportional to surface area, high hydraulic conductivities would result in slower reaction rates.

The rate of contaminant degradation that occurs within the treatment barrier is an important factor. The actual retention time in the treatment zone is obtained by dividing the pore volume of the treatment zone by the discharge through it. For degradation processes that are first order reactions, the retention time necessary is given by the formula

$$N_{1/2} = [\ln (C_{eff}/C_{inf})]/\ln (1/2)$$

where, $N_{1/2}$ = number of half lives required; C_{eff} = concentration of the desired effluent; and C_{inf} = concentration of the influent.

To put this in perspective, if the chlorinated compound TCE has a half life of 13.6 hours, to reduce a plume from 1000 ppb to 5 ppb would require 7.6 half lives or a residence time of 4.3 days. Of course this is a theoretical calculation and other factors need to be considered as discussed previously.

Since a key design consideration is to increase attenuation in the treatment zone, there are two ways to achieve this. The first way is to increase the reaction rates. This would most commonly be achieved by increasing the surface area of the treatment material (which goes back to the gate [reactor media] not needing to be more than 10 times more permeable than the aquifer) or by using a higher proportion of treatment material in the treatment zone. The second way is to increase the retention time within the reactor by decreasing the discharge through the gate (add more gates) or by increasing the gate's pore volume by making it longer or wider.

CASE STUDY—REACTIVE WALL DESIGN

Background

Pentachlorophenol (PCP) and tetrachlorophenol (TCP) were detected in on-site groundwater samples at a former wood treating facility in the western United States. The groundwater system consists of a shallow aquifer containing a heterogeneous mixture of marine deposits and artificial fill that is underlain by low permeability siltstones and mudstone. The shallow aquifer ranges in thickness from 10 to 20 feet and averages 15 feet; its saturated thickness averages seven feet. Based on the results of the site investigation, it was determined that impacted groundwater had the potential to move off-site and adversely affect downstream domestic users of the resource (Figure 8). Consequently, remedial action goals were developed to protect the quality and quantity of the resource.

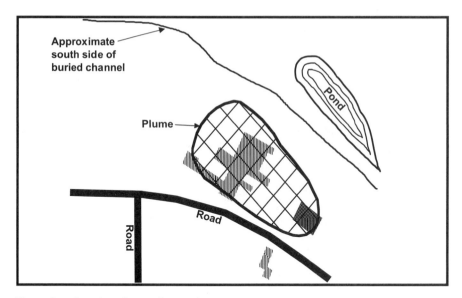

Figure 8   Location of groundwater plume.

A number of remedial alternatives were evaluated including pump and treat, containment, and *in situ* treatment. A permeable treatment barrier incorporating activated carbon was deemed the most attractive alternative and additional analyses were conducted to determine constraints on below-ground application of this technology and to evaluate its suitability for incorporation into a funnel and gate system with the funnel portion constructed using slurry wall technology. These studies focused on the following, (1) the potential for underflow beneath the funnel and gate system, (2) the spatial relationship between gates and funnels, (3) mass loadings on the gates, and (4) interferences with gate performance.

Funnel and Gate Modeling Study

Since permeable treatment barriers are generally more expensive than impermeable barriers, groundwater flow modeling was conducted to optimize the design by minimizing the number and length of gates, while still accommodating flow from the entire plume and providing adequate residence time within the gate. Groundwater modeling and particle tracking were used to delineate flow paths through and around the funnel and gate system to demonstrate the efficacy of a given configuration. A conceptual model was developed based on varying hydrogeologic conditions and groundwater flow gradients found across the site. The background transmissivity for the site was modeled at 210 ft²/day with two imbedded permeability inhomogeneities of 28 ft²/day and 3.5 ft²/day.

Groundwater streamlines were generated for several combinations of gate lengths and spacings. After several alternatives were evaluated, an L-shaped configuration was selected which contains four gates, two of them installed adjacent to the buried stream channel where flow rates are expected to be higher (Figure 9). The modeling

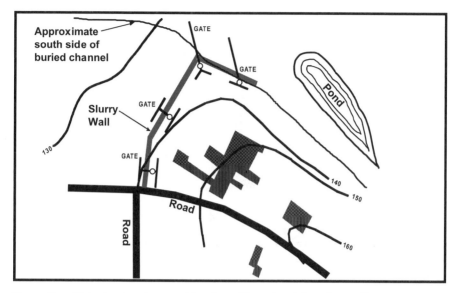

Figure 9   Water level contours and location of Funnel-and-Gate System™.

results showed that flow from a fairly broad area can readily be focused through a small number of small gates without causing groundwater seeps to occur at the surface and with minimal disruption of groundwater flow patterns.

The modeling results were also used to evaluate the volume of water expected to be captured for treatment. Based on the area encompassed within the flow lines and the hydraulic properties within this region, a flux of 22 gpm, or about 5.5 gpm per gate, was calculated.

Gradient Control

Because of the nonuniform distribution of hydraulic transmissivities across the site, groundwater may be impeded in its lateral movement toward and away from the gates. In order to minimize this effect, gravel-filled collection and distribution galleries were installed at each gate to collect water from the up-gradient side of the gate, guide it through the gate, and then redistribute it uniformly after treatment (Figures 9 and 10). The collection and distribution galleries were trenched into the aquifer to expose a large cross-sectional area to groundwater flow. This minimized the pressure required to move water from the aquifer to the carbon treatment gates. Installation of these collection and distribution galleries helped ensure that the pressure drops across the funnel and gate system were minimal so as not to affect natural groundwater gradients and flow patterns.

Underflow of Barrier

The potential for flow under the system was also evaluated by using data collected during aquifer tests conducted at the site along with results from the modeling study.

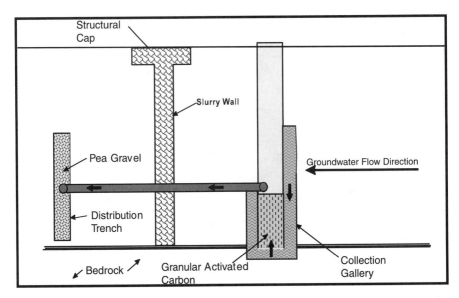

Figure 10   Profile view of treatment gate.

Aquifer tests of the upper marine terrace sediment yielded an estimated transmissivity of 210 ft²/day and the transmissivity of the lower Franciscan formation was estimated from slug tests conducted on wells completed solely within the underlying Franciscan formation. A comparison of these two values indicated that the hydraulic conductivity of the lower Franciscan formation is approximately 1/1,000 that of the overlying marine terrace sediment. The flows that are expected to travel through the treatment gates were compared with the flows that may flow under the barrier. The solution below for the flow beneath the barrier is similar to that for estimating flow under a dam, with a simplifying assumption that flow lines in the Franciscan are concentric semicircles when viewed in section from one end of the system.

k = 0.004   Hydraulic conductivity of Franciscan, in ft/day
i = 0.05   Hydraulic gradient
T = 210   Transmissivity of marine terrace, in ft²/day
w = 2   Width of barrier, in ft
h = 2.0   Average head difference across barrier, in ft
x = 400   Distance up-gradient from barrier, in ft, through which water
            permeates the Franciscan

Let q be the flow through a 1 foot thick slice under the slurry wall emanating from the shallow aquifer between the wall and a point x feet up-gradient of the wall

$$q = \frac{2kix}{\pi} + \frac{k(h - wi)}{\pi} \ln \frac{(\pi x + \pi x)}{2}$$

$$q = 0.07 \text{ ft}^3/\text{day}$$

Let Q be the flow through a 1 foot thick slice of the marine terrace deposits

$$Q = TI$$

For the parameters as defined above, the flow rate is

$$Q = 10.5 \text{ ft}^3/\text{day}$$

Since the K value for the Franciscan is a horizontal conductivity, the actual flow (q) under the wall will be less than the computed value. This is because the effective conductivity is a weighted average of the horizontal and vertical K's, and the vertical conductivity is likely much less than the horizontal conductivity. The maximum flow under the wall represents approximately 0.7 percent of the total flow through the system.

Gate Design

The gate design incorporates several factors including contaminant concentration, flow rate, and time between carbon changeout. Using an estimated flow rate of 5.5 gpm, a carbon change-out of 3 years, a concentration of 200 ppb, and a 1 percent adsorptive capacity, approximately 48 ft^3 of carbon were required for each gate.

To minimize flow velocity and maximum residence time, each gate was designed using 4 foot diameter corrugated metal pipe filled with 4 feet of activated carbon. Figure 11 shows a photo of the installation of a gate. This configuration allows water to move easily through the gates with a design pressure loss of only 2 inches of water.

Groundwater samples are collected from the monitoring points located before each of the treatment gates to measure influent concentrations entering the gate. A sample is also collected from the monitoring point within the treatment gate to verify that the treatment gate is effective in removing these compounds prior to the water exiting the gate. This monitoring point is shown in Figure 11 and is the PVC pipe located in the middle of the corrugated metal pipe. A treatment buffer zone exists downstream of this mid-gate monitoring point to ensure that impacted water does not exit the gate prior to full removal of these compounds. Upon detection of a compound of concern at a concentration above water quality objectives at the mid-gate measuring point, the carbon will be removed and replaced. It is expected that biological activity will also be a part of the removal mechanism for the contaminants.

System Performance

The system was installed in September 1995 and has been operated since that time. Flow rates though each of the gates were found to vary dependent on geology. The installation of the collection and distribution galleries which guide groundwater to and from the gates worked well. The spatial relationship between the gate, the slurry wall, and the distribution gallery is shown in Figure 12. In the photo the bulldozer is installing the distribution gallery. Figure 13 is a photo showing a close

Figure 11   Photo showing gate installation.

up of installation of the slurry wall. A portion of the slurry wall was installed as shown in the photo between the gate and the distribution gallery. In addition, the

Figure 12   Photo showing relationship between gate, slurry walls, and installation of distribution gallery.

Figure 13 Photo showing slurry wall installation.

actual pressure drops across the gates have been only a couple of inches (the system was designed based on 1 foot) and have not changed during operation of the system. No changes in local groundwater flow patterns have occurred as a result of system installation, and the yields from shallow water supply wells immediately down-gradient of the barrier have not been impacted because of some unforeseen effect of the system.

The concentrations of TCP and PCP reaching the gates have been in the low parts per billion. Low levels of TCP and PCP were detected in samples collected from the middle of the carbon beds in the third quarter of 1998. These detections were sporadic and were only slightly above the detection limit. Detections, below action levels, have also occurred in a monitor well located down-gradient of Gate No. 2. No TCP or PCP has been detected in the middle sample point of the carbon bed within Gate No. 2. In addition, no TCP or PCP was detected in follow up sampling at the top of the carbon bed in Gate No. 2 prior to the water leaving the system. It was concluded that the detections in the down-gradient monitoring well were from residual contamination that was present prior to system installation.

The first carbon change out was successfully completed in September 1999. High groundwater levels due to heavy rains prevented draining the gates during an earlier attempt to change out the carbon. Change out was accomplished by excavating the soil above the gates with a backhoe and then dewatering the gates using a submersible pump. The pumped water was transferred to one of the other gates for processing so no water was hauled off-site. The carbon was then vacuumed out using standard vacuum equipment and hauled off-site as nonhazardous waste. As designed, the system minimized changes in dissolved oxygen levels and REDOX and minimal biofouling and iron precipitation occurred within the beds. While the gate was empty a visual inspection was made of the gate and no leaks or corrosion were identified in the seals and welds. New carbon was poured into the gates after the inspection.

REFERENCES

Bianchi-Mosquera, G.C., Allen-King, R.M., and Mackay, D.M., "Enhanced Degradation of Dissolved Benzene and Toluene Using a Solid Oxygen-Releasing Compound," *Ground Water Monitoring and Remediation,* 1993.

Blowes, D.W. and Ptacek, C.J., "Geochemical Remediation of Groundwater by Permeable Reactive Walls: Removal of Chromate by Reaction with Iron-Bearing Solids," Proceedings, Subsurface Restoration Conference, Dallas, Texas, June 21-24, 1992.

Devlin, J.F. and Barker, J.F., "A Semi-Passive Injection System for *In Situ* Bioremediation," Proceedings, 1993 National Conference on Hydraulic Engineering and International Symposium on Engineering Hydrology, San Francisco, California, July 25-30, 1993.

Environmental Protection Agency, Field Applications of *In Situ* Remediation Technologies: Permeable Reactive Barriers, Solid Waste and Emergency Response, EPA 542-R-99-002, April 1999.

Gillham, R.W., *"In Situ* Remediation of VOC-Contaminated Groundwater Using Zero-Valent Iron: Long-Term Performance," Proceedings, 1999 Contaminated Site Remediation Conference "Challenges Posed By Urban & Industrial Contaminants," Centre for Groundwater Studies, Fremantle, Western Australia, 1999.

Gillham, R.W., *"In Situ* Treatment of Groundwater: Metal-Enhanced Degradation of Chlorinated Organic Contaminants," *Advances in Ground Water Pollution Control and Remediation,* Kluwer Academic Publishers, The Netherlands, 1996.

Gillham, R.W. and O'Hannesin, S.F., "Enhanced Degradation of Halogenated Aliphatics by Zero-Valent Iron," *Groundwater,* 32 (6), 1994.

Gillham, R. W. and Burris, D. R., Recent Development in Permeable *In Situ* Treatment Walls for Remediation of Contaminated Groundwater, 1994.

Gillham, R.W., O'Hannesin, S.F., and Orth, W.S., "Metal Enhanced Abiotic Degradation of Halogenated Aliphatics: Laboratory Tests and Field Trials," Proceedings, 1993 HazMat Central Conference, Chicago, Illinois March 9-11, 1993.

Hocking, G., Wells, S. L., and Ospina, R. I., 1st International Conference on Remediation of Chlorinated and Recalcitrant Compounds, Monterey, CA, May 1998.

Johnson, T.L., Sunita, S.B., and Tratnyek, P.G., "Kinetics of Halogenated Organic Compound Degradation by Iron Metal," *Environmental Science and Technology,* 30(8), 1996.

Matheson, L.J. and Tratnyek, P.G., "Reductive Dehalogenation of Chlorinated Methanes by Iron Metal," *Environmental Science and Technology,* 28(12), 1994.

Phifer, M.A., Sappington, F.C., Nichols, R.L., Ellis, W.N., and Cardoso-Neto, J.E., "Geosiphon/Geoflow Treatment Systems," Waste Management 99 Symposia, Tucson, Arizona, 1999.

Schwartz, F.W. and Xu, Y., "Modeling the Behavior of a Reactive Barrier System for Lead," Proceedings, Modern Trends in Hydrogeology, 1992 Conference of the Canadian National Chapter, International Association of Hydrogeologists, Hamilton, Ontario, 1992.

Starr, R. C. and Cherry, J. A., *In Situ* Remediation of Contaminated Groundwater: The Funnel-and-Gate System, 1993.

Tratnyek, P.G., Johnson, T.L., Scherer, M.N. and Eykholt, G.R., Remediating Ground Water with Zero, 1997.

Valent Metals, "Chemical Considerations in Barrier Design," *Ground Water Monitoring and Remediation,* XVII (4).

CHAPTER 12

Continuing Problems in Groundwater—MTBE, 1,4-Dioxane, Perchlorate, and NDMA

Evan K. Nyer, Kathy Thalman, Pedro Fierro, and Olin Braids

CONTENTS

1-56670-528-2/01/$0.00+$.50

INTRODUCTION

We have shown in Chapters 1 through 11 that strong progress has been made in the knowledge and experience for remediation of soils and groundwater. We know how to get most of the mass out, and then how to follow up with natural and enhanced biological systems to destroy the organic contaminants controlled by the geological limits of the aquifer. While we know how to take care of 99 percent of the organic chemicals, several organic compounds have been discovered in the last few years that are not remediating by these techniques.

The reader must be aware of the limitations of *in situ* remediation techniques when dealing with the following organic compounds: MTBE; 1,4 dioxane; perchlorate; and NDMA.

Several problems occur with these compounds. First, most of them have not been part of the list of organic compounds that we requested be analyzed by the laboratory. Some show up on the TIC list, some are not recognized by our standard analytical methods, and some have only recently had an analytical method for low ppb detection. Second, these compounds are generally very soluble, have low retardation in the aquifer, and are relatively slow degraders. This combination of properties creates large plumes of organic contaminants that move close to the speed of the groundwater.

These two problems have combined to create situations in which organic compounds are just recently being regulated, and have not been detected in our standard groundwater analysis, but are present. This means that on some new sites as well as sites that have already been cleaned and closed, there are large plumes of newly regulated organic compounds.

The final problem with these compounds is that their physical, chemical, and biochemical properties prevent the use of most of the remediation methods discussed in this book. Most sites that have found these compounds have had to rely on pump and treat for their control and remediation method. Even when brought above ground, the treatment methods for these compounds have proven expensive.

Let us review each of these compounds for history, properties, and remediation methods. This will prepare the reader on the type of sites that should be tested for the presence of the compounds.

METHYL TERTIARY-BUTYL ETHER—MTBE

$$
\begin{array}{c}
\quad\quad\quad O \diagup^{CH_3} \\
\quad\quad\quad | \\
H_3C \text{—} C \text{—} CH_3 \\
\quad\quad\quad | \\
\quad\quad\quad CH_3
\end{array}
$$

Background and History

Methyl tertiary-butyl ether (MTBE) is a compound that has been adopted to serve two purposes as a gasoline additive, (1) to enhance the octane rating for gasoline, and (2) to provide oxygen to boost combustion efficiency. MTBE's octane enhancing quality led to its approval as an oxygenate in 1979. It substituted for the alkylated lead compounds that were being phased out. Oxygen contained in MTBE also increases combustion efficiency, reducing carbon monoxide emissions. In 1981, the U.S. EPA approved MTBE's use in gasoline up to 10 percent by volume. It was used in higher percentages in premium gasoline than in regular gasoline. The first winter oxygenated gasoline program in the nation was implemented in Denver, Colorado in 1988 (Jacobs, Guertin, and Herron 1999). This trend expanded to other states as regulatory agencies attempted to reduce cold weather vehicular emissions.

Starting in 1992, in fifteen or more states, the Clean Air Act now requires that gasoline contain oxygen to reduce carbon monoxide emissions in nonattainment areas. Up to 17 percent MTBE may be used to achieve the required oxygen content. Although other oxygenates may be used, MTBE was used the most because of its compatible qualities. The federal reformulated gasoline (RFG) program, starting in January 1995, resulted in up to 15 percent MTBE (11 percent in California) being added to gasoline to provide 2.7 percent oxygen (California Environmental Protection Agency 1998). With the expansion of oxygenate use under these programs, 30 or more percent of the gasoline sold in the United States contains oxygenates (U.S. EPA 1998).

Federal RFG has been adopted in 28 metropolitan districts throughout the United States. Its use is primarily intended to decrease production of ozone, and it is required year round. Federal RFG must contain at least 2 percent oxygen from any oxygenate during the summer season. The demand for MTBE has made it unprecedented in its rate of production growth. In the early 1970s, its production was about 12,000 barrels per day, or the 39th highest produced organic chemical in the United States. By 1998, it had become the fourth highest and was produced at about 250,000 barrels per day (California Environmental Protection Agency 1998). In 1998, more than 10.5 mgd (million gallons per day) were being consumed in the United States and 4.2 mgd in California (Johnson et al. 2000).

MTBE Characteristics

In addition to its functional characteristics being a benefit to gasoline, MTBE's chemical and physical characteristics make it a desirable additive. MTBE is miscible in gasoline and is not hygroscopic. This is not true for alcohols, as they tend to absorb water and have limited anhydrous solubility. When alcohols absorb water, the tendency to break phase is increased. MTBE is also relatively inexpensive to produce as it is synthesized from isobutene, a product of petroleum refining. About 80 percent of the MTBE used in the nation comes from local sources, with the remainder being imported.

Table 1 summarizes the physical and chemical characteristics of MTBE. The boiling point of MTBE, 53.6° to 55.2° C, and its vapor pressure of 254 mm Hg @ 25° C are compatible with the mixture of hydrocarbons in gasoline (benzene vapor pressure 76 to 95 mm Hg). Thus, it can be successfully transported by tank or pipeline. The aqueous solubility of pure MTBE is higher than any other gasoline component, 43,000 to 54,000 ppm (Jacobs, Guertin, and Herron 1999). When it is present in gasoline in contact with water, MTBE has a tendency to dissolve into the water, but also to stay dissolved in the gasoline. Under these conditions, its solubility has been determined to be 4700 ppm from RFG with 11.1 percent MTBE and 6300 ppm from oxyfuel with 15 percent MTBE. These concentrations compare with 18 ppm for benzene and 25 ppm for toluene dissolving from gasoline in equilibrium with water.

Table 1   Physical and Chemical Properties of MTBE

Characteristic/Property	Data
Formula	$C_5H_{12}O$
Molecular Weight	88.15
Specific Gravity	0.745
Water Solubility	51,000 ppm
Log K_{ow}	1.24
K_{oc}	11.2
Henry's Law Constant (atm)	32.6
Vapor Pressure	254 mm Hg at 25° C

MTBE is chemically stable and because it is an ether and the tertiary-butyl structure provides steric hindrance, it is also resistant to biodegradation. Acclimated bacterial populations under controlled conditions, *ex situ,* show some promise with biodegradation, and *in situ* intrinsic attenuation has had some success. Partial degradation products such as tertiary-butyl alcohol are also contaminants in their own right. The high solubility of MTBE, coupled with its resistance to chemical destruction or biodegradation, result in it being highly mobile in groundwater and subject to little retardation. The observed retardation factors for MTBE in soils with 0.1 percent organic carbon and 0.4 percent organic carbon are 1.09 and 1.38, respectively. This compares to 1.75 and 3.99 for toluene under the same conditions. These factors lead to the observation that MTBE is the compound most likely to be found at the leading edge of a contaminant plume originating with gasoline (Nichols, Einarson, and Beadle 2000).

MTBE is difficult to treat because it resists air stripping, adsorbs poorly on activated carbon, and resists biodegradation. These treatment pathways are relatively effective for the hydrocarbon compounds that act as gasoline contaminants, predominantly benzene, toluene, ethylbenzene, and xylenes (BTEX). Thus, MTBE may survive standard treatment for hydrocarbons and cause the treated effluent to be out of compliance with discharge standards.

Regulatory Framework

After MTBE was authorized as a gasoline additive by the U.S. EPA, its use was slowly increased in its application as an octane enhancer. The 1990 Amendments to the federal Clean Air Act mandated that oxygenates be incorporated in gasoline sold in regions that failed to comply with federal air quality standards. For reasons previously discussed, MTBE became the oxygenate of choice for the petroleum refining industry.

The positive air quality benefits accruing from MTBE's use have been somewhat offset by public and regulatory concerns over its presence as a groundwater contaminant and its potential threat to public health. Both the public and regulatory agencies have held differing views of the risks to public health and the environment from MTBE. Some regard it as a toxic contaminant of concern and others as a low risk contaminant. As a result, regulations covering MTBE from state to state are fractured.

Three general trends seem apparent in the way states are regulating MTBE (Jacobs, Guertin, and Herron 1999). Category 1 includes those states that have set a rigorous cleanup or advisory level for MTBE based on EPA guidelines or public pressure. The concentration goals may be somewhat flexible if approached through a risk assessment.

Category 2 includes states that do not consider MTBE as a contaminant of concern, but as one of lower toxicity and one that causes esthetic concerns of taste and odor. Cleanup levels have not been established, but are arrived at from site-specific risk assessments. The general attitude is that MTBE is an indicator of the presence of more toxic hydrocarbons components of gasoline that will be cleaned up in conjunction with the hydrocarbon cleanup.

Category 3 includes states that have developed cleanup or advisory standards developed independently from EPA's health studies. Guidance in these states is derived from state-sponsored health and risk studies. The resulting concentration values have remained relatively constant over the past several years. Table 2 provides examples of the different categories and values derived.

Table 2   MTBE Guidance For Regulations

State	Cleanup or Advisory Concentration
Category 1	
Florida	50 ppb until September 1997
	35 ppb after September 1997 based on potential carcinogenicity
	risk based cleanup depending on conditions with levels 20-40 ppb
New Jersey	700 ppb until February 1997
	70 ppb after February 1997 based on health studies
Category 2	
Minnesota	risk based, MTBE not considered a "contaminant of concern" or significantly toxic
Oregon	risk based and MTBE will be remedied with cleanup of hydrocarbons
Texas	no cleanup level in groundwater and no initial MTBE testing required
	well contamination by MTBE regarded as indicator of hydrocarbon contamination
Alaska	MTBE not considered serious groundwater threat
	MTBE introduced in 1995 reformulated gas program, but abandoned in two months because of public complaint
	Department of Health banned MTBE
California	cleanup level based on oganoleptic threshold, 13 ppb
Category 3	
New York	guidance level set among general site closure criteria is 50 ppb
	guidance level in effect for 10 years
Wisconsin	60 ppb guidance set by Health Department's state-based toxicological formula

It is evident that state response to MTBE both as a potential and active contaminant ranges from banning the compound, as is the case for Maine, to regarding it as a low toxicity substance indicative of contamination by more toxic compounds.

In regard to federal regulations, the U.S. EPA, under the Safe Drinking Water Act (SDWA) as amended in 1996, published a list of contaminants that were not subject to any proposed or promulgated national primary drinking water regulation at the time of publication, that are known or anticipated to occur in public drinking water systems, and which may require regulations under the SDWA. This list, the contaminant candidate list (CCL), was published in final form March 2, 1998 (U.S. EPA 1998b). The CCL is to be republished every 5 years. The CCL included contaminants identified as priorities for drinking water research, contaminants that need additional data on frequency of occurrence, and contaminants for which development of future drinking water regulations and guidance is justified. The CCL includes fifty chemical and ten microbiological contaminants/contaminant groups. However, the SDWA limits the contaminants to thirty in any 5 year cycle. The present regulation will require monitoring for only twelve contaminants in List 1.

Related to the CCL, EPA revised the Unregulated Contaminant Monitoring Rule to evaluate and prioritize contaminants for possible new drinking water standards. The CCL contaminants are divided into three lists. List 1, including twelve contaminants for which analytical methods are established, includes MTBE. Assessment monitoring will be required beginning in 2001. Large public water systems (PWS) numbering 2,800 and 800 of 66,000 small PWS will be performing the monitoring. Surface water systems will monitor quarterly for 1 year and groundwater systems will monitor semi-annually for 1 year in a 3 year window. The EPA is presently advising that MTBE concentrations be limited to a range of 20 to 40 ppb (parts per billion) to prevent taste and odor problems and to protect human health (U.S. EPA 1999).

MTBE is odoriferous, with a reliable organoleptic threshold of 5 ppb. Heating contaminated water for cooking or bathing increases the odor intensity with increased vaporization.

Environmental Behavior and Fate

Characterizing groundwater contaminant plumes originating with MTBE containing gasoline requires some aspects that are common to all groundwater contamination investigations and some aspects specific to dealing with MTBE or oxygenates in general. Fundamentally, the groundwater regime must be characterized in regard to groundwater flow direction, velocity, and tendency for an upward or downward component of flow. The migration of MTBE will follow the flow regime at close to the flow velocity.

MTBE sources and fuel hydrocarbon sources, as evidenced by dissolved plumes, may not be the same. Because hydrocarbons are subject to relatively high biodegradation and retardation in the subsurface, whereas MTBE is not, they require more mass of hydrocarbons to create a contamination zone than it does MTBE. For example, a small leak in the dispensing system at a service station could release MTBE containing gasoline that would produce an MTBE plume without accompanying hydrocarbons. Vapor phase transport of MTBE in soil might also result in displacement of the apparent location where the MTBE enters the groundwater system. The light hydrocarbon compounds are not as likely to exhibit this behavior, as their vapor pressures are lower and their aqueous solubilities are even lower (see *MTBE Characteristics*) (Nichols, Einarson, and Beadle 2000).

MTBE is a unique component of gasoline because of its relatively high aqueous solubility and resistance to biodegradation. Natural attenuation is a process that is almost universally applied to the chemical components of gasoline when it is released into the groundwater environment. Bacteria that are present in the subsurface, particularly when conditions are aerobic, are able to metabolize the hydrocarbon compounds, thereby reducing their concentrations through biodegradation (Mackay et al. 2000). Many locations have been documented where the migration of hydrocarbons from the source area reach a point down-gradient where their migration rate and degradation rate balance and the plume stabilizes. Presence of MTBE complicates the plume management because it either does not biodegrade, or biodegrades so much more slowly than the other hydrocarbons that it persists in creating a contaminant plume of its own. Some public supply wells have become contaminated with MTBE in the absence of other gasoline hydrocarbon components.

As noted in the *MTBE Characteristics* section, MTBE's solubility is 20 or more times that of benzene, the next most soluble component. MTBE's solubility and nonpolarity result in its moving relatively unimpeded by adsorption or ionic attraction to the aquifer solid matrix when it occurs in groundwater. It therefore acts essentially as a conservative substance, moving at the velocity of groundwater. This flow characteristic allows MTBE to flow ahead of the other gasoline hydrocarbon components as they disperse into groundwater. The resulting plume, after moving a distance down-gradient, will show MTBE at the leading edge. Because its biodegradation is slow, MTBE may be found as the sole component in the advancing plume. At this stage, the other hydrocarbon compounds will have reached a stabilized position of input and degradation behind the MTBE.

MTBE Treatment Options

Activated Carbon

Many contaminants can be effectively removed from water and air by granulated activated carbon (GAC). GAC is manufactured from a variety of carbon sources including bituminous and lignite coal, wood, and coconut shells. The goal in manufacturing an activated carbon is to create a pore structure within the carbon particle that provides a large adsorption surface. Additionally, the more high energy pores that are produced, the more effective the carbon will be toward weakly adsorbing compounds. Weakly, moderately, and strongly absorbing compounds are a function primarily of the compounds' aqueous solubility and concentration. Relatively soluble compounds such as MTBE fall into the weakly adsorbed category.

GAC with the most high energy pores is derived from coconut shell. Lignite coal and wood based GAC typically have a low percentage of high-energy pores. Therefore, they do not perform well in removing MTBE from water.

Recently, a bituminous coal GAC was introduced that is manufactured from select grades of coal and is optimized with a high percentage of high energy pores.

Testing has shown it is capable of sorbing 0.24 g/100 mL at 100 ppb MTBE versus standard bituminous coal at 0.12 g/100 mL or coconut at 0.09 to 0.20 g/100 mL (Calgon Carbon Corporation 2000). The bed life for this product should be longer, reducing carbon exchanges and downtime.

Air Stripping

Air stripping is a standard treatment technique for volatile organic compounds in water. Numerous groundwater contamination problems have been caused by petroleum products containing volatile compounds or synthetic organic solvents that are volatile. Air stripping efficiency is based on Henry's Law constant. This is a mathematical value based on a combination of the compound's volatility and aqueous solubility. High volatility gives a high Henry's Law constant, whereas high aqueous solubility reduces the Henry's Law constant. For example, benzene has a high Henry's Law constant of 230 atmospheres fraction, when MTBE, with a higher vapor pressure, has a value of 32.6. This relationship requires that a much higher air-to-water ratio of 4 to 10 times be used to strip MTBE than is needed for the BTEX hydrocarbons. Since air stripping is not a destructive treatment, the resulting effluent may require additional treatment for MTBE.

Biodegradation

Ethers in general, and MTBE specifically, are not readily biodegraded. Acclimated bacteria, under controlled conditions, have been observed to mineralize MTBE. Controlling conditions for biodegradation requires that water containing MTBE be withdrawn from the aquifer and subjected to the treatment conditions. This implies pump and treat methodology that may be restricted in flow rate according to the efficiency of the biological degradation. The biological process has certain inherent uncertainties, which are a function of biomass, nutrient availability, and chemical byproduct production. These factors detract from this process as a treatment method for a public water supply.

In situ biodegradation is uncertain and slow. It has been shown to produce tertiary-butyl alcohol, a compound also regarded as a contaminant (Mackay et al. 2000). Recent research has found that MTBE degrades under highly reducing conditions (methaneogenisis) or under highly aerobic conditions. Under reducing conditions, MTBE may degrade at the same rate as benzene, but due to the increased mass (higher solubility) a plume still can form. Natural or enhanced oxygen must be present for bacteria to degrade the MTBE once it has separated from the highly reducing portion of the plume. Even under these conditions, biodegradation does not occur at every site.

The successful use of ORC in reducing MTBE concentration in groundwater from 800 ppb to less than 2 ppb at a service station spill site was reported by Koenigsberg (2000). The ORC was injected into the aquifer where BTEX and MTBE were present in dissolved form. The reported reduction in concentration was achieved over a nine-month period. Oxygen transfer by sparging has also been successful at removing MTBE.

Oxidative Processes

Advanced oxidation technologies chemically oxidize MTBE through generation of the hydroxyl radical (OH•). One means of generating OH• is with high energy, medium pressure ultraviolet (UV) light to photodissociate hydogen peroxide into two hydroxyl radicals in a UV reactor. The OH• is a powerful, rapid oxidizer, hence the term advanced to denote the rate constant that is one million to one billion times as fast as chemical oxidizers such as ozone. The UV/H_2O_2 process has been implemented in more than 250 sites worldwide for routine removal of organic compounds in drinking water (Cater, Dussert, and Megonnell 2000).

Another process that produces OH• is ozone coupled with hydrogen peroxide. This system is used in Europe because of the common use of ozone as a water sterilant. If bromide ion is present in the water, bromate ion, a suspected carcinogen, may be formed with this process.

Fenton's reagent generates OH• in a solution with ferrous iron and hydrogen peroxide at an acid pH. The reaction regenerates ferrous iron, so it actually acts as a catalyst. When Fenton's reagent reacts with an organic substrate, heat is produced. Fenton's reagent destroys BTEX, MTBE, and TPH-gasoline in water under controlled conditions as cited above. When reacted in Tedlar bags enabling off-gases to be collected, MTBE was completely oxidized to carbon dioxide (Schreier 2000). It poses problems when it is applied *in situ* as optimum conditions for the reaction are hard to maintain.

A peroxy-acid process utilizing glacial acetic acid and hydrogen peroxide also generates the OH• radical. In controlled laboratory tests, MTBE concentration was reduced by 30 percent in two hours (Halverson et al. 2000). Other recalcitrant compounds including tetrahyrofuran and 1,4-Dioxane were also degraded by this process.

An inherent problem with any oxidative approach to destruction of MTBE or any other contaminant is that of competitive reactions. The subsurface environment in most places is one having some degree of reducing conditions. Residual organic matter in geologic materials, reduced iron, manganese, and sulfur chemical species, and dissolved natural organic matter all react with oxidants that they contact. This phenomenon is called the reductive poise of a location. The amount of oxidant required to react with any or all of these substances must be introduced in order to have excess oxidant available for the target compounds. With moderate to large zones of contamination, overcoming the reductive poise may require thousands of dollars worth of reagents and hardware.

Phytoremediation

Tree roots develop a symbiotic population of bacteria, fungi, and actinomycetes in their immediate vicinity that is termed the rhizosphere (Chapter 9). Tree root exudates are organic compounds that are released from the tree roots into the soil. The associated microorganisms metabolize these compounds, and in turn, produce other compounds that are beneficial to the tree. This system becomes a biologically active zone capable of decomposing synthetic organic compounds through co-metabolism. The tree roots also may absorb organic compounds where they are fixed in the tree tissue, transported and transpired with water through the leaves, or otherwise metabolized as part of the tree's metabolic activity.

Phytoremediation is performed by planting trees (particularly hybrid poplars as they are rapid growing and transpire a large quantity of water) to interact with soil or shallow groundwater contaminants. When roots reach the capillary zone or the water table, they will transpire a large volume of water. Using radiocarbon labeled MTBE, poplar trees were shown to remove over 80 percent of MTBE from hydroponic solution within 11 days. When planted in soil, poplars removed more than 55 percent of the MTBE through transpiration, with only 4.75 percent remaining in soil (McMillan et al. 2000).

1,4-DIOXANE

Background and History

1,4-Dioxane is a high volume chemical with production exceeding 1 million pounds annually in the U.S. In 1990, the total U.S. production volume of 1,4-Dioxane was between 10,500,000 and 18,300,000 pounds (U.S. EPA 1995). In 1992, there were three producers of 1,4-Dioxane in the United States (U.S. EPA 1995).

1,4-Dioxane is used as a solvent for various applications, primarily in the manufacturing sector. 1,4-Dioxane is used as a solvent for cellulose acetate, ethyl cellulose, benzyl cellulose, lacquers, plastics, SBR latex, varnishes, paints, dyes, resins, oils, fats, waxes, greases, and polyvinyl polymers (NSC 1997). It is used as a reaction medium solvent in organic chemical manufacture, as a reagent for laboratory research and testing, as a wetting agent and dispersing agent in textile processing, as a solvent for specific applications in biological procedures, as a liquid scintillation counting medium, in the preparation of histological sections for microscope examination, in paint and varnish strippers, and in stain and printing compositions (NSC 1997). 1,4-Dioxane is also used in shampoo, deodorant, fumigants, cleaning and detergent preparations, and automotive coolant liquids. 1,4-Dioxane was also used as a solvent for coatings, sealants, adhesives, cosmetics, and pharmaceuticals, but these uses have been discontinued due to the potential carcinogenicity of the compound (NSC 1997)

In 1985, 90 percent of 1,4-Dioxane produced in the United States was used as a stabilizer for chlorinated degreasing solvents such as TCA and TCE (U.S. EPA 1995). A chlorinated degreasing solvent is a mixture of one or more chlorinated hydrocarbons plus additives that act as stabilizers and inhibitors (Jackson and Dwarakanath 1999). 1,4-Dioxane acts as an inhibitor to prevent corrosion of aluminum surfaces. TCA typically contains several percent (2 to 3.5 percent) 1,4-Dioxane and TCE may contain small quantities (<1 percent) of 1,4-Dioxane (Jackson and Dwarakanath 1999). Chlorinated degreasing solvents are used in the process of vapor degreasing which removes oil and grease from the surface of machined metal and nonmetal parts. The oil and grease originates with the machining and other fabrication operations which leave machining oils, lubricants, and soldering flux on the surface of the part being cleaned (Jackson and Dwarakanath 1999). Wastes generated by the degreasers (still bottoms or sludge) may contain solvent, additives, oil, grease, solids, and water. For example, sludge generated by the degreasers in the aerospace industry typically are composed of 70 to 80 percent solvent and 20 to 30 percent oil, grease, and solids with traces of water (Jackson and Dwarakanath 1999). Evidence suggests that soluble additives such as 1,4-Dioxane tend to concentrate in the still bottoms generated by the degreaser (Archer 1984).

It was common practice until the 1970s that the sludge from the distillation process could usually be poured on dry ground well away from buildings, and the solvents were allowed to evaporate, assuming there were no special ordinances to prevent it (Jackson and Dwarakanath 1999). Sludge or waste solvent that was poured on dry ground and that did not evaporate or was not incinerated would constitute a DNAPL upon entering the subsurface. Since the sludge from the degreasing process contained 1,4-Dioxane, sites where TCA and TCE have been detected in the groundwater would most likely contain 1,4-Dioxane. Since 1,4-Dioxane is not included on the U.S. EPA target compound lists (TCL) or standard laboratory analytical lists, it is likely that this compound may have not have been included on the chemical analyte list for the site.

Chemical Composition

The physical and chemical properties of 1,4-Dioxane are presented in Table 3. 1,4-Dioxane ($C_4H_8O_2$) is an ether. The functional group of an ether is an atom of oxygen bonded to two carbon atoms (Brown 1998). 1,4-Dioxane is a cyclic ether, which is a heterocyclic compound in which the ether oxygen is one of the atoms in a ring. Ethers are polar molecules in which oxygen bears a partial negative charge and each attached carbon bears a partial positive charge (Brown 1998). However, only weak dipole-dipole interactions exist between their molecules in the pure state. Consequently, boiling points of ethers are much lower than those of alcohols of comparable molecular weight and are close to those of hydrocarbons of comparable molecular weight (Brown 1998). Because the oxygen atom of an ether carries a partial negative charge, ethers form hydrogen bonds with water. Therefore, they are more soluble in water than hydrocarbons of comparable molecular weight and shape (Brown 1998). 1,4-Dioxane is miscible in water, which means that this compound is capable of being mixed with water in all proportions. Ethers are resistant to chemical reaction. They do not react with oxidizing agents such as potassium dichromate or potassium permanganate (Brown 1998). They are stable toward even very strong bases and most ethers are not affected by most weak acids at moderate

Table 3   Physical and Chemical Properties of 1,4-Dioxane

Characteristic/Property	Data
Formula	$C_4H_8O_2$
Molecular Weight	88.10
Specific Gravity	1.034
Water Solubility	Miscible
Log K_{ow}	-0.27
K_{oc}	1.23
Henry's Law constant (atm)	0.27
Vapor Pressure	38 mm Hg at 25°C

temperatures. Because of their good solvent properties and general inertness to chemical reaction, ethers are excellent solvents in which to carry out many organic reactions (Brown 1998).

The chemical and physical properties of a compound determine the fate and transport of a chemical in the environment.

Behavior in the Environment

1,4-Dioxane that enters the atmosphere is expected to degrade fairly quickly. Photo-oxidation by atmospheric hydroxyl radicals appears to be the most rapid degradation process for 1,4-Dioxane in the atmosphere (Howard et al. 1991). Howard (1991) reported estimated high and low half-lives for 1,4-Dioxane in the atmosphere of 3.4 days and 0.34 days, respectively.

No adsorption or volatilization data are available for 1,4-Dioxane in soil. However, based on this compounds infinite solubility and low estimated log soil-adsorption coefficient (K_{oc}) of 1.23 (compounds with a K_{oc} of this magnitude are mobile in soil), 1,4-Dioxane released to the soil is expected to leach to groundwater. The estimated Henry's Law constant suggests that volatilization from moist soils will be slow. However, based on its moderate vapor pressure, volatilization from dry soils is possible (Howard 1990). 1,4-Dioxane has been found to be resistant to biodegradation and has been classified as relatively undegradable (Howard 1990). Howard et al. (1991) reported estimated high and low half lives for 1,4-Dioxane in soil of 6 months and 4 weeks, respectively.

No hydrolysis or volatilization data are available for 1,4-Dioxane in surface water (Howard et al. 1991). When released to surface water, 1,4-Dioxane is not expected to hydrolyze significantly since ethers are generally resistant to hydrolysis (Howard 1990). The low estimated Henry's Law constant (0.27 atm) for 1,4-Dioxane and its miscibility in water suggest that volatilization will be slow (Howard 1990). From its estimated K_{oc} of 1.23, 1,4-Dioxane is not expected to significantly absorb to suspended sediments. The MITI test confirms that 1,4-Dioxane either is not degraded or is degraded slowly (Howard et al. 1991). It is expected, therefore, that 1,4-Dioxane will not biodegrade extensively in the aquatic environment.

The mobility of 1,4-Dioxane in groundwater is directly related to its solubility because very hydrophilic compounds are only weakly retarded by sorption to the aquifer matrix during groundwater transport. Additionally, 1,4-Dioxane is not expected to volatilize into the soil above the aquifer and it does not readily biode-grade. Howard (1991) reported estimated high and low half lives for 1,4-Dioxane in groundwater of 12 months and 8 weeks, respectively.

Examples of 1,4-Dioxane's behavior in groundwater has been documented at two sites contaminated by chlorinated solvents, the Seymour Superfund site in Seymour, Indiana and the Gloucester landfill site near Ottawa, Canada. Retardation factors of 1.0 and 1.1 (average) were estimated for 1,4-Dioxane at the Seymour Superfund site and Gloucester landfill site, respectively (Nyer 1991 and Priddle 1991). A retardation factor of one indicates that the compound is traveling at or near the groundwater flow rate. At both sites, the 1,4-Dioxane migrated farther than the plumes of other compounds detected: benzene, chloroethane, and tetrahydrofuran at the Seymour site; and benzene, 1,2-DCE, 1,2-DCA, and diethyl ether at the Gloucester landfill (Nyer 1991 and Priddle 1991). The other compounds, except tetrahydrofuran, detected at these sites generally had higher retardation factors than 1,4-Dioxane and were more amenable to biodegradation.

Analytical Methods

1,4-Dioxane is not included in the U.S. EPA's TCL, which is the list of analytes used for groundwater investigations at Superfund sites and is not included on standard laboratory analytical lists. However, 1,4-Dioxane is included in the U.S. EPA Appendix IX Groundwater Monitoring List (40 CFR Pt 264, App. IX).

A variety of analytical methods have been used to analyze for 1,4-Dioxane including analytical methods used for VOCs and SVOCs. The suggested groundwater analytical method for 1,4-Dioxane in the Appendix IX list is U.S. EPA SW-846 Method 8015, which is a gas chromatographic/flame ionization detector (GC/FID) method for nonhalogented organics. 1,4-Dioxane is also analyzed by U.S. EPA SW-846 Method 8260B and U.S. EPA Method 624, which are both gas chromatography/mass spectrometry (GC/MS) methods for VOCs. 1,4-Dioxane was previously analyzed by U.S. EPA SW-846 Method 8240B; however, this GC/MS method for VOC analysis was removed from SW-846 in 1996. A gas chromatographic/photo ionization detector (GC/PID) method was developed in 1991 to analyze for 1,4-Dioxane in groundwater at the Seymour Superfund site (Geraghty & Miller 1991). U.S. EPA SW-846 Method 8270C, a GC/MS method for SVOCs, is also currently being used to analyze for this compound.

When using these GC/MS and GC/FID VOC methods to analyze for 1,4-Diox-ane, purge efficiency can be quite low because of the high solubility of 1,4-Dioxane in water, resulting in high estimated quantitation limits. The U.S. EPA SW-846 Method 8015 practical quantitation limit listed in Appendix IX for 1,4-Dioxane is 150 ug/l. U.S. EPA SW-846 Method 8240B indicated that for very soluble compounds like 1,4-Dioxane the quantitation limits are approximately 10 times higher because of poor purging efficiency. One laboratory currently has a reporting limit of 500 ug/l for 1,4-Dioxane using U.S. EPA SW-846 Method 8260B.

A MDL of 40 ug/l and reporting limit of 200 ug/l for 1,4-Dioxane were obtained for the Seymour Superfund site using a GC/FID method developed for this compound (Geraghty & Miller 1991). A lower MDL (3 ug/l) for 1,4-Dioxane was achieved by the GC/PID method developed for the Seymour site. In both of these methods, 1,4-Dioxane was extracted from the groundwater samples prior to being injected into the gas chromatograph. Both the GC/FID and GC/PID methods are used to analyze the Seymour site groundwater samples for 1,4-Dioxane. The GC/PID method is used for samples with concentrations of 1,4-Dioxane below 50 ug/l and the GC/FID method is used for samples with concentrations greater then 50 ug/l (Geraghty & Miller 1991). Due to the sensitivity of the PID instrument, samples with concentrations greater than 50 ug/l can not be analyzed by this method.

According to the U.S. EPA, heating the sample to 85 °C and salting the sample with sodium sulphate prior to purging can boost the purge efficiency to more acceptable levels which gives you lower detection limits for Methods 8015, 8260B, and 624 (U.S. EPA 1997). The sodium sulphate helps decrease the solubility of 1,4-Dioxane in water. The U.S. EPA Region V laboratory can achieve a reporting limit of 5 ug/l for Methods 8260B and 624 when using heating and salting the samples prior to purging (Fuentes 2000). The U.S. EPA also indicates that use of the simultaneous ion monitoring (SIM) mode can be used to achieve lower method detection limit than the scan mode (U.S. EPA 1997).

Some laboratories in Florida are using U.S. EPA SW-846 Method 8270C for analysis of 1,4-Dioxane. Although 1,4-Dioxane is not included in the analyte list for this method, these laboratories have run method studies and are achieving MDLs of 1 to 1.6 ug/l and reporting limits of 10 ug/l for 1,4-Dioxane by using U.S. EPA SW-846 Method 8270C.

Human Health Consideration

1,4-Dioxane has low acute toxicity. The liquid is painful and irritating to the eyes, irritating to the skin upon prolonged or repeated contact, and can be absorbed through the skin in toxic amounts (U.S. EPA 1995). Breathing 1,4-Dioxane for short periods of time causes irritation to the eyes, nose, and throat in humans. Exposure to large amounts of 1,4-Dioxane can cause kidney and liver damage (U.S. EPA 1995). Acute inhalation exposure of high levels of 1,4-Dioxane has caused impaired neurological function and irritation of the eyes, nose, throat, and lungs in humans. These acute effects are not likely to occur at concentrations of 1,4-Dioxane that are normally found in the U.S. environment.

The U.S. EPA has not established a reference concentration (RfC) for chronic inhalation exposure or a reference dose (RfD) for chronic oral exposure for 1,4-Dioxane. No evidence of adverse effects attributable to 1,4-Dioxane exposure was found in three epidemiological studies on workers (U.S. EPA 1995). Dose-related liver and kidney damage have been observed in several species of animals chronically exposed by oral, inhalation, and dermal routes (U.S. EPA 1995).

No information is available on the reproductive and developmental effects of 1,4-Dioxane in humans (U.S. EPA 1998). No evidence of gross, skeletal, or visceral malformations was found in offspring of rats exposed via gavage (experimentally placing the chemical in the stomach) (U.S. EPA 1998). Embryotoxicity was observed only at the highest dose.

Human carcinogenicity data for 1,4-Dioxane are inadequate. In three epidemiological studies on workers exposed to 1,4-Dioxane, the observed number of cancer cases did not differ from the expected cancer deaths (U.S. EPA 1995). U.S. EPA has classified 1,4-Dioxane as a Group B2, probable human carcinogen of low carcinogenic hazard. The basis for the Group B2 classification is induction of nasal cavity and liver carcinomas in multiple strains of rats, liver carcinomas in mice, and gall bladder carcinomas in guinea pigs (U.S. EPA 1995). The cancer oral slope factor is estimated to be 1.1×10^{-2} mg/kg/day for 1,4-Dioxane (U.S. EPA 1990). The U.S. EPA calculated a drinking water unit risk of 3.1×10^{-7} ug/l (U.S. EPA 1990). U.S. EPA estimates that if an individual were to drink water containing 1,4-Dioxane at 3.0 ug/l over his or her entire lifetime, that person would theoretically have not more than a one-in-a-million increased chance of developing cancer as a direct result of drinking water containing this chemical (U.S. EPA 1998). U.S. EPA estimates drinking water concentrations providing cancer risks of 10^{-4} and 10^{-5} to be 300 and 30 ug/l, respectively (U.S. EPA 1990).

1,4-Dioxane has low toxicity to aquatic organisms, toxicity values are greater than 100 mg/L. 1,4-Dioxane is not likely to be acutely toxic to aquatic or terrestrial animals at levels found in the environment (U.S. EPA 1995).

Regulatory Framework

Federal

1,4-Dioxane is regulated by the following federal regulatory programs: the Clean Air Act, Occupational and Safety Health Act (OSHA), Resource Conservation and Recovery Act (RCRA), Superfund, and Toxic Release Inventory Chemicals (EDS 2000). The Clean Air Act Amendments of 1990 list 1,4-Dioxane as a hazardous air pollutant. The OSHA final permissible exposure limit (PEL) is 100 parts per million of air (ppm) as an 8 hour time weighted average (TWA) (29 CFR 1910.000). 1,4-Dioxane is classified as a U108 hazardous waste under RCRA.

There are no federal primary or secondary drinking water standards for 1,4-Dioxane. 1,4-Dioxane is not included in the Drinking Water CCL, a list of contaminants U.S. EPA is considering for possible new drinking water standards; however, this compound was on the Drinking Water priority list, the predecessor of the Drinking Water CCL. U.S. EPA has issued final 1 day and 10 day Drinking Water Health Advisories for 1,4-Dioxane of 4000 and 400 ug/l for a 10 kilogram child, respectively (U.S. EPA 1987). In the 1987 Drinking Water Health Advisory, the drinking water concentration associated with the 10^{-4} cancer risk was 700 ug/l (U.S. EPA 1987). In 1990, the drinking water concentration associated with the 10^{-4} cancer risk was updated to a more conservative 300 ug/l (U.S. EPA 1990).

Since there is no Primary MCL for 1,4-Dioxane, concentrations detected in groundwater are generally compared to the U.S. EPA Region III Risk Based Concentration (RBC) for tap water (6.1 ug/l) or the Region IX Preliminary Remediation Goal (PRG) for tap water (6.1 ug/l) when assessing a site for impacted groundwater. The U.S. EPA Region III RBCs and Region IX PRGs are chemical concentrations corresponding to fixed levels of risk (i.e., hazard quotient of one, or a lifetime cancer risk of 10^{-6}, whichever occurs at a lower concentration). The RBCs were developed by taking toxicity constants (reference doses and carcinogenic potency slopes) and combining these constants with standard exposure scenarios.

State

1,4-Dioxane is regulated under several California State Regulatory Programs including the California Air Toxics "Hot Spots" Information and Assessment Act of 1987 (Assembly Bill 2588), California Law Assembly Bill 1807, and the Safe Drinking Water and Toxic Enforcement Act (California Proposition 65). Proposition 65 includes a requirement that a list of chemicals known to the state to cause cancer or reproductive toxicity be published. Listed chemicals cannot be discharged into sources of drinking water, and warnings must be provided before exposing the public to any significant amount of a listed chemical. The state of Florida has established a groundwater guidance concentration of 5 ug/l for 1,4-Dioxane.

Treatment Options

1,4-Dioxane is one of the most recalcitrant toxic contaminants in subsurface environments. This compound's persistence and mobility presents a challenge to site remediation. All current methods rely on pump and treat methods to remove the compound from the aquifer for treatment above ground. Treatment technologies such as air stripping and carbon absorption are not viable for 1,4-Dioxane. 1,4-Dioxane is not amenable to removal by air stripping because of its hydrophilic nature and low volatility. Carbon absorption is not a viable treatment because of this compound's low carbon absorption capacity (0.5 to 1.0 milligrams of 1,4-Dioxane/gram of carbon at 500 ppb) (Nyer 1991). The state-of-the-art treatment technology for 1,4-Dioxane is UV oxidation with hydrogen peroxide.

UV-Oxidation

Pumping and treating 1,4-Dioxane with ultraviolet (UV)-oxidation and hydrogen peroxide (UV/Peroxide) is the state-of-the-art technology for the remediation of 1,4-Dioxane in groundwater. While hydrogen peroxide is a strong oxidizing agent, its effectiveness increases dramatically when stimulated by UV light. UV/Fenton and UV/Visible/Peroxide treatments can also be used to treat 1,4-Dioxane; however, they are typically not cost effective except in high concentrations (CCC 1996). UV/Fenton, which is a patented Calgon Carbon Oxidation Technologies (CCOT) process, involves the addition of a small amount of iron (II) ENOX 510 catalyst (10 ppm) to the water, adjustment of pH to between 2 and 4, followed by treatment with UV (CCC 1996). The UV/Visible/Peroxide treatment is used when the contaminated water has a chemical oxygen demand of about 1000 ppm (CCC 1996). This process uses a patented photocatalyst (ENOX 910) that strongly absorbs both UV and visible light from 200 to 500 nm wavelengths making use of significantly more of the lamp energy available to generate hydroxyl radicals.

The two primary design variables that must be optimized in sizing a UV/Peroxide system are the UV dose (the total lamp electrical energy applied to 1,000 gallons of water (kWH/1,000 gallons) and the concentration of hydrogen peroxide (CCC 1996). According to the AOT Handbook, the electrical energy per order (EE/O) values for UV/Peroxide treatment of 1,4-Dioxane typically range from 2 to 6 kWh/1,000 gal/order (CCC 1996). The EE/O is the number of kilowatt hours of electricity required to reduce the concentration of a contaminant in 1,000 gallons by one order of magnitude (or 90 percent). The UV/Oxidation treatment rate is affected by the initial concentration of 1,4-Dioxane. Compounds like 1,4-Dioxane with several carbon atoms show a significant relationship between initial concentration and treatment performance. For example, treatment of 100 ppm of 1,4-Dioxane with 200 ppm H_2O_2 requires and EE/O of 11.6 whereas treatment of 200 ppm of 1,4-Dioxane with 500 ppm of H_2O_2 requires an EE/O of 17.9 indicating the treatment efficiency is lower (CCC 1996).

Calgon Carbon Corporation has installed UV/Oxidation treatment systems at 10 sites where 1,4-Dioxane is the main contaminant (Cater 2000). Concentrations of 1,4-Dioxane being treated by UV/oxidation ranged from 20 to 103,000 ppb. Capital costs for a UV/Peroxide treatment unit ranges from $80,000 to $500,000 with O&M costs ranging from 20 cents to $1.50 per 1,000 gallons (Cater 2000). At a chemical manufacturer site in North Carolina, concentrations of 1 to 2.5 ppm 1,4-Dioxane are being treated with UV Oxidation (CCC 1997). The treatment system consists of a 3 x 90 kW Rayox[R] reactor system designed to treat up to 615 gpm from influent concentrations up to 2.5 ppm to below 10 ppb (99.96 percent destruction). O&M costs for this system are $0.76/1000 gallons.

Biodegradation of 1,4-Dioxane

Although 1,4-Dioxane is considered recalcitrant, its biodegradation has been reported under certain conditions, particularly in cultures capable of degrading tetrahydrofuran (THF) and morpholine (Grady, Sock, and Cowan 1997). Research conducted by Sock (1993) using activated sludge from a number of wastewater treatment plants receiving 1,4-Dioxane was conducted in two different types of bioreactors. One bioreactor employed a suspended growth culture and was operated as a sequencing batch reactor, and the other reactor employed a submerged attached growth culture and was operated in a continuous manner. The reactors received a complex feed containing biogenic compounds plus 1,4-Dioxane and THF. The concentrations of the biogenic organic compounds, 1,4-Dioxane, and THF were varied over time and eventually the organic compounds and THF were eliminated. In the attached growth reactor complete removal of 10 mg/l each of 1,4-Dioxane and THF to <1 mg/l (the detection limit) was achieved in the growth reactor in 4 weeks (Grady 1997). By the seventh week, the concentration of 1,4-Dioxane and THF were increased to 80 mg/l and complete removal was achieved by week 10. The suspended growth reactor required a longer time for development of the community perhaps because microbial retention was less effective (Grady 1997). When the feed concentration was 80 mg/l in the suspended growth reactor, 20 weeks were required before complete removal of both compounds was achieved (Grady 1997). After the complex feed and THF were eliminated in week 29, the attached growth bioreactor operated alone on 1,4-Dioxane for 5 weeks. However, following removal of the biogenic substrates in the suspended growth reactor in week 27, significant discharge of 1,4-Dioxane occurred (Grady 1997).

In addition to the results of the attached growth reactor summarized above, the results of additional suspended growth reactors seeded from the attached growth continuous bioreactor, which consistently removed 1,4-Dioxane without having other carbon or energy sources, showed that 1,4-dixaone can serve as a sole carbon and energy source for microbial growth (Grady 1997). This conclusion was confirmed by kinetic experiments based on microbial growth. The kinetic experiments indicated that the growth rate of the culture on 1,4-Dioxane was relatively slow and appeared to be sensitive to temperature. The growth rate of the culture on 1,4-Dioxane was highest at a temperature of 35°C (Grady 1997). The culture degrading 1,4-Dioxane was a complex bacterial community with several genera present. Researchers were unable to isolate a pure culture capable of growth on 1,4-Dioxane alone and were not able to reconstruct the mixed culture from isolated organisms (Morin 1995).

The kinetic results suggested that the failure of conventional wastewater treatment plants to remove 1,4-Dioxane from wastewater is kinetically based and not due to the inability of bacteria to degrade 1,4-Dioxane or to any inhibitory toxic characteristics (Grady 1997). A single completely mixed bioreactor is not a feasible configuration for removal of 1,4-Dioxane to below effluent standards since solids retention time of less than 20 days typical for these systems is not long enough to maintain a culture capable of biodegradation of 1,4-Dioxane (Grady 1997). Additionally, the temperatures of most activated sludge systems drops below 20°C during winter, and at that temperature the minimum solids retention time would be longer.

Theoretical modeling studies were conducted using the known kinetics of biodegradation to investigate two potential alternative configurations: the use of completely mixed tanks in series as the bioreactor configuration and the use of pretreatment of segregated 1,4-Dioxane containing streams (Grady 1997). The results indicated that using a bioreactor configuration of two or more tanks in series with bioreactor temperatures of 35°C should be used for the best removal of 1,4-Dioxane and that benefits would accrue from longer retention times (more than 9 days) (Grady 1997). Although it would be impossible to segregate all of the 1,4-Dioxane containing wastewater streams, pretreatment of 1,4-Dioxane would reduce the mass flow rate into the activated sludge system of the main bioreactor. The study indicated that this would have a large benefit if the activated sludge system were configured in a tank in series arrangement; however, it would not have much beneficial effect on the activated sludge if it contained only a single completely mixed bioreactor (Grady 1997). The simplest form of pretreatment would be to use a simple continuous stirred tank reactor with all biomass growth being discharged from it to the activated sludge (Grady 1997). Under that condition, the activated sludge system becomes a bioreactor that receives in the influent significant quantities of biomass capable of degrading a slowly degradable compound. The magnitude of the benefit to such a bioreactor depends on the mass input rate of capable biomass relative to the mass rate of capable biomass growth on the specific substrate actually applied to it (Grady 1997).

An activated sludge system for treatment of the main wastewater flow and the pretreated esterification wastewater was configured as three tanks in series with 50 percent of the system volume in tank 1 and 25 percent each in tanks 2 and 3 (CH2M Hill 1994). These would be operated at ambient temperature because of the large flow rates. CH2M Hill (1994) tested the proposed flow scheme, as well as several other schemes in pilot scale on-site at a fiber manufacturing facility. The system routinely reduced the 1,4-Dioxane to less than 40 ug/l.

This study indicates that 1,4-Dioxane can be biodegraded in properly configured treatments systems using biological processes. Lab scale studies using the kinetics of 1,4-Dioxane must be conducted to determine the proper configuration of the treatment system followed by pilot scale tests on real wastewater or groundwater influent to scale up the proposed system.

Research conducted at North Carolina State University by Zenker (2000) investigated the ability of subsurface microorganisms from contaminated soil to intrinsically biodegrade 1,4-Dioxane in microcosm assays designed to simulate aquifer conditions. Sediment from six locations contaminated with 1,4-Dioxane was used to construct sets of microcosms—four of which were incubated aerobically and two under anaerobic conditions (Zenker 2000). Each set contained five treatments that broadly examined the effects of 1,4-Dioxane concentration, temperature, presence of THF (a co-substrate), and nutrient availability on biodegredation. Treatments were incubated for approximately 1 year, during which time each was sampled periodically for 1,4-Dioxane concentrations. With the exception of one treatment at one site, no detected biodegradation of 1,4-Dioxane occurred in any microcosms (Zenker 2000). In the one microcosm, complete biodegradation of 1,4-Dioxane was demonstrated within 100 to 300 days under enhanced conditions using soil with previous exposure to 1,4-Dioxane (Zenker 1999). The enhanced conditions included addition of THF, incubation at 35°C, and the addition of nitrogen, phosphorous, and trace minerals. A subsequent readdition of 1,4-Dioxane did not yield further biodegradation until THF was added. Microcosims incubated under ambient groundwater conditions (16°C) exhibited no biodegradation of 1,4-Dioxane (Zenker, Borden, and Barlaz 1999). The sediments studied did not demonstrate that intrinsic biodegradation of 1,4-Dioxane is occurring. However, 1,4-Dioxane degradation by soil microorganisms is possible under certain conditions.

As part of Zenker's (2000) research, a mixed culture from a 1,4-Dioxane contaminated aquifer with the ability to aerobically biodegrade 1,4-Dioxane in the presence of THF was enriched. No biodegradation of 1,4-Dioxane was observed in the absence of THF and the measured cell yield was similar during degradation of 1,4-Dioxane with THF or with THF alone. However, when the consortium was grown in the presence of ^{14}C-1,4-Dioxane plus THF, 2.1 percent of the radiolabeled 1,4-Dioxane was present in the particulate fraction. The majority of the ^{14}C (78.1 percent) was recovered as $^{14}CO_2$, while 5.8 percent remained in the liquid fraction (Zenker 2000).

The biodegradation kinetics of a mixed culture with the ability to cometabolically degrade 1,4-Dioxane in the presence of THF were also studied using a model capable of incorporating the effects of product toxicity, competitive inhibition, depletion of cellular energy, and enhancement of nongrowth substrate biodegradation in the presence of growth substrate (Zenker 2000). Unlike cometabolism of other chemicals, there was no evidence that the biodegradation of 1,4-Dioxane produces toxic byproducts. The presence of 1,4-Dioxane did not inhibit the biodegradtion of THF, however, THF may inhibit 1,4-Dioxane biodegradation (Zenker 2000). The model adequately simulated biodegradation of 1,4-Dioxane and THF at molar concentrations of 0.9-3.3 (mole 1,4-Dioxane/mole THF).

The ability of both a laboratory scale trickling filter and rotating biological contractor (RBC) to biodegrade 1,4-Dioxane was also investigated by Zenker (2000). Both reactors received a feed solution designed to mimic 1,4-Dioxane concentrations typically encounted in contaminated groundwater. The feed solution contained THF and 1,4-Dioxane. The reactors were operated for approximately 1 year and were capable of biodegrading 1,4-Dioxane at 0.2 to 25 mg/l in the obligate presence of THF as the growth substrate (Zenker 2000). Removal rates ranging from 95 to 98 percent were measured for 1,4-Dioxane in the trickling filter. A simple tank in series hydraulic model combined with a kinetic model that incorporated cometabolism was utilized to model the effluent concentration of THF and 1,4-Dioxane from the trickling filter. Model predictions for THF removal were satisfactory for all loading rates analyzed. However, the model generally over predicted the amount of 1,4-Dioxane removal (Zenker 2000). This research demonstrated the ability to treat low concentrations of 1,4-Dioxane through a cometabolic process in attached growth reactors.

Phytoremediation

Although 1,4-Dioxane's half life in soils and groundwater is in the order of years, its half life in the atmosphere in the presence of NO and hydroxyl radicals is only 6.7 to 9.6 hours (Schnoor and Alvarez 1996). Researchers hypothesized that if 1,4-Dioxane could be volatized from the soil and groundwater to the atmosphere by vegetation with a large leaf index and transpiration rate, then 1,4-Dioxane could be remediated (Schnoor and Alvarez 1996). In the atmosphere, the 1,4-Dioxane would be rapidly photodegraded. Using this premise, research was conducted at the University of Iowa to assess the viability of using phytoremediation to remediate 1,4-Dioxane in groundwater and soils.

The results of the research conducted by the University of Iowa suggests that phytoremediation of 1,4-Dioxane may be a viable cleanup alternative. Plants can enhance the removal of xenobiotics by at least two mechanisms, (1) direct uptake and, in some cases, in-plant transformations to less toxic metabolites; and (2) stimulation of microbial activity and biochemical transformations in the rhizosphere (root zone) through the release of exudates and enzymes (Aitchison et al. 1997). Although the latter mechanism is not effective for removing recalcitrant xeonbiotics like 1,4-Dioxane, experiments indicate that this mechanism could be enhanced by inoculation of poplar tree roots with Actinomycete sp. CB1190 (Aitchison et al. 1997).

Hydroponic studies were conducted to assess the capacity of hybrid poplar trees for uptake and translocation of ^{14}C-labled 1,4-Dioxane (Aitchison et al. 1997). In the hydroponic studies, hybrid poplar cuttings removed 1,4-Dioxane rapidly from a solution containing 23 mg/l 1,4-Dioxane. Within 8 days, between 30 percent and 79 percent removal of the 1,4-Dioxane was achieved, with an average removal of 54 percent. Poplars removed 1,4-Dioxane more slowly from spiked soil containing 10 mg of 1,4-Dioxane per kg of dry soil, with 24 percent removal within 18 days. In both experiments, the primary pathway was uptake by transpiration and volatilization from leaf surfaces (Aitchison et al. 1997).

Soil microcosm experiments were also conducted to study the potential to enhance 1,4-Dioxane biodegradation in the rhizosphere through bioaugmentation with an Actinomycete (CB1190). The CB1190 used in the microcosm experiments was grown on a medium of 1000 mg/l tetrahydrofuran, a similar ether that is preferred by CD1190 (Aitchison et al. 1997). CB1190 degraded 100 mg/l 1,4-Dioxane in incubations without soil within 1 month. Poplar root extract (ground roots) stimulated 1,4-Dioxane microbial degradation in soil of unknown previous exposure to 1,4-Dioxane, with 100 percent removal within 45 days. 1,4-Dioxane was not removed in sterile controls or in viable microcosims not amended with CB1190 within 100 days. However, some 1,4-Dioxane removal occurred after 120 days, indicating that the indigenous soil population may have adapted and developed 1,4-Dioxane degrading abilities. All microcosms amended with CB1190 degraded 1,4-Dioxane faster suggesting the feasibility of bioaugmentation to degrade this compound (Aitchison et al. 1997).

Additional bioaugmentation experiments using the 1,4-Dioxane degrading Actinomycete, sp. CB1190 were conducted at the University of Iowa (Kelly et al. 1999). The objectives of this research were to determine if poplar trees could enhance the removal and mineralization of 1,4-Dioxane from contaminated soil and determine if bioaugmentation with CB1190 could enhance the cleanup process in both planted and unplanted soil (Kelly et al. 1999). Reactors planted with hybrid poplar trees removed more 1,4-Dioxane within 26 days than in unplanted reactors, regardless of whether CB1190 was added. Bioaugmentation with CB1190 enhanced mineralization of 1,4-Dioxane in all experiments. CB1190 increased the percentage of 1,4-Dioxane mineralized in unplanted soil from 9 percent to 26 percent (Kelly et al. 1999). Experiments were also conducted with excised tree reactors that offer a root zone, but do not remove 1,4-Dioxane through plant evapotranspiration. More 1,4-Dioxane was mineralized in the excised tree reactor with CB1190 (35 percent) than in one without CB1190 (17 percent) (Kelly et al. 1999). Although CB1190 also enhanced 1,4-Dioxane mineralization in planted soil, this enhancement was not statistically significant because plant uptake reduces the availability of 1,4-Dioxane for microbial mineralization (Kelly et al. 1999).

The research summarized above indicates using phytoremediation to remove 1,4-Dioxane from groundwater and soil is a remediation alternative which should be explored further. Further research into the final metabolites and endproducts will be necessary to fully assess the effects of full-scale phytoremediation (Atchison et al. 1997). To date, phytoremediation of 1,4-Dioxane is not being utilized at any 1,4-Dioxane contaminated sites.

PERCHLORATE

$$O = \overset{\overset{\displaystyle O}{\|}}{\underset{\underset{\displaystyle O}{\|}}{C}} - O^{-}$$

During the last decade, perchlorate has been identified as a contaminant of concern for investigations dealing with soil and groundwater contamination. Present in a variety of commercial formulations, perchlorate was brought to the forefront with the development of a low level laboratory detection method in 1997. Since that time, a variety of sites have been identified as being contaminated with perchlorate resulting in the need to respond to mitigate the effects on human health and the environment. A systematic survey of the United States has not occurred to know how prevalent this constituent is in the environment.

History and Use of Perchlorate

Perchlorate salts are utilized by a variety of industries: rocket solid propellant, electronic tubes, additives to lubricating oils, tanning and finishing leather, aluminum refining, rubber manufacturing, and production of paints and enamels. Ammonium perchlorate, however, has been the focus of increased scrutiny due to documented groundwater contamination associated with the handling of this material. The compound is used in the propellant for rockets, missiles, and fireworks. Additionally, ammonium perchlorate is being utilized as a component of air bag inflators. Some data suggest that ammonium perchlorate may also be associated with fertilizer, however, the literature references only one known occurrence of perchlorate contamination resulting from fertilizer that was imported from South America (TRC Environmental Corporation 1998).

As a solid, ammonium perchlorate is highly reactive and is used as an oxidizer for solid rocket fuel. The ammonium perchlorate is part of a solid composite propellant that includes a binder, an oxidizer, aluminum powder, a plasticizer, a ballistic modifier, a bonding agent, and an antioxidant. The solid propellant has a limited shelf life and as such requires periodic replacement with fresh material and disposal of the aged material. In the United States, large-scale production of perchlorate compounds began in the mid 1940s. Starting in the 1950s, large quantities of his material has been replaced and disposed.

The aged propellant has been land disposed as well as disposed by open pit burns and detonation. The waste material and noncombusted propellant has infiltrated into the soil and groundwater resulting in contamination of both medias (US EPA 1999).

As of June 2000, 14 states have confirmed releases of perchlorate and a total of 44 states have confirmed the presence of perchlorate manufacturers or users. Based upon usage, additional locations may be confirmed once a systematic sampling and monitoring program is instituted at other sites across the country. Based upon available information, locations where perchlorate has been manufactured or used are likely candidates for historical releases.

Chemistry

Perchlorate (ClO_4^-) is an anion generated by the dissolution of ammonium, potassium, magnesium, or sodium salts. Table 4 provides a summary of some of the

Table 4   Physical and Chemical Properties of Perchlorate

Characteristic/Property	Data
Formula	ClO_4^-
Molecular Weight	99.45
Specific Gravity	NA
Water Solubility	Miscible
Log K_{ow}	NA
K_{oc}	NA
Henry's Law Constant (atm)	NA
Vapor Pressure	NA

basic physical and chemical properties of the anion and their role in the behavior of perchlorate in the environment.

Solubility

The solubility of ammonium perchlorate is reported to be 20.2 g/100 g of solution at 25°C. This value indicates that ammonium perchlorate will dissolve and that the perchlorate anion will persist in solution. Ammonium perchlorate is stable when dissolved in water at concentrations as high as 1000 mg/l. Perchlorate as an anion is miscible.

Standard Potential

The oxidation/reduction potential of a compound is represented by the standard potential value. The value suggests that reduction of the perchlorate anion is thermodynamically favorable; however, available studies do not suggest that this reduction occurs spontaneously. The standard potential also indicates that anaerobic and anoxic conditions will promote reduction of the perchlorate anion.

Vapor Pressure

Vapor pressure of a compound indicates its propensity to partition to a gaseous phase. Consequently, volatility of a compound is reflected by a higher vapor pressure at a given temperature. The available literature does not provide a vapor pressure for ammonium perchlorate; however, looking at a similar compound, ammonium chloride, suggests that ammonium perchlorate will not be volatile at ambient temperatures.

Density

The density of ammonium perchlorate is greater than the density of water at similar temperatures indicating that the material will sink through a column of water. Likewise, a concentrated solution of ammonium perchlorate is expected to be denser than water.

Analytical Method, Detection Limits and when they were Developed

Although analytical procedures have existed for some time to detect perchlorate, the methodologies were not suitable for detection at low levels. Prior to 1997, colorimetric and ion chromatographic (IC) procedures have been employed to detect the presence or absence of perchlorate and, in the case of the IC, quantitatively measure concentrations in excess of 400 ppb. Further refinements dropped the detection limit to 100 ppb however toxicological issues suggested a detection level at 4 ppb would be necessary. In March 1997, the California Sanitation and Radiation Laboratory Branch (SRLB) in conjunction with an analytical equipment manufacturer developed an IC method that had a method detection limit of approximately 1 ppb and a reporting limit of 4 ppb.

Behavior in Environment

Behavior of the perchlorate anion is controlled by its basic chemical properties. As an oxidant, the reduction of the chlorine atom occurs slowly. Retardation to soil is not expected to be significant since perchlorate has a low affinity. As discussed, the compound is soluble in water and does not interact with the soil matrix in the aquifer. In addition, the half life of perchlorate in the environment seems to be long. Therefore, perchlorate travels at close to the speed of the groundwater, and the resulting plumes can be large.

Human Health Considerations Including History

In 1985, Region 9 of the EPA utilizing colorimetric procedures encountered perchlorate concentrations in 14 wells ranging from 0.11 to 2.6 ppm. The Center for Disease Control (CDC) was contacted to offer assistance relative to the potential health effects of perchlorate at these concentrations. The CDC recommendation was to validate the colorimetric results, but they were unable to assist with the potential toxicity issues due to lack of data. EPA Region 9 focused on other chemicals at these wells since a suitable analytical procedure and relevant toxicity data was unavailable.

EPA Region 9, however, encountered perchlorate concentrations again at a California Superfund site in excess of 1 mg/l in monitor wells. Consequently, efforts were intensified to establish a human health based reference dose (RfD). A request to the EPA Superfund Technical Support Center resulted in the release of provisional RfD values in 1992 which were revised in 1995. Groundwater cleanup guidance levels ranging from 4 to 18 ppb was calculated based upon existing information and utilizing standard assumptions relative to exposure pathways, ingestion, and body weight.

After development of a suitable analytical method for low level detection of perchlorate in 1997, the existing toxicological database was consulted to determine the adequacy of the existing information relative to perchlorate. At that time, the existing data was found to be inadequate for the purposes of performing quantitative exposure assessments for humans. Since that time, a number of studies have been performed and the database has increased; however, additional data is still needed and being gathered.

The principal exposure route for perchlorate is through oral uptake. Once consumed, perchlorate is readily absorbed from the intestinal tract. Neither dermal exposure nor inhalation is a probable exposure pathway. Once in the body, perchlorate competes with iodide in the thyroid causing a reduction in the hormones produced by the thyroid. The consequence of this behavior raises concerns relative to potential carcinogenic, neurodevelopmental, developmental, reproductive, and immunotoxic effects. Additionally, limited information is available to address potential ecotoxicological effects from perchlorate.

Regulatory Framework

Federal

Perchlorate is presently not regulated by the U.S. Environmental Protection Agency. The 1996 Safe Drinking Water Act (SDWA) mandated EPA to identify a list of contaminants to be examined for development of regulatory controls if appropriate. The Office of Water (OW) placed perchlorate on the Contaminant Candidate List in March 1998 indicating that additional research and information would be required before regulatory determinations could be made.

State

In the absence of federal regulation, regulatory control of perchlorate concentrations by the individual states is somewhat limited. Presently, only California and Nevada have adopted action levels for perchlorate. Both states have adopted 18 ppb as the recommended guideline.

Treatment Options

A viable treatment option is needed to enable removal of perchlorate from groundwater. Perchlorate does not naturally attenuate in soils or an aquifer and since the compound is not organic, bacteria will not use it as a food source. The only use the bacteria would have for perchlorate would be as a final electron acceptor. Other *in situ* methods will also not work due to the chemical properties of the compound. Soil vapor extraction, and air sparging both rely on the compound to transfer to air as the carrier. Since perchlorate is not volatile, it will not transfer to an air stream.

The current treatment methods for perchlorate require the compound to be brought above ground. A pump and treat system will have to be employed to capture the plume and bring the perchlorate to a treatment system. Conventional filtration and sedimentation are not effective and activated carbon appears to be only slightly effective in removing perchlorate. Above ground, the following treatment methods are currently being explored in an attempt to identify a method that is cost effective in addressing low level concentrations of perchlorate.

Ion Exchange

Ion exchange has a demonstrated successful history of treating water for the removal of nitrate, an anion similar to perchlorate and has received considerable attention relative to its use to treat perchlorate contaminated waters. The process involves the substitution of an innocuous anion (e.g., chloride) for the perchlorate anion.

Calgon Carbon Corporation successfully demonstrated the use of an ion exchange technology to remove perchlorate from groundwater. In a study funded by the Main San Gabriel Watermaster for the Big Dalton well, Calgon conducted a pilot study of the ISEP® system in 1998. Calgon's ISEP® process is a continuous counter-current ion exchange system that removes perchlorate contamination from groundwater. The process uses an ion exchange resin that is continuously regenerated. The system demonstrated effective removal of perchlorate (18-76 ppb) to non-detectable (<4 ppb) levels as well as removing nitrate and sulfate in the treated water. The brine generated from the process consequently has levels of these constituents present at a much greater concentration than the influent water. However, the brine volume represents only 0.75 percent of the feed water.

In a separate pilot demonstration, Calgon successfully demonstrated the use of a perchlorate and nitrate destruction module (PNDM) in the overall treatment train. The PNDM module contains a catalytic reactor for perchlorate and nitrate reduction and a nanofiltration unit for sulfate separation. Funded by National Aeronautics and Space Administration (NASA), groundwater at the Jet Propulsion Laboratory in Pasadena, California was contaminated with trichlorethene (TCE), 1,2-dichloroethane (DCA), carbon tetrachloride (CCL_4) and perchlorate (up to 1,200 ppb influent). GAC was used to remove the organic contaminants from the groundwater prior to be treated by the ISEP® process. For this pilot test, Calgon constructed a treatment platform that included the ISEP® module and the PNDM module. The resulting platform is referred to as an integrated ISEP+™ system. The pilot demonstrated the successful removal of all the constituents to nondetectable concentrations as well as the subsequent destruction of perchlorate and nitrate and removal of sulfate (exceeding 96 percent). After treatment by the PNDM module, the brine was suitable for resin regeneration.

Reverse Osmosis

Preliminary studies indicate that reverse osmosis (RO) can remove low levels of perchlorate from contaminated water. RO utilizes a fine filter/membrane that contaminated water is forced through. As the water passes through the membrane, perchlorate is removed and perchlorate free water exits the filter. Similar to the ion exchange process, a perchlorate laden brine is generated that must be properly managed. The overall cost and the actual cost effectiveness of a full-scale RO system is unknown.

U.S. Filter assisted the Jet Propulsion Laboratory in identification of remedial technologies that were capable of treating groundwater contaminated with perchlorate. U.S. Filter identified RO as one of the potential remediation technologies. RO was identified as having no chemical requirements with the exception of chemical usage for controlling scale development. The process has demonstrated success for the removal of perchlorate and other constituents. Performance results were provided for a thin film composite membrane and a cellulose acetate membrane. The influent water had a perchlorate concentration of 800 ppb and the resulting permeate had concentrations of approximately 4 to 18 ppb and approximately 600 ppb, respectively. Demonstrating the higher performance capability of the thin film composite.

Biodegradation

Use of microbes to convert perchlorate to a less toxic or innocuous substance has received considerable attention and success. As discussed earlier, the standard potential of perchlorate suggests that the anion will be reduced in anaerobic conditions. To date, efforts have focused on development of a suitable biochemical reduction process to promote the anaerobic reduction of perchlorate. In one study, the bacterium, perclace, was cultivated on acetate in an anaerobic environment and then placed in celite-packed columns for use in a flow through system. Experimentation showed that the system was capable of reducing perchlorate concentrations from 738 ppb to less than detectable levels at a flow rate of 1 ml per minute. Doubling the flow rate reduced the effectiveness of the system to 92 to 95 percent of the perchlorate. Further experimentation demonstrated that use of a circulating pump and multiple passes through the media allowed flow rates to increase to 3 ml per minute and still achieve a 95 percent removal rate (Giblin et al. 2000).

Full-scale bioreactor systems developed in late 1990s have demonstrated the ability of this technology to successfully reduce perchlorate concentrations from over 5000 ppb to the low hundreds of ppb. Of particular note is pilot scale testing that was performed at the San Gabriel Valley Superfund site in California. A bioreactor system was successful in treating groundwater with approximately 150 ppb perchlorate to nondetectable levels. The expected outcome of this work is to develop a system capable of treating 20,000 gpm. The treated water will be supplied to local drinking water utilities for additional treatment and distribution. Public perception and acceptance however may require modifications to this plan. Historically, the public will not accept a biological treatment method for drinking water.

An anoxic fluidized bed reactor (FBR) has been successfully employed at the Aerojet site (Rancho Cordova, California) as part of the Jet Propulsion Laboratory Remediation Project. U.S. Filter designed and operated a 4000 gpm FBR with an influent containing 6 mg/l perchlorate and 1.5 mg/l nitrate (NO_3-N) for treatment.

Effluent concentrations for perchlorate and nitrate were less than 4 ppb and less than 0.1 ppb, respectively.

Phytoremediation

Initial studies have shown that trees and plants can also create the environment where perchlorate is degraded. Research utilizing willows and eucalyptus species successfully reduced perchlorate concentrations of 10, 20, and 100 mg/l to below 2 ug/l concentrations. (Nzengung et al. 1999) Additionally, spinach, French tarragon, and *Myriophyllum* (an aquatic plant) also removed perchlorate from water. The phytoprocesses identified as instrumental to the remediation of perchlorate contaminated water are, (1) uptake and phytodegradation in the plant organs; and (2) rhizodegradation as a result of associated microorganisms. The experimental approach suggests that contaminated groundwater could be remediated by irrigation of the intensively cultivated parcels utilizing the successful species. However, the success of this approach will be limited by the capacity of the soil and the plantings to handle potentially large volumes of water.

In Situ and the Future

The above-ground use of biodegradation has shown that perchlorate will act as a final electron acceptor under reducing conditions. Several studies are now planned to use the IRZ technologies (Chapter 8) to change the environment to reducing conditions in order to degrade the perchlorate *in situ*. Efforts are underway to isolate the enzymes from the microorganism responsible for perchlorate reduction that may support an *in situ* approach.

N-NITROSODIMETHYLAMINE—NDMA

Background and History

N-Nitrosodimethylamine is commonly known as NDMA. It is a yellow liquid that has no distinct odor. NDMA is formed by the nitrosation of dimethylamine (DMA) with a nitrite ion (NO_2^-). Short-term or long-term exposure of animals to water or food containing NDMA is associated with serious effects, such as liver disease and death.

NDMA's use is primarily in research, but it has had prior use in the production of 1,1-dimethylhydrazine for liquid rocket fuel, and a variety of other industrial uses:

a nematocide, a plasticizer for rubber, in polymers and copolymers, a component of batteries, a solvent, an antioxidant, and a lubricant additive. NDMA was reported to be present in a variety of foods, beverages, and drugs, and in tobacco smoke; it has been detected as an air pollutant, and in treated industrial wastewater, treated sewage in proximity to a 1,1-dimethylhydrazine manufacturing facility, deionized water, high nitrate well water, and chlorinated drinking water (NTP 1998). NDMA is unintentionally formed during various manufacturing processes at many industrial sites, and in air, water, and soil from reactions involving other chemicals called alkylamines. Alkylamines are both natural and man-made compounds that are found widely distributed throughout the environment.

Optimization of drinking water treatment to minimize formation of NDMA is recommended. In particular, the suitability of the use of the specific preblended polyamine/alum water treatment coagulant identified to be contributing to levels of NDMA in drinking water.

NDMA does not persist in the air environment. When NDMA is released into the atmosphere, it breaks down in sunlight in a matter of minutes. When released to soil surfaces, NDMA may evaporate into air, break down upon exposure to sunlight, or sink into deeper soil. The rate of breakdown in water is not known.

Analytical Method

Several approaches are available for analysis of NDMA at levels at or below 2 ng/l (0.002 ppb). Continuous liquid-liquid extraction or similar techniques capable of reproducible recovery (<20 percent RSD precision) such that the GC/MS signal intensity is sufficient to achieve the level at or below 2 ng/l is required. Gas chromatographic/mass spectrometric methods offer the most sensitive and definitive measurement systems for analysis of NDMA in the low ng/l range. High resolution electron impact mass spectrometry, and low resolution chemical ionization (using ammonia, methanol, etc.) or other mass spectrometric techniques with equivalent sensitivity are acceptable.

Fate and Transport

Table 5 summarizes the chemical properties for NDMA. Once again, this compound is very soluble in water and does not interact with the soil in the aquifer. This compound will travel at close to the speed of groundwater and does not seem to naturally attenuate in the aquifer. The added problem with this compound is that low concentrations (ng/l) are considered harmful to human health. This will lead to regulatory action at very low levels of the compound, and resulting investigation and remediation to these levels.

Regulation Framework

Because NDMA historically has not been considered a common drinking water contaminant, it has no state or federal maximum contaminant level (MCL). In the

Table 5   Physical and Chemical Properties of *N*-Nitrosodimethylamine, NDMA

Characteristic/Property	Data
Formula	$C_2H_6N_2O$
Molecular Weight	74.08
Specific Gravity	1.0048 at 20/4C
Water Solubility	Miscible
Log K_{ow}	0.57
K_{oc}	12
Henry's Law Constant (atm)	0.02
Vapor Pressure	2.7 mm Hg at 20°C

absence of drinking water standards, California's DHS uses a drinking water "action level" for the protection of public health.

Risk assessments from the California's OEHHA and U.S. EPA identify lifetime *de minimis* (i.e., 10^{-6}) risk levels of cancer from NDMA exposures as 0.002 ppb and 0.0007 ppb, respectively. A *de minimis* risk is considered to be below regulatory concept 10^{-6} risk level and corresponds to up to one excess case of cancer per million per drinking 2 liters of water per day for a 70 year lifetime.

State

OEHHA lists NDMA as a chemical known to the State of California to cause cancer (Title 22, California Code of Regulations (22 CCR), Section 12000). OEHHA's 10^{-5} level corresponds to an exposure of 0.04 µg per day (2 CCR Section 12705(b)(1), which equates to 0.02 µg/l.

U.S. EPA

U.S. EPA classifies NDMA as B2, probable human carcinogen, based on the induction of tumors at multiple sites in laboratory animals exposed by various routes. EPA's 10^{-6} risk level in drinking water corresponds to a concentration of NDMA of 0.0007 µg/l (U.S. EPA 1997).

DHS initially established (in April 1998) an action level for NDMA of 0.02. However, analytical capabilities did not enable detection at that concentration. Detectable quantities of NDMA exceeded the action level.

In the Fall of 1999, coincidental with more sensitive analytical methods be available, DHS began working with utilities in the state to investigate the production of NDMA during drinking water treatment processes. During this investigation phase, DHS established a temporary action level of 0.02 µg/l for NDMA, effective 1999. In November 1999, DHS initiated studies with drinking water utilities to investigate the occurrence of NDMA in raw, treated and distributed water, the role water quality and treatment processes may play in the production of NDMA, and the possible extent of NDMA production at various steps in the water treatment process. These studies are anticipated to continue into the summer of 2000.

Treatment

The natural attenuation rate of NDMA seems to be very slow. One problem may be that the extremely low concentrations that we are working with. Biological and abiotic reactions may be limited due to the low concentrations. The natural attenuation that may occur with NDMA at ppb levels may shut off at ppt levels.

Above-ground NDMA can be treated using photolysis or carbon adsorption. Photolysis involves the interaction of light with molecules to bring about their dissociation into fragments. Light is composed of tiny energy packets called photons, whose energy is inversely proportional to the wavelength of the light. Thus, shorter wavelengths have higher energy. The optimal wavelength for photo-oxidation of NDMA is between 220 and 240 nanometers (Solarchem 1994). The UV light range is considered to be wavelengths less than 400 nanometers.

One vendor of UV oxidation systems for destruction of organics designs systems based on a UV dose, which is defined as the amount of UV lamp power in kWh applied to 1000 gallons of water. Specifically, this design parameter is known as the electrical energy per order (EE/O) with units in kWh per 1000 gallons per order, and is used to scale up of UV systems from the pilot to full-scale. Based on vendor data shown in Figure 1, 90 percent removal of ppt and ppb levels of NDMA can be achieved at an EE/O (UV dose) of 2.4 per 1000 gallons of water (Solarchem 1994). This UV dose required a power consumption of 2.5 kWh per 1000 gallons which at a typical rate of 7.5 cents per kWh results in a reasonable treatment cost of approximately 19 cents per 1000 gallons for energy alone. Results from a pilot test conducted by ARCADIS Geraghty & Miller in 1995 on a groundwater containing 230 µg/l NDMA achieved 99.6 percent destruction of NDMA using 12 kWh per 1000 gallons.

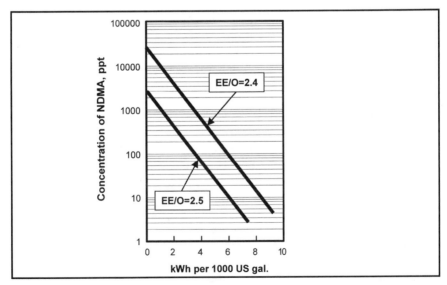

Figure 1   Treatment of groundwater containing 3,000 and 30,000 ppt of N-nitrosodimethylamine (NDMA).

As an alternative to UV photolysis systems, sunlight can also be used to photo-degrade NDMA. As described in the Background section above, NDMA breaks down in a matter of minutes when exposed to sunlight. In parts of the United States, solar treatment ponds have been used. Successful implementation of such a remedy requires careful evaluation of the amount of sunlight available, precipitation, evaporation, and daily temperatures. This method of treatment has been implemented in California. Other areas of the country where this method may be applicable include Arizona, Nevada, and New Mexico.

NDMA can be adsorbed onto GAC, however, vendor literature suggests the removal is highly dependent on the influent concentration and this method of treatment is only practically applied to concentrations in the 5 to 10 mg/l range. At a concentration of 10 mg/l, a saturated loading rate of 16 percent (16 pounds of NDMA per 100 pounds of GAC) may be achieved based on vendor isotherm data (Westates 1988). At concentrations below 5 mg/l, the saturated loading rate decreases to less than 0.1 percent (Westates 1988). Effluent concentrations achievable under these conditions were not provided.

REFERENCES

Aitchison, E.W., Kelly, S.L., Schnoor, J.L., and. Alvarez, P.J.J., "Phytoremediation of 1,4-Dioxane by Hybird Poplars," *Proc. Water Environment Federation 70th Annual Meeting and Exposition,* North Chicago, IL, October 18-22, 1997.

Archer W.L., "A Laboratory Evaluation of 1,1,1-Trichloroethane-metal-inhibitor Systems," *Werkstoffe und Korrosion, 35,* 1984.

Brown, W. H. with Foote, C. S., *Organic Chemistry,* 2nd ed., Saunders College Publishing, Fort Worth, 1998.

Calgon Carbon Corporation (CCC), "Continuous Removal and Destruction of Perchlorate and Other Contaminants Using ISEP+™ Technology," submitted to California Department of Health Services, Berkeley, CA, February 2000.

Calgon Carbon Corporation (CCC), Product Bulletin Filtrasorb® 600, 1999a.

Calgon Carbon Corporation (CCC), "Removal of Perchlorate and Other Contaminants from Groundwater at JPL: A Pilot Study," submitted to Jet Propulsion Laboratory, Pasadena, CA, June 1999b.

Calgon Carbon Corporation (CCC), "Big Dalton Perchlorate Removal Pilot Study," submitted to The Main San Gabriel Basin Water Master, October 1998.

Calgon Carbon Corporation (CCC), "UV/Oxidation Treatment of 1,4-Dioxane in Groundwater, Chemical Manufacturer, North Carolina," February 1997.

Calgon Carbon Corporation (CCC), "The AOT Handbook," October 1996.

California Department of Health Services, "Perchlorate in Drinking Water," Division of Drinking Water and Environmental Management, May 1997.

California Environmental Protection Agency, "MTBE (methyl tertiary butyl ether) Briefing Paper," April 24, 1997, updated September 3, 1998.

Cater, S. (Calgon Carbon Corporation), Telephone Interview, May 23, 2000.

Cater, S.R., Dussert, B.W., and Megonnell, N., "Reducing the Threat of MTBE-Contaminated Groundwater," *Pollution Engineering,* pp. 36-39, May 2000.

CH2M Hill, "1,4-Dioxane Treatment Guidance Document," Report to Hoechst Celanese Corporation, Charlotte, NC, 1994.

Cox, E.E., Allan, J., and Neville, S.L., "Rapid Bioremediation of Perchlorate in Soil and Groundwater," Division of Environmental Chemistry Preprints of Extended Abstracts: Vol.39(2), August, 1999.

Environmental Defense Scorecard (EDS), "Chemical Profile for 1,4-Dioxane (CAS Number 123-91-1)," "hhtp://www.scorecard.org/chemical profiles/summary.tcl?edf_substance_id =123%2d91%2d1," Search date -May 10, 2000.

Feuntes, N. (U.S. EPA Region V, Central Regional Laboratory, Chicago, Illinois), Telephone Interview, May 2000.

Geraghty & Miller, Inc, "Long-term Sampling Groundwater Monitoring Plan for the Seymour Site," January 1991 (Revised June 1991).

Giblin, T., Herman, D., Deshusses, M.A., and Frankenberger Jr., W.T., "Removal of Perchlorate in Ground Water with a Flow-Through Bioreactor," *Journal of Environmental Quality* 29, 2000.

Grady, C.P.L., Sock, S.M., and Cowan, R.M., "Biotreatability Kinetics: A Critical Component in the Scale-up of Wastewater Treatment Systems," *Biotechnology in the Sustainable Environment,* Ed. Gary S. Sayler, Plenum Press, 1997.

Halverson, J.E., Dutkus, K., Leister, M., Nyman, M., and Komisar, S. "Advanced oxidation of MTBE and ETBE using a peroxy-acid process," Pre-Prints Environmental Chemistry Division, American Chemical Society, pp. 236-237, March 2000.

Herman, D.C., and Frankenberger Jr., W.T., "Microbial-Mediated Reduction of Perchlorate in Groundwater," *Journal of Environmental Quality* 27, 1998.

Howard, P. H., Boethling, R. S., Jarvis, W. F., Meylan, W. M., and Michalenko, E. M., "Handbook of Environmental Degradation Rates," Lewis Publishers, Inc., Chelsea, Michigan, 1991.

Howard, P. H. (ed.), "Volume II, Solvents," *Handbook of Environmental Fate and Exposure Data for Organic Chemicals,* Lewis Publishers, Inc., Chelsea, Michigan, 1990.

Keith, L.H. and Walters, D.B., "Compendium of Safety Data Sheets for Research and Industrial Chemicals, Part II," VCH Publishers, Deerfield Beach, 1985.

Kelly, S. L., Aitchison, E. W., Schnoor, J. L., and Alvarez, P. J.J., "Bioaugmentation of Poplar Roots with *Amycolata* sp. CB1190 to Enhance Phytoremediation of 1,4-Dioxane," *Phytoremediation and Innovative Strategies for Specialized Remedial Applications*, Eds. B.C. Alleman and A.L. Leeson, Battelle Press, Columbus, 1999.

Jacobs, J., Guertin, J., and Herron, C., Eds. "Report & Recommendations on the Effect of MTBE on Soil and Ground Water Resources," Point Richmond, CA: Independent Environmental Technical Evaluation Institute, 1999.

Jackson, R. E. and Dwarakanath, V., "Chlorinated Degreasing Solvents: Physical-Chemical Properties Affecting Aquifer Contamination and Remediation," *Ground Water Monitoring Remediation* 19, No. 4, Fall 1999.

Johnson, R., Pankow, J., Bender, D., Price, C., and Zogorski, J. "MTBE to What Extent Will Past Releases Contaminate Community Water Supply Wells?" *Environ. Sci. Technol.*, vol. 34, pp. 210A-217A, 2000.

Koenigsberg, S., "The use of ORC in the bioremediation of MTBE," Pre-Prints Environmental Chemistry Division, American Chemical Society, pp. 289-291, March 2000.

Loehr, R.C., Katz, L.E., and Opdyke, D.R., "Fate and Transport of Ammonium Perchlorate in the Subsurface," Environmental and Water Resources Engineering Program, The University of Texas at Austin, Austin, Texas, April 1998.

Mackay, D., Wilson, R., Durrant, G., Scow, K., Smith, A., Chang, D., and Fowler, B, Pre-Prints Environmental Chemistry Division, American Chemical Society, pp. 284-286, March 2000.

Morin, M.D., "Degradative Characterization of a Mixed Bacterial Culture Capable of Mineralizing 1,4-Dioxane and Preliminary Identification of Its Isolates," Master of Science Thesis, Microbiology, Clemson University, Clemson, SC, 1995.

National Safety Council (NSC), "Environmental Writer - 1,4-Dioxane (C4H8O2) Chemical Backgrounder," Environmental Health Center, "hhtp//www.nsc.org/ehc/ew/chems/dioxane.htm," July 1, 1997.

National Toxicology Program, "NTP Chemical Repository, 1,4-Dioxane," "hhtp://ntp-db.niehs.nih.gov/NTP_Reports/NTP_Chem_H&S/NTP_Chem1/Radian123-91-1.txt," Search Date -May 10, 2000.

National Toxicology Program (NTP), *"N-Nitrosodimethylamine CAS No. 62-75-9,"* *Eighth Report on Carcinogens, 1998 Summary,* Public Health Service, US Department of Health and Human Services, 1998.

Nichols, E.M., Einarson, M.D., and Beadle, S.C. "Strategies of Characterizing Subsurface Releases of Gasoline Containing MTBE," API Publication No. 4699, American Petroleum Institute, February 2000.

Nyer, E., Boettcher, G., and Morello, B., "Using the Properties of Organic Compounds to Help Design a Treatment System," *Ground Water Monitoring Review,* 11, No. 4, Fall 1991a.

Nyer, E. K., Kramer, V., and Valkenburg, N., "Biochemical Effects on Contaminant Fate and Transport," *Groundwater Monitoring Review* 11, Spring 1991b.

Nzengung, V.A., Wang, C., Harvey, G., McCutcheon, S., and Wolfe, L. "Phytoremediation of Perchlorate Contaminated Water: Laboratory Studies," *Phytoremediation and Innovative Strategies for Specialized Remedial Applications*, April 1999.

Priddle, M.W. and Jackson, R.E., "Laboratory Column Measurement of VOC Retardation Factors and Comparison with Field Values," *Groundwater,* 29, No. 2, 1991.

Schnoor, J. and Alvarez, P. J., "Feasibility of Phytoremediation for Treatment of 1,4-Dioxane at Sites Using Hybrid Poplar Trees," A Research Proposal Submitted to Hoechst Celanese Corporation, 1996.

Schreier, C.G., "Removal of MTBE and other petroleum hydrocarbons from water using Fenton's reagent," Pre-Prints Environmental Chemistry Division, American Chemical Society, pp. 242-243, March 2000.

Sock, S.M., "A Comprehensive Evaluation of Biodegradation as a Treatment Alternative for Removal of 1,4-Dioxane," M.S. Thesis, Clemson University, Clemson, SC, 1993.

Solarchem Environmental Systems, *The UV/Oxidation Handbook*, 1994.

SRC Environmental Fate Data Base (EFDB) Information, "SRC Chemfate Search Results, CAS# 000123-91-1, Name: 1,4-Dioxane," "http://esc.syrres,com/scripts/CHFcgi.exe," Search date - April 28, 2000.

The Merck Index. An Encyclopedia of Chemicals, Drugs, and Biologicals. 11th ed. Ed.S.Budavari, Merck and Co. Inc., Rahway, NJ., 1989.

TRC Environmental Corporation, "Chemical fertilizer as a potential source of perchlorate," Burbank, CA: Lockheed Martin Corporation; November, 1998.

U.S. EPA. "Perchlorate" Office of Water "http://www.epa.gov/ogwdw/ccl/perchlor/perchlo.html," January 2000a.

U.S. EPA, "Superfund Contract Laboratory Program Target Compounds and Analytes," "hhtp://www.epa.gov/oerrpage/superfund/programs/clp/target.htm. Search date - May15, 2000b.

U.S. EPA "Perchlorate Environmental Contamination: Toxicological Review and Risk Characterization Based on Emerging Information," National Center for Environmental Assessment "http://www.epa.gov/ncea/perch.htm" April 1999a.

U.S. EPA, "Groundwater Monitoring List," 40 CFR Ch. 1, Pt. 264, App. IX, July 1, 1999b.

U.S. EPA, "Method 624-Purgeables," 40 CFR Pt. 136, App A, Meth. 624, Chapter 1, July 1, 1999c.

U.S. EPA. "Final Revisions to the Unregulated Contaminant Monitoring Regulation," Office of Water, EPA 815-F-99-005, August 1999d.

U.S. EPA, "Risk Based Concentration Table," U.S. EPA Region III, October 7, 1999e.

U.S. EPA, "PRG Tables: Air-Water, U.S. EPA Region IX, "http://www.epa.gov/region09/waste/sfund/prg," November 9, 1999f.

U.S. EPA. "MTBE Fact Sheet No. 3, Use and Distribution of MTBE and Ethanol," EPA 510-F-97-016, January 1998a.

U.S. EPA. "Announcement of the Drinking Water Contaminant Candidate List; Notice," *Federal Register*, p. 10274, March 2, 1998b.

U.S. EPA, "1,4-Dioxane 123-91-1'" Technology Transfer Network (TTN) Web, Office of Quality Planning and Standards (OAQPS), "hhtp://www.epa.gov/ttnuatw1/hlthef/dioxane.html," May 18, 1998c.

U.S. EPA, N-nitrosodimethylamine; CASRN 62-75-9 (04/01/97), Integrated Risk Information Service (IRIS) Substance File, Internet download, 1997a.

U.S. EPA, "CRL Method 624VOC-Dioxane, Standard Operating Procedure for Measurement of Purgeable 1,4-Dioxane in Water by Wide-Bore Capillary Column Gas Chromatography/Mass Spectrometry," CRL SOP# 624VOC-Dioxane, U.S. EPA Region V, Central Regional Laboratory, Chicago, Illinois, March 25, 1997b.

U.S. EPA, "OPPT Chemical Fact Sheets - 1,4-Dioxane Fact Sheet: Support Document (CAS No. 123-9-1)," EPA 749-F-95-010, February 1995.

U.S. EPA, "Integrated Risk Information System (IRIS) Substance File – 1,4-Dioxane, CASRN 123-91-1," "http://www.epa.gov/ngispgm3/iris/subst/0326.htm," On-line August 22, 1988, Revised September 1, 1990.

U.S. EPA, "Health Advisory p-Dioxane," D-288, March 31, 1987.

U.S. Department of Health and Human Services, Hazardous Substances Data Bank (HSDB, online database). National Toxicology Information Program, National Library of Medicine, Bethesda, MD, 1993.

Verschueren, K., *Handbook of Environmental Data on Organic Chemicals,* 2nd Ed. VanNostrand Reinhold Co., New York, 1983.

Westates Carbon, Technical Bulletin #132. Aqua-Scrub Capacity Calculations Using Carbon Adsorption Isotherms for Toxic Organics, Los Angeles, CA. July 30, 1988.

Zenker, M. J., "Biodegradation of Cyclic and Alkyl Ethers in Subsurface and Engineered Environments," Diss., North Carolina State University, Raleigh, NC, 2000.

Zenker, M. J., Borden, R. C., and Barlaz, M. A., "Investigation of the Intrinsic Biodegradation of Alkyl and Cyclic Ethers," *Natural Attenuation of Chlorinated Solvents, Petroleum Hydrocarbons, and Other Organic Compounds,* Eds. B.C. Alleman and A.L. Leeson, Battelle Press, Columbus, 1999.

Index

N

O

W